METHODS IN CELL BIOLOGY
VOLUME XX

Contributors to This Volume

LAWRENCE E. ALLRED

ABRAHAM AMSTERDAM

HANNA BARANOWSKA

NATHAN A. BERGER

M. J. BERRY

DANIEL BILLEN

HORST BINDING

R. H. J. BROWN

ERNESTO BUSTAMANTE

S. M. CLARKE

GLENN L. DECKER

DONALD DETERS

C. S. DOWNES

ANNA EJCHART

L. M. FINK

CHARLES GORENSTEIN

JON C. GRAFF

JOHN W. GREENAWALT

L. KARIG HOHMANN

HENRI HOMBERGER

ROLLIN D. HOTCHKISS

JOANNE HULLIHEN

JAMES D. JAMIESON

R. T. JOHNSON

LEONARD KEAY

P. F. KOHLER

DAVID KRAM

CRAIG R. LANDGREN

R. MICHAEL LISKAY

RICHARD MARZ

URS MÜLLER

A. M. MULLINGER

M. NAGY

ANN C. OLSON

DAVID PATTERSON

PETER L. PEDERSEN

PETER G. W. PLAGEMANN

WIESLAWA PRAZMO

ALEKSANDRA PUTRAMENT

MARTIN C. RECHSTEINER

BALTAZAR REYNAFARJE

GENE`A. SCARBOROUGH

PIERRE SCHAEFFER

ROBERT A. SCHLEGEL

EDWARD L. SCHNEIDER

J. SCHWENCKE

RICHARD W. SHULMAN

C. H. SISSONS

MARTIN L. SLATER

TRAVIS E. SOLOMON

JOHN W. SOPER

CONJEEVARAM E. SRIPATI

RAYMOND R. TICE

R. C. VON BORSTEL

JONATHAN R. WARNER

HANS JÜRGEN WEBER

RICHARD L. WEISS

ROBERT M. WOHLHUETER

WILLIAM C. WRIGHT

Methods in
Cell Biology

Edited by

DAVID M. PRESCOTT

DEPARTMENT OF MOLECULAR, CELLULAR AND
DEVELOPMENTAL BIOLOGY
UNIVERSITY OF COLORADO
BOULDER, COLORADO

VOLUME XX

1978

ACADEMIC PRESS • New York   San Francisco   London
A Subsidiary of Harcourt Brace Jovanovich, Publishers

ACADEMIC PRESS, INC.
111 Fifth Avenue, New York, New York 10003

United Kingdom Edition published by
ACADEMIC PRESS, INC. (LONDON) LTD.
24/28 Oval Road, London NW1 7DX

LIBRARY OF CONGRESS CATALOG CARD NUMBER: 64–14220

ISBN 0–12–564120–6

PRINTED IN THE UNITED STATES OF AMERICA

This volume is dedicated to the memory of John W. Greenawalt

CONTENTS

LIST OF CONTRIBUTORS

Numbers in parentheses indicate the pages on which the authors' contributions begin.

LAWRENCE E. ALLRED (237), Department of Molecular, Cellular, and Developmental Biology, University of Colorado, Boulder, Colorado

ABRAHAM AMSTERDAM (361), Department of Hormone Research, The Weizmann Institute of Science, Rehovot, Israel

HANNA BARANOWSKA (25), Institute of Biochemistry and Biophysics, Polish Academy of Sciences, Warsaw, Poland

NATHAN A. BERGER (325), The Hematology-Oncology Division of The Department of Medicine, The Jewish Hospital of St. Louis, Washington University School of Medicine, St. Louis, Missouri

M. J. BERRY (309), Department of Zoology, University of Cambridge, Cambridge, England

DANIEL BILLEN (315), The University of Tennessee, Oak Ridge Graduate School of Biomedical Sciences, Oak Ridge, Tennessee

HORST BINDING (113), Botanisches Institut, Christian-Albrechts-Universität, Kiel, West Germany

R. H. J. BROWN (309), Department of Zoology, University of Cambridge, Cambridge, England

ERNESTO BUSTAMANTE[1] (411), Laboratory for Molecular and Cellular Bioenergetics, Department of Physiological Chemistry, Johns Hopkins University School of Medicine, Baltimore, Maryland

S. M. CLARKE (483), Departments of Pathology and Medicine, University of Colorado Medical Center, Denver, Colorado

GLENN L. DECKER (411), Laboratory for Molecular and Cellular Bioenergetics, Department of Physiological Chemistry, Johns Hopkins University School of Medicine, Baltimore, Maryland

DONALD DETERS (107), Biozentrum der Universität Basel, Basel, Switzerland

C. S. DOWNES (255), Department of Zoology, University of Cambridge, Cambridge, England

ANNA EJCHART (25), Institute of Biochemistry and Biophysics, Polish Academy of Sciences, Warsaw, Poland

L. M. FINK (483), Departments of Pathology and Medicine, University of Colorado Medical Center, Denver, Colorado

CHARLES GORENSTEIN (45), Departments of Biochemistry and Cell Biology, Albert Einstein College of Medicine, New York, New York

JON C. GRAFF (211), Department of Microbiology, University of Minnesota Medical School, Minneapolis, Minnesota

JOHN W. GREENAWALT (411), Department of Biochemistry, University of Tennessee, Knoxville, Tennessee

L. KARIG HOHMANN (247), Department of Experimental Biology, Roswell Park Memorial Institute, Buffalo, New York

HENRI HOMBERGER (107), Biozentrum der Universität Basel, Basel, Switzerland

ROLLIN D. HOTCHKISS (149), Rockefeller University, New York, New York

[1] *Present address:* Department of Physiological Sciences, Universidad Peruana Cayetano, Herdia, Lima, Peru.

JOANNE HULLIHEN (411), Laboratory for Molecular and Cellular Bioenergetics, Department of Physiological Chemistry, Johns Hopkins University School of Medicine, Baltimore, Maryland

JAMES D. JAMIESON (361), Section of Cell Biology, Yale University School of Medicine, New Haven, Connecticut

R. T. JOHNSON (255, 309), Department of Zoology, University of Cambridge, Cambridge, England

LEONARD KEAY (169), Center for Basic Cancer Research, Washington University School of Medicine, St. Louis, Missouri

P. F. KOHLER (483), Departments of Pathology and Medicine, University of Colorado Medical Center, Denver, Colorado

DAVID KRAM (379), Laboratory of Cellular and Comparative Physiology, Gerontology Research Center, National Institute on Aging, National Institutes of Health, Public Health Services, Department of Health, Education and Welfare, Baltimore, Maryland

CRAIG R. LANDGREN[2] (159), Department of Biology, University of Oregon, Eugene, Oregon

R. MICHAEL LISKAY (355), Department of Molecular, Cellular and Developmental Biology, University of Colorado, Boulder, Colorado

RICHARD MARZ (211), Department of Microbiology, University of Minnesota Medical School, Minneapolis, Minnesota

URS MÜLLER (107), Biozentrum der Universität Basel, Basel, Switzerland

A. M. MULLINGER (255), Department of Zoology, University of Cambridge, Cambridge, England

M. NAGY (101), Laboratoire d' Enzymologie, Centre National de la Recherche Scientifique, Gif-sur-Yvette, France

ANN C. OLSON (315), The University of Tennessee, Oak Ridge Graduate School of Biomedical Sciences, Oak Ridge, Tennessee

DAVID PATTERSON (355), Eleanor Roosevelt Institute for Cancer Research and Department of Biophysics and Genetics, University of Colorado Medical Center, Denver, Colorado

PETER L. PEDERSEN (411), Laboratory for Molecular and Cellular Bioenergetics, Department of Physiological Chemistry, Johns Hopkins University School of Medicine, Baltimore, Maryland

PETER G. W. PLAGEMANN (211), Department of Microbiology, University of Minnesota Medical School, Minneapolis, Minnesota

WIESLAWA PRAZMO (25), Institute of Biochemistry and Biophysics, Polish Academy of Sciences, Warsaw, Poland

ALEKSANDRA PUTRAMENT (25), Institute of Biochemistry and Biophysics, Polish Academy of Sciences, Warsaw, Poland

MARTIN C. RECHSTEINER (341), Department of Biology, University of Utah, Salt Lake City, Utah

BALTAZAR REYNAFARJE (411), Laboratory for Molecular and Cellular Bioenergetics, Department of Physiological Chemistry, Johns Hopkins University School of Medicine, Baltimore, Maryland

GENE A. SCARBOROUGH (117), Department of Pharmacology, School of Medicine, The University of North Carolina at Chapel Hill, Chapel Hill, North Carolina

PIERRE SCHAEFFER (149), Institute of Microbiology, University of Paris-Sud, Orsay, France

ROBERT A. SCHLEGEL (341), Department of Microbiology and Cell Biology, The Pennsylvania State University, University Park, Pennsylvania

[2] *Present address:* Biology Department, Middlebury College, Middlebury, Vermont.

EDWARD L. SCHNEIDER (379), Laboratory of Cellular and Comparative Physiology, Gerontology Research Center, National Institute on Aging, National Institutes of Health, Public Health Service, Department of Health, Education and Welfare, Baltimore, Maryland

J. SCHWENCKE (101), Laboratoire d' Enzymologie, Centre National de la Recherche Scientifique, Gif-sur-Yvette, France

RICHARD W. SHULMAN (35), Department of Biochemistry, Albert Einstein College of Medicine, New York, New York

C. H. SISSONS[3] (83), Department of Cell Biology, University of Auckland, Auckland, New Zealand

MARTIN L. SLATER (135), Laboratory of Biochemistry and Metabolism, National Institute of Arthritis, Metabolism, and Digestive Diseases, National Institutes of Health, Bethesda, Maryland

TRAVIS E. SOLOMON (361), Veterans Administration Center, Los Angeles, California

JOHN W. SOPER (411), Laboratory for Molecular and Cellular Bioenergetics, Department of Physiological Chemistry, Johns Hopkins University School of Medicine, Baltimore, Maryland

CONJEEVARAM E. SRIPATI (61), Institut de Biologie Physico-chimique, Paris, France

RAYMOND R. TICE (379), Medical Department, Brookhaven National Laboratories, Upton, New York

R. C. VON BORSTEL (1), Department of Genetics, The University of Alberta, Edmonton, Alberta, Canada

JONATHAN R. WARNER (45, 61), Departments of Biochemistry and Cell Biology, Albert Einstein College of Medicine, New York, New York

HANS JÜRGEN WEBER (113), Department of Bacteriology and Immunology, University of California, Berkeley, California

RICHARD L. WEISS (141), The Biological Laboratories, Harvard University, Cambridge, Massachusetts

ROBERT M. WOHLHUETER (711), Department of Microbiology, University of Minnesota Medical School, Minneapolis, Minnesota

WILLIAM C. WRIGHT (499), Human Tumor Cell Laboratory, Walker Laboratory, Sloan-Kettering Institute, Rye and New York, New York

---

[3] *Present address:* Ruakura Animal Research Station, Hamilton, New Zealand.

PREFACE

Volume XX of this series continues to present techniques and methods in cell research that have not been published or have been published in sources that are not readily available. Much of the information on experimental techniques in modern cell biology is scattered in a fragmentary fashion throughout the research literature. In addition, the general practice of condensing to the most abbreviated form materials and methods sections of journal articles has led to descriptions that are frequently inadequate guides to techniques. The aim of this volume is to bring together into one compilation complete and detailed treatment of a number of widely useful techniques which have not been published in full detail elsewhere in the literature.

In the absence of firsthand personal instruction, researchers are often reluctant to adopt new techniques. This hesitancy probably stems chiefly from the fact that descriptions in the literature do not contain sufficient detail concerning methodology; in addition, the information given may not be sufficient to estimate the difficulties or practicality of the technique or to judge whether the method can actually provide a suitable solution to the problem under consideration. The presentations in this volume are designed to overcome these drawbacks. They are comprehensive to the extent that they may serve not only as a practical introduction to experimental procedures but also to provide, to some extent, an evaluation of the limitations, potentialities, and current applications of the methods. Only those theoretical considerations needed for proper use of the method are included.

Finally, special emphasis has been placed on inclusion of much reference material in order to guide readers to early and current pertinent literature.

DAVID M. PRESCOTT

METHODS IN CELL BIOLOGY

VOLUME XX

Chapter 1

Measuring Spontaneous Mutation Rates in Yeast

R. C. VON BORSTEL

*Department of Genetics, The University of Alberta,
Edmonton, Alberta, Canada*

I. Introduction

Luria and Delbrück (1943) devised the first reasonably accurate method for measuring spontaneous mutation rates in microorganisms. Their fluctuation test is based on variance in mutant number from clone to clone, which is caused by rare mutations early in the growth of a clone yielding

1

more mutants in the clone than the more frequent mutations occurring late in clonal growth. They devised the fluctuation test to determine whether phage-resistant strains of *Escherichia coli* arise as random spontaneous mutations or as an adaptive response, such as the depletion of phage receptors among bacterial progeny by a massive phage attack on parental phage receptors (Luria, 1966). The fluctuation test came to be regarded as the preeminent quantitative method which provided respectability to the budding science of bacterial genetics.

Lea and Coulson (1949) developed mathematical expressions to deal more effectively with the boundary conditions of fluctuation experiments, introducing, for example, the method of the median. Ryan (cf. 1963) made many careful comparisons of theoretical distributions with observed distributions of mutants in clones, taking into account such phenomena as delayed nuclear segregation (Witkin, 1951) and phenotypic lag (Ryan, 1955). The validity and usefulness of the fluctuation test is now established, and some technical variant of this test is used for most measurements of spontaneous mutation rates in microorganisms.

The fluctuation test has been modified in order to measure the spontaneous mutation rates in a variety of strains and species of yeast. Yet another method was developed especially for yeast by Ogur and his collaborators (1959); it depends upon the selective elimination of new mutants in a population so that the mutants present when the population is examined represent those that arose in the last cell generation. These methods are described in this chapter.

An independent method for accurately measuring spontaneous mutation rates was introduced by Novick and Szilard (1950), stemming from their work on the continuous cultivation of bacteria in a chemostat. In a chemostat, the bacteria cannot grow at a maximal rate because they are prevented from doing so by the introduction, at a slow rate, of a limiting requirement into the medium while an equal volume overflows. The density of the bacterial culture remains in a steady state.

There is a large literature on the growth and mutation of *E. coli* in chemostat experiments (cf. Kubitschek, 1970). Suffice it to say that the chemostat has been utilized for studying selection (Francis and Hansche, 1972, 1973) and mutation (McAthey and Kilbey, 1976, 1977) in yeast. The basic discoveries in chemostat experiments with *E. coli*, such as the periodic selection of strains more adapted to the chemostat environment and mutations arising as a function of time or cell generation, depending upon the growth-limiting factor, have been demonstrated in yeast. Experimental methods for research on yeast in the chemostat are not dealt with here.

II. Measuring the Spontaneous Mutation Rate in Mitotic Cells by Limiting a Required Nutrient

A. The 1000-Compartment Fluctuation Test

The most accurate method for measuring the spontaneous mutation rate in yeast is based on the continued growth of prototrophic revertants after a required nutrient in the medium has been exhausted. A synthetic complete medium is used in which the relevant nutrient is low enough in concentration so that the titer of cells in the medium is kept well below the saturation level. This nutrient medium is called the limiting medium. For the method described here (von Borstel et al., 1971), the relevant titer of cells is usually limited to somewhere between 10^5 and 10^7 cells/ml. The titer is regulated by altering the concentration of the limiting nutrient so that a reasonable number of revertants can be scored.

Limiting medium is inoculated with 2000 to 5000 cells/ml, and the medium is then distributed in 1-ml aliquots by a Brewer Model 60453 automatic pipetting machine (Baltimore Biological Laboratory) into 10 compartmented culture boxes each having 100 compartments. (The boxes were developed by F. J. de Serres, unpublished, and made on special order by Falcon Plastics). The boxes are then sealed and incubated. The yeast cells settle out on the bottom of each compartment, and revertant colonies begin to appear in about 3 days when incubation is at 26°C. The boxes are incubated for 12 days before colonies are counted, picked, and analyzed.

At the time the colonies are counted, the cell number is determined by hemocytometer counts from compartments without colonies. Two compartments per box are sampled. Since cell growth has ceased, cell counts at this time contribute to the extreme accuracy of this method. That is, residual cell growth, the limiting factor in the accuracy of some methods for measuring the spontaneous mutation rate, is not a limiting factor for accuracy with the 1000-compartment fluctuation test.

Special care must be taken with two items in order to ensure maximum accuracy: First, evaporation from the compartments must be minimal. This can be achieved either by tightly sealing the boxes with masking tape or by incubating them in a chamber where the humidity is kept very high. Second, small variations in temperature can alter measurably the spontaneous mutation rate; with some strains a threefold depression in rate takes place with each 10°C decrease in temperature (R. C. von Borstel and C. M. Steinberg, unpublished data).

Examples of the appearance of the revertants in the boxes can be seen in Fig. 1. The yeast strain in the box on Fig. 1B has a greatly enhanced spon-

(A)

(B)

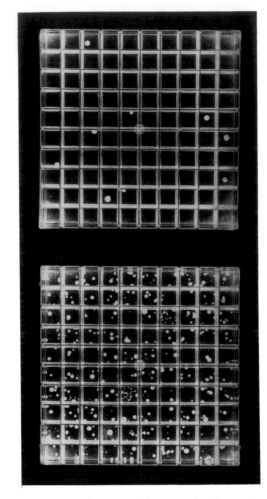

FIG. 1. Hundred-compartment boxes used for measuring the spontaneous mutation rate. Each box measures 15 cm on each side and is 2 cm deep; each compartment holds 3 ml. (A) box containing a control strain, with few revertants present. (B) box containing a mutator stock, a double mutant of *mut1-2* and *mut2-1*, with many revertants present. Both boxes are lysine-limiting with the same amount (1 μg/ml) of lysine present, and the mutant colonies are revertants of *lys1-1* (locus revertants and ochre suppressor mutants) von Borstel *et al.*, 1973 .

taneous rate (the concentration of the limiting nutrient and the mutant allele reverted are the same for both strains). Two items are worth noting: First, the even distribution of cell growth from 2×10^3 up to about 2×10^6 cells/ml can be seen in the indirectly lighted corner compartments in the box on Fig. 1A. Second, up to at least 20 colonies per compartment can be counted accurately. For example, one of the compartments in the box on Fig. 1B has 19 colonies in it. Although the spontaneous mutation rate is calculated by the P_0 method (i.e., by the number of compartments containing no mutants), the rate also can be measured by taking multiple colonies into account. We do not do this routinely because revertant colonies can deplete the medium for another nutrient that then becomes limiting, or they sometimes release the limiting nutrient into the medium, i.e., they are feeders. See the Appendix for estimation by P_0, P_1, P_2, and P_3.

The usefulness of the 1000-compartment fluctuation test has been extended by Leung (1974). She measured the spontaneous mutation rate for resistance by adding the drug and an additional carbon source after growth of the cells was complete. This can be done also by growing the cells in liquid instead of the soft agar as she did, and adding the drug and carbon source in a liquid medium.

The spontaneous mutation rate for the 1000-compartment test can be computed by the following analytical procedure, taken from von Borstel *et al.* (1971):

Let N be the number of compartments in an experiment, and N_0 the number of compartments without revertants. From the zeroth term of a Poisson distribution we have

$$e^{-m} - \frac{N_0}{N} \qquad (1)$$

where m equals the average number of mutational events (not mutants) per compartment. Most of these mutational events are due to new mutations arising during the growth of the cells in the limiting medium, but some are due to mutants present in the inoculum. We correct for this "background" by

$$m_g = m - m_b \qquad (2)$$

where m_b is the average number of mutants per compartment in the inoculum (as determined by direct plating), and m_g is the corrected average number, i.e., the mutational events occurring during the growth in the compartments. This can be converted to the mutational events per cell per generation, M, by

$$M = \frac{m_g}{2C} \qquad (3)$$

where C is the number of cells per compartment after growth has ceased in the limiting medium. The factor of two in the denominator is necessary because the number of cell generations in the history of a culture is approximately twice the final number of cells. Actually, the proper value for this numerical factor depends upon the point(s) in the cell cycle at which growth is terminated in the limiting medium and also upon the distribution of mutation production over the cell cycle. Since it enters only as a scale factor in all mutation rate calculations, relative mutation rates are unaffected by the value used.

This method for determining mutation rates is due to Luria and Delbrück (1943). The principal advantage of the method is that the results are not affected by many types of selection. Since we only score the presence or absence of a mutational event in a culture, it is clearly irrelevant whether the mutants grow faster or slower than non-mutants.

In those experiments where the mutants are further analyzed into categories, the mutation rate may be partitioned by

$$M_i = f_i M \tag{4}$$

where M_i is the mutation rate (per cell per generation) for the ith category and f_i is the fraction of the mutants tested which were found to be in the ith category.

The factor of 2 in the denominator is not used by some investigators. For instance, Magni and von Borstel (1962) preferred not to amplify the difference between the spontaneous mutation rates in mitosis and meiosis; therefore they did not use the factor of 2.

Ordinarily, the revertants begin to appear on the third or fourth day after the experiment is initiated and essentially no longer appear by the seventh or eighth day. Consequently, when cumulative revertants are plotted against the duration of the experiment, there is a delay prior to a steep slope which is followed by a stable plateau.

In Fig. 2 examples are shown of the sudden appearance of revertants, followed by a plateau. The curves are for three haploid strains each con-

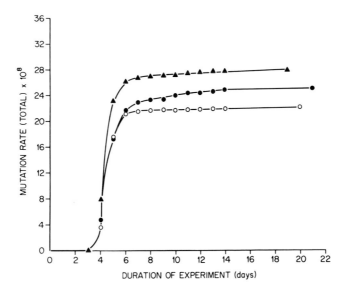

FIG. 2.   Kinetics of the appearance of *lys1-1* revertants (both locus revertants and ochre suppressor mutants) for haploid strains each bearing the radiation-sensitive allele *rad18-2*. Solid circles, KC372; open circles, XV108-4A; triangles, XV108-6A. (von Borstel *et al.*, 1971).

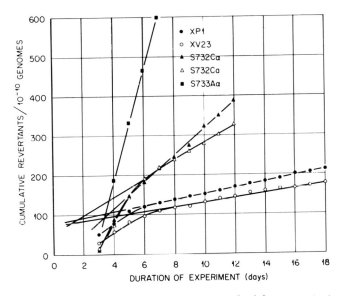

FIG. 3. Kinetics of the appearance of *trp1-1* revertants (both locus revertants and amber suppressor mutants) for two diploid strains (XP1 and XV23) and three haploid strains (S732Cα, S732Ca, and S733Aa).

taining the same radiation-sensitive mutant. It can be seen that, although genetic background causes some variation, the strains have similar mutation rates within a range of about 30%, and the plateau is established from the tenth to the twentieth day.

One major problem has been encountered occasionally with the 1000-compartment test, and that is the continuous appearance of revertant colonies in cultures where growth has been limited, in this case by tryptophan deprivation. We observed that some strains carrying *trp1-1* varied in the expression of revertant appearance. In Fig. 3, three haploid strains and two diploid strains are compared. It can be seen that the accumulation of revertants varies from strain to strain, and that a plateau is never achieved with any strain. We assume that after the first few days the appearance of new revertants comes from cell turnover—cells die and release their contents which serve as nutrient for the living cells which undergo another cell division. It is not known whether this phenomenon is related to the mutant undergoing reversion or to the genetic background of the strains being tested. Therefore the real reversion rate can be estimated by extrapolating from the long linear rise back toward the vertical axis. When this is done, all the curves intersect, in this case at between 90 and 100 × 10^{-10} reversions per genome per generation.

With most strains used in our laboratory, the nonrevertant cells die after a few days, but these dead cells tend not to serve as nutrient for the living ones, consequently the plateau is stable.

B.   The Lassie Test

A simple test has been devised for qualitatively comparing the spontaneous mutation rates in two different strains of yeast. This method was given the meaningless name "Lassie test" by C. M. Steinberg for quick reference in the laboratory. The test is a reversion test on limiting nutrient medium on a plate. From 10^5 to 10^7 cells from a mutant strain are spread

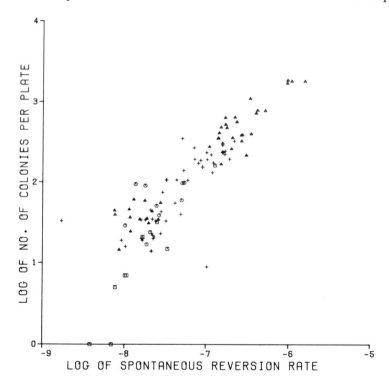

FIG. 4.   Comparison of the Lassie test (ordinate) with the box test (abscissa). Each point represents one 1000-compartment box test (10 boxes) and the average of two Lassie test dishes inoculated with the same strain on the same day; both tests represent revertants for *lys1-1* (both locus revertants and ochre suppressor mutants). The strains vary from mild antimutators to double mutators. The control strains have spontaneous revertant rates in the range of 10^{-8} to 4×10^{-8} mutations per cell per generation. The correlation between the two methods is .883 (d.f. = 107, $p < .00001$). ▲, Haploid strains with single and double mutator combinations of *mut1*, *mut2*, *mut3*, and *mut4* and their controls; +, an array of haploid strains showing weak to strong mutator activity and their controls; ▢, haploid strains showing mild antimutator activity and their controls; ☉, diploid strains.

on a plate which has enough of a limiting nutrient to allow several divisions to take place. For even distribution, we spread each plate on a turntable with 0.5 ml of the cell suspension delivered by a 1-ml pipet—the pipet is tilted, held close to the surface of the plate, and used to spread the cell suspension. A "lawn" forms, and revertant colonies continue to grow against this background. The control for the Lassie test is to plate a similar number of cells on dropout medium.

The Lassie test is a qualitative test because it is not possible to determine accurately how many cell replications have taken place. Nevertheless, it is possible to use this test for analyzing tetrads of heterozygotes of strains differing in spontaneous mutation rate.

A comparison of the 1000-compartment fluctuation test with the Lassie test is shown in Fig. 4 for an array of mutator strains and nonmutator controls strains (R. C. von Borstel and S.-K. Quah, unpublished data). If the two tests are proportional, the individual experiments should fall on a line with a slope of 1. The relation between the two tests is quite good. It is reliable enough so that Lassie test data can be used to determine the limiting concentration of medium necessary to obtain similar numbers of mutant compartments in the 1000-compartment fluctuation test when different mutator strains are under investigation (Gottlieb and von Borstel, 1976).

Examples are shown of the Lassie test for a nonmutator strain (Fig. 5) and for *rad51* a mutator mutant (Fig. 6). With the level of adenine reduced

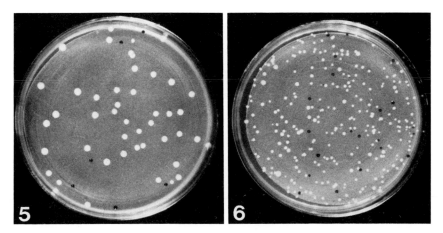

FIG. 5.   Lassie test for a nonmutator strain showing revertants of *lys1-1*, the light colonies being ochre suppressor mutants and the dark colonies revertants at the *lys1-1* locus itself.

FIG. 6.   Lassie test for a *rad51* mutant showing revertants of *lys1-1*, the white colonies being ochre suppressor mutants and the dark colonies revertants at the *lys1-1* locus itself.

to approximately one-fourth of the amount usually provided as supplement, some of the revertants shown in Figs. 5 and 6 accumulated a red pigment and appear as dark colonies. The limiting nutrient is lysine, the white colonies represent newly arisen suppressors which also suppress the adenine-requiring mutation, and the red colonies represent mutations at the lysine-requiring locus itself (Schuller and von Borstel, 1974). The mutator activity of *rad51* is quite evident for both the suppressors and the lysine-requiring locus.

Strains are occasionally tested which show no revertants. It has been found with these strains that, besides the mutant requirement being tested, one of the other requirements becomes limiting. When this happens, the problem is usually overcome by increasing all the requirements by a factor of 4, except the one under test.

The Lassie test is useful for rapid analysis of tetrads in following a mutator mutant in meiosis. But the test may not work well with different genetic backgrounds. For example, when mutator mutants are outcrossed to multiple-marker strains in order to locate them on a genetic map, the mutator phenotype may disappear when the Lassie test is used. This can usually, but not always, be remedied by altering the medium as described above.

This type of qualitative assay was used for *E. coli* by Demerec and Cahn (1953) and Kondo *et al.* (1970); Naumov and Yurkevich (1968) observed that it could be used for yeast in liquid culture. In order to make the test at least semiquantitative, Kondo *et al.* cut out the colonies or a piece of the agar without colonies, washed off the cells, spread them on complete medium, and counted the colony formers. Yeast cells can be counted in a hemocytometer.

If very few cells are plated on limiting medium, they will be seen as separate small colonies rather than as a lawn. Nevertheless, revertants will show up as very large colonies or as larger outgrowths from the small colonies. This procedure has been used by Devin and Arman (1970) to obtain a crude measure of mutation rates, because the small nonmutant colonies can be picked up and the cell numbers estimated.

Another qualitative assay related to the Lassie test has been developed for examining strains with differing spontaneous mutation rates. Golin and Esposito (1977) and Gottlieb (1973) showed that tetrads of a mutator strain segregate for high spontaneous mutation frequency when streaked and show colonies resistant to canavanine or reversions from auxotrophy to prototrophy.

The spore products from dissected asci are streaked onto complete medium, allowed to grow, and then replica-plated onto omission (or canavanine-containing) media as well as complete medium. The next day, another replication of the culture replicated onto the complete medium raises the observed reversion frequency substantially, especially for mutator strains.

C. The Leningrad Test

The genetics group at the University of Leningrad has devised an ingenious and accurate method for measuring the spontaneous reversion rates for alleles of *ade1* and *ade2* of *Saccharomyces* (Khromov-Borisov, 1973). In the method of "ordered plating," a concentrated suspension of yeast cells (ca. 10^6–10^8 cells/ml) is taken up by a pen point used for graphic work. The tip of the pen point is touched lightly, quickly, and with care to the agar surface, following a diagram laid under the plate. The small drops, about 1 mm in diameter, which are deposited on the plate grow into colonies which are quite uniform in size (Figs. 7 and 8). As an alternative to this procedure, a special replicator has been devised with 150 blunt pegs on it (N. N. Khromov-Borisov, personal communication). A lawn of yeast is grown on a plate of complete medium, and 5 ml of sterile water is placed on it. The replicator is dipped into this suspension and inverted, and a plate of limiting medium is stamped lightly against it. The medium contains a limiting concentration of adenine (2 mg/liter), and the thickness of the agar must be more than is normally poured onto a plate. With an adenine insufficiency, the *ade1* and *ade2* strains accumulate red pigment, which tends to slow their growth. Reversions have slightly faster growth which shows up as

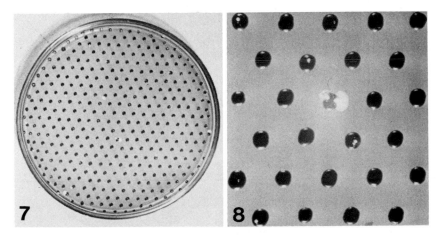

FIG. 7. Leningrad test, the method of ordered plating, showing red, adenine-requiring colonies with some of the colonies exhibiting white, adenine-independent outgrowths. (Photograph by courtesy of N. N. Khromov-Borisov.)

FIG. 8. Magnification of a portion of Fig. 7.

white papillae, sectors, or outgrowths. Numerous colonies with outgrowths can be seen in Fig. 7; the magnified portion of the plate (Fig. 8) shows one colony with white outgrowths and four colonies with one white papilla each.

For determination of the spontaneous mutation rate, the P_0 is determined from the colonies without papillae, sectors, or outgrowths. These colonies are quite uniform in size, and the number of cells follows a normal distribution with a relative standard error in the range of 0.1–1.0%. In order to determine the cell number, several colonies can be carefully cut out of the agar and shaken into separate measured amounts of water, and the cells counted in a Coulter counter or the equivalent. In principle, the method can be made to be as accurate as the 1000-compartment fluctuation test. Also, as in the case of the 1000-compartment fluctuation test, the Leningrad test has been used to measure intragenic recombination, mutator effects, and the effects of exceedingly low levels of mutagens.

III. Measuring the Spontaneous Mutation Rate in Mitotic Cells Grown in Complete Medium

A. The Fluctuation Test

Ever since the experiments of Luria and Delbrück (1943), the standard procedure for determining the spontaneous mutation rate has been to grow cells in complete medium in a large number of tubes. Cell number is assayed by using some of the tubes for dilution and by the plating of cells onto complete medium. The mutant frequency is determined in the rest of the tubes by plating the contents of each tube on selective medium, sometimes directly and sometimes after centrifugation, washing, and resuspension.

From this general procedure, different steps are taken in order to measure the spontaneous mutation rate by the P_0 method or by the method of the median (see also Lea and Coulson, 1949).

1. THE PROBLEM OF RESIDUAL GROWTH

When cells are grown on complete medium and the cell population then tested for the presence of mutants, the cells often undergo growth and division on the selection medium. The amount of residual growth usually varies from none to twofold, depending upon the strain of yeast.

The method for determining the residual growth index used by Magni and von Borstel (1962) is quite satisfactory and is now routinely carried out with two petri dishes containing the selection medium. Approximately 10^7 cells are spread on a dish. When the suspension has dried, the plate is placed under a microscope for examination. The cells occur singly or in groups of two or more attached cells. Eight hundred to one thousand groups are counted and then another 800 to 1000 groups are counted again after 48 hours of incubation at 30°C, when residual growth is complete. An example of the group counts is shown in Table I. The ratio between the second and first scores is a factor of the increase in the number of cells (the residual growth index in Table I).

2. THE P_0 METHOD

An account of this procedure is given by Moore and Sherman (1975) who used it to measure the rate of spontaneous mitotic recombination. They adjust the number of cell divisions for each cross by varying the concentration of the complete growth medium. For example, Moore and Sherman ordinarily use the standard yeast growth medium YEPD used by most investigators, which is 1% Bacto yeast extract, 2% Bacto peptone, and 2% glucose; for strains which show relatively high prototrophic frequencies from intragenic recombination, they decrease the concentration of growth medium to as little as 0.00167% Bacto yeast extract, 0.0033% Bacto peptone, and 0.0033% glucose.

In an experiment, freshly grown cells are suspended in the YEPD medium at a concentration of 15 to 60 cells/ml. One-milliliter aliquots are dispensed by a Brewer automatic pipet into 170 to 250 tubes. The tubes are then shaken at 30°C until the culture is saturated. Moore and Sherman use 10 of the cultures for cell density determinations by diluting them and growing them on medium in which glycerol substitutes for glucose. The remaining cultures are cultured on lactate plates in their entirety for determining the number of cultures with recombinants from *cyc1* to *CYC1+*. (Respiratory-deficient mutants do not grow on lactate or glycerol medium, but *cyc1* strains grow on glycerol, not lactate, medium.) They then calculate the spontaneous recombination (or mutation) rate by a standard P_0 analysis.

Their method can be adapted to reversions and forward mutations, e.g., drug resistance, by determining the cell density, after dilution, on medium composed of yeast extract, peptone, and glucose, or by counting the cells in a hemocytometer if the viability is high.

3. THE METHOD OF THE MEDIAN

This method is a convenient way to measure the spontaneous mutation rate when most or all of the cultures contain mutants.

Yeast is grown in 10 to 30 culture tubes in 10 ml of yeast extract, peptone, and glucose. From an initial inoculum of 100 to 3000 cells/ml, Magni and

TABLE I

EXAMPLE OF DATA SCORED FOR ESTIMATION OF RESIDUAL GROWTH IN STRAIN 5207[a]

Scoring time (hours)	Number of cells per group													Total number of groups	Average number of cells per group[b]
	1	2	3	4	5	6	7	8	9	10	11	12	13		
0	299	415	58	44	11	4	—	—	—	—	—	—	—	831	1.87
48	123	325	312	100	86	35	19	14	7	—	1	6	—	1028	3.05

[a] Magni and von Borstel, 1962.
[b] Residual growth index = 3.05/1.87 = 1.63.

von Borstel (1962) grew the yeast to a titer of 5×10^7 cells/ml before centrifuging and resuspending them in 1 ml of water. The suspension can be used as follows: 0.5 ml is transferred into 10 ml of sporulation medium (see Section IV), 0.5 ml is spread on five dishes of selection medium (either spread as 0.1 ml per dish or resuspended in 2 ml to spread at 0.5 ml per dish), and 0.1 ml is diluted for counting the cells in a hemocytometer and for plating efficiency and residual growth. Variations in the above procedure are often used. For example, Magni grows the cells to a titer of 5×10^7 cells/ml because the frequency of sporulation is depressed at higher titers, and his protocol provides for the study of mutation rates during meiosis. Other investigators normally grow the cells to saturation, since they are studying mitotic cells only.

The mutation frequency in each tube is measured by counting revertant or resistant colonies growing on the selection medium. The number of revertants per culture tube (or per milliliter or per 10^8 cells) is ranked, and the median number is selected for estimating the mutation rate of the strain. (If an even number of growth tubes was used, the average of the middle two tubes in the rank is taken.) The median number of mutants r and the number

TABLE II
PRELIMINARY ESTIMATION OF m FROM THE MEDIAN VALUE OF r [a]

r_0	r_0/m	r_0	r_0/m	r_0	r_0/m	r_0	r_0/m
1.4	1.3	15.3	2.9	117	4.5	787	6.1
1.6	1.4	17.4	3.0	132	4.6	884	6.2
1.9	1.5	19.9	3.1	150	4.7	993	6.3
2.3	1.6	22.7	3.2	169	4.8	1115	6.4
2.7	1.7	25.9	3.3	190	4.9	1251	6.5
3.2	1.8	29.5	3.4	215	5.0	1404	6.6
3.7	1.9	33.5	3.5	242	5.1	1575	6.7
4.3	2.0	38.1	3.6	273	5.2	1767	6.8
5.0	2.1	43.3	3.7	307	5.3	1981	6.9
5.7	2.2	49.2	3.8	346	5.4	2221	7.0
6.6	2.3	55.8	3.9	389	5.5	2490	7.1
7.7	2.4	63.2	4.0	438	5.6	2791	7.2
8.8	2.5	71.6	4.1	493	5.7	3127	7.3
10.1	2.6	81.1	4.2	554	5.8	3503	7.4
11.6	2.7	91.7	4.3	623	5.9	3924	7.5
13.3	2.8	104	4.4	700	6.0	4395	7.6

Thus if the middle culture of the series has 50 mutants, interpolation in the table between $r_0 = 49.2$ and $r_0 = 55.8$ gives $r_0/m = 3.81$, so that $m = 50/3.81 = 13.1$. This is the mean number of mutations per culture.

[a] Table 3 from Lea and Coulson, 1949.

of cells per culture are used to estimate the revertant frequency. The revertant frequency can be used as r_0, and r_0/m is interpolated from Table II.
Then m, the average mutation frequency, is estimated by

$$m = \frac{r_0}{r_0/m} \tag{5}$$

This can be converted to mutational events per cell per generation M by

$$M = \frac{m}{2CW} \tag{6}$$

where C is the number of cells per culture (or per milliliter) and W is the
residual growth index. The factor of 2 in the denominator is necessary for
the same reason given for its use in Eq. (3), although, as noted, some investigators prefer to omit this factor.

B. Direct Estimate of Mutation Rate from Mutant Frequency

This method requires selective elimination or cessation of growth of new
mutants during growth of the culture. When the population is growing, and
new mutants arise during a cell generation while the mutant cells from the
previous generation die, then the mutation rate M is

$$M = \frac{m}{T} \tag{7}$$

where m is the mutant frequency and T is the number of cells in the population. If the mutant cells simply cease growing every generation, the mutation
rate from Eq. (7) will be approximately twice the real rate.

This method for measuring the mutation rate is particularly useful for
mutation rates in excess of 10^{-3} mutations per cell per generation, although
it can be used to measure mutation rates an order of magnitude lower. Ogur
and his collaborators (1959) used this method for measuring the rate of
formation of petite from grande strains by spreading cells grown in glycerol
medium onto dishes where glucose was the carbon source, but see Crosby
et al. (1978) for a method for handling mutation frequencies in excess of 2
or 3%.

This method can be used to measure intragenic recombination rates for
drug-resistant alleles, or to measure reversion rates of unstable mutants.
Residual growth results in sectored colonies and, if sectoring is common,
should be distinguished from clumpiness in any computation of rates. Under
conditions in which the mutant can be selected against during growth of the
culture, and selected for during mutant frequency estimation, the method
can be used to estimate mutation rates occurring at low frequency (Ogur
et al., 1959).

IV. Measuring the Spontaneous Mutation Rate during Meiosis

A suitable method for measuring the spontaneous mutation rate during meiosis was developed by Magni and von Borstel (1962). It is essential to measure the spontaneous mutation frequency in cells just prior to meiosis, so that this fraction of mutants can be subtracted from the number of mutants arising during meiosis. Therefore cells must be grown vegetatively under conditions where sporulation frequency is high, where nonsporulating cells are either killed or their viability measured and, ideally, where sporulation takes place without premeiotic cell divisions. These restrictions are formidable for *Saccharomyces cerevisiae* but are not a problem with *Schizosaccharomyces pombe*. Many strains of *S. cerevisiae* have low sporulation frequencies, and the best sporulation media (with a nitrogen source) permit cells to divide once or twice before entering meiosis. Therefore it has been essential to devise a sporulation medium (without a nitrogen source) which will not permit residual growth to exceed 0.2 to 0.3 (see Section IV, A). However, *S. pombe* cells sporulate immediately after zygotes form (cf. Friis *et al.*, 1971).

A. Sporulation

G. E. Magni (personal communication) uses cells in the exponential phase of growth (cf. Section III,A,3) and sporulates them in a sporulation medium which does not contain glucose or nutrients. The sporulation medium (VB medium) contains sodium acetate (8.2 gm), KCl (1.9 gm), $Mg(SO_4)_2$ (0.35 gm), NaCl (1.2 gm), and distilled water (1 liter).

After sporulation is complete, ascus walls are digested with an enzyme preparation from the digestive tract of the snail *Helix pomatia* (Glusulase). The suspension is washed, resuspended, and gently sonicated to separate the spores. In Magni's laboratory, the frequencies of spores and diploid cells are determined microscopically, and the survival of the diploid cells and the spores is determined by picking up 200 of each with the aid of a micromanipulator to observe the frequency of colony formation.

B. Selective Killing of Unsporulated Cells

One of the two simple methods for preferentially destroying unsporulated cells in mixtures of spores and cells can eliminate the micromanipulation control used by Magni.

1. Diethyl Ether

Dawes and Hardie (1974) showed that treatment of a mixture of spores and cells with diethyl ether for 5 minutes causes a millionfold reduction in cells in that time, while reducing spore survival by about 50%. Their protocol is:

> For liquid cultures, organisms were centrifuged and resuspended in the same volume of 0.2 M potassium acetate buffer (pH 5.5). An equal volume of diethyl ether was added to each sample and the mixture contained in a 1 oz. vial was emulsified by rolling on a Voss roller for an appropriate time at ambient temperature. The two phases were then allowed to separate and the lower aqueous layer removed by pipette. For cultures sporulated on plates, ether treatment was carried out by replica plating onto YEPD in glass petri dishes, exposing these to ether vapour for 2.5 h at 23°C or 90 min at 30°C in a sealed vessel followed by incubation to detect survivors. It should be noted that the boiling point of diethyl ether under standard conditions is 34.5°C. Precautions must be taken to ensure that vessels containing ether at 30°C are perfectly sealed and capable of withstanding some positive internal pressure.

Care must be taken that segregation ratios are normal after ether treatment, and this may vary from strain to strain.

2. French Pressure Cell

DiCaprio and Hastings (1976) found that cells are destroyed preferentially after Glusulase treatment when a mixture of spores and cells is placed under high pressure. Their protocol is:

> Sporulated cultures were washed and resuspended in 0.3 ml .5 M sodium thioglycolate in .05 M Tris (pH 8.8) plus 0.05 ml undiluted glusulase. The mixture was incubated for a minimum of one hour at 37°. This suspension was washed and resuspended in 5 ml water. A French Pressure Cell (Aminco, rapid-fill and manual-fill models) was sterilized either by passing steam through the cylinder for 10 minutes and allowing to cool or by leaving a 1% Roccal solution in the cell for 20–30 minutes, then rinsing thoroughly with sterile distilled water. The suspension was put into the cell, then collected by drops while a pressure of 15,000 PSI was maintained. If by haemocytometer count there remained greater than 1% vegetative cells or groups of spores, the suspension was passed through the pressure cell repeatedly until a count of 99% single spores was achieved.

C. Estimation of the Spontaneous Mutation Rate

If the diploid, unsporulated cells have not been killed by either of the methods listed in Section V,B, then the mutant frequency of spores m_T is estimated by

$$m_T = r - mN_c \qquad (8)$$

where r is the number of revertants on the median dish from a sporulated culture, m is the mutant frequency from Eq. (5), and N_c is the number of viable cells spread on the dish as determined from the hemocytometer counts of cell frequency and the cell survival V_c. If the unsporulated cells were destroyed preferentially by either of the methods described in Section V,B, then N_c is taken to be zero.

A certain fraction of the mutant spores is a reflection of mutations already present in the population before sporulation began. This fraction m_s is determined by

$$m_s = \frac{mN_s}{2} \qquad (9)$$

where N_s is the number of viable spores spread on a dish as determined from hemocytometer counts of spores and the spore survival V_s. If either of the methods in Section V,B was used, then V_s must be estimated by dilution and spreading the spores on a dish containing complete medium. The factor 2 in the denominator is necessary because the spore mutants already present in the culture can be represented in but one-half of the spores because the diploid cells are certainly heterozygous for the spontaneous mutations.

The number of mutants arising during meiosis m_m is estimated by

$$m_m = m_T - m_s \qquad (10)$$

The meiotic mutation rate M_m is estimated by

$$M_m = \frac{2m_m}{N_s} \qquad (11)$$

The factor of 2 in the numerator is used because a mutational event during meiosis yields at most two mutant spores and two auxotrophic spores. This factor can be omitted from the equation if it is assumed that meiotic mutational events are nonreciprocal. It must be left in if meiotic recombinational rates are being measured, and, in that case, the frequency of gene conversion events can be subtracted from the factor of 2 in the numerator.

V. Conclusion

It has become apparent from studies of mutants that repair metabolism comprises recombination, mutation, meiosis, and assorted aspects of DNA metabolism. Many of the mutants affecting DNA repair processes have alterations in phenotype, which can be measured by the procedures we have described in this chapter. Some of the roles of repair-deficient mutants and mutator mutants in repair processes in yeast can be deduced thereby (Hastings *et al.*, 1976).

ACKNOWLEDGMENTS

I am grateful to Charles M. Steinberg and Siew-Keen Quah for data (Figs. 3 and 4, respectively) and to Dorothy Woslyng for typing the manuscript for this chapter. Nikita N. Khromov-Borisov and Giovanni E. Magni generously forwarded their most recent protocols, and Carol W. Moore provided helpful hints. T. Mary Holmes analyzed the data in Fig. 4.

Appendix: Biometrical Aspects of Measuring Mutation Rates

N. N. KHROMOV-BORISOV

Department of Genetics, Leningrad State University,
Leningrad, U.S.S.R.

The methods described in the preceding discussion involve the counting of "particles" that are revertant colonies randomly distributed over "zones" which are equal in volume, as is the case in the 1000-compartment test, or equal in area, as in the Leningrad test.

Let N be the number of such zones in the study and let N_i be the observed number of zones each containing a different number of i particles. The parameter to be estimated is m, the average number of mutational events (the number of revertant colonies) per zone.

Before a method of statistical estimation of m is used in the study, it is obviously necessary to test the null hypothesis that the Poisson distribution is an adequate statistical model for the observed data. This can easily be done by means of visualization of the Poisson distribution as a straight line on a graph with coordinates $N_i \times i!$ on the logarithmic coordinate versus i on the linear coordinate (Schenck, 1968), or by using any other Poisson probability plot (Hahn and Shapiro, 1967). Visualizing the Poisson probability plot on a straight line is sufficient, because the human eye works in agreement with the least-squares principle. For other tests and a bibliography, the handbook by Haight (1967) should be consulted.

Only when the Poisson model is accepted as adequate can one use a simple and quick estimation method designed particularly for the Poisson parameter and its standard deviation and/or standard error.

The most commonly used method is the P_0 method, in which

$$m \pm \text{S.E.} = -\ln \frac{N_0}{N} \pm \sqrt{\frac{1}{N_0} - \frac{1}{N}} \tag{12}$$

Unfortunately, this method gives a large standard error $\sqrt{(1/N_0) - (1/N)}$ with the minimum at $P_0 \cong 0.20$ and $m \cong 1.59$, (Tippett, 1932; Lea and Coulson, 1949; Khalizev, 1969).

So, when $P_0 \geq 0.2$ and m is small, it is simple to provide an estimation of m from inspection of the complete sample. When $P_0 \leq 0.2$, it is better to use more precise procedures. One of them—the method of the median—is described in Section III,A,3. A simpler method, but still precise enough, was deduced and elaborated by Tippett (1932, p. 435).

> The method consists in making counts only on zones with few particles, those zones with more than a certain number of particles being classed together. . . . This method saves time, not only because relatively high concentrations of particles can be used, rendering comparatively few zones necessary for any given accuracy, but also because the numbers of particles in zones with only a few can be counted at a glance, and hence, very quickly. A further advantage arises from the fact that a greater range of particle densities can be investigated than might otherwise be possible.

It is easy to estimate m from such an incomplete (classed or grouped) sample from Fig. 9. This figure is more compact and convenient to use

TABLE III

ESTIMATION OF m AND ITS STANDARD ERROR FROM
THE MUTATOR STRAIN SHOWN IN FIG. 1 AND
BY THE USE OF FIG. 9.

i	N_i	N_t	P_t	m	S.E.
0	1	1	0.01	—[a]	—[a]
1	5	6	0.06	4.5	0.40
2	13	19	0.19	4.4	0.26
3	15	34	0.34	4.5	0.24
.	.	.	.		
.	.	.	.		
.	.	.	.		
k		$N = 100$	1.00	4.5[b]	0.21[b]

[a] These values are not provided because P_t is too small to read from $t = 0$ in Fig. 9.

[b] These values are estimated from the complete sample by $m \pm \text{S.E.} = R/N \pm \sqrt{R}/N$, where $R = \Sigma_1^k i N_i$ is the total observed number of revertant colonies.

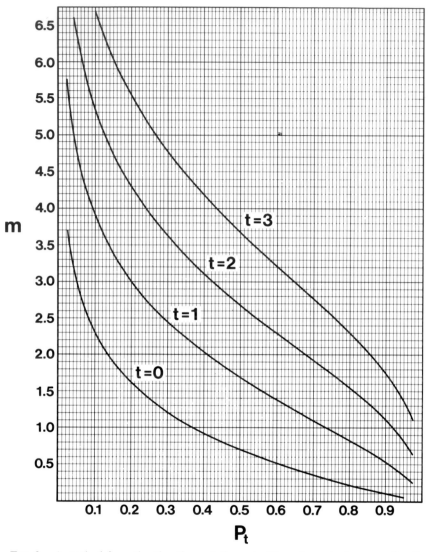

FIG. 9. A method for estimating the mutation rate. The value m is shown as a function of P_t for different values of t (see text).

than the nomograms of Tippett (1932) and curves used by others (Hahn and Shapiro, 1967).

For example, let us estimate m from the revertant colonies seen in the lower box in Fig. 1, assuming that the numbers of colonies per compartment follow a Poisson distribution. The data taken from Fig. 1 are presented in the first two columns of Table III; i.e., the first column i lists the variants of the compartments containing 0, 1, 2, 3... k revertant colonies, and the second column N_i lists the number of compartments with i colonies.

N_t in column 3 of Table III is simply the accumulated number of each class, P_t in column 4 is the corresponding relative frequency, this is, $P_t = N_t/N$, and N is the total number of compartments. The value m is estimated by taking a given P_t on the abscissa of Fig. 9, going up to the appropriate t curve, and then reading the value of m on the ordinate.

The standard error, S.E., of m estimated from observed P_t may be approximated by

$$\text{S.E.} = \pm \frac{1}{P_t - P_t - 1} \sqrt{\frac{P_t(1 - P_t)}{N}} \tag{13}$$

or

$$\text{S.E.} = \pm \frac{1}{N_t - P_t - 1} \sqrt{\frac{N_t(N - N_t)}{N}} \tag{14}$$

Here P_{t-1} (or N_{t-1}) can be obtained from the same sample data or with the aid of Fig. 9. In the last case the estimated value m on the ordinate is taken over to the appropriate $t - 1$ curve, and the value of P_{t-1} is read on the abscissa. For $t = 0$, formula (12) must be used.

The values of m, and the standard error of each estimate of m, are shown in columns 5 and 6 of Table III. The agreement is good.

In order to estimate the mutation rate M, it is necessary to obtain a cell count. Obviously, this will have its own standard error. Consequently, the method of counting cells must be precise enough so that the standard error of the counts is much less than the standard error of m. In that case the standard error of m will suffice.

For an estimate of the significance of difference between the values of m obtained from two different experiments, the Kastenbaum and Bowman (1966, 1970) tables may be used, but with two precautions. First, the same t must be used for each experiment. Second, the two-sided test must be used; therefore the level of significance in the Kastenbaum and Bowman tables must be multiplied by 2.

It is essential to bear in mind that in some cases the Poisson model may not be adequate. If the distribution is something other than a Poisson distribution, the above treatment of data is not sufficient.

REFERENCES

Crosby, W. L., Colson, A.-M., Briquet, M., Moustacchi, E., Goffeau, A., and Tyteca, D. (1978). *Mol. Gen. Genet.* (in press).
Dawes, I. W., and Hardie, I. D. (1974). *Mol. Gen. Genet.* **131**, 281–289.
Demerec, M., and Cahn, E. (1953). *J. Bacteriol.* **65**, 27–36.

Gottlieb, D. J. C., and von Borstel, R. C. (1976). *Genetics* **83**, 655–666.

Hahn, G. J., and Shapiro, S. S. (1967). "Statistical Models in Engineering." Wiley, New York.

Haight, F. A. (1967). "Handbook of the Poisson Distribution" (Operations Research Society of America, Publications in Operations Research, 11). Wiley, New York.

Hastings, P. J., Quah, S.-K., and von Borstel, R. C. (1976). *Nature (London)* **265**, 719–722.

Kastenbaum, M. A., and Bowman, K. O. (1966). "The Minimum Significant Number of Successes in a Binomial Sample", ORNL = 3909, Oak Ridge National Laboratory, Oak Ridge, Tennessee.

Kastenbaum, M. A., and Bowman, K. O. (1970). *Mutat. Res.* **9**, 527–549.

Khalizev, A. E. (1969). *Genetika* **5**, 157–168.

Khromov-Borisov, N. N. (1973). *Conf. Genet. Ind. Microorg., 1973,* Conference Abstracts, p. 43.

Kondo, S., Ichikawa, H., Iwo, K., and Kato, T. (1970). *Genetics* **66**, 187–217.

Kubitschek, H. E. (1970). "Introduction to Research with Continuous Culture." Prentice-Hall, Englewood Cliffs, New Jersey.

Lea, D. E., and Coulson, C. A. (1949). *J. Genet.* **49**, 264–285.

Leung, E. S. (1974). *Heredity* **33**, 417–419.

Luria, S. E. (1966). *In* "Phage and the Origins of Molecular Biology" (J. Cairns, G. S. Stent, and J. D. Watson, eds.), pp. 173–179. Cold Spring Harbor Lab., Cold Spring Harbor, New York.

Luria, S. E., and Delbrück, M. (1943). *Genetics* **28**, 491–511.

McAthey, P., and Kilbey, B. J. (1976). *Biol. Zentralbl.* **95**, 415–421.

McAthey, P., and Kilbey, B. J. (1977). *Mutat. Res.* **44**, 227–234.

Magni, G. E., and von Borstel, R. C. (1962). *Genetics* **47**, 1097–1108.

Moore, C. W., and Sherman, F. (1975). *Genetics* **79**, 397–418.

Naumov, B. I., and Yurkevich, W. W. (1968). *Proc. Tech. High Sch., Sect. Biol. Sci., 1968* Vol. 10, pp. 88–91.

Novick, A., and Szilard, L. (1950). *Proc. Natl. Acad. Sci. U.S.A.* **36**, 708–719.

Ogur, M., St. John, R., Ogur, S., and Mark, A. M. (1959). *Genetics* **44**, 483–496.

Ryan, F. J. (1955). *Am. Nat.* **89**, 159–162.

Ryan, F. J. (1963). *In* "Methodology in Basic Genetics" (W. J. Burdette, ed.), pp. 39–82. Holden-Day, San Francisco, California.

Schenck, H., Jr. (1968). "Theories of Engineering Experimentation," 2nd ed. McGraw-Hill, New York.

Schuller, R. C., and von Borstel, R. C. (1974). *Mutat. Res.* **24**, 17–23.

Tippett, L. H. C. (1932). *Proc. R. Soc. London, Ser. A* **137**, 434–446.

von Borstel, R. C., Cain, K. T., and Steinberg, C. M. (1971). *Genetics* **69**, 17–27.

von Borstel, R. C., Quah, S.-K., Steinberg, C. M., Flury, F., and Gottlieb, D. J. C. (1973). *Genetics* **73**, Suppl., 141–151.

Witkin, E. M. (1951). *Cold Spring Harbor Symp. Quant. Biol.* **16**, 357–372.

Chapter 2

Manganese Mutagenesis in Yeast

ALEKSANDRA PUTRAMENT, HANNA BARANOWSKA,
ANNA EJCHART, AND WIESLAWA PRAZMO

Institute of Biochemistry and Biophysics,
Polish Academy of Sciences,
Warsaw, Poland

I. Introduction

Manganese was known long ago to be strongly mutagenic toward bacteria (Demerec and Hanson, 1951; Bohme, 1961) and bacteriophage T4 (Orgel and Orgel, 1965). There was also evidence that the cation decreased the fidelity of DNA polymerases *in vitro* (Berg *et al.*, 1963; Hall and Lehman, 1968; for later data, see Chang, 1973; Sirover and Loeb, 1976). However, mutator properties of some strains of bacteriophage T4 carrying temperature-sensitive mutations in the gene coding for DNA polymerase suggested that the polymerase takes an active part in selecting nucleotides during DNA replication (Speyer, 1965; Speyer *et al.*, 1966; Drake and Greening, 1970). Thus it can be assumed that manganese acts mutagenically by decreasing the fidelity of DNA polymerases.

Tests for the mutagenicity of manganese toward yeast seemed to be useful for two reasons. First, the results would answer the question whether or not the cation is mutagenic toward eukaryotes. Second, in *Saccharomyces cerevisiae*, not only nuclear but also mitochondrially inherited mutations conferring either a respiratory deficiency or resistance to antibiotics can be easily scored. The search for the latter type of mitochondrial mutations was of particular interest, since a few years ago there was no evidence that any factor could induce such mutations.

The assumption seemed justified that, if manganese can induce mutations at all, it must act on replicating DNA, therefore the treatment should be performed in growth medium (see Bohme, 1961).

II.   Induction of Mitochondrial Mutations

A.   Mutations Conferring Resistance to Antibiotics

Two sets of conditions proved to favor induction of this class of mutations: 6–8 mM manganese added to YEPGlu (see legend for Fig. 1) medium and inocula of up to 5×10^5 cells/ml (Fig. 1), or 10 mM manganese in the same medium (standard or diluted) and inocula of 2–5×10^7 cells/ml. We are unable to explain why 8 mM manganese acted very weakly if at all when the inocula were increased to 10^6–10^7 cells/ml (Prazmo *et al.*, 1975; Putrament *et al.*, 1977). In the majority of strains definite induction of an *ery-r* mutation was observed after 3–4 hours' incubation in the mutagenic medium. After 18–24 hours of incubation the highest frequencies of mutations were obtained (100 to 1000 times the spontaneous level; Putrament *et al.*, 1975a,b, and unpublished results).

The yields of mutants recovered were always higher when, after mutagenic treatment, the cells resuspended in YEPGlu or in 2% glucose were plated on selective medium, as compared with yields from plated cells suspended in glucose-free media (Fig. 1). It seems likely that traces of glucose, by allowing one or two cell divisions on the selective media, facilitated phenotypic expression of the newly induced mutations, as is the case in, for example, streptomycin-resistant mutants of *Escherichia coli* (Demerec *et al.*, 1949).

Mutations conferring resistance to antibiotics were not induced by manganese when the cells were mutagenized in the presence of hydroxyurea (Putrament *et al.*, 1975b), a known inhibitor of DNA synthesis (Slater, 1973), or when the treatment was performed under nongrowth conditions (Putrament *et al.*, 1977). This supports our assumption that the cation acts mutagenically only on replicating DNA.

A total of 134 manganese-induced *ery-r, cap-r,* and *oli-r* mutants was

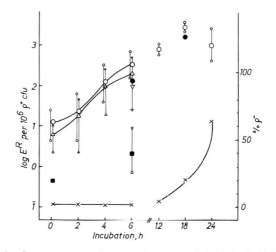

Fig. 1. Induction by manganese of *ery-r* mutations in diploid strain 26-DP after different periods of treatment. The medium contained 1% yeast extract, 1% Bacto peptone, 2% glucose, pH adjusted to 6 (YEPGlu), and 8 mM MnSO$_4$. It is referred to as the mutagenic medium. At pH 6.5 or higher manganese often precipitated. Late stationary-phase cells were used as inocula, at a final density of about 5×10^5/ml. The points for 0–6 hours' incubation in the mutagenic medium represent mean values from five experiments. The lowest and highest mutation frequencies obtained are marked by vertical bars. The *ery-r* frequencies after 12, 18, and 24 hours represent mean values from 2, 3, and 4 experiments respectively. The spontaneous mutation frequencies in stationary cells varied from 0.34 to 1.4 per 10^6 *rho*$^+$ cfu, while those in log-phase cells varied from 1×10^{-6} to 4×10^{-6} in most experiments. Higher *ery-r* frequencies in logphase cells could be spurious; stationary-phase cells are single, while the log phase produces many chains consisting of up to three cells. In experiments with 0- to 2-hour mutagenesis, the number of *ery-r* mutants varied from 10 to 100, thus decreasing the reliability of the data. All the other experiments resulted in one to several hundred. The mutagenized cells were plated on selective medium suspended in: ○, mutagenic or standard YEPGlu medium; ●, YEP plus 0.2% glucose and 2% glycerol; ▽, Ringer solution. ■, Control (plated in YEPGlu or YEPGlycerol); x, percentage of *rho*$^-$ mutants.

tested (Putrament *et al.*, 1973, 1975a; G. Kergastel, A. Touze, and A. Nicolas, unpublished results, courtesy of Dr. L. Clavillier); 129 of them proved to be mitochondrially inherited.

Haploid yeast cells contain at least 20 copies of mitochondrial DNA (mtDNA). Antibiotic-resistant clones must have an identical mutational change in all or at least in the majority of mtDNA molecules within each cell. Such mutant clones originate through intracellular selection of mitochondria (or parts of mitochondria?) carrying mutant mtDNA (Birky, 1973; Dujon *et al.*, 1976; Putrament *et al.*, 1976), as is the case with *Paramecium* (Beale *et al.*, 1972; Queiroz and Beale, 1974).

Manganese induces mitochondrial mutations in *Schizosaccharomyces pombe* also (Colson *et al.*, 1976).

B. *rho*⁻ Mutations Conferring a Respiratory Deficiency

Most experiments were performed using 6–8 m*M* manganese and small inocula. Under these conditions, in all strains tested, the percentage of *rho*⁻ mutants remained unchanged during the first 6–8 hours of incubation in the mutagenic medium. The increase in *rho*⁻ mutants started after 12–24 hours, depending on the strain tested, and after 48–72 hours up to 99% of the cells developed a respiratory deficiency.

The suppressiveness of manganese-induced *rho*⁻ primary clones varied from 0 to over 50%; that is, it was similar to that usually observed in spontaneous *rho*⁻ mutants.

C. *mit*⁻ and *syn*⁻ Mutations

mit⁻ mutations were defined by Slonimski and Tzagoloff (1976) as those that are mitochrondrially inherited. Strains with such mutations are unable to grow on nonfermentable substrates but retain mitochondrial protein synthesis. The assay of the latter can be performed routinely in the presence of cycloheximide, according to Tzagoloff *et al.* (1975). In *syn*⁻ mutants the ability to carry out mitochondrial protein synthesis is also impaired or lost (Bolotin-Fukuhara *et al.*, 1976; Faye *et al.*, 1976; Tzagoloff *et al.*, 1976). However, both types of mutants differ from *rho*⁻ mutants in their ability to restore a standard phenotype through (1) reversion due to back- or suppressor mutation, (2) recombination with strains carrying other mutations of the same type, (3) recombination with certain *rho*⁻ mutants. Therefore, in contrast to the *rho*⁻ mutation, *mit*⁻ and *syn*⁻ mutations are nonlethal for the *rho* factor (i.e., mtDNA or mitochondrial genophore).

The *mit*⁻ and *syn*⁻ mutants so far isolated were obtained by one of the following procedures.

1. A standard strain was grown until the stationary phase in YEPGlu with 7 m*M* manganese. The cells were plated, and petite clones which were capable of mitochondrial protein synthesis but which did not show a 2:2 segregation in tetrads (i.e., were not nuclear *pet* mutants) were classified as *mit*⁻ mutants (Tzagoloff *et al.*, 1975). A detailed genetic analysis (Slonimski and Tzagoloff, 1976) led to the unequivocal conclusion that *mit*⁻ mutations are indeed located in mtDNA.

2. Mutagenic treatment of a standard strain was performed as in method 1, and *mit*⁻ and *syn*⁻ mutants were identified as those in which the *mit*⁺ (standard) phenotype was restored in crosses with certain *rho*⁻ mutant testers (Bolotin-Fukuhara *et al.*, 1976; Faye *et al.*, 1976).

3. A standard strain was grown for 24 hours in YEPGlu with 10 m*M* manganese, *mit*⁻ mutants were identified as in method 2, and mitochondrial protein synthesis was additionally tested (Rytka *et al.*, 1976).

4. A standard strain carrying *cap-r*, *ery-r*, *oli-r*, and *par-r* mitochondrial mutations was incubated in YEPGlu with 10 m*M* manganese for 5 hours, and *mit⁻* mutants were screened as those petites which in crosses with *rho⁺* antibiotic-sensitive strains transmitted all mitochondrial markers to their diploid progeny (Mahler *et al.*, 1976).

5. A strain carrying the nuclear mutation *opl* was used. The *opl* mutant strains cannot grow on nonfermentable substrates (Kovac *et al.*, 1967), and *rho⁻* mutations are lethal for them (Kovacova *et al.*, 1968). The cells were grown in YEPGlu with 8 m*M* manganese for 24–30 hours. Colonies from the treated cells were crossed with a *rho⁰* tester. Diploids originating from *OP rho⁰* × *opl mit⁺* crosses could grow normally on glycerol medium, whereas those from *OP rho⁰* × *opl mit⁻* crosses could not grow on glycerol. The *mit⁻* mutants thus identified were crossed with *rho⁻* testers (Kotylak and Slonimski, 1976). In contrast with the *mit⁻* mutants newly isolated from standard strains (Rytka *et al.*, 1976) the *mit⁻* mutants isolated from the *opl* strain were stable even when transferred to an *OP* nuclear background (Z. Kotylak, personal communication).

The *mit⁻* and *syn⁻* mutants resemble *rho⁻* mutants as regards the lack of selective advantage over the standard mitochondrial genophore, but they also resemble antibiotic-resistant mutants as regards the specificity of mutational lesions. For the expression of a particular mutational lesion, for instance, the absence of cytochrome oxidase, all or at least a majority of mtDNA molecules within each cell of the mutant clone must carry an identical mutational lesion. Therefore the origin of pure *mit⁻* and *syn⁻* clones is so far unknown.

D. Mutational Lesions Induced by Manganese in mtDNA

In bacteriophage T4 manganese induces transitions and transversions (Orgel and Orgel, 1965; Ripley, 1975), as can be expected from its effects on DNA polymerases *in vitro*. So far nothing is known about mutational changes in yeast mtDNA leading to antibiotic resistance. However, genetic (Slonimski and Tzagoloff, 1976) and biochemical evidence (Rabinowitz *et al.*, 1976; Faye *et al.*, 1976) indicates that many *mit⁻* and *syn⁻* mutations are due to deletions. One of them (Rabinowitz *et al.*, 1976) consists of a loss of about 7% of the mitochondrial genophore. These data, together with the fact that manganese can induce *rho⁻* mutations, indicate that the mutagenic action of manganese toward yeast mtDNA is more complex than previously assumed. One of the possibilities is that manganese inhibits the activity of mtDNA polymerase as described by Iwashima and Rabinowitz (1969), while the remaining proteins involved in mtDNA replication act abnormally, producing deletion mutations with an increased frequency, as is the case with *polA⁻* strains of *E. coli* (Coukell and Yanofsky, 1970).

III. Induction of Nuclear Mutations

The reversion of nuclear auxotrophic mutations was tested in 15 strains. The highest mutation frequencies were obtained using 8 mM manganese in YEPGlu medium, and an 18-hour treatment. Depending on the strain used, reversion of the *thr2-1* (*hom-2*) mutation varied from 1.7 to 95 times the spontaneous level. Similar results were obtained for the reversion of *thr2-2*, *arg4-2*, and *arg4-17*. In heteroallelic *thr2-1/thr2-2* and *arg4-2/arg4-17* diploids the highest induction of revertants was 34 times the spontaneous level; i.e., it was not higher than mutational reversion of either of the mutants involved. Therefore there is no reason to suppose that manganese induces gene conversion in yeast (Baranowska *et al.*, 1977).

It seems likely that manganese is weakly mutagenic toward nuclear DNA, because it strongly inhibits its replication (see Section IV,B).

IV. The Uptake of Manganese and Its
General Effects on Yeast Cells

A.   Uptake and Interaction with Magnesium

The uptake of Mn^{2+} was strongly stimulated by glucose (Fuhrmann, 1973). Other components of complete medium further increased the uptake in the presence of glucose, but not in glycerol medium (Fig. 2). Magnesium, even in a concentration 10 times lower than that of manganese, strongly inhibited the uptake of the latter. When the cells were incubated for 3 hours in manganese medium and magnesium was then added in an equimolar concentration, the uptake of manganese was immediately inhibited, mutation induction was arrested, and cell division resumed (Putrament *et al.*, 1977). Thus magnesium apparently not only inhibits the uptake of manganese but also causes the release of Mn^{2+} from the cells. Alternatively, Mg^{2+} replaces Mn^{2+} at the sites essential for error-proof mtDNA replication and cell growth.

B.   Inhibition of Protein and DNA Synthesis

Under the experimental conditions used (see legend for Fig. 2) the cells did not divide, and their viability decreased. Therefore the percentages of the inhibition by manganese of protein and DNA synthesis (Table I) do not apply to conditions routinely used for mutation induction. The results show,

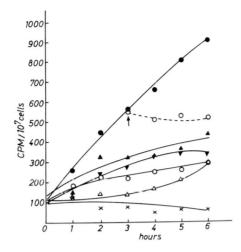

FIG. 2. The effects of media on manganese uptake by the yeast strain S/M13-2D. Cells grown to mid log phase in YEPGlycerol medium were collected by centrifugation and resuspended to a final density of about 2×10^7 cells/ml in a medium containing 0.25% yeast extract, 0.25% Bacto peptone, and 1% glucose, pH 6 (diluted YEPGlu), and diluted YEP-Glycerol, or in 1% glucose. All samples contained 10 mM $MnSO_4$ and 1 μCi/ml $^{54}MnCl_2$. The cells were incubated at 30°C in a shaker. After the times indicated 1-ml samples were taken, filtered through nitrocellulose Millipore filters Ha 0.45 μm, $\emptyset$ 25 mm, and washed twice with 10 ml volumes of water or 2% glucose solution. The filter disks were transferred to scintillation vials and dissolved in 0.5 ml of dioxan Bray scintillator mixture. ●, Diluted YEPGlu with manganese; ○ (broken line), the same, with 10 mM $MgCl_2$ added after 3 hours; ○ (solid line), the same, with 1 mM $MgCl_2$ added at time 0; x, the same, with 10 mM $MgCl_2$ added at time 0; ▽, the same, with 10 mM $ZnCl_2$ added at time 0; ▼, 1% glucose plus 10 mM manganese; △, diluted YEPGlycerol plus 10 mM manganese.

however, that manganese inhibits more strongly cytosolic than mitochondrial protein synthesis. The inhibition of nuclear DNA replication may be due to the direct action of manganese on this process, or it may be an after-effect of the inhibition of protein synthesis.

mtDNA replication seems to be resistant to manganese. This result was rather unexpected, since the only known mtDNA polymerase (Iwashima and Rabinowitz, 1969; Wintersberger and Wintersberger, 1970) is *in vitro* very sensitive to the cation (Iwashima and Rabinowitz, 1969).

C. Toxicity

During the first 24 hours of incubation in YEPGlu with 8 mM manganese no cell lethality was ever observed. Division rates were from about 4 hours to more than 12 hours, depending on the strain. Between 48 and 96 hours of incubation the turbidity of the cultures usually increased, but the number of viable cells decreased to 1–10% (unpublished results). The same seems to be true for *S. pombe* (Colson *et al.*, 1976).

When cells from the logarithmic phase were incubated in diluted YEPGlu with 10 m*M* manganese, or in 1% glucose solution with 8 m*M* manganese, the survival rate decreased after 4–6 hours' incubation (Prazmo *et al.*, 1975; Putrament *et al.*, 1977). Upon incubation in 1% glucose, magnesium at an equimolar concentration with manganese did not prevent cell death.

V.   Other Mutagens Inducing Mutations Nonlethal for *rho* Factor

Mitochondrial *ery-r* mutations were induced by nitrosoguanidine (Dawes and Carter, 1974). Temperature-sensitive mitochondrially inherited mutations were produced by treatment with nitrosoguanidine or 2-aminopurine (Lancashire, 1976; Storm and Marmur, 1975). A *mit⁻* mutant was obtained after treatment with ethyl methanesulfonate (Tzagoloff *et al.*, 1975). These mutagens, however, also induce nuclear mutations efficiently.

In *S. pombe* strains carrying an extranuclear *ant-r* mutation, *mit⁻* mutations were induced by ethidium bromide (Wolf *et al.*, 1976).

TABLE I

INHIBITION BY MANGANESE OF PROTEIN AND DNA SYNTHESIS[a,b]

Protein synthesis		DNA synthesis	
Total	Mitochondrial	Total	Mitochondrial
97%	47%	84%	12%

[a]Putrament *et al.* (1977).

[b]Cells of strain S/M13D were incubated in manganese medium as described in the legend for Fig. 2. For measurements of protein synthesis the cell samples, after a 3-hour treatment with manganese, were harvested by centrifugation and resuspended in the incubation mixture containing [¹⁴C]leucine (Bilinski and Jachymcyzk, 1973). Inhibition of mitochondrial protein synthesis was measured in the presence of cycloheximide (200 µg/ml) which inhibited cytosolic protein synthesis. DNA labeling was carried out according to Hatzfield (1973). Control samples were incubated for 6 hours in manganese-free medium with [¹⁴C]adenine (1.5 µCi/ml); inhibition of total DNA synthesis by manganese was measured in the same medium containing 10 m*M* MnSO₄, while for inhibition of mtDNA synthesis the incubation mixture was supplemented with cycloheximide (200 µg/ml) added 45 minutes before the label (Grossman *et al.*, 1969).

ACKNOWLEDGMENTS

The work described here was supported by the Polish Academy of Sciences under Project 09.7.2.1., and in part by PHS Grant No. 05-001-0. The authors are very grateful to Professor R. K. Mortimer for critical reading of the manuscript.

REFERENCES

Baranowska, H., Ejchart, A., and Putrament, A. (1977). *Mutat. Res.* **42**, 343–348.
Beale, G. H., Knowles, J., and Tait, A. (1972). *Nature (London)* **235**, 396–397.
Birky, C. Q. (1973). *Genetics* **74**, 421–432.
Bilinski, T., and Jachymczyk, W. (1973). *Biochem. Biophys. Res. Commun.* **52**, 379–387.
Bohme, H. (1961). *Biol. Zentralbl.* **80**, 5–32.
Bolotin-Fukuhara, M., Faye, G., and Fukuhara, H. (1976). *In* "The Genetic Function of Mitochondrial DNA" (C. Saccone and A. M. Kroon, eds.), pp. 243–250. North-Holland Publ., Amsterdam.
Colson, A. M., Labaille, F., and Goffeau, A. (1976). *Mol. Gen. Genet.* **149**, 101–109.
Coukell, M. B., and Yanofsky, C. (1970). *Nature (London)* **228**, 633–635.
Dawes, I. W., and Carter, B. L. A. (1974). *Nature (London)* **250**, 709–712.
Demerec, M., and Hanson, J. (1951). *Cold Spring Harbor Symp. Quant. Biol.* **16**, 215–228.
Demerec, M., Wallace, B., Witkin, E. M., and Bertani, G. (1949). *Carnegie. Inst. Wash., Yearb.* **48**, 154–166.
Drake, J. W., and Greening, E. O. (1970). *Proc. Natl. Acad. Sci. U.S.A.* **66**, 823–829.
Dujon, B., Bolotin-Fukuhara, M., Coen, D., Deutsch, J., Netter, P., Slonimski, P., P., and Weill, L. (1976). *Mol. Gen. Genet.* **143**, 131–165.
Faye, G., Bolotin-Fukuhara, M., and Fukuhara, H. (1976). *In* "Genetics and Biogenesis of Chloroplasts and Mitochondria" (T. Bücher *et al.*, eds.), pp. 547–555. North-Holland Publ., Amsterdam.
Fuhrmann, G. F. (1973). "European University Papers," Vol. 3. Lang, Bern.
Grossman, L. I., Goldring, E. S., and Marmur, J. (1969). *J. Mol. Biol.* **46**, 367–376.
Hall, R. M., and Brammar, J. (1973). *Mol. Gen. Genet.* **121**, 271–276.
Hall, Z. W., and Lehman, I. R. (1968). *J. Mol. Biol.* **36**, 321–333.
Hatzfield, J. (1973). *Biochim. Biophys. Acta* **299**, 34–42.
Iwashima, A., and Rabinowitz, M. (1969). *Biochi. Biophys. Acta* **178**, 283–293.
Kotylak, Z., and Slonimski, P. P. (1976). *In* "The Genetic Function of Mitochondrial DNA" (C. Saccone and A. M. Kroon, eds.), pp. 143–154. North-Holland Publ., Amsterdam.
Kovac, L., Lachowicz, T. M., and Slonimski, P. P. (1967). *Science* **158**, 1564–1567.
Kovacova, V., Irmlerova, J., and Kovac, L. (1968). *Biochim. Biophys. Acta* **162**, 157–163.
Lancashire, W. E. (1976). *In* "Genetics and Biogenesis of Chloroplasts and Mitochondria" (T. Bücher *et al.*, eds.), pp. 481–490. North-Holland Publ., Amsterdam.
Mahler, H. R., Bilinski, T., Miller, D., Hanson, D., Perlman, P. S., and Demko, C. A. (1976). *In* "Genetics and Biogenesis of Chloroplasts and Mitochondria" (T. Bücher *et al.*, eds.), pp. 857–863. North-Holland Publ., Amsterdam.
Orgel, A., and Orgel, L. E. (1965). *J. Mol. Biol.* **14**, 453–457.
Prazmo, W., Balbin, E., Baranowska, H., and Putrament, A. (1975). *Genet. Res.* **26**, 21–29.
Putrament, A., Baranowska, H., and Prazmo, W. (1973). *Mol. Gen. Genet.* **126**, 357–366.
Putrament, A., Baranowska, H., Ejchart, A., and Prazmo, W. (1975a). *J. Gen. Microbiol.* **90**, 265–290.
Putrament, A., Baranowska, H., Ejchart, A., and Prazmo, W. (1975b). *Mol. Gen. Genet.* **140**, 339–347.
Putrament, A., Polakowska, R., Baranowska, H., and Ejchart, A. (1976). *In* "Genetics and

Devin, A. B., and Arman, I. P. (1970). *Genetika* **6**, 186–190.
DiCaprio, L., and Hastings, P. J. (1976). *Genetics* **84**, 697–721.
Francis, J. C., and Hansche, P. E. (1972). *Genetics* **70**, 59–73.
Francis, J. C., and Hansche, P. E. (1973). *Genetics* **74**, 259–265.
Friis, J., Flury, F., and Leupold, U. (1971). *Mutat. Res.* **11**, 373–390.
Golin, J. E., and Esposito, M. S. (1977). *Mol. Gen. Genet.* **150**, 127–135.
Gottlieb, D. J. C. (1973). Ph.D. Dissertation, University of Tennessee, Knoxville.
   Biogenesis of Chloroplasts and Mitchondria" (T. Bücher *et al.*, eds.), pp. 415–418. North-Holland Publ., Amsterdam.
Putrament, A., Baranowska, H., Ejchart, A., and Jachymcyzyk, W. (1977). *Mol. Gen. Genet.* **151**, 69–76.
Queiroz, C., Beale, G. H. (1974). *Genet. Res.* **23**, 233–238.
Rabinowitz, M., Jakovcic, S., Martin, N., Hendler, F., Halbreich, A., Lewin, A., and Morimoto, R. (1976). *In* "The Genetic Function of Mitochondrial DNA" (C. Saccone, and A. M. Kroon, eds.), pp. 219–230. North-Holland Publ., Amsterdam.
Ripley, L. S. (1975). *Mol. Gen. Genet.* **141**, 23–40.
Rytka, J., English, K. J., Hall, R. M., Linnane, A. W., and Lukins, H. B. (1976). *In* "Genetics and Biogenesis of Chloroplasts and Mitochondria" (T. Bücher *et al.*, eds.), pp. 427–442. North-Holland Publ., Amsterdam.
Sirover, M. A., and Loeb, L. A. (1976). *Biochem. Biophys. Res. Commun.* **70**, 812–817.
Slater, M. L. (1973). *J. Bacteriol.* **113**, 263–270.
Slonimski, P. P., and Tzagoloff, A. (1976). *Eur. J. Biochem.* **61**, 27–41.
Speyer, J. F. (1965). *Biochem. Biophys. Res. Commun.* **21**, 6–8.
Speyer, J. F., Karam, J. D., and Lenny, A. B. (1966). *Cold Spring Harbor Symp. Quant. Biol.* **31**, 693–697.
Storm, E. M., and Marmur, J. (1975). *Biochem. Biophys. Res. Commun.* **64**, 752–759.
Tzagoloff, A., Akai, A., and Needleman, R. B. (1975). *J. Bacteriol.* **122**, 826–831.
Tzagoloff, A., Foury, F., and Akai, A. (1976). *In* "The Genetic Function of Mitochondrial DNA" (C. Saccone and A. M. Kroon, eds.), pp. 155–161. North-Holland Publ., Amsterdam.
Wintersberger, U., and Wintersberger, E. (1970). *Eur. J. Biochem.* **13**, 20–27.
Wolf, K., Lang, B., Burger, G., and Kaudewitz, F. (1976). *Mol. Gen. Genet.* **144**, 75–81.

Chapter 3

Yeast Cell Selection and Synchrony: Density Gradients and Mating Factor Block

RICHARD W. SHULMAN

Department of Biochemistry,
Albert Einstein College of Medicine,
New York, New York

I. Introduction

The desire to examine events during the cell cycle has generated a variety of methods for selecting homogeneous populations of cells from heterogeneous cultures. These homogeneous populations are either examined directly or allowed to grow in a synchronous manner to provide cells at all stages of the cell cycle. A general discussion of the yeast cell cycle appeared in Vol. XI of this series (Chapter 11, Mitchison and Carter) and included some specific methods. The purpose of this chapter is to detail methods which are easy to perform and do not require special rotors or equipment. Furthermore, these techniques can be applied equally well to small or large quantities of cells.

Two methods are presented, one physical and the other physiological. The physical method utilizes the cell density shifts observed during the cell cycle of *Saccharomyces cerevisiae* (Hartwell, 1970; Wiemkin *et al.*, 1970). Two procedures for deriving homogeneous populations based on this observation are detailed. The Ludox density method (Shulman *et al.*, 1973) has the advantage of producing self-generating gradients, while the Renographin density method utilizes ordinary gradient markers. Both techniques are recommended for diploid cells.

Haploid cells, because of asynchronous bud growth (L. Hartwell, personal communication) and separation, are more difficult to synchronize by physical methods. Thus a physiological method, involving mating factor block of cells in G_1, is presented for both *a* and *α* cells. Methods for the purification of *a* and *α* mating factors are detailed.

II.   Isopycnic Density Gradients

Isopycnic equilibrium banding depends upon the periodic density fluctuations yeast cells undergo during the cell cycle. *Saccharomyces cerevisiae* is least dense just prior to cell separation (ρ_{min}, 0.95 through cell cycle) and most dense just after the end of DNA synthesis (ρ_{max}, 0.45 through cell cycle) (Fig. 1). Cells at these points in the cell cycle can be isolated from homogeneous populations by separation on density gradients. Such gradients have the advantage of exposing cells to minimal physiological perturbation. Selection by the cells' inherent density characteristics requires no metabolic manipulation. Also, the materials used to generate the gradients are metabolically inert.

A.   Ludox Gradients

Ludox HS-30 or -40 (E. I. DuPont de Nemours and Company, Industrial and Biochemicals Department, Wilmington, Del.; Van Waters and Rogers, 4000 First Avenue South, Seattle, Wash.) is a colloidal glass suspension. It is inexpensive and, because of its range of particle sizes, capable of producing self-generating gradients in a centrifugal field. (Ludox HS-30 or -40 usually can be obtained *gratis* from breweries, soft-drink bottlers, cardboard box factories, or other industries which use it as a coating on cardboard boxes to prevent skidding during palleting operations.) The concentration of Ludox used for the gradient depends upon the cell strain, the medium in which the cells are grown, and the phase of growth of the cells. Mid-log-phase cells separate optimally, while late log- or stationary-phase cells are generally unsuitable.

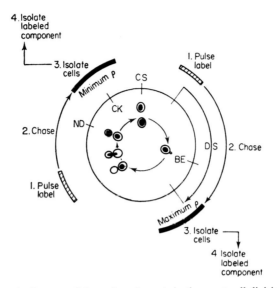

FIG. 1.   Schematic diagram of the order of events in the yeast cell division cycle. CS, Cell separation; DS, nuclear DNA synthesis; BE, bud emergence; ND, nuclear division; CK, cytokinesis; ρ, density. The numbered events refer to the four steps of the labeling protocol. (From Shulman *et al.*, 1973).

To determine the appropriate Ludox concentration for the yeast strain and condition of growth utilized, cells are initially displayed on a preformed stepwise Ludox gradient. A few milliliters of Ludox and distilled water ranging in concentration from 12 to 19% (v/v) are layered in 2% steps in a transparent test tube. A small quantity of mid-log-phase cells which have been removed quickly from medium by centrifugation and resuspended thoroughly in distilled water are layered on top. The tube is centrifuged in a Sorval centrifuge (preferably in the HB-4 swinging-bucket rotor) at 2000–3000 rpm for 5 minutes. The percent Ludox solution in which the cells are found is the appropriate concentration with which to proceed.

To the solution of Ludox HS-30 or -40 determined above, add sufficient 2 N HCl to bring to pH 7.5. (If the solution is titrated below pH 7.1, the Ludox will precipitate out.) If the cells are to be recultured, up to 0.1 volume of water can be replaced by medium. Lesser amounts of medium must be used if the Ludox precipitates out. Subsequent growth is further facilitated by the addition of 0.02 M glutathione. Dextran T40 (0.25 gm/10 ml, Pharmacia Fine Chemicals, Piscataway, N. J.) is then added, and the Ludox solution chilled to 4°C. (Spinco SW 27 No. 331220 long-bucket cellulose nitrate tubes are suitable for 14-ml gradients. These tubes fit the Sorvall No. 402 test-tube holder.) Mid-log-phase cells are collected by centrifugation, and the medium removed thoroughly by pouring off and aspiration.

The cell pellet is resuspended in a small volume of the chilled Ludox solution and clumps are dispersed by pipetting up and down with a 10-ml pipet. The dispersed cells are mixed with the remaining Ludox solution and centrifuged at 19,000 rpm for 10 minutes at 4°C in a Sorvall SS-34 rotor. The minimum-density cells at the top edge of the top band can be removed sequentially with a bent pasteur pipet, or pumped off. The maximum-density cells at the bottom edge of the bottom band can be removed by puncturing from the side or bottom with a needle. This is facilitated by using the ISCO 1801 G clamp seal assembly (Instrumentation Specialties Company, Lincoln, Nebr.) from which the inside ridge has been removed. The degree of homogeneity can be determined microscopically. Minimum-density cells can be obtained which are almost entirely ($\sim 90\%$) composed of cells with very large buds, while the remaining cells are unbudded. The maximum-density cells are composed mostly of cells with small buds ($\sim 70\%$), with the remainder consisting of equal numbers of unbudded cells or cells with medium-sized buds. The cells may be cultured directly without removal of the Ludox (Fig. 2). In a 14-ml gradient 1.0×10^{10} diploid yeast cells have been partitioned, and larger gradients containing more cells can be run.

B.   Renographin Gradients

Isopycnic equilibrium banding can also be achieved using Renographin, an iodinated polysaccharide (E. R. Squibb and Sons, Inc., New York, N.Y.). In this method a preformed 20–27% Renographin gradient containing 1% polyvinylpyrrolidone (PVP 360, Sigma Chemical Company, St. Louis, Mo.) is made using an ordinary chilled two-chamber gradient maker (Hartwell, 1970). Mid-log-phase cells are collected by centrifugation and resuspended thoroughly in chilled 20% Renographin. This is the light component, and an equal volume of 27% Renograph comprises the heavy component. The preformed gradient is centrifuged at 12,000 rpm for 1–2 minutes in a Sorvall HB-4 or SS-34 rotor at 4°C. The minimum- and maximum-density cells are removed and cultured as described for Ludox gradients.

III.   Isolation of Cells Labeled at Specific Periods of the Cell Cycle

Although density gradients select homogeneous populations from two points in the cell cycle, it is still possible to observe synthetic periods in other parts of the cell cycle. The limitation of the method to be described is that the

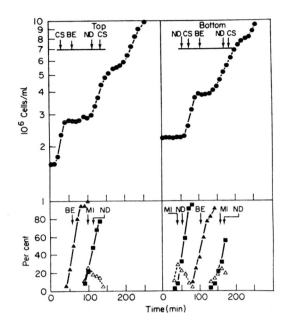

FIG. 2. Timing of cell cycle events in synchronous cultures of *S. cerevisiae* strain A364A D5. Cells were removed from the top and bottom bands of a Ludox density gradient and resuspended in synthetic medium with 0.1% yeast extract at 23°C. CS, Cell separation (from cell number, solid circles); BE, bud emergence (solid triangles); MI, mitotic index (open triangles); ND, nuclear division, (solid squares). Arrows indicate the time at which 50% of the cells completed a particular event. (From Shulman *et al.*, 1973).

component of interest must be stable. There are four steps in the approach, which are presented diagrammatically in Fig. 1. First, an asynchronous culture is pulse-labeled with a radioactive precursor. Second, a chase is imposed with an excess of nonradioactive precursor. During this time the cells continue their traverse of the cell cycle. Third, the cells of maximum and minimum density are isolated by isopycnic density banding. Fourth, the component of interest is purified from the two homogeneous populations of cells isolated on the density gradient.

Thus, if cells with a 120-minute cell cycle are pulse-labeled for 1 minute, chased for 30 minutes, and then isolated by density banding, the cells isolated will be those which at the time of labeling were at 0.25 of the cell cycle prior to the minimum-density (0.9 through the cell cycle) and maximum-density (0.45 through the cell cycle) collection points.

For the pulse, cells are concentrated to one-fifth the culture volume over a filter funnel with immobilized filters such as those manufactured by Falcon,

Nalgene, Seitz, or Statlich, fitted with a vacuum shutoff valve. A propeller is used to aerate the cells during the pulse. The chase is effected by returning the cells to the original culture volume plus the chase precursor.

IV. a- and α-Factor Selection and Synchronization of Haploid Yeast Cells

Constitutive a and α mating factors cause haploid $S.$ $cerevisiae$ cells of the opposite mating type to arrest as unbudded cells at a point just prior to the initiation of DNA synthesis (Buckring-Throm et $al.$, 1973; Wilkinson and Pringle, 1974). Removal of factor by endogenous inactivation or by washing out permits a synchronous round of growth. For α-factor block, approximately 90% of an originally heterogeneous population accumulate as single cells, and when these blocked cells are released 70–80% recover in synchrony (Bucking-Throm et $al.$, 1973). a-Factor block produces 96% unbudded cells (Wilkinson and Pringle, 1974).

The following method applies to either a or α synchrony blocks. Cells in early or mid log phase are collected by filtration and resuspended at a density of 5×10^6 cells/ml in fresh, well-aerated YM-1 (see Section IV, A) containing factor. Minimally sufficient factor from the opposite mating type cell (empirically determined) should be employed to give 90–95% single cells within the time span of one cell cycle. At this time the cells are removed by filtration, washed with fresh medium, and resuspended in well-aerated, fresh medium for subsequent synchronous growth.

A. a-Factor Partial Purification

$Saccharomyces$ $cerevisiae$ strain $\times$ 2180-1a (obtained from R. R. Mortimer, Donner Laboratory, Berkeley, Calif.) is grown in 12 liters of YNB (7 gm per liter yeast nitrogen base without amino acids plus 20 gm per liter glucose) or YM-1 (YNB plus 10 gm per liter succinic acid and 6 gm per liter NaOH) at 30°C in a New Brunswick Microferm fermentor or carboys to a density of about 2×10^8 cells/ml (stationary phase). The cells are removed by centrifugation in a Sharples continuous-flow centrifuge. The chilled, clear supernatant is subjected to ultrafiltration in an Amicon Model TC-1B recirculating system utilizing a 15-cm-diameter UM-05 membrane and a 20-liter reservoir (Amicon Corporation, Lexington, Mass.; Wilkinson and Pringle, 1974). The ultrafiltration rate is adjusted to 2 ml per minute under 75 psi nitrogen, with the recirculation pump set at an intermediate setting. When the volume has been reduced about 20-fold, the slightly turbid concentrate is partitioned into 10 to 20-ml aliquots and stored at $-20°C$. Further concentration can be obtained by subjecting the centrifuge-clarified concentrate to further ultra-filtration in an Amicon Model 50 cell equipped with a UM-05 or UM-10 membrane, at 45 psi nitrogen. When stored at $-20°C$, activity remained undiminished for at least 3 weeks.

B. α-Factor Purification

Saccharomyces cerevisiae strain X2180-1Bα is grown as described for a-factor purification. The use of minimal medium (described in Section IVA) is particularly important because of the interference of unwanted polypeptides present in undefined medium. The culture medium, freed of cells by centrifugation or filtration, is passed through Amberlite CG-50, 100 to 200-mesh (20-ml column, 1.5 × 11.5 cm; Bio-Rad Laboratories, Richmond, Calif.). To prepare the poured column, wash successively with 100 ml of 1 N NaOH, distilled water, 100 ml of 3 N HCl, distilled water, and 100 ml of 0.01 N HCl in 80% ethanol, and equilibrate with 0.1 M acetic acid. The cleared medium is passed through the column at room temperature at a flow rate of 50 ml per minute. The column is then washed with 200 ml of ethanol–water (1:1, v/v). This eluate is discarded, and the α factor is quantitatively eluted with 200 ml 0.01 N HCl in 80% ethanol. The resultant eluate is concentrated by rotary evaporation at 40°C to about 40 ml, diluted threefold with distilled water, and again evaporated. This procedure is repeated once more to remove most of the ethanol.

The resultant acidic solution is brought to pH 5.5 with concentrated ammonium hydroxide and left overnight at 4°C. The precipitate formed is removed by centrifugation and discarded. The remaining clear, brown supernate is then lyophilized.

The residue is suspended in 25 ml of methanol and stirred for 2 hours at room temperature. The insoluble material is removed by centrifugation and extracted twice more with 25 ml of methanol. The resultant supernatants are pooled and concentrated by rotary evaporation to about 15 ml. Water is added to permit freezing, and the solution is lyophilized. This lyophilized material is again extracted with 20 ml of methanol, and then twice with 10 ml of methanol. The dark-brown extract is concentrated to 10 ml by rotary evaporation, and the white precipitate formed is discarded.

The clear, dark-brown extract is then applied to a 5 × 60 cm column of Sephadex LH-20 (Pharmacia Fine Chemicals, Piscataway, N. J.) equilibrated with methanol. The column is eluted with methanol at a flow rate of 30 ml per hour. Two bands develop, a fast-moving brown one and a somewhat slower-moving faintly blue one. It is the blue band which contains the bulk of the factor and the highest-specific-activity material. When 7.5-ml fractions are collected, the blue-band material is found in fractions 70 to 80. When following absorbancy at 260 or 280 nm, the α factor is found in the major optical density peak (see Fig. 1; Duntze *et al.*, 1973).

The pooled fractions are dried in a rotary evaporator (40°C), resuspended in distilled water, and again evaporated to dryness. The resultant material is stored in this form or in 20 ml of 0.05 sodium acetate, pH 5, at 4°C. In the latter storage form α-factor activity has remained undiminished for over a year.

Further fractionation of α factor into four closely related oligopeptides has been achieved (Stotzler and Duntze, 1976). However, for the purpose of inducing synchrony, there seems to be no advantage for further fractionation, given the properties of the oligopeptides so far reported.

V. Discussion

A major concern in employing cell separation and synchrony techniques is the degree of physiological perturbation induced by these techniques. Physical methods such as isopycnic density gradients appear to have little or no effect on cell growth (unpublished results) or ribosomal protein synthesis (Shulman *et al.*, 1973).

Where the mating factor block has been employed, there is evidence that the method produces some metabolic perturbation. Protein and RNA synthesis continues, although DNA synthesis ceases (Throm and Duntze, 1970; Wilkinson and Pringle, 1974). The longer the block, the greater the macromolecule imbalance. Furthermore, cell wall changes occur in the presence of extended hormone treatment (Lipke *et al.*, 1976). It is clear that each application of any cell synchrony method requires an assessment of the effect of the method on the parameter to be measured.

ACKNOWLEDGMENT

I thank Vivian McRay and W. Duntze for their expert comments on mating factor isolation and characteristics.

REFERENCES

Bucking-Throm, E., Duntze, W., Hartwell, L. H., and Manney, T. R. (1973). *Exp. Cell Res.* **76**, 99–110.
Duntze, W., Stotzler, D., Bucking-Throm, E., and Kalbitzer, S. (1973). *Eur. J. Biochem.* **35**, 357–365.

Hartwell, L. H. (1970). *J. Bacteriol.* **104**, 1280–1285.
Lipke, P. N., Taylor, A., and Ballou, C. E. (1976). *J. Bacteriol.* **127**, 610–618.
Shulman, R. W., Hartwell, L. H., and Warner, J. R. (1973). *J. Mol. Biol.* **73**, 513–525.
Stotzler, D., and Duntze, W. (1976). *Eur. J. Biochem.* **65**, 257–262.
Throm, E., and Duntze, W. (1970). *J. Bacteriol.* **104**, 1388–1390.
Wiemkin, A., Matile, P., and Moore, H. (1970). *Arch. Microbiol.* **70**, 89–103.
Wilkinson, L. E., and Pringle, J. R. (1974). *Exp. Cell Res.* **89**, 175–184.

Chapter 4

The Ribosomal Proteins of
Saccharomyces cerevisiae

JONATHAN R. WARNER AND CHARLES GORENSTEIN

*Departments of Biochemistry and Cell Biology,
Albert Einstein College of Medicine,
New York, New York*

While studies on the ribosomal proteins of *Escherichia coli* have made great strides in the past few years (Stoffler and Wittman, 1977; Nomura, 1976), eukaryotic ribosomal proteins have been relatively neglected. Now that methods for purifying ribosomal proteins from rat liver have been developed (Tsurugi *et al.*, 1976), further work in this area should proceed more rapidly.

We have been studying the ribosomal proteins of *Saccharomyces cerevisiae* for several years because we are interested in the coordinated regulation of their synthesis (Gorenstein and Warner, 1976; Warner and Gorenstein, 1977) and because it is possible to obtain mutants which are likely to involve ribosomal proteins (Mortimer and Hawthorne, 1966; Skogerson *et al.*, 1973; Grant *et al.*, 1976).

This chapter describes methods of preparing and analyzing the ribosomal proteins of *S. cerevisiae* and points out certain properties that may be shared by the ribosomal proteins of all eukaryotes.

I. Preparation of Ribosomes

A 500-ml culture of *S. cerevisiae* strain A364A (Hartwell, 1967; Warner, 1971) growing in synthetic medium (Warner, 1971; Sripati and Warner, this volume) at a concentration of 2 to 4 × 10^7/ml is harvested by centrifugation and washed twice with water. The cells are suspended in 10 ml of TMN (50 mM Tris–acetate, pH 7.0, 50 mM NH$_4$Cl, 12 mM MgCl$_2$, 1 mM dithiothreitol), added to 5 ml of glass beads (0.1 mm, VWR Scientific No. 34007-088), and shaken in Bronwill homogenizer (VWR Scientific No. 34006-008) for two 30-second bursts. The glass beads are allowed to settle, the supernate removed, and the beads rinsed with 4 ml of TMN. The combined supernatants are centrifuged at 10,000 g for 10 minutes. The supernatant is removed, and the centrifugation repeated one or two times until there is no gross turbidity. Five milliliters of extract is then layered over 4 ml of 10% sucrose in HKB (500 mM KCl, 5 mM MgCl$_2$, 20 mM Tris–HCl, pH 7.4) in a polycarbonate screw-cap tube (Beckman No. 339574) and centrifuged for 120 minutes at 60,000 rpm in a Beckman 75 Ti or 65 rotor. The high concentration of KCl used in the sucrose cushion strips loosely bound proteins from the ribosomes (Sherton and Wool, 1974). The supernatant is carefully sucked off, and the ribosomes are resuspended by stirring in 1.5 ml of TMN.

If it is desired to separate subunits, the pellet is suspended in 1.5 ml of HKB and incubated at 30°C for 15 minutes with 1 mM puromycin (Sherton and Wool, 1974). The sample is layered on 10–25% (w/w) sucrose gradients in HKB and centrifuged at 15°C for 20 hours at 18,000 rpm in the large bucket of a SW 27 rotor. Each tube accommodates about 50–75 A_{260} units of ribosomes. Centrifugation at 15°C reduces the dimerization of 40S subunits. The gradients are collected through a recording spectrophotometer, and the 60 and 40S fractions are pooled, diluted with an equal volume of HKB, and centrifuged for 16 hours at 4°C at 50,000 rpm. The pellet is suspended in 1–1.5 ml of TMN. To obtain subunits free of contamination at least two cycles of sucrose gradients are necessary (see Fig. 2).

II. Preparation of Ribosomal Proteins

The preparation of ribosomal proteins is essentially as developed by Hardy *et al.* (1969) for *Escherichia coli*. To the ribosome suspension stirring in ice is added 0.1 volume of 1 M MgCl$_2$, 2.5 volumes of glacial acetic acid, and 0.1 volume of 0.1 M dithiothreitol. Eukaryotic ribosomal proteins appear to be more susceptible to disulfide-induced aggregation than prokaryotic ribosomal proteins. The acetic acid solubilizes the protein and causes the RNA to precipitate. After 30–60 minutes, the sample is centrifuged at 20,000 g for 10 minutes, and the supernatant carefully removed into Spectrapor No. 3 membrane tubing (Spectrum Medical Industries, Los Angeles, Calif.). This is a low-molecular-weight cutoff tubing. Ribosomal proteins are small, and high osmotic pressure builds up during the dialysis. Using conventional tubing we often suffered appreciable losses. After extensive dialysis against 1% acetic acid, the proteins can be stored at $-20°C$ indefinitely.

In some cases it is necessary to prepare ribosomal proteins directly from a whole-cell lysate (Gorenstein and Warner, 1976). If so, the lysate is treated with $MgCl_2$, dithiothreitol, and acetic acid as described above for the ribosome suspension and, after dialysis, subjected to two-dimensional gel electrophoresis (see Section III,B,2).

III. Analysis of Ribosomal Proteins

A. One-Dimensional Gels

Ribosomal proteins can be analyzed directly on one-dimensional SDS gels without removing RNA. Figure 1 shows such a pattern of 60 and 40S subunits. Cells were labeled for several generations with 15 amino acids uniformly labeled with ^{14}C. Subunits were prepared as described in Section 1. Samples from a sucrose gradient were treated with 10% trichloroacetic acid for 30 minutes at 0°C and collected by centrifugation at 20,000 g for 15 minutes. A sample was suspended in 0.2 ml of 0.01 M NaPO$_4$, pH 7, 1% SDS, and 1% mercaptoethanol, heated in boiling water for 1 minute, and analyzed on a cylindrical gel as described by Maizel (1969). The gel was fractionated and counted (Warner, 1971). Although a few of the larger proteins of the 60S subunit are resolved, most peaks represent a mixture of proteins (Warner, 1971). To obtain clear a resolution of ribosomal proteins, two-dimensional methods are necessary.

B. Two-Dimensional Gels

We have employed two systems of two-dimensional polyacrylamide gel analysis to separate yeast ribosomal proteins. The first is based on the Kaltschmidt–Wittman (1970; Wittman, 1974) system and employs 6 M urea at pH 8.6 in the first dimension and at pH 4.5 in the second dimension. The second is a modified version of the Mets and Bogorad (1974) system and

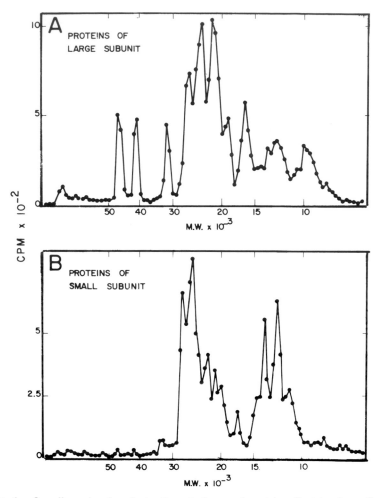

FIG. 1.   One-dimensional analysis of yeast ribosomal proteins. Proteins from ribosomal subunits were analyzed on SDS gels as described in the text. Protein markers in parallel gels provided the molecular-weight scale. Reprinted from Warner (1971) by permission.

employs 8 M urea at pH 5.0 in the first dimension and sodium dodecyl sulfate (SDS) in the second dimension. The two systems are described separately, and the results then compared. For both gel systems the importance of using proteins free of RNA cannot be overemphasized. RNA binds to ribosomal proteins even in urea solutions and causes poor resolution and decreased yields.

Parenthetically, several attempts to adapt to ribosomal proteins the O'Farrell (1975) two-dimensional system, based on isoelectric focusing, have failed because the isoelectric points of most of the ribosomal proteins are higher than the working range of presently available ampholytes.

1. Method 1: pH 8.6 and pH 4.5 in 6 M Urea

This method is a slightly modified version of the method described by Wittman (1974).

The solutions needed are:

Solution A: Sample buffer
    100 ml 8 M urea
    0.085 gm Na_2EDTA
    0.32 gm boric acid
    Store frozen in small aliquots.
    Add 1% mercaptoethanol just before use.

Solution B: First-dimension gel
    6.0 gm acrylamide, recrystallized
    0.225 gm bisacrylamide, recrystallized
    0.8 gm Na_2EDTA
    3.2 gm boric acid
    4.85 gm Tris
    10 ml 3% linear polyacrylamide (BDH 29788)
    75 ml 8 M urea
    Water to 100 ml
    For polymerization of 10 ml, use 30 μl 10% ammonium persulfate
        and 15 μl TEMED.
    A 5 × 100 mm gel requires ~2 ml.

Solution C: First-dimension running buffer
    2.4 gm Na_2EDTA
    9.6 gm boric acid
    14.55 gm Tris
    Water to 1000 ml
    Degas

Solution D: Soaking gel after first dimension
    (1) 2.5 ml concentrated HCl (first soak for 30 minutes)
        100 ml 8 M urea (Avital and Elson, 1974)
    (2) 0.074 ml acetic acid
        0.24 ml 5 N KOH (second soak for 30 minutes)
        100 ml 8 M urea

Solution E: Second-dimension gel
   180 gm acrylamide
   5 gm bisacrylamide
   52.3 ml acetic acid
   9.6 ml 5 N KOH
   750 ml 8 M urea
   Water to 1000 ml
   Use 80 ml per slab; add 1.32 ml 10% ammonium persulfate and
   0.45 ml TEMED.

Solution F: Second-dimension running buffer
   14 gm glycine
   1.5 ml acetic acid
   Water to 1000 ml

Note: All 8 M urea is deionized, filtered, and stored in a cold room.

a. First-Dimension Gel. A 5 × 140 mm glass tube is sealed at one end with Parafilm and taped to the side of a bench. It is filled to 100 mm with degassed solution B to which TEMED and ammonium sulfate have been added, and overlaid with water. After polymerization, which occurs in 30–60 minutes, the Parafilm is removed, and the tube is placed in the apparatus and filled with running buffer containing 300 mg/ml urea. The urea prevents the precipitation of any proteins which migrate upward. The sample (100–300 μg of ribosomal protein) is dissolved in 100 μl of solution A and layered under the running buffer using microsyringe. For the separation of *basic proteins* run toward the cathode at 110 V for 20 hours. For the separation of *acidic proteins* run toward the anode at 60 V for 20 hours.

b. Equilibration for the Second Dimension. Remove gels with a syringe containing glycerol. Mark the bottom by inserting a No. 27, $\frac{1}{2}$-inch needle. Equilibrate for 30 minutes in solution D1 with gentle shaking and then for 30 minutes in solution D2.

Notes: (1) We found that polymerizing the sample in the middle of the first-dimension gel, as described by Kaltschmidt and Wittmann (1970), gave poor and irreproducible yields of yeast ribosomal protein. Therefore we analyze acidic and basic ribosomal proteins on separate gels. The improved yield and pattern more than compensate for the duplicate sample required. (2) The linear polyacrylamide in the first dimension provides strength and dimensional stability to the gel during the several steps in which it must be handled. (3) Touch the first-dimensional gel only with gloves during equilibration. An electrophoresed fingerprint can complicate the analysis of the gel.

c. Second-Dimension Gel. The apparatus for the second-dimension gel, shown in Studier (1973), has been modified as described in Stewart and Crouch (1977). The separation gel is 140 mm wide, 130 mm high, and 3 mm thick. One glass plate is beveled, and the first-dimension gel is placed in the bevel between the two plates.

One hundred milliliters of solution E, containing TEMED and ammonium persulfate, is degassed and poured around and finally over the first-dimension gel. The apparatus can be tilted to avoid the formation of bubbles at the seal between the two gels. A good seal between the gels is essential to keep the spots from spreading. The gel is overlaid with water and allowed to polymerize. The second-dimension running buffer, solution F, is placed in both wells, and electrophoresis toward the cathode carried out for 20 hours at 85 V.

The gels are stained in 0.2% coomassie brilliant blue in 50% methanol for 4–6 hours and destained in 30% methanol containing a mesh bag of washed charcoal (Sherton and Wool, 1974). The gels are stored in plastic bags (No. F-13178, Bel Art Products, Pequannock, N.J.), sealed with a T-bar plastic sealer (Harwill Company, Santa Monica, Calif.), and keep indefinitely.

d Results of Method 1. The results of the Wittman gel analysis of yeast ribosomal proteins are described in Zinker and Warner (1976), from which Figs. 2–4 have been extracted. Similar but not identical patterns have been presented by Kruiswijk and Planta (1974) and by Ishiguro (1976). There are

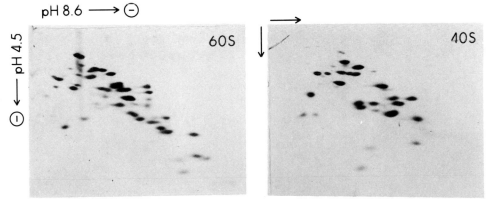

pH 8.6 ⟶ ⊖

pH 4.5 ⟶ ⊕

60S   40S

FIG. 2. Separation of basic yeast ribosomal proteins by method 1 (pH 8.6 and pH 4.5). Proteins of 60 and 40S subunits were subjected to two-dimensional electrophoresis as described for basic proteins, and the gels stained. A small amount of cross-contamination permits alignment of the two gels; e.g., L1 and L2 can be seen faintly above S1; S3 and S4 can be seen faintly above L8. Reprinted from Zinker and Warner (1976) by permission.

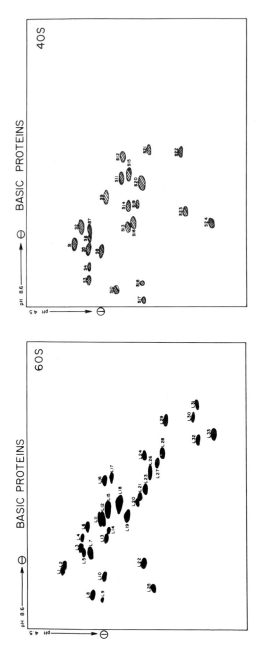

FIG. 3. Numbering system of yeast basic ribosomal proteins separated by method 1. Compare with Fig. 2. Reprinted from Zinker and Warner (1976) by permission.

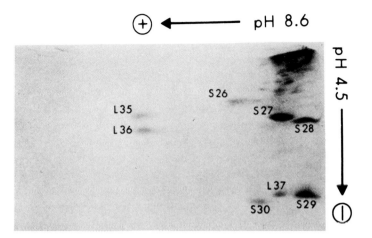

FIG. 4. Separation of acidic yeast ribosomal proteins by method 1 (pH 8.6 and pH 4.5). Total ribosomal proteins were subjected to two-dimensional electrophoresis as described for acidic proteins, and the gels stained. The attribution of proteins to individual subunits was determined from the analysis of separated subunits. Proteins L34 and S25 are sometimes seen in the upper right corner which, however, is where most contaminating proteins run. Reprinted from Zinker and Warner (1976) by permission.

sufficient differences that a compatible numbering system must await further comparisons.

As is the case with bacteria (Wittman, 1974) and mammalian cells (Sherton and Wool, 1972), nearly all yeast ribosomal proteins are very basic, migrating toward the cathode at pH 8.6. A comparison between yeast and mammalian ribosomal proteins reveals that:

1. The 60 S subunits of all eukaryotes have two proteins larger than all the rest. This is particularly clear in one-dimensional SDS gels, e.g., Fig. 1. It is likely therefore that proteins L1 and L2 in Fig. 1 correspond to proteins L3 and L4 of rat liver (Sherton and Wool, 1972).

2. Proteins S5 and S6 are two isomers of a phosphorylated protein whose unphosphorylated form is probably S7 (Zinker and Warner, 1976). Because of their comigration in a Wittman gel and in a gel containing SDS (see Section III.B.2) these proteins are undoubtedly the functional counterparts of the phosphorylated protein, S6, of rat liver (Gressner and Wool, 1974).

3. Proteins L35 and L36 are very acidic proteins, which are also phosphorylated (Zinker and Warner, 1976). They appear to be analogous to L40 and L41 of rat liver, which may have a functional relationship with proteins L7 and L12 of *E. coli* (Wool and Stoffler, 1974). It remains to be seen whether or not L35 and L36 have the same amino acid sequence as L7 and L12.

2. METHOD 2: pH 5.0 IN 8 M UREA AND SDS

This method is a modification (Gorenstein and Warner, 1976) of the method of Mets and Bogorad (1974). Briefly, in the first dimension the proteins are separated based on their charge at pH 5.0. In the second dimension they are separated on the basis of their molecular weight in the presence of SDS.

a. First Dimension
Solution A: First-dimension gel mix
    4% (w/v) acrylamide
    0.1% (w/v) bisacrylamide
    8 M urea
    0.057 M bis Tris adjusted to pH 5.0 with acetic acid
Prior to polymerization this solution is degassed and 3 μl of 10% ammonium persulfate and 1 μl of TEMED are added per milliliter of solution A.

Solution B: Upper electrode solution
    0.01 M bis Tris adjusted to pH 4.0 with acetic acid

Solution C: Lower electrode solution
    0.179 M potassium acetate adjusted to pH 5.0 with acetic acid

Solution D: First-dimension sample buffer
    10% (v/v) β-Mercaptoethanol
    10% (v/v) glycerol
    1% (v/v) acetic acid
    0.01% (w/v) basic fuschin
    8 M urea
This solution is stored in aliquots at $-20°$C.

Solution E: equilibration buffer
    0.5 M Tris, 1% SDS, adjusted to pH 6.8 with HCl
b. Second Dimension
Solution F: Resolving gel
    17% (w/v) acrylamide
    0.45% (w/v) bisacrylamide
    0.1% (w/v) SDS
    0.175 M Tris adjusted to pH 8.8 with HCl
If the gel is to be dried for autoradiography, inclusion of 0.3% linear polyacrylamide in solution F reduces the tendency of the gels to crack during drying. Prior to polymerization this solution is degassed, and 6 μl of 10% ammonium persulfate and 0.5 μl of TEMED are added per milliliter of solution F.

Solution G: Stacking gel
  4% (w/v) acrylamide
  0.11% (w/v) bisacrylamide
  0.1% (w/v) SDS
  0.0625 M Tris adjusted to pH 6.8 with HCl
The gel is degassed and polymerized with 10 μl 10% ammonium persulfate
and 0.5 μl TEMED per milliliter of solution G.

Solution H: Electrode buffer
  0.05 M Tris
  0.38 M glycine
  0.05% (w/v) SDS

  c. *First Dimension.* Six-millimeter (I.D.) 150-mm glass tubes, washed
with chromic acid, methanolic KOH, and 1% Siliclad (Beckman Instruments), are sealed at the bottom with Parafilm, filled to 110 mm with gel mix
solution A, and overlayered with water. The gels are allowed to polymerize
for 1 hour. The gel tubes are placed in the apparatus, and the upper reservoir
filled with electrode solution B and the lower reservoir with electrode
solution C.

  The sample, consisting of up to 400 μg of lyophilized protein is dissolved
in 100 μl of sample buffer D. Occasionally difficulties are encountered in
solubilizing total yeast extracts. They can be overcome by increasing the
acetic acid concentration in sample buffer D to 10% (w/v) without affecting
the resolution of the gel. The sample is layered on the surface of the gel with
a microsyringe.

  The gels are electrophoresed toward the cathode at constant voltage for
900 volt-hours or until the tracking dye, basic fuschin, has reached the
bottom of the gel.

  d. *Equilibration.* The first-dimension gels are removed from the glass
tubes by rimming them with a 0.1% SDS solution. Each gel is then placed
in a 125-ml Erlenmeyer flask containing 25 ml of solution E and shaken for
45 minutes. This procedure changes the pH of the first-dimensional gel to
that of the second-dimension stacking gel and introduces the detergent
SDS.

  e. *Second Dimension.* A 100 × 3 mm second-dimension slab gel is polymerized between two glass plates in the apparatus described in Section
III,B,1. Solution F is added and overlayered with 0.1% SDS, using an
atomizer.

  After the gel has polymerized, the excess unpolymerized material is removed and the equilibrated first-dimension gel is laid across the apparatus,
resting between the beveled edge and the straight glass plate. The cylindrical
gel is cemented into place by means of the stacking gel solution G. Care

must be taken to see that no bubbles are trapped under the cylindrical gel. If a reference sample or a tracking protein is to be used, a well can be made by placing a small piece of Plexiglas near one edge of the apparatus and allowing the stacking gel to polymerize around it. After polymerization, the reservoirs are filled with electrode buffer, and electrophoresis is conducted at constant voltage toward the anode for 1100 volt-hours or until a cytochrome-c marker is within 1 cm of the bottom of the gel.

The gels are stained, destained, and stored as described in Section III,B,1.

f. Quantitation. The amount of radioactively labeled protein in a spot can be quantitated from an autoradiograph by comparing the intensity of the protein spot with a series of known standards. A more accurate means of measuring the amount of radioactively labeled protein is to excise the spot from the gel and count the radioactivity in a scintillation counter (Gorenstein and Warner, 1976). Capillary stainless-steel tubing (Small Parts, Inc.), sharpened at one end, make very convenient borers. The excised spots are dried in glass scintillation vials at 60°C for 2 hours; then 0.5 ml of freshly made 30% H_2O_2 (made by diluting 50% H_2O_2) is added. The vials are tightly capped and incubated at 60°C overnight. After cooling, 10 ml of Aquasol (New England Nuclear) or Readisolv GP (Beckman Instruments) is added. Because of the large amount of peroxide present, problems with fluorescence may be encountered. We avoid this problem by storing the vials in a cool, dark place for several hours before counting in a refrigerated counter.

g. Results of Method 2. The analysis of yeast ribosomal proteins by method 2 is shown in Fig. 5. Most of the proteins have been attributed to one of the two subunits (Table I). In only a few cases has it been possible to cross-identify proteins in the two systems:

1. Because of their size, proteins 1 and 2 are likely to L1 and L2. The identification of L2 with protein 2 is confirmed by the finding that both are methylated (Cannon *et al.*, 1977; D. Barton, unpublished).

2. Protein 9 is the unphosphorylated form of the 40S protein S5, S6, S7 (P2 of Zinker and Warner, 1976). The phosphorylated form runs slightly to the left of protein 9.

3. Protein 14 is the phosphorylated protein S27 (P3 of Zinker and Warner, 1976).

4. Protein 16 is the exchangeable 60S protein L7 (E2 of Zinker and Warner, 1976).

Although method 1 probably gives better resolution of more proteins, method 2 has three distinct advantages:

1. It is more reliable.

2. It gives better separation of marginally basic ribosomal proteins, e.g., protein 14.

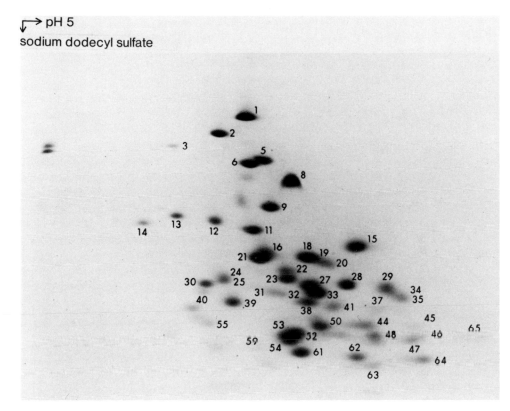

FIG. 5. Separation of yeast ribosomal proteins by method 2 (pH 5 0 SDS). Total ribosomal proteins were subjected to two-dimensional electrophoresis as described, and the gels stained. Reprinted from Gorenstein and Warner (1976) by permission.

3. Proteins are more soluble in its sample buffer. Therefore method 2 is preferable when isolating ribosomal proteins from a total cell extract (Gorenstein and Warner, 1976).

IV. Conclusion

A. What is a Ribosomal Protein?

There are no covalent bonds between the components of a ribosome. Since ribosomes are well known to bind many proteins non- or quasi-specifically, e.g., RNase of *E. coli* (Waller, 1964) and initiation factors in both prokaryotes and eukaryotes (Weissbach and Ochoa, 1976), it has been a

TABLE I

ASSIGNMENT TO SUBUNIT OF PROTEINS
SEPARATED BY METHOD 2—FIG. 5[a]

60S				40S	
1	22	38	61	5	37
2	23	39	62	9	40
3	24	44	64	12	41
6	25	45	65	13	42
8	27	47		14	50
10	28	48		19	52
11	29	49		20	55
15	31	51		21	60
16	32	57		30	61
18	33	58		36	63

[a] Protein 61 is found in both subunits. Whether this represents two unresolved proteins or a single protein that can be purified with either subunit is unclear at present.

continuous problem to determine which proteins should be considered truly ribosomal. The question is complicated by the fact that the salt washes necessary to remove contaminating proteins often remove some ribosomal proteins (Hardy, 1975). Clearly certain proteins in Figs. 2 and 5 are much less distinct than others; it is unlikely that the variation is due only to different affinities for the coomassie stain. Furthermore, since only a small portion of the ribosomes is active in most *in vitro* systems, a functional assay is seldom feasible.

Thus, while it is likely that most of the spots in Figs. 2–5 are ribosomal proteins, a few may not be. To distinguish between the two it may be useful to develop criteria based on the assembly of ribosomal particles or on the regulation of the synthesis of these components, as we have proposed for mammalian cells (Warner, 1966) and for *S. cerevisiae* (Gorenstein and Warner, 1976).

B.   Uniformity of Eukaryotic Ribosomal Proteins

From an evolutionary point of view ribosomal proteins appear to be very stable, as first demonstrated by Nomura *et al.* (1968) when they reconstituted active 30S subunits from 16S RNA of *E. coli* and ribosomal proteins of *Bacillis stearothermophilus*. It is not surprising therefore that the ribosomal proteins of all mammals appear to be very similar (Delaunay *et al.*, 1973; S. Morgan and J. R. Warner, unpublished).

Nevertheless, it is interesting to note that the two phosphorylated yeast proteins revealed on a pH 5 SDS gel, proteins 9 and 14, migrate in positions essentially identical to those of the two phosphorylated proteins of hamster 40S ribosomes (Schubart *et al.*, 1977). Furthermore, the three exchangeable proteins of yeast (Warner and Udem, 1972) migrate in positions essentially identical to those of the three exchangeable proteins of HeLa (Warner, 1966) on a pH 5 SDS gel. We suggest that for most of the ribosomal proteins there is likely to be a 1:1 correlation between eukaryotes as distant as yeast and humans.

ACKNOWLEDGMENTS

Work from this laboratory was supported by grants from the American Cancer Society No. NP 72 G, the National Science Foundation No. PCM 7503938 and the National Institutes of Health No. P 30 CA 13330. J.R.W. was a Faculty Career Awardee of the American Cancer Society.

REFERENCES

Avital, S., and Elson, D. (1974). *Anal. Biochem.* **57**, 287–292.
Cannon, M., Schindler, D., and Davies, J. (1977). *FEBS Lett.* **75**, 187–191.
Delaunay, J., Creusot, F., and Schapira, G. (1973). *Eur. J. Biochem.* **39**, 305–312.
Gorenstein, C., and Warner, J. R. (1976). *Proc. Natl. Acad. Sci. U.S.A.* **73**, 1547–1551.
Grant, P., Schindler, D., and Davies, J. (1976). *Genetics* **83**, 667–673.
Gressner, A. M., and Wool, I. G. (1974). *J. Biol. Chem.* **249**, 6917–6925.
Hardy, S. J. (1975). *Mol. Gen. Genet.* **140**, 253–274.
Hardy, S. J., Kurland, C. G., Voynow, P., and Mora, G. (1969). *Biochemistry* **8**, 2897–2905.
Hartwell, L. (1967). *J. Bacteriol.* **93**, 1662–1670.
Ishiguro, J. (1976). *Mol. Gen. Genet.* **145**, 73–79.
Kaltschmidt, E., and Wittman, H. G. (1970). *Anal. Biochem.* **36**, 401–407.
Kruiswijk, T., and Planta, R. J. (1974). *Mol. Biol. Rep.* **1**, 409–414.
Maizel, J. V., Jr. (1969). *In* "Fundamental Techniques in Virology" (K. Habel and N. P. Salzman, eds.), pp. 334–357. Academic Press, New York.
Mets, L., and Bogorad, L. (1974). *Anal. Biochem.* **57**, 200–207.
Mortimer, R., and Hawthorne, D. (1966). *Genetics* **53**, 165–173.
Nomura, M., Traub, P., and Bechmann, H. (1968). *Nature (London)* **219**, 793–795.
O'Farrell, P. H. (1975). *J. Biol. Chem.* **250**, 4007–4013.
Schubart, U.K., Shapiro, S., Fleischer, N., and Rosen, O. M. (1977). *J. Biol. Chem.* **252**, 92–101.
Sherton, C., and Wool, I. G. (1972). *J. Biol. Chem.* 4460–4467.
Sherton, C., and Wool, I. G. (1974). *In* "Methods in Enzymology" (L. Grossman and K. Moldave, eds.), Vol. 30, pp. 506–526. Academic Press, New York.
Skogerson, L., McLaughlin, C., and Wakatama, E. (1973). *J. Bacteriol.* **116**, 818–822.
Stewart, M., and Crouch, R. (1977). *Anal. Biochem.* (in press).
Stöffer, G., and Wittman, H. G. (1977). *In* "Molecular Mechanisms of Protein Biosynthesis" H. Weissbod and S. Pestka, eds.), pp. 117–202. Academic Press, New York.
Studier, F. W. (1973). *J. Mol. Biol.* **79**, 237–244.
Tsurugi, K., Collatz, E., Wool, I. G., and Lin, A. (1976). *J. Biol. Chem.* **251**, 7940–7946.
Waller, J. P. (1964). *J. Mol. Biol.* **10**, 319–336.

Warner, J. R. (1966). *J. Mol. Biol.* **19**, 383–398.
Warner, J. R. (1971). *J. Biol. Chem.* **246**, 447–454.
Warner, J. R., and Gorenstein, C. (1977). *Cell* **11** (201–212).
Weissbach, H., and Ochoa, S. (1976). *Annu. Rev. Biochem.* **45**, 191–216.
Wittman, H. G. (1974). *In* "Methods in Enzymology" (L. Grossman and R. Moldave, eds.), Vol. 30, pp. 497–518. Academic Press, New York.
Wool, I. G., and Stoffler, G. (1974). *In* "Ribosomes" (M. Nomura, A. Tissieres, and P. Lengyel, eds.), p. 417. Cold Spring Harbor Lab., Cold Spring Harbor, New York.
Zinker, S., and Warner, J. R. (1976). *J. Biol. Chem.* **251**, 1799–1807.

NOTE ADDED IN PROOF

Using as a third dimension the analysis of proteolytic digests of proteins separated on 2D gels (Cleveland, D., Fischer, S., Kirschner, M., and Laemmli, U. (1977). *J. Biol. Chem.* **252**, 1102–1106), P. J. Wejksnora has cross-identified a number of ribosomal proteins as separated by the two 2D gel methods (see Table II). These results demonstrate the identity of the phosphorylated proteins S5 and S7 with No. 9. Also No. 61 is shown to consist of two proteins, one from each subunit.

TABLE II

CROSS IDENTIFICATION OF PROTEINS AS SEPARATED BY
METHODS 1 AND 2

40S		60S	
System 1	*System 2*	*System 1*	*System 2*
S1	12	L1	2
S2	5	L2	1
S3	30	L3	6
S4	40	L4	11
S5	9	L6	8
S7	9	L8	25
S8	21	L11	16
S9	19	L13	18?[a]
S10	55	L16	15?[a]
S12	41	L17	22
S13	42	L18	33
S14	52	L20	62?[a]
S16	50	L21	31?[a]
S20	61?[a]	L24	29?[a]
S21	37		
S22	45		
S27	14		
S28	13		

[a] Protein pairs marked (?) are likely matches, but not confirmed unquestionably by either method.

Chapter 5

Isolation, Characterization, and Translation of mRNA from Yeast

CONJEEVARAM E. SRIPATI

*Institut de Biologie Physico-chimique,
Paris, France*

AND

JONATHAN R. WARNER

*Departments of Biochemistry and Cell Biology, Albert Einstein College of Medicine,
New York, New York*

Many of the methods for the preparation and analysis of RNA and ribosomes from *Saccharomyces cerevisiae* were described by Rubin (1975) in Vol. XII of this series. This chapter deals more specifically with mRNA, its isolation in undegraded form from cells and polysomes, the characterization of the 5'- and 3'-structures unique to mRNA, and its translation in an extract of wheat germ.

I.   Media, Cells, and Spheroplasts

Most of the following procedures have been carried out with cells of strain A364A (a, gal_1, $ade_{1,2}$, ura_1, his_7, lys_2, tyr_1) (ATC No. 22244) grown in synthetic complete medium (SC) which contains, per liter, 6 gm of yeast nitrogen base without amino acids (Difco), 10 gm of succinic acid, 6 gm of NaOH, 20 mg each of adenine and uracil, and 40 mg each of histidine, lysine, and tyrosine (Hartwell, 1967; Warner, 1971). Log-phase cells grown in synthetic medium are more easily disrupted and more easily converted to spheroplasts than cells grown in broth medium.

Because of *Saccharomyces'* tough cell wall, fragile structures such as poly-ribosomes can only be isolated from spheroplasts. For the preparation of spheroplasts (Hutchison and Hartwell, 1967; Udem and Warner, 1972) a culture of logarithmically growing cells at a concentration of less than 2.5×10^7 cells/ml is collected either by filtration on a Millipore filter or by centrifugation, washed with sterile water, and suspended in 1 M sorbitol at one-tenth the original volume. One-thousandth of the original volume of Glusulase (Endo Laboratories) is added, and the mixture incubated at room temperature with occasional swirling. Spheroplasts are usually formed with-in 10–20 minutes, as can be checked by diluting an aliquot into 0.1% sodium dodecyl sulfate (SDS). The culture is then diluted 10-fold with SC containing 0.4 M $MgSO_4$ (SCM) as an osmotic support and incubated with gentle aeration. Within 60 minutes macromolecular synthesis returns to normal and continues for 3–6 hours, although no cell division occurs (Hutchison and Hartwell, 1967). Spheroplasts are harvested by pouring the culture over frozen, crushed 1 M sorbitol. To prevent ribosomes from running off poly-somes as the temperature falls, cycloheximide is added to 100 μg/ml just before the culture is chilled.

II.   Isolation of Polysomes, Single Ribosomes, and Ribosomal Subunits

A working definition of mRNA, in the absence of demonstrable template activity, is that fraction of cytoplasmic RNA, excluding rRNA and tRNA, which sediments with polysomes and can be released from polyribosomes by EDTA or puromycin (Penman *et al.*, 1968). Thus, for many purposes, the preparation of mRNA necessarily requires the isolation of polysomes, which must be carried out in such a way as to prevent nuclear lysis during the disruption of spheroplasts and to inhibit nuclease activity. The optimization of the lysis medium with respect to pH, ionic strength, detergent, and other constituents, as well as the method of disruption of spheroplasts without damaging cell nuclei have been described (Udem, 1972; Udem and Warner, 1973). RNase activity is suppressed by the use of diethylpyrocarbonate (DEPC, diethyloxydiformate, Eastman No. 9793). The use of DEPC as a nuclease inhibitor has been evaluated by Ehrenberg *et al.* (1976). We have found it to protect mRNA from degradation without affecting its template activity in the synthesis of ribosomal proteins *in vitro*. (See Section VIII).

The following operations are carried out at 2 to 4°C using autoclaved glassware and sterile solutions. Washed spheroplasts are suspended at a concentration of 2×10^8 cells/ml in buffer A (10 mM NaCl, 10 mM PIPES, 5 mM MgCl$_2$, 1 mM CaCl$_2$, 1 mM dithiothreitol, 0.01% spermidine–HCl, pH 6.5) by repeated drawing and blowing gently with the aid of a pipet. The spheroplasts swell in this hypotonic medium. After 2 minutes the swollen spheroplasts are lysed by adding 0.1 volume of a 1% (v/v) solution of saponin. Lysis is allowed to proceed for about 3 minutes with frequent mixing. To complete solubilization of the cytoplasmic membranes Brij-58 is added to a final concentration of 0.3%. The lysate is centrifuged for 10 minutes at 12,000 *g*. The slightly opalescent supernatant which constitutes the cytoplasmic fraction is removed and treated with DEPC at a final concentration of 2% (v/v) to inhibit nuclease activity. An aliquot of 1.2 ml is layered over an 11-ml linear 7–47% (w/w) sucrose gradient in buffer A and centrifuged for 100 minutes in a Beckman SW 41 rotor at 36,000 rpm at 4°C. The gradients are pumped from the bottom, and the absorbance at 260 nm is recorded continuously. A typical polysome profile is shown in Fig. 1. Nearly 95% of the ribosomes are in the polysomes and only about 4% in the 80S monosomes. The polysome profile is apparently not altered by treatment of the cytoplasmic fraction with DEPC. When EDTA is present in the gradient, the polysomes dissociate and sediment as 60 and 40S subunits. Under these conditions the mRNA sediments at about 60S. DEPC treatment also does not alter these properties. However, if DEPC treatment of the cytoplasmic fraction is omitted, much less poly-A-containing RNA is recovered from the polysomes (see Section IV). This is even more likely if the polysomes are dissociated and analyzed on gradients containing EDTA, where the mRNA sediments near the top of the gradient. Treatment of the cytoplasmic fraction with DEPC as described appears to suppress nuclease activity effectively.

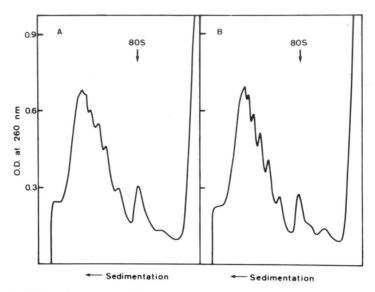

FIG. 1. Effect of treatment of cytoplasm with DEPC on the sedimentation profile of polysomes. Cytoplasm prepared from spheroplasts (derived from approximately 3×10^8 cells) was divided into two fractions: One fraction was treated with DEPC as described, and each fraction was centrifuged on a 7–47% linear sucrose gradient. Optical density scanning of the gradients is shown. (A) Untreated cytoplasm. (B) Cytoplasm treated with DEPC.

III. Detection of mRNA in Polysomes and Subunits

For studying the sedimentation pattern of pulse-labeled total and poly-A-containing RNA, 1-ml fractions of the polysome gradient are collected in 15-ml Corex tubes containing 5 μl of 10% SDS. LiCl and EDTA are added to a final concentration of 0.5 M and 10 mM, respectively. RNA and protein are precipitated at –20°C with 95% ethanol. The precipitate is collected by centrifugation at 15,000 g for 20 minutes. The supernatant is sucked off, and the pellets dissolved in 1 ml of buffer (10 mM Tris–HCl, pH 7.4, 10 mM EDTA, 0.1 M NaCl, 0.2% SDS). Proteins are digested by the addition of 100 μg of proteinase (proteinase K, fungal, lyophilized, 15 mAnson units/mg, E. M. Laboratories, Elmsford, N.Y.) to each sample and incubation for 30 minutes at 37°C; a 0.1 volume of 5 M LiCl is added to each sample, and RNA is precipitated with ethanol at –20°C. RNA is pelleted by centrifugation at 15,000 g for 20 minutes and dissolved in 0.5–1 ml of binding buffer (0.12 M NaCl, 0.01 M, Tris–HCl pH 7.4); 50- to 100-μl portions are assayed for poly-A-containing RNA by binding to poly U immobilized on glass filters as described by Kates (1973) (see also Section VI,B). For the determination of total RNA, a measured volume of the RNA solution is treated with 1 ml of cold 10% trichloroacetic acid (TCA), and 50 μg of bovine serum albumin is added as carrier. After 10 minutes on ice the precipitate is collected on glass filters (Whatman GF/C), washed with cold 5% TCA, and dried; radioactivity is then determined in a toluene-based scintillant mixture.

IV. Preparation of RNA

A. Preparation of RNA from Whole Cells

All glassware and solutions are autoclaved or sterilized by filtration. A culture of 3–75 ml is chilled by pouring over crushed ice. It is centrifuged at 10,000 g for 2 minutes, and the cells resuspended in water and resedimented twice. The cells are then suspended in 3 ml of LETS (0.1 M LiCl, 0.01 M EDTA, 0.01 M Tris, pH 7.4, 0.2% SDS) supplemented with SDS to a final concentration of 1%. The suspension is transferred to a 13 × 100 mm plastic screw-cap tube (Falcon Plastics No. 2027) containing 2.5 ml of glass beads (0.1 mm, VWR Scientific No. 34007-088) which have been washed in LETS. This is fit into the microchamber (VWR Scientific No. 34007-011) of a Bronwill mechanical cell homogenizer (VWR Scientific No. 34006-008) and shaken for two 30-second periods. This generally suffices to break log-phase cells, but more extensive disruption may be necessary for late log- and stationary-phase cells.

The contents of the tube, including the glass beads, are poured into a 20 × 150 mm screw-cap tube containing 5 ml of phenol–chloroform–isoamyl alcohol (50:48:2). The disruption tube is rinsed with 2.5 ml of LETS. The extraction is carried out by vigorous shaking for 5 minutes at room temperature, and the emulsion broken by centrifugation at 1000 g for 5 minutes. The aqueous layer is removed and reextracted twice with fresh organic phase. The advantage of this method is that none of the sample is lost by being intermixed with the glass beads, which fall to the bottom of the organic phase. The final aqueous phase is centrifuged at 25,000 g for 15 minutes to remove insoluble material. Two volumes of ethanol are added, and the RNA can then be stored at $-20°$C under ethanol for years without degradation. The yield from 1 × 10^7 cells is roughly 0.6 A_{260} units of RNA with a A_{260}/A_{280} ratio of 2.0–2.1.

B.   Preparation of RNA from Spheroplasts

Washed spheroplasts pelleted in a glass tube are lysed in LETS (10 mM Tris–HCl, pH 7.4, 10 mM EDTA, 0.1 M LiCl, 0.2% SDS) containing 1% SDS (1 ml buffer/10^8 cells) by shaking on a vortex mixer. Immediately following lysis, DEPC can be added to a final concentration of 2% (v/v) and mixed on the vortex for 1 minute. The lysate is then deproteinized with phenol–chloroform–isoamyl alcohol (50:48:2) as described in Section IV,A.

The final material is essentially free of endogenous nucleases but may contain traces of proteins, as well as DNA, polysaccharides, and other high-molecular-weight material. However, the mRNA and rRNA can be separated from these contaminants by precipitation with 2 M LiCl (Baltimore,

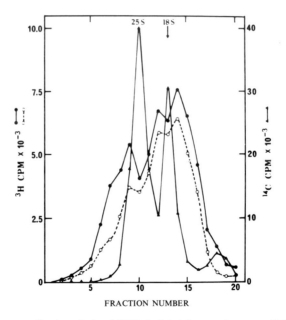

FIG. 2.   Sucrose gradient analysis of RNA isolated from polysomes. Cells grown for six generations in SC containing 0.05 μCi/ml of [^{14}C]uracil were converted to spheroplasts and cultured for 2.5 hours in label-free SCM before being exposed to [^{3}H]adenine (10 μCi/ml) for 5 minutes. The cytoplasm was isolated and treated with DEPC, and polysomes were displayed on sucrose gradients as described in Section II. RNA was extracted from fractions containing polysomes (>90S) and centrifuged on a 20-ml linear 15–30% (w/w) sucrose gradient in 0.2 M NaCl, 0.01 M EDTA, 0.01 M Tris, 0.2% SDS at 23,000 rpm for 18 hours at 25°C. RNA in 1-ml fractions of the gradient was precipitated with ethanol. Radioactive poly-A-containing RNA and total RNA were assayed in each fraction as described in Section III. Solid circles ^{3}H-pulse-labeled total RNA; open circles, ^{3}H-pulse-labeled poly-A-containing RNA; triangles, ^{14}C-labeled stable RNA.

1966; Kates, 1973). The final RNA pellet is dissolved in 1–2 ml of LETS without SDS and mixed with an equal volume of 4 M LiCl, and RNA allowed to precipitate for 12–24 hours at 2°C. Under these conditions all RNAs precipitate as lithium salts except for tRNA which remains in solution along with DNA and the other contaminants. The precipitated RNA is spun out at 15,000 g for 20 minutes and reprecipitated with 2 M LiCl. The final pellet is dissolved in LETS, mixed with 2 volumes of ethanol, and stored at −20°C.

C.   Preparation of RNA from Polysomes and Ribosomal Subunits

Polysomes, 80S monosomes, ribosomal subunits, and the top of the gradient are collected separately in flasks containing SDS. The concentration of SDS is adjusted to 0.5%, and LiCl and EDTA are added to a final concentration of 0.5 M and 10 mM, respectively. The RNA and proteins are precipitated at −20°C with 2.5 volumes of 95% ethanol. The precipitates are collected by centrifugation at 10,000 g for 20 minutes and dissolved in LETS containing 1% SDS. DEPC is added to a final concentration of 1%, the mixture is shaken for 1 minute, and RNA is extracted and purified by deproteinization with phenol–chloroform–isoamyl alcohol as described for spheroplasts.

The high recovery and sedimentation profile of mRNA in the RNA extracted from polysomes is shown in Fig. 2. It can be seen that most of the long-labeled [^{14}C] RNA sediments as 25 and 18S rRNA, whereas the ^{3}H-pulse-labeled RNA, mostly mRNA, shows a broad heterogeneous sedimentation. The poly-A-containing mRNA, 70–80% of the pulse-labeled RNA, shows a similar heterogeneous profile of sedimentation. The poly-A-containing molecules in the RNA extracted from the 80S, and the top region of the polysome gradient which contains the subunits and free mRNA, also show a similar sedimentation profile. Moreover, the poly-A-containing RNA isolated from the two regions of the polysome gradient have blocked 5'-termini (Section VII; Sripati et al., 1976). Thus the mRNA in the different regions of the polysome gradient is recovered intact.

V.   Analysis of RNA on Polyacrylamide Gels

Polyacrylamide gel analysis of high-molecular-weight RNA is carried out essentially according to Kadowaki and Halvorson (1971). The following solutions are prepared:

20 × PD Buffer (PDB)
    108 gm Tris
    9.3 gm Na_2 EDTA
    55 gm boric acid, to 1000 ml water

Acrylamide monomer mix
    10 gm acrylamide (recrystallized from warm chloroform)
    0.5 gm N, N'-methylenebis (acrylamide) (recrystallized from acetone)
    to 100 ml with water
    Filter through a 0.45 μm filter.

Enough solution for six to eight gels of 2.75% acrylamide is prepared by mixing 1.0 ml of PDB, 5.5 ml of acrylamide mix, 0.2 ml of 10% SDS, 2 ml of 50% glycerol, 11.1 of ml water, 0.2 ml of 10% ammonium persulfate, and 20 μl of N,N,N',N'-tetramethylethylenediamine (TEMED). The mixture is swirled, degassed, and pipetted into gel tubes.

The gels are 70 × 6 mm, cast in 110-mm tubes. The tubes are sealed at one end with Parafilm and taped vertically to the edge of a bench. The gel solution is carefully pipetted in and overlaid with running buffer. Polymerization requires 10–15 minutes. Running buffer consists of 75 ml of 20 × PDB, 10 ml of 10% SDS, and 915 ml of water.

After polymerization the tubes are mounted in a vertical apparatus, and the Parafilm is removed and replaced with a damp fragment of Handiwipe or pantyhose, attached with a rubber band. The overlaid running buffer is replaced with fresh solution. The gel surface is fragile and must be handled with care.

1. Sample: The RNA is usually stored as a precipitate under ethanol. It is collected by centrifugation, the supernate carefully sucked off, and the tube exposed briefly to a vacuum to remove traces of ethanol. Twenty to forty micrograms of RNA is dissolved in 50–100 μl of sample buffer consisting of 0.5 ml of 20× PDB, 0.5 gm of sucrose, 10 mg of SDS, 0.5 mg of bromphenol blue, and 9.5 ml of water and deposited on the surface of the gel using a microsyringe.

2. Electrophoresis: The gels are electrophoresed toward the anode for 20 minutes at 0.5 mA per gel and then for 120–140 minutes at 5 mA per gel.

3. Analysis: The gels are removed by reaming with a fine syringe needle filled with 0.1% SDS and then gentle blowing. The gels are easily deformed and stick avidly to anything dry such as paper or fingers. They can be scanned at 260 nm in a spectrophotometer. If the RNA is labeled, the gels are frozen on a weighing scoop and transferred to a Mickle gel slicer (Brink-

mann Instruments) on which 1-mm sections are cut. Each slice is placed in a Minivial (Nuclear Associates) with 0.75 ml of a solution containing 300 ml of 0.5 M Protosol (New England Nuclear), 175 ml of toluene, and 25 ml of water (Udem and Warner, 1972). The vial is shaken on a vortex mixer vigorously and covered. After 2–3 hours at room temperature, during which the RNA is hydrolyzed, 5 ml of toluene-based scintillation fluid is added and the samples counted.

4. Notes: (a) The glycerol in the gels has two roles. It makes the gel solution denser, facilitating overlayering. It prevents the formation of large ice crystals, permitting smooth slicing of frozen gels. (b) The RNA becomes very concentrated as it enters the gel. As a result, the gels are sensitive to overloading; no more than 20–40 μg of RNA should be applied to a gel of 6-mm diameter. (c) In our experience, yeast 18S RNA sometimes runs as a double peak, because of variations in conformation. If, immediately before layering, the sample is briefly dipped in boiling water and then chilled, the 18S RNA runs as a sharp peak (E. Falke, personal communication).

VI.   Isolation of Poly-A-Containing RNA

A.   Chromatography on Oligo-dT-Cellulose

In *S. cerevisiae*, most of the mRNA molecules contain relatively homogeneous polyriboadenylic acid (poly A) sequences of about 50 residues covalently linked to the 3'-end (McLaughlin *et al.*, 1973; Reed and Wintersberger, 1973; Groner *et al.*, 1974). These molecules can be readily separated from total RNA by affinity chromatography on oligo-dT-cellulose columns (Edmonds and Caramela, 1969; Aviv and Leder, 1972; Kates, 1973). Columns are prepared and run at room temperature.

Oligo-dT-cellulose (Collaborative Research, Waltham, Mass.) is suspended in binding buffer (10 mM Tris–HCl, pH 7.4, 10 mM EDTA, 0.5 M LiCl, 0.2% SDS) and, after eliminating the fines, the suspension is made 0.1 gm (dry weight)/ml. Pasteur pipet columns are plugged with glass wool to allow a flow rate of about 0.5 ml per minute. The oligo-dT-cellulose suspension, 30–50 mg for 10–20 A_{260} units of total RNA, is added to the column containing the binding buffer and mixed to eliminate air bubbles. The column is washed with 2 volumes of water, 5 volumes of 0.1 M KOH, 5 volumes of water, 5 volumes of binding buffer, and 5 volumes of 10 mM Tris–HCl, pH 7.4, and finally equilibrated with the binding buffer. RNA samples are dissolved in buffer (10 mM Tris–HCl, pH 7.4, 10 mM EDTA, 0.2% SDS) at approximately 20 A_{260} units/ml, mixed with 0.1 volume of 5 M LiCl, and applied to the column in small portions (about 0.2 ml) with an interval of 10 minutes between additions. The samples flowing through the column are collected and applied again to the column. About 20 minutes after the second passage, the column is washed with 2 volumes of binding buffer. The nonabsorbed material and the washings are pooled, and the RNA precipitated with 2.5 volumes of ethanol. The column is then washed with 5 volumes of binding buffer containing only 0.15 M LiCl in order to remove traces of nonabsorbed RNA. If the poly-A-containing RNA is to be used for translation in a cell-free system or is to be treated with enzymes, SDS must be omitted from the buffer in this washing step. The bound RNA is eluted with 0.5-ml portions of 10 mM Tris–HCl, pH 7.4. Fractions containing RNA are pooled and mixed with 0.1 volume of 2 M sodium or potassium acetate, pH 5.0, and the RNA precipitated at $-20°$C with 2.5 volumes of ethanol.

Figure 3 is a typical elution profile of RNA isolated from spheroplasts. Stable RNA was labeled with [^{14}C]adenine, and newly synthesized RNA was labeled with [^{3}H]adenine. About 96% of the stable RNA and 75% of the pulse-labeled RNA is excluded in the void volume of the binding buffer. Very little labeled RNA is released when the salt concentration of the washing buffer is lowered to 0.15 M (A in Fig. 3), but the RNA (about 20% ^{3}H- and 2% ^{14}C-labeled) retained by the column is almost completely eluted in a single fraction of the salt-free elution buffer (B in Fig. 3).

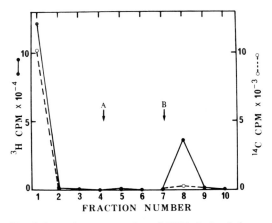

FIG. 3. Oligo-dT-cellulose chromatography of RNA isolated from spheroplasts. Cells were grown for five to six generations in SC containing 0.2 μCi/ml of [^{14}C]adenine. They were converted to spheroplasts and incubated in SCM as described. After 90 minutes, a 10-ml portion was exposed to 200 μCi of [^{3}H]adenine for 15 minutes. The spheroplasts were harvested, and the RNA was extracted and subjected to chromatography on oligo-dT-cellulose. Fractions 1 to 4 were eluted with 0.5 M LiCl, 0.01 M EDTA, 0.01 M Tris–HCl, (pH 7.4, 0.1% SDS. Fractions 5 to 7 (A) were eluted with the same buffer, except that LiCl was at 0.15 M. Fractions 8 to 10 (B) were eluted with 0.01 M Tris, pH 7.4, 0.1% SDS.

B. Absorption to Poly-U–Glass-Fiber Filters

The affinity chromatography on oligo-dT-cellulose described above separates the bulk of the poly-A-containing molecules from the total RNA and thus is highly suitable for preparing large quantities of poly-A-containing RNA. However, analysis for methylated components shows that the preparations contain appreciable rRNA contamination which persists even after repeated binding to oligo-dT-cellulose. The only methyl group in yeast mRNA is m^7G at the 5'-end (Sripati et al., 1976). However, 25 and 18S rRNAs together contain 67 methyl groups occurring as ribose and base-methylated residues (Klootwijk and Planta, 1973). Hence quantitative analysis for methylated components in poly-A-containing RNA preparations provides a highly sensitive tool for estimating the degree of contamination by rRNA. Therefore the criteria we have employed to evaluate the purity and yield of poly-A-containing RNA are: the absence of methylated nucleosides and the recovery of nuclease- and alkaline phosphatase-resistant 5'-termini in the final fractions (see Section VII).

More complete purification of yeast mRNA has been obtained using poly U immobilized on glass-fiber filters. These are prepared by applying 300 μg of poly U per filter (Whatman GF/C, 2.4 cm) and ultraviolet-irradiating the dried filters as described by Kates (1973). Two of these filters are mounted on a Millipore filter support for microanalysis. Unbound poly U is washed off the filters with 100 ml of distilled water, and then the filters are equilibrated with the binding buffer (0.01 M Tris–HCl, pH 7.5, 0.12 M NaCl). The RNA eluted from oligo-dT-cellulose is precipitated with ethanol and dissolved in 1–2 ml binding buffer. An equal volume of binding buffer containing 40% formamide is added, and the solution warmed at 50°C for 5 minutes. The concentration of formamide is then reduced to 6% by mixing 3 volumes of binding buffer, and the resulting solution is immediately passed through the poly-U filters at a rate of approximately 0.5 ml per minute using a reduced vacuum. Unbound RNA is washed off by passing 10 ml of binding buffer through the filters. The bound RNA is eluted with 0.5-ml portions of a neutral solution of 90% formamide containing 3 mM EDTA. Fractions containing RNA are pooled, 0.1 volume of 2 M sodium acetate, pH 5.0, is added, and the RNA precipitated at −20°C with ethanol.

The rRNA contamination of the oligo-dT-purified poly-A-containing RNA and its removal by rechromatography on poly-U filters are illustrated in Table I and Fig. 4 which show that most of the molecules containing methylated nucleosides which bound to poly-dT columns did not bind to poly-U filters, especially in the presence of formamide. Since the molar ratio of methyl groups in mRNA to those in rRNA is ~1:33 (averaging 25 and 18S rRNA), the molar ratio of mRNA to rRNA in the sample analyzed

TABLE I

RATIO OF METHYLATED NUCLEOSIDES TO BLOCKED 5'-TERMINI OF OLIGO-dT-
CELLULOSE-BOUND RNA RECHROMATOGRAPHED ON POLY U–GLASS-FIBER FILTERS[a]

RNA fraction from poly-U filters	[³H]methyl (cpm × 10⁻³)		Ratio, 5'-termini/ Nucleosides
	Nucleosides	5'-Termini	
Not bound in the absence of formamide	11.40	1.04	0.09
Bound in the absence of formamide	3.07	2.80	0.91
Bound in the presence of 6% formamide	1.00	4.15	4.15

[a] [³H]-methyl-labeled RNA bound to oligo-dT-cellulose was eluted and chromatographed on poly-U–glass-fiber filters with and without formamide in the binding buffer. The fraction of RNA that did not bind and the fraction that bound to poly-U filters were analyzed for methylated nucleosides and the blocked 5'-termini as described in Section VII. The radioactivity in the spots corresponding to the nucleosides and the spots I and II corresponding to the 5'-termini were determined. The third line is data from Fig. 4.

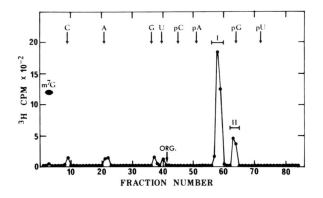

FIG. 4. Analysis of the methylated constituents of oligo-dT-purified RNA rechromatographed on poly-U filters in the presence of formamide. RNA isolated from spheroplasts labeled with [³H]methyl methionine for 10 minutes was purified by oligo-dT-cellulose chromatography, warmed in binding buffer containing 20% formamide, and then diluted with binding buffer to 6% formamide and applied to poly-U filters. Bound RNA was eluted and digested with P_1 nuclease followed by bacterial alkaline phosphatase, and the products were separated by paper electrophoresis as described. The radioactive peaks represent the spots of methylated constituents of the RNA. The position of the spots of ³H standards are indicated by arrows. The spot of standard m⁷G was identified under ultraviolet light. Peaks I and II are spots of the blocked 5'-termini of mRNA, m⁷GpppA and m⁷GpppG, respectively. ORG, Origin.

in Fig. 4 is about 100:1. Even this contamination could probably be eliminated by a second passage through poly-U filters. There is very little loss of 5′ termini during the rechromatography, therefore the poly-A-containing mRNA is recovered intact.

C. Isolation of Poly-A Segments

The poly-A sequences are resistant to pancreatic and T_1 RNases in the presence of moderate concentrations of salts. Hence they can be isolated from poly-A-containing RNA molecules and analyzed by gel electrophoresis (Edmonds and Caramela, 1969; Kates, 1973).

The ethanol-precipitated RNA samples (about $1-2 A_{260}$ units) are centrifuged at 20,000 g for 15 minutes, the supernatants sucked off, and traces of ethanol in the pellet removed under vacuum. The RNA is dissolved in 0.5 ml of 0.01 M Tris–HCl, pH 7.4, 0.1 M NaCl, 0.01 M EDTA; 0.025 μg of RNase A and 2.5 units of RNase T_1 are added to each sample which is then incubated at 36°C for 30 minutes. The reaction is stopped by the addition of 0.5 ml of a solution containing 1% SDS, 1 M NaCl, and about 5 A_{260} units of tRNA per milliliter followed by 200 μg of proteinase K, and the incubation is continued for 30 minutes in order to digest the RNases. The reaction mixture is deproteinized with a phenol–chloroform–isoamyl alcohol mixture, and the RNA precipitated as described earlier. The precipitate is collected by centrifugation at 30,000 g for 30 minutes, dissolved in 1 ml of buffer (10 mM Tris–HCl, pH 7.4, 10 mM EDTA, 0.2% SDS), mixed with 0.1 volume of 5 M LiCl, and passed through 50 mg of oligo-dT-cellulose packed in a pasteur pipet. The column is washed with 10 ml of the same buffer containing 0.5 M LiCl. The bound poly-A tracts are eluted and precipitated with ethanol as described for the oligo-dT-cellulose chromatography of RNA.

Poly-A sequences are analyzed on 12.5% polyacrylamide gels in the same manner described for high-molecular-weight RNA (McLaughlin et al., 1973).

VII. Analysis of 5′-Termini of mRNA

A. Isolation of 5′-Termini of mRNA

The 5′-termini of yeast mRNA, like those of other eukaryotic cells and their viruses, are of the form $m^7G(5′)ppp(5′)A$ and $m^7G(5′)ppp(5′)G$ (Sripati et al., 1976). There are no other methyl groups in yeast mRNA.

Since the $5'-5'$ linkage in the structure is inverted with respect to the normal $3'-5'$ phosphodiester bond in the RNA chain and, since there is a pyrophosphate linkage between the two $5'$-nucleotides, these dinucleotide structures are resistant to nucleases and phosphatases. Therefore they can be easily isolated and analyzed (Furuichi et al., 1975a,b; Lavi and Shatkin, 1975; Groner and Hurwitz, 1975; Sripati et al., 1976). In principle the RNA is digested with appropriate nucleases to mononucleotides which are then converted to nucleosides by alkaline phosphatase. The blocked $5'$-termini, which remain intact and negatively charged, are separated from the positively charged nucleosides by paper electrophoresis of the digest. Although the modified $5'$-termini are also resistant to alkali, alkaline digestion of RNA should be avoided, since these conditions provoke opening of the imidazole ring at the $8-9$ bond, leading to conversion of m^7G to the ring-opened derivative lacking a positive charge. The two nucleases generally employed are RNase T_2 (Sankyo Company, Ltd.) which cleaves ribonucleotide sequences, except those containing $2'-O$-methyl derivatives, to $3'$-nucleotides, and P_1 nuclease (Yamasa Shoyu Company), a nonspecific nuclease which splits polynucleotides, including sequences containing $2'-O$-methyl residues, to give $5'$-mononucleotides (Fujimoto et al., 1974).

Either methyl- or purine-labeled RNA preparations are used for analysis. The RNA must be freed of salts and SDS. This is done by collecting the RNA precipitates in conical tubes, dissolving the pellet in water, and reprecipitating the RNA at $-20°C$ after the addition of 0.1 volume of 1 M sodium acetate, pH 5.0, and 2.5 volumes of ethanol. Reprecipitation is repeated two or three more times. The final pellet, after the supernatant is carefully sucked off, is dried in a vacuum desiccator and then dissolved in 100 μl of water; 1 M sodium acetate, pH 6.0, is added to a final concentration of 10 mM, and the solution incubated for 60 minutes at $37°C$ with 1 mg/ml of P_1 nuclease. The pH of the mixture is then carefully raised to 8 by adding $1-2$ μl of 1 M Tris–base. The pH must be verified with test papers and should never be allowed to exceed 9. Bacterial alkaline phosphatase (Worthington Biochemicals Corporation) is added, and incubation continued for 60 minutes. Since alkaline phosphatase is sensitive to the inorganic phosphate formed in the reaction, 1 unit of enzyme is added every 15 minutes. The reaction mixture is used directly for analysis of the products by paper electrophoresis.

Since alkaline phosphatase is sensitive to inorganic phosphate, not more than 15–20 A_{260} units of RNA can be digested as described above. Furthermore, ^{32}P-labeled RNA cannot be used directly in the above procedure, since the large amounts of $^{32}P_i$ released will interfere with electrophoretic analysis of the products. Therefore ^{32}P-labeled or large amounts of RNA ($20-100$ A_{260} units) are first digested with RNase-T_2, and the oligonucleotides separated from the mononucleotides by DEAE-cellulose chromatography in the presence of urea (Tomlinson and Tener, 1963). The advantage of using RNase -T_2 instead of P_1 nuclease in this step lies in the fact that the products of T_2 digestion are 3'-nucleotides. Hence the 5'-dinucleotide termini released from the yeast mRNA will have a net charge of -4.5, instead of -2.5 as in the case of P_1 nuclease digestion, and can be conveniently separated from the large quantities of mononucleotides of charge -2.

Salt- and SDS-free RNA (methyl-, purine-, or ^{32}P-labeled) is dissolved in 100–200 μl of 0.05 sodium acetate (pH 4.5) and incubated with 5 units of RNase-T_2 for 15 hours at 37°C. The digestion mixture is brought to 1 ml with 50 mM Tris–HCl, pH 7.6, containing 7 M urea and 5 mM NaCl, mixed with tRNA hydrolyzate (1 mg of tRNA digested with 100 μg of RNase A for 7 hours at 37°C), applied to a DEAE-cellulose (Whatman, DE 52) column (0.7 × 22 cm) prepared as described by Tomlinson and Tener (1963), and equilibrated with 50 mM Tris–HCl, pH 7.6, containing 7 M urea and 5 mM NaCl. The nucleotides are eluted at a flow rate of about 0.3 ml per minute, using a linear salt gradient (50 ml of 0.005 M NaCl and 50 ml of 0.3 M NaCl) in the same buffer. One-milliliter fractions are collected, and absorbance at 260 nm is measured; 50- to 100-μl portions are used for determining radioactivity in Aquasol (New England Nuclear, Boston, Mass.) The bulk of the T_2-digested product, which consists of 3' nucleotides, elutes with the optical density peak corresponding to a charge of -2, whereas the blocked 5'-termini elute as a minor peak between the two optical density peaks of charge -4 and -5. The fractions eluted between these two peaks are pooled and desalted on a small DEAE-cellulose column as described by Tener (1967). The resulting product is digested with P_1 nuclease followed by bacterial alkaline phosphatase as described earlier.

High-voltage electrophoresis is carried out on Whatman 3 MM paper strips, 105 × 23 cm, in an "up-and-over" ionophoresis tank (Brownlee, 1972). A pencil line is drawn across the strip, 50 cm (20 cm for running ^{32}P-labeled samples) from one end marked for the cathode, and spaces for the origin are marked on the line at 3 cm, followed by 5.5, 1, 5.5, 3, and 2.5 cm. Samples are spotted in the middle of each of these spaces: 2 μl of the dye mixture (1% xylene cyanol, 1% acid fuchsin, 1% methyl orange) are spotted in the first 3-cm space. The enzyme-digested RNA samples are applied as a 2-cm band in each of the 5.5 cm spaces. About 10 μl is spotted at a time and dried by blowing slightly warm air from a hair dryer. Two microliters of a dilute solution of m^{7}G is spotted in the space next to the samples. ^{3}H-labeled ribonucleosides and ribonucleotides, each containing about 1 × 10^4 cpm/μl are mixed separately, 1–2 μl of the nucleoside mixture is spotted next to m^{7}G, and the nucleotide mixture is similarly spotted in the last space. The spots are thoroughly dried, and the paper mounted on an electrophoresis rack. Two strips can be run at a time. The strips are carefully wetted with pyridine–acetate buffer, pH 3.5 (pyridine–glacial acetic acid–water, 1:10:89), on either side of the origin, and the buffer fronts are allowed to advance slowly and meet at the origin. The buffer is applied to the rest of the paper, excess buffer is quickly wiped off, the rack is placed in the tank with the paper ends dipping into the pyridine–acetate buffer, pH 3.5, contained in the electrode chambers, and the rest of the paper is completely immersed in the water-cooled Varsol coolant in the tank. Electrophoresis is run at 60 V/cm until the blue dye has migrated about 20 cm from the origin (about 2.5 hours at 4 KV). The strips are dried by placing the racks in a hood for 8–10 hours. Vertical lines are drawn through the pencil marks separating the spaces containing samples. Parallel lines, first 0.5 cm on either side of the origin, and then 1 cm apart, are drawn across the paper. The m^7G spot is located by its blue fluorescence under ultraviolet light. Then 1-cm strips are cut, and the radioactivity determined in a toluene scintillation mixture. The counting efficiency for ^3H is 10–15% as compared to the spots eluted from strips and counted in Aquasol (New England Nuclear).

A typical electrophoretic analysis of P_1 nuclease and bacterial alkaline phosphatase-digested $[^3H]$methyl-labeled RNA is shown in Fig. 4. The migration of the methyl-labeled components of the enzyme-digested RNA is shown as radioactive peaks, and the position of the spots of ^3H-methyl-labeled standards are indicated by arrows at the top. The modified 5'-termini, m^7GpppA and m^7GpppG, derived from the RNA, migrate as peaks I and II, respectively, between the markers pA and pG.

B. Constituents of the 5', 5'-Dinucleotides

For further analysis, the strips containing the radioactive peaks of the nuclease and bacterial alkaline phosphatase-resistant 5'-dinucleotides are removed from the vials, and the scintillant washed off with toluene. The spots are eluted from the dry strips with 0.5–1 ml water into small conical tubes and dried in a desiccator. The product is dissolved in 100 μl of buffer (Tris–HCl, pH 7.5, 2 mM MgCl$_2$) and incubated with 0.03 units of nucleotide pyrophosphatase (Sigma Chemical Corporation) at 37°C for 60 minutes. Since commercial preparations of this enzyme contain traces of phosphomonoesterase activity, higher concentrations of the enzyme and longer periods of incubation must be avoided. The enzyme splits the pyrophosphate linkages in the 5'-dinucleotide structures and liberates the mononucleotides and P_i. If the 5'-termini were derived from ^{32}P-labeled RNA, the products of nucleotide pyrophosphatase digestion can be analyzed by paper electrophoresis (Sripati *et al.*, 1976) as described earlier. The m^7GMP migrates very little from the origin during electrophoresis. If the 5'-termini were derived from methyl- or purine-labeled RNA, the nucleotide pyrophosphatase digestion can be followed by incubation with bacterial alkaline phosphatase as described earlier and the nucleosides liberated analyzed by paper electrophoresis. The m^7G migrates ahead of cytosine and can be identified as a blue fluorescent spot under ultraviolet light.

VIII.   Translation of Yeast mRNA

A.   The Extract

Wheat germ extract actively translates yeast mRNA (Gallis et al., 1975; Mager et al., 1975; Warner and Gorenstein, 1977). While the characteristics of each batch of wheat germ vary slightly, we have found the following conditions, based on those of Roberts and Paterson (1973), effective.

All procedures are carried out at 0–4°C using sterile solutions and materials. Six grams of wheat germ was ground for 60 seconds with 6 gm of autoclaved glass fragments (prepared by crushing pasteur pipets) in a mortar. Twenty-five milliliters of a solution containing 20 mM HEPES, pH 7.6, 100 mM KCl, 1 mM MgCl$_2$, 2 mM CaCl$_2$, and 6 mM β-mercaptoethanol was added, and grinding continued until the slurry was thoroughly mixed. The slurry was poured into a sterile tube and centrifuged at 30,000 g for 10 minutes at 0°C. Fifteen to twenty milliliters of yellowish supernate was removed, avoiding the large brownish pellet and the thin layer of fat on the meniscus. The extract was passed through a 100-ml column of Sephadex (G-25, coarse) which had been equilibrated with 20 mM HEPES, pH 7.6, 120 mM KCl, 5 mM MgAc$_2$, 6 mM β-mercaptoethanol. Within 15–20 minutes, a light-brown, turbid material began to emerge from the column. About 15 ml of this was collected, frozen by dripping slowly into liquid Nitrogen, and stored in a liquid-nitrogen freezer. This extract had an A_{260} of 70, a A_{260}/A_{280} ratio of 1.68, and a protein concentration of 18.5 mg/ml. It was stable for many months.

B.   The Incubation Mix

The final incubation mix contains 28 mM HEPES, pH 7.8, 2.25 mM MgAc$_2$, 80 mM KAc, 24 mM KCl, 25 μM each of 19 amino acids, 1.2 mM ATP, 0.03 mM GTP, 2.4 mM dithiothreitol, 40 μM spermine, 9.6 mM creatine phosphate, and 8 μg/ml creatine phosphokinase. An incubation mix of 25 μl contains 5 μl of extract (92.5 μg of protein) and 5 μCi of [^{35}S] methionine (New England

Nuclear) or 2 μCi of [^{3}H]leucine. Incorporation is carried out at room temperature, 20–23°C, and is linear for about 90 minutes when using [^{35}S]methionine and for more than 200 minutes when using [^{3}H]leucine. Incorporation is measured by spotting 5- or 10-μl samples on halves of Whatman 3 MM filter paper disks impregnated with methionine or leucine and washing them successively in cold 10% TCA containing 0.5 mg/ml methionine or leucine, boiling 5% TCA, cold 5% TCA, and 70% ethanol. The dried disks are counted in a toluene-based scintillation fluid.

C. The Response to Yeast RNA

The response of this system to added yeast RNA is shown in Fig. 5. There is an optimum concentration of RNA, which is quite independent of the batch. The cause of the inhibition by excess RNA is unknown. In studying the synthesis of ribosomal proteins we found that, as a function of the concentration of added RNA, the synthesis of individual proteins was parallel to that of total incorporation, as illustrated in Fig. 6. Thus the wheat germ extract appears not to discriminate among different mRNAs, as observed in other systems, e.g., Kabat and Chappell (1977). While poly A (+) RNA is very active as a template in this system, the sum of the template activities of RNA bound and not bound to a poly-dT column is less than that of the input RNA. Mixing the two RNAs does not restore the activity. The reason for this is not known. Although the template activities (per microgram) of total, poly A (−), and poly A (+) RNA differ greatly, the spectra of proteins synthesized by the three classes of RNA are very similar (Warner and Gorenstein, 1977). In particular, there are no detectable proteins synthesized in response to poly A (−) RNA that are not synthesized to a greater extent

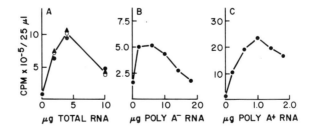

FIG. 5.   Response of wheat germ extract to yeast RNA. To the wheat germ extract described in Section VIII,A was added total RNA prepared from three different cultures (A), RNA not bound to oligo-dT-cellulose (B), and RNA bound to and eluted from oligo-dT-cellulose (C) (Section VI,A). After incubation for 60 minutes at room temperature aliquots were prepared for counting as described in Section VIII, B.

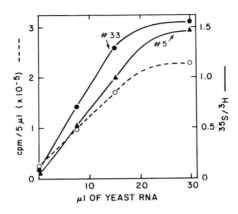

FIG. 6. Lack of discrimination among mRNAs by the wheat germ extract. To incubation mixtures of 150 μl (final volume) were added increasing amounts of total yeast RNA. After 90 minutes aliquots were removed to measure total incorporation of [^{35}S]methionine, and the rest was analyzed for incorporation into individual ribosomal proteins as described in Section VIII,D. Open circles, Total incorporation; triangles, incorporation into protein 5, a 60S protein; solid circles, incorporation into protein 33, a 40S protein. Similar curves were obtained for 20 other ribosomal proteins.

in response to poly A (+) RNA. Whole-cell RNA isolated in the absence of in the presence of DEPC was equally active as a template. However, RNA from polysomes isolated in the presence of DEPC was considerably more active than RNA from polysomes isolated without DEPC.

D. Analysis of the Products

For analysis of the products on a one-dimensional gel, 20 μl of reaction mix containing [^{35}S]methionine is added to 20 μl of sample buffer containing 1% SDS and 1% mercaptoethanol, immersed in boiling water for 1 minute, and analyzed on a one-dimensional slab gel as described by Studier (1973).

For analysis of the synthesis of ribosomal proteins a reaction mixture of 150 μl is diluted with 0.85 ml of 10 mM HEPES, pH 7.5, 100 mM NaCl, 10^{-5} M MgCl$_2$, 1 mg/ml methionine, and then total protein is extracted with 67% acetic acid as described in Chapter 4 (Warner and Gorenstein, 1977, also this volume). The dialyzed product is lyophilized, run on a pH 5 × SDS two-dimensional gel (Gorenstein and Warner, 1976; Warner and Gorenstein, this volume). The gel is dried onto a piece of filter paper and autoradiographed. The result is shown in Fig. 7, which demonstrates the synthesis of many ribosomal proteins in vitro (compare to Fig. 5 in Chapter 4).

FIG. 7. Acrylamide gel analysis of products of cell-free translation. Total yeast RNA was added to a wheat germ extract containing $[^{35}S]$methionine, which was then incubated at 23°C for 90 minutes. The products were extracted and analyzed as described in Section VIII,C, and the gel autoradiographed. This pattern should be compared with Fig. 5 in Chapter 4 to identify ribosomal proteins, a few of which have been numbered. In the absence of added yeast RNA there are no detectable spots on the autoradiograph.

For quantitative analysis of ribosomal proteins, the dialyzed product is mixed with protein extracted from yeast uniformly labeled with $[^3H]$methionine, lyophilized, and run on a pH 5 × SDS two-dimensional gel. The stained spots are excised and counted as described in Chapter 4.

ACKNOWLEDGMENTS

Work from this laboratory was supported by grants from the American Cancer Society No. NP 72 G, the National Science Foundation No. PCM 7503938 and the National Institutes of Health No. P 30 CA 13330. J.R.W. was a Faculty Career Awardee of the American Cancer Society. C.E.S. is supported by Centre National de la Recherche Scientifique, France.

References

Aviv, H., and Leder, P. (1972). *Proc. Natl. Acad. Sci. U.S.A.* **69**, 1408–1412.

Baltimore, D. (1966). *J. Mol. Biol.* **18**, 241–428.

Brownlee, G. G. (1972). "Determination of Sequences in RNA-Laboratory Techniques in Biochemistry and Molecular Biology" (T. S. Work and E. Work, eds.), pp. 54–66. North-Holland Publ., Amsterdam.

Edmonds, M., and Caramela, M. G. (1969). *J. Biol. Chem.* **244**, 1314–1324.

Ehrenberg, L., Fedorcsak, I., and Solymosy, F. (1976). *Prog. Nucleic Acid Res. Mol. Biol.* **16**, 189–262.

Fujimoto, M., Kuninaka, A., and Yoshino, H. (1974). *Agric. Biol. Chem.* **38**, 1555–1561.

Furuichi, Y., Morgan, M., Muthukrishnan, S., and Shatkin, A. J. (1975a). *Proc. Natl. Acad. Sci. U.S.A.* **72**, 362–366.

Furuichi, Y., Morgan, M., Shatkin, A. J., Jelinek, W., Salditt-Georgieff, M., and Darnell, J. E. (1975b). *Proc. Natl. Acad. Sci. U.S.A.* **72**, 1904–1908.

Gallis, B. M., McDonnell, J. P., Hopper, J. E., and Young, E. T. (1975). *Biochemistry* **14**, 1038–1046.

Gorenstein, C., and Warner, J. R. (1976). *Proc. Natl. Acad. Sci. U.S.A.* **73**, 1547–1551.

Groner, B., Hynes, N., and Phillips, S. (1974). *Biochemistry* **13**, 5378–5383.

Groner, Y., and Hurwitz, J. (1975). *Proc. Natl. Acad. Sci. U.S.A.* **72**, 2930–2934.

Hartwell, L. H., (1967). *J. Bacteriol.* **93**, 1662–1670.

Hutchison, H. T., and Hartwell, L. H. (1967). *J. Bacteriol.* **94**, 1697–1705.

Kabat, D., and Chappell, M. R., (1977). *J. Biol. Chem.* **252**, 2684–2690.

Kadowaki, K., and Halvorson, H. D., (1971). *J. Bacteriol.* **105**, 826–830.

Kates, J. (1973). *Methods Cell Biol.* **7**, 53–65.

Klootwijk, J., and Planta, R. J. (1973). *Eur. J. Biochem.* **39**, 325–333.

Lavi, S., and Shatkin, A. J. (1975). *Proc. Natl. Acad. Sci. U.S.A.* **72**, 2012–2016.

McLaughlin, C. S., Warner, J. R., Edmonds, M., Nakazato, H., and Vaughan, M. H. (1973). *J. Biol. Chem.* **248**, 1466–1471.

Mager, W. H., Hoving, R., and Planta, R. J. (1975). *FEBS Lett.* **58**, 219–221.

Penman, S., Vesco, C., and Penman, M. (1968). *J. Mol. Biol.* **34**, 49–69.

Reed, J., and Wintersberger, E. (1973). *FEBS Lett.* **32**, 213–217.

Roberts, B. E., and Paterson, B. M. (1973). *Proc. Natl. Acad. Sci. U.S.A.* **70**, 2330–2334.

Rubin, G. M. (1975). *Methods Cell Biol.* **12**, 45–64.

Sripati, E. C., Groner, Y., and Warner, J. R. (1976). *J. Biol. Chem.* **251**, 2898–2904.

Studier, F. W. (1973). *J. Mol. Biol.* **79**, 237–248.

Tener, G. M. (1967). *In* "Methods in Enzymology" (L. Grossman and K. Moldave, eds.), Vol. 12, Part A, pp. 398-404. Academic Press, New York.

Tomlinson, R. V., and Tener, G. M. (1963). *Biochemistry* **2**, 697–702.

Udem, S. A. (1972). Ph.D. Thesis, Albert Einstein College of Medicine, New York.

Udem, S. A., and Warner, J. R. (1972). *J. Mol. Biol.* **65**, 227–242.

Udem, S. A., and Warner, J. R. (1973). *J. Biol. Chem.* **248**, 1412–1416.

Warner, J. R. (1971). *J. Biol. Chem.* **246**, 447–454.

Warner, J. R., and Gorenstein, C. (1977). *Cell* **11**, 201–212.

Chapter 6

Method for Yeast Protein Synthesis in a Cell-Free System

C. H. SISSONS[1]

Department of Cell Biology,
University of Auckland,
Auckland, New Zealand

I. Introduction

Yeast has become increasingly important as an eukaryotic model system. At present, more intensive genetic and biochemical analysis of yeast is possible than of any other eukaryote. This has led to many attempts over the last 15 years to obtain a cell-free protein-synthesizing system programed by exogenous natural mRNA. The early systems (see Section II) were success-

---

[1] Present address: Ruakura Animal Research Station, Hamilton, New Zealand.

ful in obtaining polyphenylalanine synthesis programed by poly U but had very low activity. Every attempt to translate mRNAs other than artificial polynucleotides has failed. It seems not only that yeast has unusual elongation factor requirements (Skogerson and Engelhardt, 1977) and delicate ribosomes even for elongation reactions (Sissons, 1974), but also that the system for the initiation of protein synthesis is exceptionally labile in extracts (see, for example, Gallis and Young, 1975). Further, the ribosome cycle found *in vivo* (Kaempfer, 1969; Branes and Pogo, 1975) does not occur in cell extracts, so that stable "runoff" ribosomes (Schreier and Staehelin, 1973) for initiation on exogenous mRNA are not generated by incubating extracts.

The development of *in vitro* yeast protein-synthesizing systems has proceeded in two directions. The first involved highly purifying ribosomes and ribosome and elongation factors and studying (1) elongation requirements (Albrecht *et al.*, 1970; Skogerson and Wakatama, 1976; Skogerson and Engelhardt, 1977), (2) factor specificity of mitochondrial versus cytoplasmic systems (Perani *et al.*, 1971; Richter and Lipmann, 1970), (3) initiation on poly U using *N*-acetylphenylalanyl-tRNA as an initiator tRNA (Ayuso and Heridia, 1968; Torano *et al.*, 1972; Pranger *et al.*, 1974; Somasundaran and Skogerson, 1976), (4) Met-tRNA initiation (Richter *et al.*, 1971), and (5) the genetics of ribosomal antibiotic resistance (e.g., Rao and Grollman, 1967; Skogerson *et al.*, 1973; Grant *et al.*, 1974; Schindler *et al.*, 1974; Bucher and Skogerson, 1976; Somasundaran and Skogerson, 1976). Some of the more recent work on highly purified ribosomes has generated systems with a high activity in polyphenylalanine formation (e.g., van der Zeijst *et al.*, 1972; Skogerson *et al.*, 1973).

The second approach was to develop and characterize systems based on crude extracts analogous to the cell-free systems obtained from *Escherichia coli* and mammalian cells. The major advantage of crude extract systems is that their very simplicity minimizes the damage to labile components by handling, thus enhancing the probability of obtaining, for example, a functioning initiation system.

We have developed a yeast *in vitro* system (Sissons, 1974) analogous to the *E. coli* S-30 systems and it is described here. It is much more active than the early cell-free systems. If the various precautions described herein are taken and details adhered to, it should prove reliable and relatively easy to prepare and handle. It provides a useful base for studies on tRNA function, inhibitors, ionic conditions, and supplementation with heterologous components, and hopefully for further development of the system. Polyphenylalanine synthesis operates with very high fidelity and does not show some of the elongation defects of some other systems (Halvorson and Heridia, 1967; Bretthauer and Golichowski, 1968), nor the dependence on high Mg^{2+}

concentrations that can occur in highly purified systems (van der Zeijst *et al.*, 1973). As pointed out by Barbacid *et al.* (1975), the results from studies on, for example, inhibitor mechanisms in highly purified systems do not always agree with those obtained *in vivo* or in crude cell-free systems (see the review by Vazquez, 1974). Thus crude cell-free systems have a role in confirming the results obtained from better resolved systems, particularly where confirmation *in vivo* is not feasible.

Further fractionating the S-30 system reduces activity (Sissons, 1974), and we have not described fractionation techniques here. For references to techniques, see those on purified systems mentioned above and those in Section II.

II. Early Yeast Cell-Free Systems

Much useful information is in the early articles on the conditions for and characteristics of yeast *in vitro* protein synthesis. Since they are of such help to anyone trying to use or develop a yeast system, some of them are listed here.

The first report of *in vitro* protein synthesis on yeast ribosomes was probably by Webster (1957), and in 1962 two preliminary papers appeared (Barnett *et al.*, 1962; Cooper *et al.*, 1962). In 1963 three separate groups described yeast cell-free systems in some detail (Bretthauer *et al.*, 1963; Marcus *et al.*, 1963; So and Davie, 1963; Maeda and Imahori, 1963), followed by two more groups in 1964 (Lucas *et al.*, 1964; Siegel and Sigler, 1964). Some of these groups then published a series of articles on, particularly, polysomes (Marcus *et al.*, 1965, 1967) and the characterization of fractionated systems in aminoacyl transfer from aminoacyl-tRNA to polypeptides (Downey *et al.*, 1965; Dietz *et al.*, 1965; Heridia and Halvorson, 1966; Tanner, 1967). This early work was reviewed by Halvorson and Heridia (1967). In 1967–1970, work centered on the characterization of elongation factors (Richter and Klink, 1967; Richter *et al.*, 1968; Albrecht *et al.*, 1970; Ayuso and Heridia, 1967, 1968) and the specificity of mitochondrial and cytoplasmic systems (Lamb *et al.*, 1968; Richter and Lipmann, 1970; Perani *et al.*, 1971). Work also started on the nature of initiation on poly U with *N*-acetylphenylalanyl-tRNA (Ayuso and Heridia, 1968; Vazquez *et al.*, 1969; Torano *et al.*, 1972). More recent studies are described in Section I. Work on yeast protein synthesis *in vivo*, yeast tRNA structures and coding specificity, and yeast ribosome structure also proceeded. Further useful papers on yeast ribosomes are those by Bruening and Bock (1965),

Ghosh (1969), Pêtre (1970), and Walters and van Os (1970). See also the methods and references in Rubin (1975) in Vol. XII of this series and the isolation method for ribosomes in Skogerson and Wakatama (1976) and for polysomes in Martin and Hartwell (1970) and Barbacid *et al.* (1975).

III.   The S-30 Extract

Yeast strain, growth and harvesting conditions, breakage method, buffers, protective agents, and freezing technique all have critical requirements for the successful preparation of an active cell-free system.

A.   Yeast

The major relevant difference among yeast strains lies in their ease of breakage, which increases with increasing size and cell wall fragility. There may also of course be unexpected strain differences in nuclease or protease activity and stability of components of the protein-synthesizing system, leading to changed activity in the *in vitro* system. A large polyploid, *Saccharomyces cerevisiae* (isolated by P. L. Bergquist from a New Zealand Breweries, Ltd. yeast), was used in the work described here. This yeast was originally chosen because of its size and because its tRNAs had been characterized in considerable detail (Bergquist *et al.*, 1968). Nevertheless, fragility can be maximized by high growth rates and harvesting at an appropriate stage of culture (see Section B,2), and polyploidy thus avoided. A fully active cell-free system has also been prepared by the method described here, from genetically well-characterized haploid *Schizosaccharomyces pombe* strain 972h⁻ (C. H. J. Berry and A. Coddington, personal communication).

B.   Growth and Harvesting

1.   METHOD

The yeast was maintained on slants containing 2% glucose, 2% Bacto peptone, 1% Difco yeast extract, 1.5% agar, 10μg/ml adenine, 10 μg/ml uracil, 40 μg/ml lysine, and 10 μg/ml tryptophan. Starter cultures were prepared by inoculating 100 ml of growth medium (in 500-ml conical flasks) from a slant and growing for 24 hours at 25°C (mid-stationary phase). Starter cultures could be stored at 4°C for up to 4 days. The growth medium contained, per liter, 5 gm Difco yeast extract, 0.5 gm sodium chloride,

0.5 mg ferric chloride, 1 gm potassium dihydrogen phosphate and, added after sterilization as concentrated sterile solutions, 0.4 gm calcium chloride and 6 gm glucose (Linnane, 1965).

Six flasks (4-liter conical), each containing 2 liters of growth medium, were inoculated with 1 ml of stationary-phase starter culture per liter of growth medium and grown in a 29°C room on a rotary shaker at 125 rpm until an E_{650} of 1.2 (10^7 cells/ml) was reached, taking about 20 hours. The final yield was 2 gm wet weight of yeast per liter of culture. The culture was poured onto $\frac{1}{3}$ volume of crushed ice and then harvested at 4°C either in a Lourdes continuous-flow centrifuge or batchwise.

The packed cells were resuspended in a mixture of ice-cold 0.01 M Tris–HCl (pH 7.6) and 0.01 M MgCl$_2$ (15 ml per liter of original culture) and centrifuged at 16,000 g_{av} for 10 minutes (Sorval RC-2, GSA head). This wash was then repeated. Next the yeast was resuspended in 5 volumes of isolation buffer containing 50 mM Tris–HCl (pH 7.6), 8 mM magnesium acetate, 60 mM NH$_4$Cl, 0.5 mM spermidine, 0.5 dithiothreitol, and 10% (v/v) glycerol, and recentrifuged at 16,000 g_{av}. The pellet, in 0.5- to 1-gm lots on a spatula, was plunged into liquid nitrogen. The liquid nitrogen was decanted, and the yeast stored at −80°C, where it remained stable indefinitely.

2. CRITICAL FEATURES

Several aspects of these procedures are important for high activity in protein synthesis.

The phase of the growth cycle in which the cells are harvested affects the *in vivo* state of the protein-synthesizing system, the ionic state of the cells, or both (Lucas *et al.*, 1964; Dietz *et al.*, 1965). This variation is caused by changing batch culture conditions (Dietz *et al.*, 1965) and thus varies with different media. The cell walls, as the cultures near stationary phase, thicken and cross-link, resulting in increased breakage difficulties.

Maximal growth rates also favor the isolation of active systems. For the yeast used in these studies, the temperature giving fastest growth was 28–30°C (100 minutes, measured by nephelometry). The yeast grew at the maximum rate for one generation after a shift from 28 to 33°C and then slowed to 125 minutes; 7 hours after the shift, the rate suddenly decreased even more (202 minutes). This behavior indicates the existence of temperature-sensitive functions in this yeast affected by a temperature increase as small as 3°C, which strongly suggests that each new yeast strain should have its growth temperature carefully optimized (cf. Rose, 1975).

The washing procedure is vital. All growth medium must be removed, and adequately protective ionic conditions for ribosomes, and so on, must be present in the frozen yeast pellet, probably because this procedure not only presaturates the cell walls but is the major source of protection during break-age. The final isolation buffer contains the ionic components of the *in vitro* reaction mixtures, minus amino acids and energy sources, except for a higher concentration of dithiothreitol, a lower concentration of spermidine, and 10% glycerol. All these components are necessary to protect labile amino-acyl-tRNA synthetases and ribosomes against degradation and freezing. It is extremely likely that further study will improve the protective conditions.

Freezing the yeast pellet in liquid nitrogen caused it to crack and fragment, which facilitated subsequent breakage. This helped avoid overgrinding with alumina (see Section III,C,2), which gave inactive extracts. Yeast, frozen as above, gave extracts as active as those from unfrozen yeast in endogenous and poly U-stimulated protein synthesis. Freezing should probably be omitted if one is trying to isolate the study uncharacterized components of the system.

C.   Yeast Breakage

1.   Procedure

Frozen yeast (10–20 gm) was placed in a precooled ($-15°C$) 4-inch mortar on ice, and powdered. Twice the yeast wet weight of Bacto alumina was divided into three equal portions, and one portion was added to the mortar and gently mixed with the yeast. When an even powder was obtained, the mortar was placed at room temperature, and the yeast ground vigorously for about 5 minutes. As soon as the mixture was slightly moist, the mortar was placed on ice. After 5 minutes of grinding the mixture was usually quite wet and crackled audibly on being stirred. The second portion of alumina was added, and the mixture again ground until a semiliquid paste formed. The third alumina portion was added gradually such that the paste stayed slightly liquid. The total grinding process took 15 minutes. The paste was washed into 12-ml glass centrifuge tubes with isolation buffer equal in amount to the wet weight of the yeast, and the slurry thoroughly mixed.

2.   Critical Features

Success in following the exact grinding procedure was the most important factor in obtaining active extracts. Longer grinding than 15 minutes gave relatively inactive preparations. When the alumina was added too fast, so that the yeast did not reach the viscous "snap-crackle-pop" stage, breakage dropped from 80 to 30%. An attempt to scale up the grinding process was unsuccessful. A five-fold increase in yeast and a 6-inch mortar took an hour to grind to paste, giving an inactive extract. Probably because the larger mortar did not warm sufficiently fast, grinding at this scale was less efficient.

3.   ALTERNATIVE BREAKAGE METHODS

Breaking open the yeast wall and cell membrane without destroying fragile organelles with consequent delocalization of a variety of possible inhibitors, the mitochondrial protein-synthesizing system, and so on, is akin to trying to obtain intact eggs from a modern safe with explosives. Because it is so critical to the whole procedure, it is almost certainly worth considering alternatives for specific purposes.

Although slow and labourious, of only moderate scale, and with somewhat variable yields, alumina grinding was superior to the other breakage techniques we tried. Sand grinding is possibly equally effective. A French pressure cell required for complete breakage over twice the pressure previously reported to inactivate yeast ribosomes (Bretthauer et al., 1963). A French pressure cell has, however, been used to prepare high-salt-washed ribosomes (e.g., Schindler et al., 1974). Free and membrane-bound polysomes have been prepared successfully by Schneider et al. (1976) using a French pressure cell as an Eaton press (Eaton, 1962). Sonication was more destructive to yeast organelles than to the yeast cell wall. Blending with glass beads in a high-speed blender (Virtis, Model 45) was tried and ruled out as being very destructive and taking 10 minutes for adequate cell breakage. However, techniques involving the stirring of yeast suspensions with glass beads (40 mesh) using a Vibromix (Shandon Southern, Cheshire, England), or even a vortex mixer, have proved successful in breaking yeast. Such methods are surprisingly gentle, and reasonably intact mRNA (Fraser and Carter, 1976) and polysomes (Gallis et al., 1975) can be isolated from the homogenate. These methods can probably provide an alternative to alumina grinding. Breakage is dependent on the glass bead/liquid ratio, 2.5 gm of beads per milliliter of yeast suspension being optimal in dilute suspensions (C. H. Sissons, unpublished). It is important for in vitro protein synthesis to keep the buffer/cell ratio low in obtaining crude extracts. In alumina grinding the cell contents are diluted only about threefold.

One disadvantage of the alumina grinding procedure is the need for relatively high centrifugal forces to sediment the alumina from the S-30 extracts, which precludes the isolation of heavy membrane fractions. There was usually some alumina present after the first 29,000-g sedimentation during the preparation of S-30 extracts. Glass beads would help overcome this, as they can be removed by decanting (or centrifuging) the extract through bolting silk or fine stainless-steel mesh, but there is always fine broken glass left.

The major disadvantage of the standard alumina grinding procedure is that polysomes are degraded to 80S monosomes. Although the rate of endogenous protein synthesis was high, this clearly meant that most of these monosomes contained mRNA lacking termination codons, thus rendering preincubation techniques largely futile (see Section IV,C). Poly U was active on only a small proportion of the ribosome population (Sissons, 1974). Improved mechanical breakage methods, as outlined above, may yield polysomes, as grinding unfrozen yeast for 3 minutes gave a high proportion of large polysomes.

The major method for obtaining yeast polysomes is protoplast lysis. Protoplast formation (for review, see Kuo and Yamamoto, 1965) drastically changes the physiological state of the cells, is expensive on a large scale, and is reported to give ribosomes no more active than those produced by the grinding technique (Lucas et al., 1964). Where these limitations are immaterial, protoplasts potentially serve as a yeast polysome source. Incubating protoplasts after formation, in nutrient solutions, has allowed many studies on functioning polysomes (e.g., Gallis et al., 1975; Branes and Pogo, 1975; Mager and Planta, 1976). There appear to be, however, complicated controls on translation, which vary during protoplast incubation (Vessey and Keck, 1970). These may be important in some studies.

Using protoplasts as a source of all the components of the in vitro system may lead to difficulties with dilution factors. In our experience, S. pombe protoplasts have proved quite resistant to mechanical rupture unless it is preceded by significant osmotic shock.

D.   Preparation of the S-30 Extract

1.   CENTRIFUGATION

The grindates were centrifuged at 29,000 g_{av} for 15 minutes, and the supernatant was recentrifuged at 15,000 g_{av} for 30 minutes. Great care should be taken to avoid the floating lipid and membrane layers by taking only about two-thirds of the supernatant with a pasteur pipet.

2.   DIALYSIS

This supernatant, designated the S-30 extract, was dialyzed against 1 liter of isolation buffer for 4 hours to lower the amount of endogenous amino acids. Some protein precipitated, but this made no difference in protein synthesis activity or in physical state as analyzed on gradients and in an analytical ultracentrifuge (Sissons, 1974).

Other procedures such as a Sephadex-G25 chromatography have been used to remove low-molecular-weight components (e.g., So and Davie, 1963; Heridia and Halvorson, 1966; Gallis et al., 1975), however, such procedures considerably dilute the whole system. For some experiments it may be desirable to omit the dialysis or equivalent step.

3. STORAGE AND STABILITY

Portions (0.3 ml) were placed in 2-ml glass conical centrifuge tubes, frozen, and stored in liquid nitrogen. At $-15°C$ the extracts degraded rapidly. Glycerol, dimethyl sulfoxide, and polyethylene glycol are all inhibitory toward the cell-free system (C. H. Sissons, unpublished), which limits the use of high concentrations to protect against freeze-thawing. Repeated freeze-thawing of the system as prepared caused only slight loss of activity. At 25 to 30°C the S-30 extract was fairly stable until protein synthesis was started.

E. Physical Features of the S-30 Extract

The S-30 extracts contained 300–400 E_{260} units per milliliter and a biuret protein concentration of about 25–30 mg/ml. The λ_{max} was 260 nm, the λ_{min} was 240 nm, and the E_{260}/E_{240} ratio was 1.39–1.40 while the E_{260}/E_{280} ratio was 1.70–1.71. The ribosome content was 3–4 nmoles of 80S ribosomes per milliliter.

IV. Assay of Protein Synthesis

A. Incubation Procedure

1. CONSTRUCTION OF REACTION MIXTURES

The standard reaction mixture contained 25 μl of S-30 extract appropriately diluted or preincubated (see Section IV,C), 30 μl of incubation mixture (Section IV,B,1), mRNA, tRNA, and variable components to make up 100 μl. The incubation mixture contained all the components necessary to give the final desired reaction mixture after allowing for contributions from the other components (see Section IV,B,2).

To minimize degradation, all components except the cell extracts were added, using glass constriction pipets, to 12-ml conical centrifuge tubes placed in ice, in the order: water, inhibitors or other compounds (e.g., cations, buffer to balance different amounts of S-30 extract), tRNA, and, as soon as possible before starting a set of reactions, mRNA. Sets of 18 reactions were started at 20- or 30-second intervals by adding the S-30 extract prewarmed to room temperature to avoid ribosome precipitation. Gentle mixing of the S-30 extracts every second assay was carried out in a Super Mixer (Lab-line Instrument Company No. 1291) at low speed to avoid foaming. This constant mixing is essential to disperse ribosome aggregates. Strict adherence to appropriate techniques is a prerequisite for reliable results.

2. ANALYSIS OF PRODUCTS

Routinely, the reactions were stopped with 0.2 ml of 0.1 M KOH and the mixture incubated at 30°C for 10 minutes to hydrolyze aminoacyl-tRNA (Capecchi, 1966). Three milliliters of 7% (v/v) trichloroacetic acid was then added, and the precipitate immediately filtered off under vacuum on Whatman GF/C filters. The reaction tube was rinsed twice, and the filter holder rinsed twice more with 3 ml of 7% (v/v) trichloroacetic acid. The filters were dried at 80°C for 45 minutes, and the radioactivity measured on a liquid scintillation counter using 5 ml of toluene scintillator containing 0.5% 2,5-diphenyloxazole and 0.1% 1,4-bis(5-phenyloxazole-2-yl) benzene. It is necessary to prewash each filter in the holder twice with 3 ml of 7% trichloroacetic acid to reduce a high variable background binding of radioactive decay products. A time-zero reaction involving the complete reaction mixture kept 20 seconds on ice before KOH addition is usually the most appropriate blank. Polypeptide plus aminoacyl-tRNA radioactivity can be measured by omitting the KOH hydrolysis.

As poly U is the main exogenous mRNA still available in yeast, techniques analyzing phenylalanine incorporation are particularly useful. Partial hydrolysis followed by the Sephadex G-25 columns of Bretthauer and Golichowski (1968) is a method for analyzing polyphenylalanine chain length we have found to be more reliable (Sissons, 1974) than paper or thin-layer chromatography, which gave problems with streaking after partial hydrolysis. For some purposes paper chromatography can be used (e.g., van der Zeijst et al., 1972). Carboxyl-terminal phenylalanine estimation by hydrazinolysis may prove useful (van zer Zeijst et al., 1973). N-Acetyl-phenylalanyl-tRNA, an artificial "initiator tRNA," has had considerable use in initiation studies (Ayuso and Heridia, 1968; Torano et al., 1972; van der Zeijst et al., 1972; Pranger et al., 1974). For other potential mRNAs, a large number of possible analytical techniques now exists, including gel electrophoretic characterization of endogenous yeast mRNA (Gallis and Young, 1975; Gallis et al., 1975).

B.   Reaction Conditions

1. REACTION TEMPERATURE

The temperature optimum was 20°C measured by total protein synthesis. The standard incubation was at 20°C for 30 minutes during which the reaction rate was virtually constant. At 25°C the initial polymerization rate was faster but terminated sooner, indicating limited stability of some component, possibly the aminoacyl-tRNA synthetase (Sissons, 1974). The temperature optimum may well vary between yeast strains and depend on whether labeled amino acid or aminoacyl-tRNA is supplied. It is probably better to sacrifice a fast initial reaction rate to enhanced stability in yeast reaction mixtures until proper preservation of initiation function is achieved.

2. S-30 EXTRACT CONCENTRATION

The optimum concentration of S-30 extract was determined for each S-30 preparation, as there is an inhibitor present, localized in the 100,000-g supernatant fraction (Sissons, 1974). Usually a 1:3 or 1:4 dilution was optimal.

3. IONIC CONDITIONS

The compositions of the standard reaction and incubation mixtures are shown in Table I. The techniques for making up and storing the various components are shown also.

These cation concentrations, probably the most important ionic variables, are near optimal for both poly-U-dependent and endogenous mRNA-dependent protein synthesis. With poly U, the Mg^{2+} optimum is 8–10 mM and the spermidine optimum 3 mM. For endogenous protein synthesis, the Mg^{2+} optimum varied between 4 and 8 mM for different S-30 extracts, probably depending on the intrinsic polyamine levels. Polyamines increase RNA secondary structure, so 1 mM, which gives 75% of the maximal poly-phenylalanine synthesis, is more desirable for routine use.

These optima are those in the described concentrations of cation-buffering compounds, most notably the nucleotides and phosphoenolpyruvate. The free Mg^{2+} is between 1 and 3 mM (Manchester, 1970). Changing the concentration of chelators of cations changed the Mg^{2+} equilibria. Using 15 mM phosphoenolpyruvate instead of 5 mM raised the poly-U Mg^{2+} optimum 5 mM and enhanced polyphenylalanine synthesis by 20%, showing that probably the exact state of the Mg^{2+} chelates in the extracts is an important variable. Added yeast tRNA caused inhibition by changing the free Mg^{2+} concentration, an effect superimposed on a direct inhibition by high levels of tRNA (C. H. Sissons, unpublished).

Polyamines especially, show complex interrelationships with Mg^{2+}. The stimulation by polyamines varies with Mg^{2+} concentration, and there is evidence for limited replacement of Mg^{2+} by polyamines and competition between Mg^{2+} and polyamines, but both cations have independent functions (C. H. Sissons, unpublished). The high degree of dependence on polyamines is nearly unique to yeast cell-free systems (but cf. E. coli, Fuchs, 1976). Magnesium and probably spermidine optima should be established

TABLE I

STANDARD REACTION AND INCUBATION MIXTURE COMPOSITIONS

Component	Reaction concentration[a]	S-30 contribution (mM)	Incubation mix concentration[a]	Stock concentration[a,b]	Stock per 20 assay incubation mix (μl)
Tris–Cl, pH 7.6	50	12.5	125	1000	75
NH$_4$Cl	60	15	150	1000	90
Mg(OAc)$_2$	8	2	20	200	60
ATP	1.33	—	4.3	100[c]	—
GTP	0.167	—	0.55	10[c]	—
PEP	5.0	—	16.67	100[c]	—
Glutathione	8.0	—	26.67	400	40
Spermidine	1.125	0.125	3.	100	20
Dithiothreitol	0.125	0.125	—	—	—
19[12C]amino acids[a]	0.04	—	0.1333	1	80
[14C]amino acid[e]	50 nCi/assay	—	1.67 μCi/ml	50 μCi/ml	20
[12C]amino acid of [14C]amino acid[f]	0.01[g]	—	0.0333	1	20
Water	—	—	—	—	20
Total volume	—	—	—	—	30
					595[h]
PEP kinase	5 units/assay	—	—	Stock	6[i]

[a]Millimolar unless otherwise noted.
[b]All stocks are frozen and stored at −20°C, except the phosphoenolpyruvate (PEP) kinase.
[c]Stored together as one solution.
[d]Twenty protein amino acids each 1 mM minus the labeled amino acid(s).
[e]Usually [14C]phenylalanine at 250 μCi/μmole.
[f]Unlabeled amino acid corresponding to the unlabeled stock.
[g]The exact amount of radioactive amino acid was calculated, and radioactivity measured for each reaction mixture.
[h]The incubation mixture minus PEP kinase can be stored frozen.
[i]The PEP kinase should be added on the day of assay.

for a new system, as well as a given set of RNA and other chelator concentrations.

Although KCl is often used as the cation in reaction mixtures (Grant *et al.*, 1974; Gallis *et al.*, 1975; van der Zeijst *et al.*, 1972), we have found NH_4Cl gives greater poly-U-directed polypeptide synthesis. Both KCl and NH_4Cl have a Mg^{2+}-dependent optimum. In 8 mM Mg^{2+} the NH_4Cl optimum is 100 mM; in 4 mM Mg^{2+} it is 60 mM. For endogenous protein synthesis in 4 mM Mg^{2+}, 40 mM was optimal for both KCl and NH_4Cl. Both cations gave identical activities on endogenous mRNA. The final choice is probably always a compromise.

C. Preincubation Procedures

Sometimes, but not always, a two- to fourfold activation of the S-30 extract for poly-U-dependent polyphenylalanine synthesis was obtained by preincubating it at 0°C in low Mg^{2+} (2–5 mM; cf. Section IV,B,3) with all the components necessary for protein synthesis. This activation, although often very significant when it happened, remained variable and elusive in spite of considerable investigation. Its nature and the circumstances producing it are still unknown. Preincubation at 20°C to obtain ribosomes inactivated the system (Sissons, 1974). This inactivation by preincubation under protein-synthesizing conditions seems to be the universal experience in yeast systems, even those using polysomes (see Gallis and Young, 1975), except for two very early reports on low-activity systems (Maeda and Imahori, 1963; Bretthauer *et al.*, 1963). The failure to preincubate yeast extracts successfully is probably the major technical limitation on current yeast *in vitro* systems after the failure to maintain a complete initiation system.

D. mRNAs Other Than Poly U

Apart from poly U, various synthetic polynucleotides have been used successfully in the yeast cell-free system. We found poly (U, C) as efficient an mRNA for phenylalanine and serine incorporation into polypeptide as poly U was for phenylalanine. Poly A was functional in early yeast *in vitro* systems (Halvorson and Heridia, 1967), but the polylysine produced was larger than the total codon number in the poly A used, indicating a translational defect. Poly AUG forms polymethionine (Richter *et al.*, 1971). There seems to be little difficulty in translating these artificial mRNAs which, however, omit most of the steps in initiation.

Three of the first publications on the yeast cell-free system (Barnett *et al.*, 1962; So and Davie, 1963; Maeda and Imahori, 1963) reported that, in cell-free systems programed by endogenous mRNA, added mRNA preparations, containing mostly rRNA, stimulated amino acid incorporation into poly-peptide up to 100%. This has not been repeated since. It seems likely that in these early, low-activity extracts, instead of the enhanced incorporation being caused by initiation on new mRNA molecules, the RNA preparations protected or enhanced the translation of mRNA already bound to ribo-somes (cf. Hunt and Wilkinson, 1967).

The most recent attempt to make natural exogenous mRNA function in a crude system appears to be that reported by Gallis and Young (1975), who obtained no evidence for any initiation onto added reticulocyte mRNA or yeast polysomal RNA. They suggest that initiation factors which dis-sociate ribosomes (see Pêtre, 1970) may be inactive in their yeast extracts. We attempted R_{17}RNA translation in the cell-free system described here. A wide variety of detection techniques and supplements including $E.\ coli$ components showed no R_{17}RNA-specific translation. The defect was in the initiation, and this could indicate species specificity. Evidence was obtained that the R_{17}RNA bound to 40S subunits but not to any assembled 80S ribosomes.

V. Prospects

As discussed, the major defects in the existing yeast cell-free system are lack of initiation on added natural mRNA caused by the instability of initia-tion factors (Gallis and Young, 1975) or instability of the ribosomes (cf. Schreier and Staehelin, 1973), preventing ribosome recycling $in\ vitro$. Until some small level of initiation is detected on natural mRNAs and pre-incubation procedures can be carried out using poly U, both ribosome and initiation factor integrity must be suspect.

The suggestion by Branes and Pogo (1975) that yeast mRNA initiates protein synthesis as a membrane-bound structure, a suggestion reinforced by Schneider $et\ al.$ (1976), adds a new dimension to the initiation of protein synthesis in yeast extracts. It also reaffirms the importance of using crude rather than purified extracts for initial studies. Hopefully, purification of the proteins responsible for mRNA transport, and suitable membranes, will allow more progress.

Nevertheless, systems such as that from wheat germ and Krebs ascites cells initiate protein synthesis on exogenous mRNA, where existing yeast systems do not (see Gallis and Young, 1975; Gallis $et\ al.$, 1975) and probably should. Further attention should be paid to protecting labile components from thermal or freezing denaturation and from nuclease or protease attack.

For a review on protecting protein in yeast from protease attack, see Pringle (1975a). However, attempts by Gallis and Young (1975) to protect their system from protease attack did not result in a system which would initiate on mRNA. Replacement studies using heterologous initiation factors, ribosomes, and soluble factors should help track down the exact defects in the yeast system. Conversely, one could possibly use a highly fractionated eukaryotic system to test yeast components. These approaches have been used successfully in studying elongation reactions (e.g., Skogerson and Engelhardt, 1977). *In vivo* manipulation of the ribosomal state, such as treatment with azide of protoplasts (van er Zeijst *et al.*, 1972) or whole cells (Grant *et al.*, 1974), is another approach which is little used but may enhance the chances of success. The processes of RNA and protein synthesis are now known to have a battery of specific inhibitors (reviewed by Vazquez, 1974; Schindler and Davies, 1975) and mutants, mostly t_s (reviewed by Pringle, 1975b). Together these add up to several possible approaches to improving the yeast cell-free protein-synthesizing system.

ACKNOWLEDGMENTS

I thank Ray Ralph for wise encouragement and guidance during my work on the yeast cell-free system, Peter Bergquist for teaching me *E. coli in vitro* system technology, my colleagues in the Cell Biology Department, Auckland University, where this work was done, the Zoology Department, Edinburgh University, who endured its emergence, and Doug Wright, in whose laboratory I wrote this article. I was supported during this work by a Research Fellowship from the Auckland Division of the New Zealand Cancer Society.

REFERENCES

Albrecht, V., Prenzel, K., and Richter, D. (1970). *Biochemistry* 9, 361–368.
Ayuso, M. S., and Heridia, C. F. (1967). *Biochim. Biophys. Acta* 145, 199–201.
Ayuso, M. S., and Heridia, C. F. (1968). *Eur. J. Biochem.* 7, 111–118.
Barbacid, M., Fresno, M., and Vazquez, D. (1975). *J. Antibiot.* 28, 453–462.
Barnett, L. B., Frens, G., and Koningsberger, V. V. (1962). *Biochem. J.* 84, 89P–90P.
Bergquist, P. L., Burns, D. J. W., and Plinston, C. A. (1968). *Biochemistry* 7, 1751–1761.
Branes, L., and Pogo, A. O. (1975). *Eur. J. Biochem.* 54, 317–328.
Bretthauer, R. K., and Golichowski, A. M. (1968). *Biochim. Biophys. Acta* 155, 549–557.
Bretthauer, R. K., Marcus, L., Chaloupka, J., Halvorson, H. O., and Bock, R. M. (1963). *Biochemistry* 4, 1079–1084.
Bruening, G. E., and Bock, R. M. (1965). *Biochim. Biophys. Acta* 149, 377–386.
Bucher, K., and Skogerson, L. (1976). *Biochemistry* 15, 4755–4759.
Capecchi, M. (1966). *J. Mol. Biol.* 21, 173–193.
Cooper, A. H., Harris, G., and Neal, G. E. (1962). *Biochim. Biophys. Acta* 61, 573–582.
Dietz, G. W., Reid, B. R., and Simpson, M. V. (1965). *Biochemistry* 4, 2340–2350.
Downey, K. M., So, A. G., and Davie, E. W. (1965). *Biochemistry* 4, 1702–1709.
Eaton, N. R. (1962). *J. Bacteriol.* 83, 1359–1360.
Fraser, R. S. S., and Carter, B. L. A. (1976). *J. Mol. Biol.* 104, 223–242.
Fuchs, E. (1976). *Eur. J. Biochem.* 63, 15–22.
Gallis, B. M., and Young, E. T. (1975). *J. Bacteriol.* 122, 719–726.

Gallis, B. M., McDonnell, J. P., Hopper, J. E., and Young, E. T. (1975). *Biochemistry* **14**, 1038–1046.

Ghosh, N. (1969). *Biochem. J.* **115**, 1005–1007.

Grant, P., Sánchez, L., and Jiménez, A. (1974). *J. Bacteriol.* **120**, 1308–1314.

Halvorson, H. O., and Heridia, C. (1967). *In* "Aspects of Yeast Metabolism" (A. K. Mills and S. H. Krebs, eds.), pp. 107–132. Blackwell, Oxford.

Heridia, C. F., and Halvorson, H. O. (1966). *Biochemistry* **5**, 946–952.

Hunt, J. A., and Wilkinson, B. R. (1967). *Biochemistry* **6**, 1688–1693.

Kaempfer, R. (1969). *Nature (London)* **222**, 950–953.

Kuo, S.-C., and Yamamoto, S. (1975). *Methods Cell Biol.*, **11**, 169–183.

Lamb, A. J., Clark-Walker, G. D., and Linnane, A. W. (1968). *Biochim. Biophys. Acta* **161**, 415–427.

Linnane, A. W. (1965). *Oxidases Relat. Redox Syst. Prve. Symp., 1964* Vol. 2, pp. 1102–1128.

Lucas, J. M., Schuurs, A. H. W. M., and Simpson, M. V. (1964). *Biochemistry* **3**, 959–967.

Maeda, A., and Imahori, K. (1963). *Biochim. Biophys. Acta* **76**, 543–557.

Mager, W. H., and Planta, R. J. (1976). *Eur. J. Biochem.* **62**, 193–197.

Manchester, K. L. (1970). *Biochim. Biophys. Acta* **213**, 532–534.

Marcus, L., Bretthauer, R. K., Bock, R. M., and Halvorson, H. O. (1963). *Proc. Natl. Acad. Sci. U.S.A.* **50**, 782–789.

Marcus, L., Bretthauer, R., Halvorson, H. O., and Bock, R. M. (1965). *Science* **147**, 615–617.

Marcus, L., Ris, H., Halvorson, H. O., Bretthauer, R. K., and Bock, R. M. (1967). *J. Cell Biol.* **34**, 505–512.

Martin, T. E., and Hartwell, L. H. (1970). *J. Biol. Chem.* **245**, 1504–1508.

Perani, A., Tiboni, O., and Ciferri, O. (1971). *J. Mol. Biol.* **55**, 107–112.

Pêtre, J. (1970). *Eur. J. Biochem.* **14**, 399–405.

Pranger, M. H. Roos, M. H., van der Zeijst, B. A. M., and Bloemers, H. P. J. (1974). *Mol. Biol. Rep.* **1**, 321–327.

Pringle, J. R. (1975a). *Methods Cell Biol.* **12**, 149–184.

Pringle, J. R. (1975b). *Methods Cell Biol.* **12**, 233–272.

Rao, S. S., and Grollman, A. P. (1967). *Biochem. Biophys. Res. Commun.* **29**, 696–704.

Richter, D., and Klink, F. (1967). *Biochemistry* **6**, 3569–3575.

Richter, D., and Lipmann, F. (1970). *Nature (London)* **227**, 1212–1214.

Richter, D., Hameister, H., Petersen, H. G., and Klink, F. (1968). *Biochemistry* **10**, 3753–3761.

Richter, D., Lipmann, F., Tarragó, A., and Allende, J. E. (1971). *Proc. Natl. Acad. Sci. U.S.A.* **68**, 1805–1809.

Rose, A. H. (1975). *Methods Cell Biol.* **12**, 1–16.

Rubin, G. M. (1975). *Methods Cell Biol.* **12**, 45–64.

Schindler, D., and Davies, J. (1975). *Methods Cell Biol.* **12**, 17–38.

Schindler, D., Grant, P., and Davies, J. (1974). *Nature (London)* **248**, 535–536.

Schneider, E., Lochman, E.-R., and Lother, H. (1976). *Biochim. Biophys. Acta* **432**, 92–97.

Schreier, M. H., and Staehelin, T. (1973). *J. Mol. Biol.* **73**, 329–349.

Siegel, M. R., and Sisler, H. D. (1964). *Biochim. Biophys. Acta* **87**, 83–89.

Sissons, C. H. (1974). *Biochem. J.* **144**, 131–140.

Skogerson, L., and Engelhardt, D. (1977). *J. Biol. Chem.* **252**, 1471–1475.

Skogerson, L., and Wakatama, E. (1976). *Proc. Natl. Acad. Sci. U.S.A.* **73**, 73–76.

Skogerson, L., McLaughlin, C., and Wakatama, E. (1973). *J. Bacteriol.* **116**, 818–822.

So, A. G., and Davie, E. W. (1963). *Biochemistry* **2**, 132–136.

Somasundaran, U., and Skogerson, L. (1976). *Biochemistry* **15**, 4760–4764.

Tanner, M. J. A. (1967). *Biochemistry* **6**, 2686–2694.

Toraño, A., Sandoval, A., Sanjosé, C., and Heridia, C. F. (1972). *FEBS lett.* **22**, 11–14.

van der Zeijst, B. A. M., Kool, A. J., and Bloemers, H. P. J. (1972). *Eur. J. Biochem.* **30**, 15–25.

van der Zeijst, B. A. M., Engel, K. J. M., and Bloemers, H. P. J. (1973). *Biochim. Biophys. Acta* **294**, 517–526.

Vazquez, D. (1974). *FEBS Lett.* **40**, Suppl. 563–584.

Vazquez, D., Battaner, E., Neth, R., Heller, G., and Monro, R. E. (1969). *Cold Spring Harbor Symp. Quant. Biol.* **34**, 369–375.

Vessey, D. A., and Keck, K. (1970). *Biochemistry* **9**, 2923–2930.

Walters, J. A. L. I., and van Os, G. A. J. (1970). *Biochim. Biophys. Acta* **199**, 453–463.

Webster, G. C. (1957). *J. Biol. Chem.* **229**, 535–546.

Chapter 7

Preparation of Protoplasts of Schizosaccharomyces pombe

J. SCHWENCKE AND M. NAGY

Laboratoire d'Enzymologie,
Centre National de la Recherche Scientifique
Gif-sur-Yvette, France

I. Introduction

The so-called fission strains of yeasts are difficult to transform into protoplasts using snail enzyme preparation even under a wide range of conditions. This failure is related to the particular composition of the cell walls of the genus *Schizosaccharomyces*. Thus the wall of *Schizosaccharomyces pombe* (Bush *et al.*, 1974) is made up of a β-glucan fraction (46–54%) probably containing two separate polysaccharides: a β-1,6-linked glucan and another with β-1,3 linkages. Furthermore, unlike the situation in the majority of other yeasts there exists an important α-glucan fraction (28%) most probably consisting of a linear polymer with α-1,3 linkages. There is also a galactomannan fraction (9–14%). These data of Bush *et al.* (1974) are in general agreement with the previous work of Kreger (1954), Bacon *et al.* (1968), Gorin and Spencer (1968), Gorin *et al.* (1969), Phaff (1971, review), and Fleet and Phaff (1973).

Snail digestive juice is unable to degrade α-1,3-glucan (Kanetsuma *et al.*, 1969; de Vries and Wessels, 1972). Furthermore, it has been found that purified β-1,3-glucanase of the basidiomycete QM 806 is also unable to dissolve isolated walls of *S. pombe* (Tanaka and Phaff, 1966), though this enzyme can degrade the β–1,3 polysaccharide component.

The aforementioned characteristics of both snail enzymes and the cell wall composition of *Schizosaccharomyces* explain the difficulties encountered by previous workers in making protoplasts using snail enzymes alone (Holter and Ottolenghi, 1960; Havelkova, 1966; Svoboda, 1967; Nečas *et al.*, 1968; Rost, 1969; Shahin, 1971; Mitchison *et al.*, 1973). Alternative methods, based on the inhibitory effect of 2-deoxyglucose (2-DG) on the cell wall formation of *Schizosaccharomyces* cells (Megnet, 1965; Johnson, 1968; Poole and Lloyd, 1973; Johnson *et al.*, 1974), have been also described. One type involves wall degradation mediated by the addition of 2-DG and $MgSO_4$, to the culture medium (Berliner, 1971; MacDonald and Berlinger, 1972). The other involves the growth of cells in media containing 2-DG, and subsequent treatment with snail gut juice (Birnboim, 1971; Foury and Goffeau, 1973; Poole and Lloyd, 1973, 1976). Both methods have the disadvantage that cells are grown under conditions which increase the risk of possible metabolic perturbations. This problem has been minimized for 2-DG by running short preincubations of exponentially growing cells of *S. pombe* with low concentrations (70 μg/ml) of 2-DG (Foury and Goffeau, 1973). However, none of these methods allows complete conversion of the whole cell population into protoplasts, nor the isolation of protoplasts from cells grown in normal media, a condition necessary for the majority of biochemical studies.

In this article, we describe a method by which cells of *S. pombe* are completly and efficiently converted into protoplasts. The method is essentially that previously described by us (Housset *et al.*, 1975). It requires a mixture of α-1,3-glucanase, exo-β-1,3-glucanase, and snail enzymes. In an independent communication, Kopecka (1975), has also reported the quantitative preparation of protoplasts from two strains of *Schizosaccharomyces* by the combined action of snail gut juice and a lytic preparation from *Trichoderma viride* containing α-1,3-glucanase, as previously demonstrated by Hasegawa *et al.* (1969) and de Vries and Wessels (1973).

II.   Preparation of Protoplasts

Our previously described method (Housset *et al.*, 1975) has been subsequently modified as follows.

1. *Schizosaccharomyces pombe* cells, wild-type strain 972h⁻, are grown at 30°C in a complete medium containing in 1 liter: 5 gm of yeast extract, 5 gm of Difco peptone, and 30 gm of glucose. The addition of sulfur amino acids (Svihla *et al.*, 1961) has been found to be unnecessary.

Protoplasts are prepared from exponentially growing cells (up to 8×10^6 cells/ml), harvested by centrifugation (5000 g for 1 minute), and washed once with distilled water.

2. The yeast pellet (10^9 cells, about 350 mg wet weight) is resuspended in 5 ml of a freshly prepared solution containing 100 mM Tris, 5 mM EDTA, and 5 mM dithiothreitol brought to pH 8 with HCl (Schwencke *et al.*, 1969). After 15 minutes at 30°C, the cells are separated by centrifugation (5000 g for 1 minute) and washed once with 0.6 M sorbitol containing 20 mM (N-morpholino) ethanesulfonic acid (MES) brought to pH 6 with KOH.

3. The cells are then incubated (10^9 cells/ml) with gentle shaking at 30°C in a fresh mixture of 0.6 M KCl, 5 mM dithiothreitol, and 40 mg/ml lyophylized snail enzymes (Industrie Biologique Française), to which is added 5 mg/ml of α-1,3-glucanase (3 units/mg) and 5 mg/ml of β-1,3-glucanase (10 units/mg). (The last two enzymes were obtained from E. T. Reese, Food Microbiology Group, U.S. Army, Natick Laboratories, Natick, Mass.) This enzyme mixture is previously clarified by centrifugation at 12,000 g for 15 minutes. Complete conversion of cells to protoplasts (Fig. 1) is obtained after about 60 minutes of incubation for cells grown in complete medium

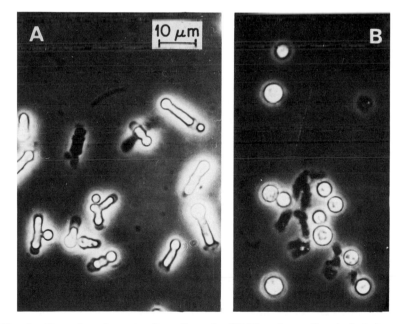

FIG. 1. Protoplasts formation from *S. pombe*. (A) The release of protoplasts from cells, leaving the cell envelopes intact. (B) Free protoplasts at the end of the incubation. The dark bodies are partially digested envelopes.

without sulfur amino acids, and after about 80 minutes for those grown in their presence.

4. Protoplasts may be purified from debris by sedimentation at 2000 g for 5 minutes through a layer of 0.8 M sucrose containing 0.2 M KCl. The pelleted protoplasts are resuspended and stored in 0.6 M sorbitol containing 0.2 M KCl and 20 mM MES (pH 5, KOH).

III. Concluding Remarks

Both α-1,3 and β-1,3-glucanase are indispensable for good results. Protoplasts obtained and stored under the conditions described above maintain good microscopic and metabolic characteristics as indicated by the study of their guanine uptake (Housset *et al.*, 1975). Early stationary cells and cells grown in mineral media also have been converted to protoplasts using a stronger pretreatment medium (Schwencke *et al.*, 1977). The method described by Kopecka (1975) should also be useful. A limitation common to both methods is the lack of a commercial source of α-1,3-glucanase. A mixture of β-1,3-glucanases is available in gram quantities under the trade name Zymolase from Kirin Brewery Co., Ltd., Takasaki, Gumma Prefecture, Japan. However, as expected, complete conversion of cells into protoplasts with this enzyme preparation alone has not been obtained.

ACKNOWLEDGMENTS

The work described here was supported by the Délégation Générale à la Recherche Scientifique et Technique (74-7-0399) and by the Centre National de la Recherche Scientifique (ERA 224). We are most grateful to Dr. E. T. Reese for the continuous supply of α- and β-glucanase.

REFERENCES

Bacon, J. S. D., Jones, D., Farmer, V. C., and Webley, D. M. (1968). *Biochim. Biophys. Acta* **158**, 313–315.

Berliner, M. D. (1971). *Mycologia* **63**, 819–825.

Birnboin, H. C. (1971). *J. Bacteriol.* **107**, 659–663.

Bush, D. A., Horisberger, M., Horman, L., and Wursch, P. (1974). *J. Gen Microbiol.* **81**, 199–206.

Fleet, G. H., and Phaff, H. J. (1973). *In* "Yeast, Mould and Plant Protoplasts" (J. R. Villanueva *et al.*, eds.), pp. 33–59. Academic Press, New York.

Foury, F., and Goffeau, A. (1973). *J. Gen. Microbiol.* **75**, 227–229.

Gorin, P. A. J., and Spencer, J. F. T. (1968). *Can. J. Chem.* **46**, 2299–2304.

Gorin, P. A. J., Spencer, J. F. T., and Magus, R. J. (1969). *Can. J. Chem.* **47**, 3569–3576.

Hasegawa, S. H., Nordin J. H., and Kirkwood, S. (1969). *J. Biol. Chem.* **244**, 5460–5470.

Havelkova, M. (1966). *Folia Microbiol.* (*Prague*) **11**, 453–458.

Holter, H., and Ottolenghi, P. (1960). *C.R. Trav. Lab. Carlsberg* **31**, 409–422.
Housset, P., Nagy, M., and Schwencke, J. (1975). *J. Gen. Microbiol.* **90**, 260–264.
Johnson, B. F. (1968). *J. Bacteriol.* **95**, 1169–1172.
Johnson, B. F., Lu, C., and Brandwein, S. (1974). *Can. J. Genet. Cytol.* **16**, 593–598.
Kanetsuna, F., Carbonell, L. M., Moreno, R. E., and Rodriguez, J. (1969). *J. Bacteriol.* **97**, 1036–1041.
Kopecka, M. (1975). *Folia Microbiol. (Prague)* **20**, 273–276.
Kreger, D. R. (1954). *Biochim. Biophys. Acta* **13**, 1–9.
MacDonald, C. E., and Berliner, M. D. (1972). *Appl. Microbiol.* **24**, 993–994.
Megnet, R. (1965). *J. Bacteriol.* **90**, 1032–1035.
Mitchison, J. M., Creanor, J., and Sartirana, M. L. (1973). *In* "Yeast, Mould and Plant Protoplasts" (J. R. Villanueva *et al.*, eds.), pp. 229–275. Academic Press, New York.
Nečas, O., Svoboda, A., and Havelkova, M. (1968). *Folia Biol. (Prague)* **14**, 80–85.
Phaff, H. J. (1971). *In* "The Yeasts" (A. H. Rose and J. S. Harrison, eds.), Vol. 2, pp. 135–210. Academic Press, New York.
Poole, R. K., and Lloyd, D. (1973). *Arch. Microbiol.* **88**, 257–272.
Poole, R. K., and Lloyd, D. (1976). *J. Gen. Microbiol.* **93**, 241–250.
Rost. K. (1969). *Z. Allg. Mikrobiol.* **9**, 289–295.
Schwencke, J., Gonzalez, G., and Farias, G. (1969). *J. Inst. Brew., London* **75**, 15–19.
Schwencke, J., Magaña-Schwencke, N., and Laporte, J. (1977). *Ann. Microbiol. (Paris)* **128**, 3–17.
Shahin, M. M. (1971). *Can. J. Genet. Cytol.* **13**, 714–719.
Svihla, G., Schlenk, F., and Dainko, J. L. (1961). *J. Bacteriol.* **82**, 808–814.
Svoboda, A. (1967). *In* "Symposium ueber Hefe-Protoplasten" (R. Mueller, ed.), pp. 31–35. Akademie-Verlag, Berlin.
Tanaka, H., and Phaff, H. J. (1966). *Abh. Dtsch. Akad. Wiss. Berlin, K. Med.* **6**, 113–129.
de Vries, O. M. H., and Wessels, J. G. H. (1972). *J. Gen. Microbiol.* **73**, 13–22.
de Vries, O. M. H., and Wessels, J. G. H. (1973). *J. Gen. Microbiol.* **76**, 319–330.

Chapter 8

Bulk Isolation of Yeast Mitochondria

DONALD DETERS, URS MÜLLER, AND HENRI HOMBERGER

Biozentrum der Universität Basel,
Basel, Switzerland

I. Introduction

Various methods have been established for isolating yeast mitochondria. However, most are not suitable for the isolation of mitochondria in gram quantities. Generally, small-scale isolation methods cannot be conveniently scaled up. For example, the treatment of yeast cells with snail gut enzyme yields fragile spheroplasts which can be gently lysed to provide fairly intact mitochondria (Ohnishi *et al.*, 1966; Schatz and Kovac, 1974). However, snail enzyme is expensive and therefore unsuitable for large-scale work. Likewise, the Braun (or Bronwill) glass-bead shaker effectively breaks yeast cells (Mason *et al.*, 1973; Salton, 1974; Tzagoloff, 1971) but accommodates only 20–30 gm of yeast cells per shaking vessel.

Faced with these limitations, researchers have generally resorted to two methods for the bulk isolation of yeast mitochondria. One involves rupturing the cells in a Waring Blendor in the presence of liquid nitrogen (Tzagoloff, 1969), and the second involves breaking the cells in a commercially available Manton-Gaulin homogenizer (Mason *et al.*, 1973). Both methods have been successfully used in many laboratories, including our own, but are relatively laborious and time-consuming, particularly when large amounts of cells need to be broken. Moreover, we have found that these methods are not always equally reliable with different yeast strains or commercial preparations of yeast.

In our experience use of the Dyno-Mill, Model KD-L, a continuous-flow cell disintegrator, is superior to the other large-scale methods of yeast cell breakage. The principle of disruption is described in publications (Rěháček *et al.*, 1969; Rěháček, 1971), patents [Czechoslovak Patent No. 115–017 (1965)], and in information provided by the manufacturer. Briefly, a cellular suspension is pumped continuously through a 600-ml chilled glass chamber containing vigorously stirred glass beads. Conditions can be readily adjusted to optimize cell breakage.

II. Materials and Methods

The Model KD-L Dyno-Mill was donated to the Biozentrum by the manufacturer, Willy A. Bachofen, Maschinenfabrik, Basel, Switzerland. Glass beads were obtained from B. Braun, Melsungen, AG, and pressed cake yeast from E. Klipfel and Company, AG, Rheinfelden, Switzerland, or Bäko, Hefevertrieb, Lörrach, West Germany, unless otherwise stated. The yeast cells were suspended in 0.9% KCl, 50 mM K$_2$HPO$_4$, and 1 mM EDTA buffer (KPE) by mixing 1 liter of KPE with each 0.5 kg of yeast cells and vigorously aerated overnight at 20°C with stirring in a 14-liter Magnaferm fermentor (New Brunswick Scientific Company, Inc., New Brunswick, N.J.). Cytochrome spectra, heme a, and protein were determined as described earlier (Mason *et al.*, 1973).

III. Breakage of Yeast Cells

We have found that 4 kg of yeast cells is a convenient amount to process in 1 day. After aeration overnight the yeast cell suspension is cooled in ice to 4°C and pumped with a variable-speed peristaltic pump (Biowerk, Biozentrum der Universität Basel) to the cold, jacketed glass breakage chamber of the Dyno-Mill (Fig. 1). The chamber is cooled continuously with dry ice-chilled ethanol circulated through the jacket of the breakage chamber from an external reservoir at 200 liters per hour. Circulation of a coolant at only −20°C was less satisfactory.

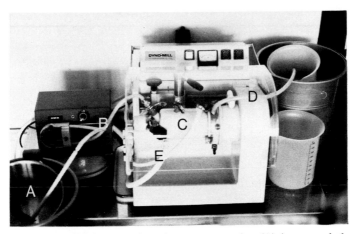

FIG. 1. The Dyno-Mill in operation. A yeast suspension (A) is pumped through the inlet tubing (B) to the breakage chamber (C). Broken cells exit through tubing (D) to a collection vessel. Tubing (E) carries cold ethanol from a dry ice–ethanol reservoir (not shown) to the jacketed breakage chamber.

The Dyno-Mill cell breakage system is quite flexible and allows adjustment of experimental conditions to achieve optimal results. We have varied the cellular concentration, glass-bead size, stirring velocity, and flow rate, and monitored the effect on cell breakage (Deters et al., 1976). We now routinely use 510 ml of 0.45- to 0.5-mm glass beads, stirred at 3000 rpm, and pump the cellular suspension through the breakage chamber at 7 liters per hour. Glass beads usually can be used for three to four preparations and are then discarded.

The efficiency of cell breakage was monitored by microscopic examination. In more than 100 separate runs carried out in the past 2 years with commercial yeasts from three different sources and with two different laboratory strains, breakage was always 80% or greater. Therefore we believe the conditions described above are optimal.

IV. Isolation of Mitochondria

The procedure has been slightly modified from that described previously (Mason et al., 1973). The broken cell suspension emerged from the Dyno-Mill at 6–8°C. All subsequent steps were carried out at 0–4°C. The cellular suspension was centrifuged immediately two times for 10 minutes at 8000 rpm (13,200 g_{max}) in a Sorvall G-3 rotor to remove unbroken cells and cell debris. The mitochondria were sedimented from the resulting supernatant using two (for 4 kg of yeast) continuous-flow-type GLT Cepa centrifuges (Carl Padberg, Lahr, Schwarzwald, West Germany), each operated at 50,000 rpm, and a flow rate of 30 ml per minute. This centrifuge is essentially identical to a Sharples continuous-flow centrifuge. The mitochondrial pellet (5–7 gm of protein per kilogram of yeast) was suspended in cold KPE at 35 mg protein/ml with a large Teflon homogenizer and resedimented by centrifugation for 90 minutes at 33,000 rpm in a No. 35 rotor (87,000 g_{av}) of a Beckman L-5 ultracentrifuge. The supernatant was discarded, and the pellet resuspended in KPE at 30 mg protein/ml. The mitochondria were stored frozen at $-80°C$ or used immediately.

V.   Properties of Isolated Mitochondria

Over the past 2 years we have processed over 300 kg of yeast cells using the Dyno-Mill, and isolated mitochondria as described above. We have used the mitochondria to prepare cytochrome c oxidase, mitochondrial F_1 ATPase and the oligomycin-sensitive ATPase complex.

TABLE I

COMPARISON OF MITOCHONDRIA PREPARED WITH THE DYNO-MILL WITH THOSE OBTAINED BY OTHER MECHANICAL METHODS[a]

Breakage method	Mitchondrial yield (gm protein/kg pressed yeast)	Submitochondrial particles (gm protein/kg pressed yeast)	Submitochondrial particles (nmoles heme a/mg protein)	Particle-bound heme a recovered (%)
Dyno-Mill	5–7	4–6	0.37–0.48	15–25
Manton-Gaulin homogenizer	0.7–1.2	0.5–0.8	0.35–0.45	15–20
Liquid nitrogen disintegration	2.24	0.81	0.48	14.8

[a]Conditions for cell breakage with the Dyno-Mill are described in the text. Ranges of values obtained from at least 50 separate preparations are reported. Results from breakage with the Manton-Gaulin homogenizer (Mason et al., 1973) are from 20 preparations. Results for the liquid-nitrogen disintegration method (Tzagoloff, 1969) are from a single experiment, although a second trial gave a similar yield of mitochondria. The last column shows the percentage yield of heme a recovered from submitochondrial particles in highly purified preparations of cytochrome c oxidase.

To prepare cytochrome c oxidase, mitochondria are generally sonicated and submitochondrial particles (SMP) collected by ultracentrifugation (Mason *et al.*, 1973). Table I compares the yield of SMP per kilogram of pressed yeast with the yield obtained using other large-scale procedures. The yield is five to eight times higher with the Dyno-Mill than with either of the other large-scale procedures. The specific cytochrome c oxidase contents of the various preparations are about equal (Table I). The cytochrome c oxidase obtained by slight modification of published procedures (Mason *et al.*, 1973) has high activity (specific first-order rate constant $= 1000$ min^{-1} mg of protein^{-1}), a high heme a /protein ratio (10–12 nmoles of heme a per milligram of protein), and the typical subunit composition on analytical sodium dodecyl sulfate polyacrylamide gel electrophoresis (Mason *et al.*, 1973). Similarly, yields of F_1 ATPase and the oligomycin-sensitive ATPase complex are generally as good or better than yields obtained from yeast mitochondria prepared by other methods (Takeshiga *et al.*, 1976; Tzagoloff and Meagher, 1971).

VI. Discussion

The Dyno-Mill has two major advantages over other large-scale methods of yeast cell breakage. It is a far more convenient and easy procedure, and breakage is very efficient, leading to a high yield of mitochondria. With the Dyno-Mill it is easily possible for one person to process 4–5 kg of yeast per day and to obtain 25 gm or more of mitochondrial protein. Since 4 kg can be broken in 2 hours or less, production is limited mainly by the volume of the yeast suspension that must be subsequently processed.

ACKNOWLEDGMENT

The authors acknowledge the support and encouragement of Dr. Gottfried Schatz, in whose laboratory the work reported here was done.

REFERENCES

Deters, D., Müller, U., and Homberger, H. (1976). *Anal. Biochem.* **70**, 263–267.
Mason, T. L., Poyton, R. O., Wharton, D. C., and Schatz, G. (1973). *J. Biol. Chem.* **244**, 1346–1354.
Ohnishi, T., Kawaguchi, K., and Hagihava, B. (1966). *J. Biol. Chem.* **241**, 1797–1806.
Řeháček, J. (1971). *Experientia,* **27**, 1103–1104.
Řeháček, J., Beran, K., and Bičík, V. (1969). *Appl. Microbiol.* **17**, 462–466.
Salton, M. R. J. (1974). *In* "Methods in Enzymology" (S. Fleischer and L. Packer, eds.), Vol. 31, Part A, pp. 653–667. Academic Press, New York.

Schatz, G., and Kovac, L. (1974). *In* "Methods in Enzymology" (S. Fleischer and L. Packer, eds.), Vol. 31, Part ·A, pp. 627–632. Academic Press, New York.
Takeshiga, K., Hess, B., Böhm, M., and Zimmerman-Telschow, H. (1976). *Hoppe-Seyler's Z. Physiol. Chem.* **357**, 1605–1622.
Tzagoloff, A. (1969). *J. Biol. Chem.* **244**, 5020–5026.
Tzagoloff, A. (1971). *J. Biol. Chem.* **246**, 3050–3056.
Tzagoloff, A., and Meagher, P. (1971). *J. Biol. Chem.* **246**, 7328–7336.

Chapter 9

Isolation, Regeneration, and Fusion of Protoplasts of Phycomyces

HORST BINDING AND HANS JÜRGEN WEBER

Botanisches Institut,
Christian-Albrechts-Universitat,
Kiel, West Germany

Department of
Bacteriology and
Immunology, University
of California, Berkeley

I. Introduction

Most cells of bacteria, fungi, and green plants are surrounded by rigid cell walls. Removal of these walls makes them susceptible to various types of biochemical and genetic manipulation. This chapter deals with experiments on the isolation of protoplasts from germinating sporangiospores of *Phycomyces*, the regeneration of isolated protoplasts, the fusion of protoplasts from different auxotrophs, and the formation of heterokaryotic regenerants. More details have been published previously (Binding and Weber, 1974).

II. Material

The following strains of *Phycomyces blakesleeanus* from M. Delbrück's collection at the California Institute of Technology were used:

113

wild-type NRRL 1555−, the color mutants C2 (carA5−), C9 (carR21−), and the auxotrophic mutants H1 (leu-51−), H3 (sul-52−), C242 (nicA101 carA5+), C244 (leu-51+), and C245 (leuA51+).

III. Preparation of Cells

Sporangiospores were heat-shocked at 48°C for 1 hour and incubated in a rich medium (Heisenberg and Cerdá-Olmedo, 1968) at a density of 10^5 spores/ml on a rotary shaker (300 rpm) at 22°C for 8 hours. The germlings were washed twice in 0.35 M sorbitol–10^{-2} M phosphate buffer before being subjected to the protoplast isolation procedure.

IV. Protoplast Isolation

The cell walls were removed by chitin and chitosan-hydrolyzing enzymes. A chitinase from Streptomyces griseus was obtained from the Nutritional Biochemical Company. A chitosanase preparation was made by a procedure modified from methods used by D. E. Eveleigh (personal communication) and Price and Storck (1974): Streptomyces strain 6 was grown with shaking at 22°C in a medium which contained 5 × 10^{-2} M glucose, 1% yeast extract (Difco), 2.8 × 10^{-3} M MgSO$_4$, 7.4 × 10^{-2} M KH$_2$PO$_4$, 8.6 × 10^{-2} M NaCl, 6.7 × 10^{-2} M KCl, and 7.6 × 10^{-2} M (NH$_4$)$_2$SO$_4$. Chitosanase production was induced after 36 hours by transfer of the cells to a similar medium in which glucose and yeast extract were replaced by 10^{-2} M D-glucosamine. The cells were removed 6 hours later and subjected to fractionation by (NH$_4$)$_2$SO$_4$. The precipitate between 70 and 100% saturation was collected, dissolved in 10^{-2} M phosphate buffer at pH 7.0, dialyzed against distilled water, and lyophilized.

The incubation mixture for protoplast isolation contained 0.35 M sorbitol, 10^{-2} M phosphate buffer, pH 7.0, 0.2% chitinase, 0.1% chitosanase, and 1–5 × 10^6 germlings/ml. Protoplasts were released after 30–60 minutes at about 25°C. The yields were close to 100% of germinated spores if the cell density was not higher than 5 × 10^6/ml. The removal of cell walls newly built during germination was indicated by the spherical shape of the protoplasts, their osmotic sensitivity, and especially their ability to fuse. The spore walls remained undigested, either sticking to the protoplasts or floating freely in the medium.

V. Reversion of Protoplasts

Free protoplasts were collected by centrifugation at 2000 g and washed twice in sorbitol solution to remove the hydrolytic enzymes. Wall regeneration occurred after transfer to a culture medium of the following composition: $2 \times 10^{-2} M$ MgSO$_4$, $1 \times 10^{-2} M$ CaCl$_2$, $5 \times 10^{-4} M$ KH$_2$PO$_4$, trace elements (Sutter, 1975), $7 \times 10^{-4} M$ thiamine, $5 \times 10^{-2} M$ glucose, 0.25 M sorbitol, and 0.5% yeast extract (Difco). The protoplasts were incubated at a cell density of about 10^6/ml at 25°C with occasional resuspension. Up to 95% of the protoplasts showed polar growth in the regeneration medium after 4–8 hours. Growth began frequently in a budding-like manner, gradually changing to normal development of hyphae. The newly formed cells were plated after 8–10 hours on agar plates containing the rich medium of Heisenberg and Cerdá-Olmedo adjusted to pH 3.3 to obtain colonial growth of the mycelia. The plating efficiency of strains NRRL 1555 and C9 reached values of about 0.8, and that of the other strains, about 0.1.

VI. Protoplast Fusion

Protoplasts with different genetic markers were used for fusion experiments. It emerged from preliminary experiments that the combination H1–H3 was most appropriate for fusion experiments, because of relatively high regeneration rates and low cross-feeding capacity on the minimal medium of Sutter (1975); no vigorously growing colony was formed at plating densities of less than 2×10^4 cells/cm^2. About equal numbers of protoplasts of both types were mixed at the end of the enzymic digestion, rapidly washed twice in 0.35 M sorbitol solution, transferred to fusion media, spun down at about 2000 g for 3 minutes to form a dense sediment, and incubated for 15 minutes. The supernatant was then replaced by solutions resembling the fusion media but adjusted to pH 7.0. Incubation was continued for a further 30–45 minutes to allow completion of the fusion. The protoplasts were then transferred to the regeneration medium. The fusion tendency of protoplasts was investigated in the following media. Natural seawater (Atlantic Ocean, Cold Spring Harbor, N.Y., and Pacific Ocean, Los Angeles, Calif.) was diluted with deionized water to two-thirds, and the pH adjusted to 9.0 with NH$_4$OH (cf. Binding, 1974). Solutions of 0.2 M Ca(NO$_3$)$_2$ of different pH values were prepared in 0.05 M glycine–NaOH buffer (cf. Keller and Melchers, 1973).

In order to calculate the frequency of heterokaryons, about 2×10^4/cm^2

regenerating protoplasts were plated on minimal agar plates, and the vigorously growing mycelia were counted after 5 days. To confirm their heterokaryotic nature, all of them were checked for segregation in the offspring of sporangiospores. The frequency of heterokaryons is given by the ratio of the number of heterokaryons to the number of viable protoplasts used for the fusion experiment. This number was calculated by regenerating aliquots of the two protoplast suspensions on rich agar medium at pH 3.3. The frequency of heterokaryons after incubation in seawater at pH 9 was 2.7×10^{-5}. Treatment by $Ca(NO_3)_2$ solutions at pH 4, 7, 9, and 10 revealed values of 5.5×10^{-5}, 7.2×10^{-5}, 1.3×10^{-4} and 3.7×10^{-5}, respectively. No heterokaryons were formed in a sorbitol solution containing about 2×10^5 viable protoplasts of each type.

VII. Discussion

High yields of isolated protoplasts and high regeneration rates have been obtained. Protoplast fusion has been induced by seawater and calcium nitrate. Significantly higher fusion frequencies were achieved by Anné and Peberdy (1975) in solutions of polyethylene glycol (MW 6000). Up to 4% of the viable protoplasts exhibited complementation of nutritional deficiencies when calcium ions and high pH were used in addition to polyethylene glycol. Somatic hybridization in fungi through fusion of auxotrophic mutants has been investigated in great detail by Ferenczy and co-workers (for references, see Ferenczy et al., 1977).

ACKNOWLEDGMENTS

The experiments described here were carried out at the Cold Spring Harbor Laboratories during the *Phycomyces* workshop in 1974, and in the laboratory of Dr. M. Delbrück at the California Institute of Technology, Pasadena.

REFERENCES

Anné, J., and Peberdy, J. F. (1975). *Arch. Microbiol.* **105**, 201–205.
Binding, H. (1974). *Z. Pflanzenphysiol.* **72**, 422–426.
Binding, H., and Weber, H. J. (1974). *Mol. Gen. Genet.* **135**, 273–276.
Ferenczy, L., Szegedy, M., and Kevei, F. (1977). *Experientia* **33**, 184–186.
Heisenberg, M., and Cerdá-Olmedo, E. (1968). *Mol. Gen. Genet.* **102**, 187–195.
Keller, W. A., and Melchers, G. (1973). *Z. Naturforsch., Teil C* **28**, 737–741.
Price, J. S., and Storck, R. (1974). *Abstr., Annu. Meet. Am. Soc. Microbiol.* p. 146.
Sutter, R. (1975). *Proc. Natl. Acad. Sci. U.S.A.* **72**, 127–130.

Chapter 10

The Neurospora Plasma Membrane: A New Experimental System for Investigating Eukaryote Surface Membrane Structure and Function[1]

GENE A. SCARBOROUGH

Department of Pharmacology, School of Medicine,
The University of North Carolina at Chapel Hill, Chapel Hill, North Carolina

I. Introduction

During the last few decades, an accelerating interest has developed in the molecular biology of the eukaryote surface membrane because of the obvious critical role of this structure in a wide range of cellular activities, including cell–cell interaction, hormone response, nerve impulse transmission, cell growth, division, and differentiation, and regulation of the entry and exit of ions, small molecules, and macromolecules, to name a few.

[1] The studies described here were supported by research grants from the National Institutes of Health (GM 19971) and the National Science Foundation (GB 38801).

Although the existence of a cell surface membrane has been recognized for most of this century, progress in understanding the molecular nature of this structure has come only recently. The most widely used system for investigating eukaryote surface membrane structure and function has been the erythrocyte plasma membrane, primarily because it is easy to isolate in a pure form. But, while much indispensable information, and indeed the entire design and direction of eukaryote plasma membrane research, has come primarily from the study of the erythrocyte membrane, the erythrocyte is nevertheless incapable of many of the above-mentioned functions. It is thus limited as a system for investigating the molecular mechanisms of cell–cell interaction, plasma membrane biogenesis, response to many hormones, active metabolite transport, and generation and maintenance of large membrane potentials. Progress on these fronts will come only after the development of new experimental systems with plasma membranes isolated from cells which perform these functions. It is the purpose of this article to acquaint the reader with the *Neurospora* plasma membrane, a new experimental system for exploring the molecular biology of a variety of important aspects of eukaryote plasma membrane structure and function. In addition, it is hoped that the description of the strategies and methodology employed in the development of this system will provide a framework which can be used by others to identify, isolate, and assess the functionality of plasma membranes from other eukaryotic cells.

II.   Isolation of *Neurospora* Plasma Membranes

A.   Problems

A variety of problems is encountered in the development of a method for the isolation of plasma membranes, in high yield and purity, from most eukaryotic cells. In plant, fungal, and certain other eukaryotic cells there exists a rigid cell wall which makes gentle lysis of the cells impossible. Without prior removal of the cell wall, the drastic homogenization methods necessary to break such cells usually lead to the destruction of most of the subcellular structures. Another problem is that eukaryotic cells, almost by definition, contain a variety of membranous structures and the specific identification of the plasma membranes in a cell homogenate is often difficult. Additional serious problems arise as a result of the almost universal tendency for plasma membranes to fragment and vesiculate upon lysis of the cells. This usually leads to vesicles of variable density, inside which are trapped most of the other components of the lysate. Such vesicles are obviously contaminated and, in addition, their nonuniform density commonly leads to extensive smearing in standard isopycnic centrifugation procedures.

B. Solutions

For *Neurospora*, the rigid cell wall problem was solved by using cells of an already available mutant, the *sl* strain (Emerson, 1963) which grows in osmotically stabilized medium as spheroplasts essentially devoid of a cell wall. In the case of other walled cells, spheroplasts must be prepared by enzymic removal of the cell wall. With most animal cells the problem does not exist, however, excessive amounts of extracytoplasmic surface material can present a problem.

In identifying plasma membranes, the most widely accepted criterion for proving that an isolated membrane fraction consists of plasma membranes is the demonstration that it is enriched in certain so-called plasma membrane marker enzymes such as alkaline phosphatase, 5'-nucleotidase, Na$^+$, K$^+$-stimulated Mg^{2+}-ATPase, and adenylate cyclase (DePierre and Karnovsky, 1973). But there is no compelling reason to expect, a priori, the presence of any of these enzymes in the plasma membrane of a previously uncharacterized organism. The literature has many examples of exceptions to the generality that these enzymes are dependable plasma membrane markers. In a previously uncharacterized free-living cell, the only predictable difference between the plasma membrane and other cellular membranes is that the plasma membrane is exposed to the external environment. Thus surface labeling of intact cells by a membrane-impermeable reagent which forms covalent bonds with components of the cell surface is the most dependable approach to identifying the plasma membranes of an uncharacterized cell. In addition to providing a method for confidently identifying plasma membranes, the surface-labeling approach allows an accurate estimation of the yield of plasma membranes in a given procedure. This is not possible with the marker enzyme approach. The reagent chosen to label the *Neurospora* plasma membranes was diazotized [^{35}S]sulfanilic acid (Berg, 1969), but many other surface-labeling reagents have also been reported (Maddy, 1964; Vansteveninck *et al.*, 1965; Pardee and Watanabe, 1968; Phillips and Morrison, 1970; Bretscher, 1971; Staros and Richards, 1974; Whiteley and Berg, 1974; Cabantchik *et al.*, 1975; Dutton and Singer, 1975; Hawkes *et al.*, 1976). Of course, surface labeling is used only in development of the isolation method. After a reproducible isolation method has been established, the use of surface labeling, which may affect important plasma membrane functions, is no longer necessary.

The serious problems which arise when plasma membranes fragment and vesiculate upon cell lysis can be solved only by stabilizing them prior to lysis so that they do not spontaneously fragment and vesiculate and can be isolated as open sheets or "ghosts" which represent intact plasma membranes freed of intracellular contents. Warren *et al.* (1966) have reported a variety of reagents that stabilize plasma membranes, including fluorescein mercuric acetate, 5,5′-dithiobis(2-nitrobenzoic acid), Zn^{2+}, acetic acid, and Tris. And Nachman *et al.* (1971) have used glutaraldehyde as a plasma membrane-stabilizing agent. Although most of these reagents are rather harsh, and other objections to their use have been pointed out (DePierre and Karnovsky, 1973), they may be worth trying. It should be mentioned that, whatever method is chosen for stabilizing plasma membranes, it is relatively easy to determine whether the treatment adversely affects any function of interest (i.e., enzyme activity, ligand binding) by comparing that activity in homogenates of treated and untreated cells. If the treatment does not affect the activity of interest, it is probably adequate for study of that system.

For stabilizing *Neurospora* plasma membranes against fragmentation and vesiculation, the most gentle and effective agent by far is the plant lectin concanavalin A (Con A). Treatment of intact cells of the *sl* strain with Con A just prior to lysis leads to remarkable stabilization of the plasma membranes, and after lysis the resulting stabilized plasma membrane sheets are visible entities in the light microscope, which can be isolated in a nearly pure state in high yield by low-speed centrifugation procedures. Importantly, Con-A stabilization is essentially reversible. Treatment of the purified plasma membrane ghosts with α-methylmannoside removes the bulk of the Con A and, after its removal, the plasma membranes fragment and vesiculate, as is their nature, giving rise to essentially pure plasma membrane vesicles. The ability to isolate *Neurospora* plasma membranes as open sheets or closed vesicles is an important technical advantage. Structural studies on the three-dimensional orientation of plasma membrane components are easier to carry out with open plasma membrane sheets than with resealed vesicles, whereas functional studies involving transmembrane movements of ions and molecules require a closed vesicle system.

Since the Con-A method for stabilizing *Neurospora* plasma membranes was reported (Scarborough, 1975), it has been shown to be suitable for stabilizing the plasma membranes of yeast protoplasts (Durán *et al.*, 1975), *Dictyostelium* (Parish and Müller, 1976), and wild-type *Neurospora* protoplasts (G. A. Scarborough, unpublished). Hopefully, the method will be useful for stabilizing the plasma membranes of other cells as well.

C. The Method

1. GROWTH OF *sl* CELLS

sl cells are grown in Vogel's N medium (Vogel, 1956) supplemented with 2% (w/v) mannitol, 0.75% (w/v) yeast extract, and 0.75% (w/v) nutrient broth. Cultures are maintained by daily transfer of 1 ml of an overnight culture into 50 ml of fresh medium, followed by rotary shaking (150 rpm) at 30°C overnight. Five-hundred-milliliter cultures are obtained by inoculating 450 ml of fresh medium with 50 ml of an overnight culture and shaking overnight as above. Stock cultures can be maintained on slants prepared from the above-mentioned liquid medium solidified with 1.5% agar. The slants should be transferred at least once a month. *sl* cells should not be stored in the refrigerator.

2. ISOLATION OF PLASMA MEMBRANE GHOSTS

The following method was developed using the surface label diazotized [^{35}S]sulfanilic acid to identify *Neurospora* plasma membranes. Although the details regarding the preparation of diazotized [^{35}S]sulfanilic acid and the surface-labeling procedure are not described here, they can be found in Scarborough (1975).

Cells from one 500-ml culture (O.D.$_{650\ nm}$ = 0.6–0.9, approximately 250 mg of cell protein) are harvested by centrifugation at 700 g for 10 minutes, resuspended in 80 ml of ice-cold buffer A (0.05 M Tris–HCl, pH 7.5, containing 0.01 M MgSO$_4$ and 0.25 M mannitol), divided into four 20-ml aliquots, and washed five times with 80 ml of ice-cold buffer A (20 ml per tube) by alternate centrifugation and gentle resuspension (50-ml plastic tubes in a swinging-bucket clinical centrifuge at 140 g for 6 minutes). The rather extensive wash procedure is necessary to remove all broken cell debris which accumulates in the culture. It is extremely important that the cell preparation be free of contamination by broken cell debris before the Con A is added, since such contaminating material can be bound to the cell surface by the Con A. The purity of the cell suspension before the Con-A treatment should be verified by observation with a light microscope. The washed cells are resuspended in a total volume of 20 ml of buffer A (at 25°C), gently mixed with 20 ml of Con-A solution (0.5 mg/ml in buffer A, 25°C), and incubated with occasional gentle agitation for 10 minutes at 25°C. The Con A agglutinates the cells during this period. Figure 1 shows the effects of Con-A treatment on the cell surface. Figure 1A is an electron micrograph of the cell surface just prior to the addition of Con A, and Fig. 1B is an electron micrograph of the cell surface after Con-A treatment. The effect of Con A is

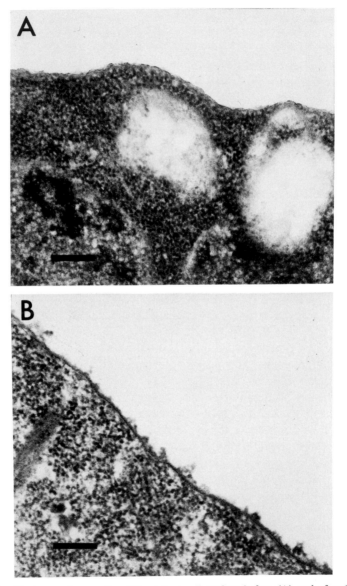

FIG. 1. Electron micrographs of the intact cell surface before (A) and after (B) Con-A treatment. The horizontal bars represent 0.25 μm. (From Scarborough, 1975).

manifested as a fuzzy appearance of the cell surface. The Con-A-agglutinated cells are then chilled, centrifuged at 140 g for 1 minute, resuspended gently in 40 ml of ice-cold buffer A, and centrifuged at 140 g for 6 minutes. The resulting cell pellet is then resuspended in 50 ml of ice-cold 0.01 M Tris–HCl, pH 7.5, containing 5 mM MgSO$_4$ and 5 mg of DNase (Worthington, Code D), and homogenized in a glass-Teflon tissue homogenizer (50 passes over a 10-minute period, clearance approximately 0.008 inch). Twelve-milliliter portions of the resulting lysate are layered over 35 ml of ice-cold buffer B (0.1 M tris–HCl, pH 7.5, containing 0.5 M mannitol), and the resulting two-phase systems are centrifuged at 140 g for 30 minutes in a swinging-bucket clinical centrifuge at 4°C. The supernatant fluids containing most of the cell contents are removed by aspiration, and the plasma membrane pellets are resuspended in a total of 20 ml of ice-cold 0.01 M Tris–HCl, pH 7.5, and again homogenized in a glass-Teflon tissue homogenizer (20 passes, 4°C). Ten-milliliter aliquots of the resulting suspension are layered over 35 ml of buffer B, and the resulting two-phase systems are centrifuged at 250 g for 30 minutes in a swinging-bucket clinical centrifuge (4°C). The supernatant fluids are removed by aspiration, and the pellet contains the plasma membrane ghosts. The yield of ghosts is about 50%. Figure 2A is an electron micrograph of the isolated ghosts. It can be seen that the plasma membrane ghosts exist primarily as extended sheets with Con-A deposits on one side only.

3. Isolation of Plasma Membrane Vesicles

The procedure for the isolation of plasma membrane vesicles is identical with the procedure for isolation of plasma membrane ghosts up to the final centrifugation through buffer B. Instead of layering over buffer B, 10-ml aliquots of the plasma membrane ghost suspension are layered over a two-phase system consisting of 30 ml of buffer B layered over 5 ml of buffer C [0.01 M Tris–HCl, pH 7.5, containing 20% (w/v) sucrose], and the resulting three-phase discontinuous gradients are centrifuged at 250 g for 30 minutes (4°C). The supernatant fluids are removed by aspiration, and the pellets resuspended in a total of 4 ml of 1 M α-methylmannoside in 0.01 M Tris–HCl, pH 7.5. The resulting plasma membrane suspension is incubated with shaking for 30 minutes at 30°C, chilled, diluted with 16 ml of ice-cold 0.01 M Tris–HCl, pH 7.5, and homogenized in a glass-Teflon tissue homogenizer (100 passes, 4°C). The mixture is then carefully layered over 30 ml of ice-cold buffer C and centrifuged at 250 g for 30 minutes. The plasma membrane vesicles, which remain on top of the buffer C, are removed with a pasteur pipet, diluted to 50 ml with ice-cold 0.01 M Tris–HCl, pH 7.5, and centrifuged at 12,000 g for 30 minutes (4°C). The supernatant fluid is decanted, and the pellet contains the plasma membrane vesicles. The yield is about

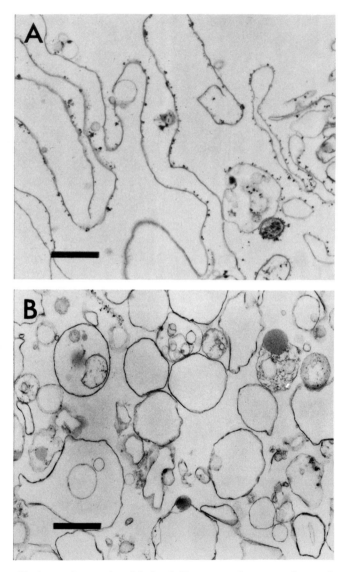

FIG. 2.   Electron micrographs of isolated *Neurospora* plasma membrane ghosts (A) and vesicles (B). The horizontal bars represent 1 μm. (From Scarborough, 1975).

25%. Figure 2B is an electron micrograph of isolated plasma membrane vesicles.

III.   Properties of Isolated *Neurospora* Plasma Membranes

A.   Chemical Composition

Table I describes the chemical composition of isolated *Neurospora* plasma membrane vesicles compared to that of the cells from which they were isolated. The vesicles consist of about 56% protein, 12% sterol, 17% phospholipid, and 14% carbohydrate (weight). The plasma membranes contain very little RNA and DNA. The molar ratio of sterol to phospholipid is about 1.3. A high sterol/phospholipid ratio is characteristic of all eukaryote plasma membranes studied thus far. When corrected for yield (approximately 25%),

TABLE I

CHEMICAL COMPOSITION OF *sl* CELLS AND PLASMA MEMBRANE VESICLES[a,b]

Component	Cells		Plasma membrane vesicles	
	I	II	I	II
Protein (mg)	216	224	2.1	1.4
Sterol (ergosterol equivalents)				
Micromoles	6.8	7.6	1.1	0.81
Milligrams	2.7	3.0	0.43	0.32
Milligrams per milligram of protein	0.013	0.013	0.20	0.23
Sterol/phospholipid ratio (mole/mole)	0.17	0.15	1.29	1.40
Phospholipid				
Micromoles	40.7	49.9	0.85	0.58
Milligrams (approx. 0.75 mg/μmole)	30.5	37.4	0.64	0.44
Milligrams per milligram of protein	0.14	0.17	0.30	0.31
Carbohydrate (glucose equivalents)				
Milligrams	52.2	45.8	0.51	0.35
Milligrams per milligram of protein	0.24	0.20	0.24	0.25
RNA				
Milligrams	38.0	40.2	0.06	0.04
Milligrams per milligram of protein	0.18	0.18	0.03	0.03
DNA				
Milligrams	2.0	2.1	0.01	0.006
Milligrams per milligram of protein	0.009	0.009	0.005	0.004

[a] From Scarborough, 1975.
[b] Roman numerals indicate separate experiments.

it can be estimated that the plasma membrane, which contains only a few percent of the total cellular mass, contains more than 60% of the total cellular sterol. This is suggestive of an essential relationship between eukaryote plasma membranes and sterols. An understanding of the biochemical basis for this striking fact would be an important contribution to our knowledge of eukaryote plasma membrane structure and function in general. Figure 3 shows the results of sodium dodecyl sulfate (SDS) polyacrylamide slab gel electrophoretic analysis of *Neurospora* plasma membrane proteins according to the method of Ames (1974). The wells contained decreasing amounts of protein from left to right. *Neurospora* plasma membranes are quite complex with respect to their protein composition, as evidenced by the large number of bands in the gel. If this complexity is a reflection of functional multiplicity, then much is yet to be learned about the functions performed by the *Neurospora* plasma membrane. Another interesting feature of the chemical composition of the *Neurospora* plasma membrane is the high iron content (G. A. Scarborough, unpublished). Isolated *Neurospora* plasma membranes contain tightly bound nonheme iron in amounts in the range of 50 nmoles/mg protein. This is at least 10 times the concentration of nonheme

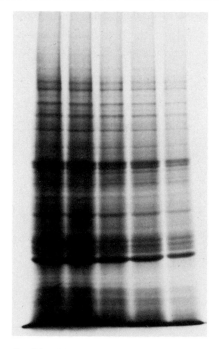

Fig. 3. SDS polyacrylamide slab gel electrophoretic analysis of *Neurospora* plasma membrane ghost proteins by the Ames method (Ames, 1974).

iron present in mitochondria (Doeg and Ziegler, 1962) which are known for their high iron content. The nature of the ligands holding the iron in the membrane and the function of the membrane-bound iron are presently unknown, but the investigation of this tightly bound iron may reveal an important new aspect of eukaryote plasma membrane biochemistry.

B. Functional Properties

1. Metabolite Transport Systems

Isolated plasma membrane vesicles possess a variety of transport systems, as evidenced by the fact that it is possible to measure the facilitated diffusion of glucose and certain amino acids, but not others, into the vesicles (G. A. Scarborough, unpublished). However, since the methodology for energizing these transport systems has not yet been developed, and the amount of uptake by facilitated diffusion is necessarily quite small, none of these transport systems have been characterized.

2. The Electrogenic Pump

Thus far, the most extensively studied enzyme in the isolated plasma membranes is the Mg^{2+}-dependent ATPase. Various biochemical properties of this enzyme have been characterized (Scarborough, 1977). The enzyme catalyzes the hydrolysis of ATP, resulting in the formation of ADP and inorganic phosphate; the pH optimum is about pH 6.0–6.5; the apparent K_m for Mg^{2+}-ATP is about $2 mM$; the ATPase requires a divalent cation for activity in the following order of preference: $Mg^{2+}, Co^2 > Mn^{2+} > Zn^{2+}$; no specific monovalent cation requirement has been demonstrated; the enzyme is quite specific for ATP compared to other nucleotides; treatment of the plasma membranes with sodium deoxycholate inactivates the ATPase, and activity is restored upon the addition of acidic phospholipids, suggesting an acidic phospholipid requirement for activity; organic mercurials are potent irreversible inhibitors, but N-ethylmaleimide does not affect activity; treatment of the plasma membranes with trypsin inactivates the ATPase, and this inhibition is prevented by Mg^{2+}-ATP, suggesting substrate-induced conformational changes in the enzyme.

Earlier electrophysiological studies by Slayman and collaborators (1970, 1973; Slayman and Gradmann, 1975) with wild-type cells of *Neurospora* suggested that ATP hydrolysis catalyzed by a plasma membrane ATPase is necessary for the generation and maintenance of the *Neurospora* membrane potential (approximately 200 mV, interior negative). Recent studies in this laboratory with isolated *Neurospora* plasma membrane vesicles have demonstrated *in vitro* that this is indeed the case; that the *Neurospora* plasma membrane ATPase is an electrogenic pump (Scarborough, 1976). These studies were possible because a significant proportion of the isolated vesicles are functionally inverted. Incubation of the isolated plasma membrane vesicles with Mg^{2+}-ATP gives rise to the generation of an electrical potential (interior positive) across the vesicle membrane. This can be demonstrated by the uptake of $[^{14}C]$thiocyanate, which can be used to monitor membrane potentials because of its ability to penetrate biological membranes and become asymmetrically distributed in the presence of a membrane potential (Schuldiner *et al.*, 1974). Figure 4 shows the results of a typical thiocyanate

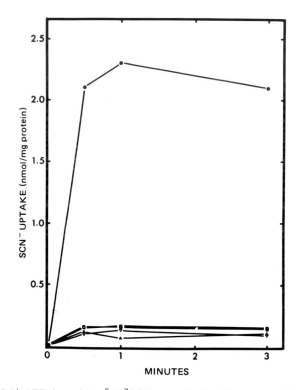

FIG. 4. Mg^{2+}-ATP-dependent $[^{14}C]SCN^-$ uptake by *Neurospora* plasma membrane vesicles. The uptake assay has been described (Scarborough, 1976). Incubations contained 50 μl of plasma membrane suspension (1.2 mg of protein per ml in 0.01 M Tris phosphate, pH 6.8), 5 μl of 1 mM $[^{14}C]$KSCN (specific activity of 60 mCi/mmole), and the indicated additions in a total volume of 60 μl. Points indicate the average of duplicate determinations. Additions: $\triangledown$, 5 μl of H_2O; $\bigcirc$—$\bigcirc$, 5 μl of 0.1 M Tris phosphate containing 0.1 M MgSO$_4$ (pH 6.8); $\triangle$, 5 μl of 0.1 M Tris–ADP containing 0.1 M MgSO$_4$ (pH 6.8); $\square$, 5 μl of 0.1 M Tris–ATP (pH 6.8); $\bullet$, 5 μl of 0.1 M Tris–ATP containing 0.1 M MgSO$_4$ (pH 6.8). (From Scarborough, 1976).

uptake experiment. The amount of thiocyanate uptake in the controls (no addition, Mg^{2+}-P_i, Mg^{2+}-ADP, and ATP) approximates equilibration of the thiocyanate between the intra- and extravesicular spaces. In the presence of Mg^{2+}-ATP, thiocyanate uptake is markedly stimulated about 20-fold over that in the controls. This concentration ratio corresponds to a membrane potential of about 75 mV (interior positive). This is a minimum estimate, since the ratio of right-side-out to inside-out vesicles is presently unknown, and whereas vesicles of either orientation can contribute to the equilibrium uptake level, only inside-out vesicles can be energized by Mg^{2+}-ATP. Mg^{2+}-ATP-dependent thiocyanate uptake requires ATP hydrolysis catalyzed by the plasma membrane ATPase and is inhibited by the metabolic poisons azide and carbonyl cyanide-*m*-chlorophenylhydrazone which dissipate membrane potentials by virtue of their ability to facilitate transmembrane proton movements (Harold, 1972).

Another independent means of monitoring the ATP-generated membrane potential involves the 1-anilinonaphthalene-8-sulfonic acid (ANS) fluorescence enhancement seen upon energizing *Neurospora* plasma membrane vesicles with Mg^{2+}-ATP. ANS fluorescence changes have been used to monitor the energized state of a variety of membrane systems (Brand and Gohlke, 1972; Reeves *et al.*, 1972; Azzi *et al.*, 1971); the generation of an interior negative membrane potential results in the quenching of ANS fluorescence, whereas generation of an interior positive membrane potential results in the enhancement of ANS fluorescence. Figure 5 shows the ANS fluorescence changes which take place upon energizing *Neurospora* plasma membrane vesicles with Mg^{2+}-ATP. The addition of Mg^{2+}-ATP leads to the immediate enhancement of ANS fluorescence (attributable to salt effects) followed by a slower fluorescence increase which follows a time course similar to that observed for Mg^{2+}-ATP-dependent $[^{14}C]SCN^-$ uptake. Furthermore, the slower component of the Mg^{2+}-ATP-dependent fluorescence increase is essentially eliminated by the addition of sodium azide, as is Mg^{2+}-ATP-dependent $[^{14}C]SCN^-$ uptake.

These experiments demonstrate that *Neurospora* plasma membrane ATPase is an electrogenic pump (it is most likely a proton pump) and establish *Neurospora* plasma membrane vesicles as an excellent experimental system for studying this form of biological energy transduction.

3. PHOSPHORYLATION BY $[\gamma$-$^{32}P]$ATP

Another interesting functional property of isolated *Neurospora* plasma membranes is the autophosphorylation of several of the membrane proteins by γ-labeled $[^{32}P]$ATP (G. A. Scarborough, unpublished). Incubation of plasma membranes with $[\gamma$-$^{32}P]$ATP for only a few seconds at 0°C results in considerable transfer of ^{32}P to membranes. SDS polyacrylamide slab

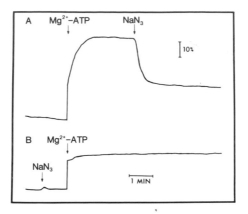

FIG. 5. Mg^{2+}-ATP-dependent ANS fluorescence enhancement in *Neurospora* plasma membrane vesicles. Incubations contained plasma membrane vesicles (0.081 mg of protein), ANS (20 nmoles), and Tris phosphate (0.01 M, pH 6.8) in a volume of 0.9 ml. In (A) the following additions were made at the times indicated by the arrows: 100 μl of 0.1 M Tris–ATP containing 0.1 M MgSO$_4$, and then 10 μl of 0.1 M NaN$_3$. In (B) the order of the additions was reversed. The vertical bar represents a 10% increase over the baseline fluorescence values. Fluorescence was measured at an angle of 90°. Excitation, 366 nm; emission, 470 nm. (From Scarborough, 1976).

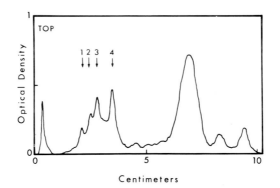

FIG. 6. Phosphorylation of *Neurospora* plasma membranes by [γ-^{32}P]ATP. Plasma membrane ghosts (0.3 mg of protein) suspended in 0.05 ml of 0.01 M Tris–succinate, pH 6.0, containing 0.02 M MgSO$_4$, were mixed with 15 μCi of [γ-^{32}p]ATP (Amersham-Searle, triethylammonium salt, specific activity 1500 Ci/mmole, 0.75 mCi/ml in H$_2$O) and incubated for 30 seconds (2 seconds is sufficient) at 0°C. The reaction was terminated by the addition of 2 ml of ice-cold 0.3 M perchloric acid, and the membranes pelleted by centrifugation at 700 g for 5 minutes. The phosphorylated membranes were washed once with ice-cold perchloric acid (0.3 M, 2 ml) and then subjected to SDS slab gel electrophoresis (110 mA, 1.5 hours, 22°C) in the S (2.4) system of Fairbanks and Avruch (1972). The resulting gel was dried and placed under x-ray film for 14 hours. The exposed film was then developed and analyzed with an Ortec scanning densitometer.

gel electrophoresis of the labeled membranes at pH 2.4, followed by auto-radiography of the dried gels, allows the delineation of four major ^{32}P-labeled proteins in *Neurospora* plasma membranes. Figure 6 shows a densitometer tracing of a typical autoradiogram. The numbered peaks in the upper half of the gel correspond to the four major ^{32}P-labeled proteins. The material at the top is an artifact of the electrophoresis and the three peaks at the bottom are ATP, an unknown compound, perhaps lipid, and inorganic phosphate, respectively, from left to right. The four bands migrate in the 100,000–200,000 molecular-weight range. The material in each of the numbered bands exhibits the characteristic pH-versus-hydrolysis profile of protein-bound acyl phosphate (Hokin *et al.*, 1965), and the bulk of the radioactivity released from the labeled membranes above pH 9 behaves like inorganic phosphate in the isobutanol–benzene procedure for separat-ing free phosphate from phosphate esters (Ernster *et al.*, 1950). It is thus likely that the radioactivity in all four labeled proteins represents protein-bound acyl phosphate. In view of the probable role of protein-bound acyl phosphate in the mechanisms of several other eukaryotic energy-trans-ducing ATPases (Dahl and Hokin, 1974; Inesi, 1972; Hasselbach, 1974) it is likely that one, several, or all of the ^{32}P-labeled *Neurospora* plasma membrane proteins are related to the electrogenic ATPase. The reason for the multiplicity of acyl phosphate-containing proteins in the *Neurospora* plasma membrane is presently unclear. There may be artifacts due to pro-teolysis or to the electrophoretic separation itself; or there may exist several ATPases of similar or different function which are kinetically undistinguish-able; or the various bands may represent posttranslational forms of the same enzyme; or the ATPase mechanism may involve a complex series of acyl phosphate transfers among proximal subunits. Investigation of these pos-sibilities is presently in progress in the hope of gaining new insight into the mechanism and overall molecular biology of *Neurospora* plasma membrane ATPase.

IV. Conclusion

The purpose of this brief chapter has been to describe the development of a new experimental system with which to explore the structure and function of eukaryote plasma membranes and to present what has been learned so far. To date, the bulk of the progress in understanding plasma membrane *structure* has come from studies of the erythrocyte plasma membrane, because these membranes are easy to isolate and because they can be isolated as open ghosts without a permeability barrier function.

The study of this system has engendered fundamental concepts of plasma membrane structure such as the asymmetric distribution of lipids and carbohydrates and the multiple modes of protein orientation in the plasma membrane (Singer, 1974; Rothman and Lenard, 1977). Most of the progress in understanding plasma membrane *function*, however, has come from studies of vesicular plasma membrane systems, primarily bacterial, in which it is possible to measure transmembrane movements of important metabolites because of the existence of permeability barrier function in such closed vesicles. The *Neurospora* plasma membrane system described here is suitable for both structural and functional studies, since these plasma membranes can be isolated as open sheets or closed vesicles. With open sheets it should be possible to confirm or deny the generality of structural concepts which have emerged from erythrocyte ghost studies; with closed vesicles it should be possible to gain new insight into the mechanisms of solute translocation across the eukaryote plasma membrane and the mechanisms of energy coupling which lead to the active transport of ions and molecules into and out of eukaryotic cells; and, with either sheets or vesicles, it should be possible to explore the mechanism of plasma membrane biogenesis. While the *Neurospora* plasma membrane system is suitable for the above-mentioned studies, it is limited in the sense that cells of *Neurospora* do not exhibit cell–cell interactions, hormone response, pino- and phagocytosis, and certain other critical functions of higher eukaryotes. But, hopefully, the information provided in this article will facilitate the isolation and study of plasma membranes from cells which do perform these functions.

REFERENCES

Ames, G. F. -L. (1974), *J. Biol. Chem.* **249**, 634–644.
Azzi, A., Gherardini, P., and Santato, M. (1971). *J. Biol. Chem.* **246**, 2035–2042.
Berg, H. C. (1969). *Biochim. Biophys. Acta* **183**, 65–78.
Brand, L., and Gohlke, J. R. (1972). *Annu. Rev. Biochem.* **41**, 843–868.
Bretscher, M. S. (1971). *J. Mol. Biol.* **58**, 775–781.
Cabantchik, I. Z., Balshin, M., Breuer, W., and Rothstein, A. (1975). *J. Biol. Chem.* **250**, 5130–5136.
Dahl, J. L., and Hokin, L. E. (1974). *Annu. Rev. Biochem.* **43**, 327–356.
DePierre, J. W., and Karnovsky, M. L. (1973). *J. Cell. Biol.* **56**, 275–303.
Doeg, K. A., and Ziegler, D. M. (1962). *Arch. Biochem. Biophys.* **97**, 37–39.
Durán, A., Bowers, B., and Cabib, E. (1975). *Proc. Natl. Acad. Sci. U.S.A.* **72**, 3952–3955.
Dutton, A., and Singer, S. J. (1975). *Proc. Natl. Acad. Sci. U.S.A.* **72**, 2568–2571.
Emerson, S. (1963). *Genetica* **34**, 162–182.
Ernster, L., Zetterström, R., and Lindberg, O. (1950). *Acta Chem. Scand.* **4**, 942–947.
Fairbanks, G., and Avruch, J. (1972). *J. Supramol. Struct.* **1**, 66–75.
Harold, F. M. (1972). *Bacteriol. Rev.* **36**, 172–230.
Hasselbach, W. (1974). *In* "The Enzymes" (P. D. Boyer, ed.), 3rd ed., Vol. 10, pp. 432–467. Academic Press, New York.

Hawkes, S. P., Meehan, T. D., and Bissell, M. J. (1976). *Biochem. Biophys. Res. Commun.* **68**, 1226–1233.

Hokin, L. E., Sastry, P. S., Galsworthy, P. R., and Yoda, A. (1965). *Proc. Natl. Acad. Sci. U.S.A.* **54**, 177–184.

Inesi, G. (1972). *Annu. Rev. Biophys. Bioeng.* 1, 191–210.

Maddy, A. H. (1964). *Biochim. Biophys. Acta* **88**, 390–399.

Nachman, R. L., Ferris, B., and Hirsch, J. G. (1971). *J. Exp. Med.* **133**, 785–806.

Pardee, A. B., and Watanabe, K. (1968). *J. Bacteriol.* **96**, 1049–1054.

Parish, R. W., and Müller, U. (1976). *FEBS Lett.* **63**, 40–44.

Phillips, D. R., and Morrison, M. (1970). *Biochem. Biophys. Res. Commun.* **40**, 284–289.

Reeves, J. P., Lombardi, F. J., and Kaback, H. R. (1972). *J. Biol. Chem.* **247**, 6204–6211.

Rothman, J. E., and Lenard, J. (1977). *Science* **195**, 743–753.

Scarborough, G. A. (1975). *J. Biol. Chem.* **250**, 1106–1111.

Scarborough, G. A. (1976). *Proc. Natl. Acad. Sci. U.S.A.* **73**, 1485–1488.

Scarborough, G. A. (1977). *Arch. Biochem. Biophys.* **180**, 384–393.

Schuldiner, S., Padan, E., Rottenberg, H., Gromet-Elhanan, Z., and Avron, M. (1974). *FEBS Lett.* **49**, 174–177.

Singer, S. J. (1974). *Annu. Rev. Biochem.* **43**, 805–833.

Slayman, C. L., and Gradmann, D. (1975). *Biophys. J.* **15**, 968–971.

Slayman, C. L., Lu, C. Y.-H., and Shane, L. (1970). *Nature (London)* **226**, 274–276.

Slayman, C. L., Long, W. S., and Lu, C. Y.-H. (1973). *J. Membr. Biol.* **14**, 305–338.

Staros, J. V., and Richards, F. M. (1974). *Biochemistry* **13**, 2720–2726.

Vansteveninck, J., Weed, R. I., and Rothstein, A. (1965). *J. Gen. Physiol.* **48**, 617–632.

Vogel, H. J. (1956). *Microb. Genet. Bull.* **13**, 42–43.

Warren, L., Glick, M. C., and Nass, M. K. (1966). *J. Cell. Physiol.* **68**, 269–287.

Whiteley, N. M., and Berg, H. C. (1974). *J. Mol. Biol.* **87**, 541–561.

Chapter 11

Staining Fungal Nuclei with Mithramycin

MARTIN L. SLATER

Laboratory of Biochemistry and Metabolism, National Institute of Arthritis, Metabolism, and Digestive Diseases, National Institutes of Health, Bethesda, Maryland.

I. Introduction

Nuclear division is a particularly useful cell cycle marker in yeast because it occurs near the middle of the cell cycle and involves a series of distinctive morphological changes (Williamson, 1966). Meiosis, which occurs during sporulation, also involves a series of morphological changes in the nucleus (Pontefract and Miller, 1962). Accordingly, observation of the position, shape, and number of nuclei in yeast cells has been useful in monitoring mitotic cell cycles and sporulation and in classifying mutants defective in these processes (Esposito *et al.*, 1970; Hartwell *et al.*, 1970). The observation of nuclei is also used in studies of mitosis (Knox-Davies, 1966) and meiosis (Singleton, 1953) in filamentous fungi.

Conventional nuclear staining methods are prolonged and difficult when applied to fungi (Knox-Davies, 1966; Robinow and Marak, 1966). This is partly due to the low DNA content of fungal nuclei in relation to the amount of cellular RNA. Recently, methods for fluorescent staining of fungal nuclei have been reported, which are more rapid but which require a critically timed hydrolysis of RNA (Laane and Lee, 1975; Lemke *et al.*, 1975). A notable exception is the one-step procedure using 4′,6′-diamino-2-phenylindole (DAPI) (Williamson and Fennell, 1975).

Here we describe the use of mithramycin as a nuclear stain. The procedure is designed primarily for observing stages of division of the nucleus rather than its internal structures. It has been used mainly with *Saccharomyces cerevisiae* (Conde and Fink, 1976; Slater, 1976) and has also been found to work with a filamentous fungus.

II. Mithramycin

Mithramycin (areolic acid) is a member of a group of antitumor antibiotics (Gause, 1965) produced by *Streptomyces* (Grundy *et al.*, 1953). Its structure (Fig. 1) (Bakhaeva *et al.*, 1968) is similar to those of oligomycin and chromomycin, the other antibiotics in the group (Ward *et al.*, 1965). Mithramycin is inert toward RNA and protein but fluoresces when bound to DNA; binding to the latter requires both Mg^{2+} and base pairing (Gause, 1965; Ward *et al.*, 1965). The fluorescence and specificity properties of mithramycin have been exploited in a microassay for DNA (Hill and Whatley, 1975) and in flow microfluorometry (Crissman and Tobey, 1974).

We obtained mithramycin as a gift from Nathan Belcher of Pfizer, Inc.,

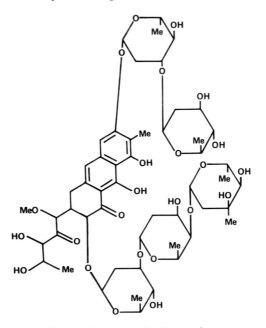

FIG. 1.   Structure of mithramycin.

Eastern Point Road, Groton, Conn. It is supplied in vials containing 2.5 mg of mithramycin and about 40 parts by weight of carrier (mannitol and sodium diphosphate).

III. Nuclear Staining Procedures

To stain vegetatively growing yeast cells, a sample of culture is added to an equal volume of 50% (v/v) aqueous ethanol containing mithramycin (0.2 mg/ml), 30 mM $MgCl_2$, and 60 μg/ml of fluorescamine. After about 10 minutes a drop is examined as a wet mount.

For staining yeast during meiosis, the cells are centrifuged and resuspended in 70% ethanol. After 20 minutes, 2 volumes of aqueous 22.5 mM $MgCl_2$ containing mithramycin (0.6 mgu/ml) and 45 μg/ml of fluorescamine are added.

To stain filamentous fungi, a small piece of mycelium is transferred to about 0.2 ml of 25% (v/v) aqueous ethanol containing mithramycin (0.2 mg/ml), 15 mM $MgCl_2$, and 30 μg/ml of fluorescamine. After the mycelial mat is frayed in a few drops of stain on a slide, the preparation is viewed as a wet mount. Spores are stained in the same solution.

The final staining solutions all contain 25% (v/v) ethanol, 15 mM $MgCl_2$, and 30 μg/ml of fluorescamine. Fluorescamine (Fluram, Hoffman La Roche), stored as a 0.03% (w/v) stock solution in acetone, acts as a counterstain for the cytoplasm.

Several alternative procedures were found satisfactory. (1) Acridine orange (AO) at a final concentration of 1 μg/ml can be used as a counterstain. In this case, however, the cells should be stained with mithramycin until the nuclei are visible before adding AO. Permount-sealed preparations stained with mithramycin and AO have been stored for over a year with no loss of intensity, whereas the fluorescamine counterstain fades in 1 or 2 days. (2) Cells can be fixed with either absolute ethanol and acetic acid (3:1 v/v) or absolute ethanol and acetone (1:1 v/v) before being stained. However, the fixative should be removed and the cells washed in 25% ethanol before staining. The fixation may be at room temperature for 10 minutes or overnight at 4°C. Cells should not be exposed to Formalin or strongly acidic fixatives before staining. (3) Cells may be either harvested and stored in 70% ethanol at −20°C for several days or stored in 25% (v/v) ethanol at 4°C before being stained.

IV. Microscopy

Preparations were examined with a Leitz vertical illumination fluorescence microscope with excitation filter KP 400, a 455-nm dichroic mirror, with suppression filter K460 (position 2). The initial fluorescence is brighter with excitation filter KP 500, a 510-nm dichroic mirror, and suppression filter K 515 (position 3). However, with this filter combination the tendency of fresh samples to fade while exposed to the light beam is increased. This tendency can be counteracted by inserting into the light beam a sunglasses lens (J. Fink, personal communication) or a 35 or 50% transmission neutral density filter.

Cells examined with a Leitz microscope at position 3 were photographed using Kodak Tri-X film. Exposures were from 10 to 20 seconds.

Examples of vegetatively growing and sporulating *S. cerevisiae* stained with mithramycin and AO are illustrated in Figs. 2 and 3. Mycelia and a

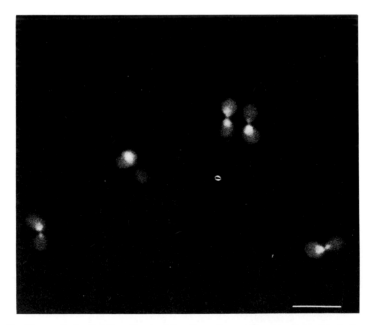

Fig. 2. Vegetatively dividing *S. cerevisiae*. Log-phase cells were exposed to hydroxyurea to accumulate cells in midnuclear division. Samples were stained with mithramycin and AO. The marker represents 10 μm. [Reprinted from Slater (1976) with permission from the American Society for Microbiology].

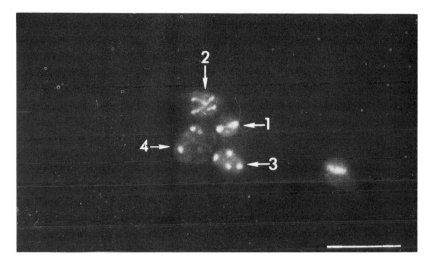

FIG. 3. Sporulating *S. cerevisiae*. Samples removed from sporulation medium were suspended in 70% ethanol and refrigerated overnight. They were then stained with mithramycin and AO. Cells 1, 2, and 3 contained one dividing nucleus, two dividing nuclei, and four nuclei, respectively. Cell 4 contained two brightly stained nuclei and two ascospores in which the stain does not penetrate easily. The marker represents 10 μm. [Reprinted Slater (1976) with permission from the American Society for Microbiology].

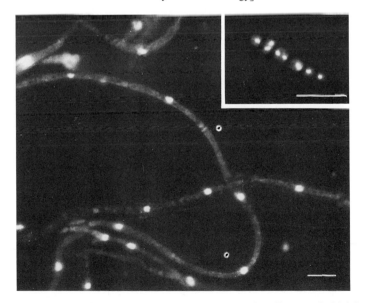

FIG. 4. *Alternaria* sp. The inset shows a spore containing six cells, two of which have two nuclei. Mycelia and spores were stained with mithramycin and fluorescamine. The marker represents 10 μ m.

spore of *Alternaria* sp. stained with mithramycin and fluorescamine are illustrated in Fig. 4.

V. Summary

The specific binding and fluorescence properties of mithramycin make it possible to stain fungal nuclei in a rapid, one-step procedure. This procedure⁻ is suitable for monitoring yeast mitotic cell cycles or sporulation as they occur rather than after they are completed. Alternatively, large numbers of samples can be fixed, stored, and processed later to screen mutants or to assess synchrony.

REFERENCES

Bakhaeva, G. P., Berlin, Y. A., Boldyreva, E. F., Chuprunova, D. A., Kolosov, M. N., Soifer, V. S., Vasiljeva, T. E., and Yartseva, I. V. (1968). *Tetrahedron Lett.* **32**, 3595–3598.
Conde, J., and Fink, J. (1976). *Proc. Natl. Acad. Sci. U.S.A.* **73**, 3651–3655.
Crissman, H. A., and Tobey, R. A. (1974). *Science* **184**, 1297–1298.
Esposito, M. S., Esposito, R. E., Arnaud, M., and Halvorson, H. O. (1970). *J. Bacteriol.* **104**, 202–210.
Gause, G. F. (1965). *Adv. Chemother.* **2**, 179–195.
Grundy, W. E., Goldstein, A. W., Rickher, G. J., Hanes, M. E., Warren, H. B., and Sylvester, J. C. (1953). *Antibiot. Chemother.* (*Washington, D.C.*) **3**, 1215–1221.
Hartwell, L. H., Culotti, J., and Reid, B. (1970). *Proc. Natl. Acad. Sci. U.S.A.* **66**, 352–359.
Hill, B. T., and Whatley, S. (1975). *FEBS Lett.* **56**, 20–23.
Knox-Davies, P. S. (1966). *Am. J. Bot.* **53**, 220–244.
Laane, M. M., and Lee, T. (1975). *Mikroskopie* **31**, 85–90.
Lemke, P. A., Ellison, J. R., Marino, R., Morimoto, B., Arous, E., and Kohman, P. (1975). *Exp. Cell Res.* **96**, 367–373.
Pontefract, R. D., and Miller, J. J. (1962). *Can. J. Microbiol.* **8**, 573–584.
Robinow, C. F., and Marak, J. (1966). *J. Cell Biol.* **29**, 129–151.
Slater, M. L. (1976). *J. Bacteriol.* **126**, 1339–1341.
Ward, D. C., Reich, E., and Goldberg, I. H. (1965). *Science* **149**, 1259–1263.
Williamson, D. H. (1966). *In* "Cell Synchrony" (I. L. Cameron and G. M. Padilla, eds.), pp. 81–100. Academic Press, New York.
Williamson, D. H., and Fennell, D. J. (1975). *Methods Cell Biol.* **12**, 335–352.

Chapter 12

Methods for Protoplast Formation in Escherichia coli

RICHARD L. WEISS[1]

The Biological Laboratories, Harvard University, Cambridge, Massachusetts

I. Introduction

The investigation of bacterial membrane structure and function has relied on the isolation of membranes from osmotically sensitive forms, protoplasts and spheroplasts. In this chapter "protoplast" refers to a gram-positive organism after complete digestion of the cell wall; "spheroplast" describes gram-negative forms surrounded by or retaining part of the outer membrane. The outer membrane of gram-negative bacteria is difficult to remove from spheroplasts. It either surrounds the cell (Kaback, 1971) or peels back, exposing the cell surface membrane (Birdsell and Cota-Robles, 1967). In both cases the outer membrane remains attached to the cell.

A method was recently developed for complete removal of the outer membrane from gram-negative cells, and this method is described here. These protoplasts can be obtained from log-phase and stationary cultures of *Escherichia coli* by enzymic and chemical alteration of the cell wall. This alteration includes rapid cell wall lysis wherein the outer membrane de-

[1] Present address: Department of Botany, San Diego State University, San Diego, California.

taches from the cell body, forming small vesicles. This detachment proceeds without restraint until the cell is bound only by the cytoplasmic or inner membrane.

II.   General Methods

Protoplasts of *E. coli* ML30 are formed by the following procedure. Cells grown in glucose mineral salts (Fiil and Branton, 1969) without iron at 37°C to the logarithmic (A_{450} = 0.9) phase of growth are centrifuged at 10,000 g for 5 minutes. The cells are washed twice in 10 mM tris–HCl buffer, pH 8, at 23°C. [Tris (hydroxymethyl) aminomethane can be obtained from Schwartz-Mann, Orangeburg, N.Y.] The pellet is resuspended in a solution containing 20% (w/w) sucrose and 0.1 M tris-HCl (pH 8) which is pipetted directly into the centrifuge tube. The product of a culture's volume and its measured absorbance at 450 nm is calculated to determine the total absorbance units (AU) contained in the culture (Osborn *et al.*, 1972). The cells are suspended in 1 ml of buffer per 10 AU. For example, in our experiments a culture with a volume of 500 ml and an absorbance of 0.9 was resuspended in 45 ml of buffer at 37°C. Cells are then transferred into a small flask, and the temperature is adjusted to 37°C. The cells are stirred with a magnetic stirrer, and within 1 minute 2.25 ml of lysozyme is added from a 2 mg/ml stock solution in distilled water to a final concentration of 100 μg/ml. The temperature is adjusted to 37°C, and stirring is continued for the next 12 minutes at 37°C. The cell suspension is diluted [1 part ethylenediaminetetraacetic acid (EDTA) per 10 parts cells] with 0.1 M K$_2$EDTA prewarmed to 37°C. Continuous stirring and slow dilution over 2.5 minutes prevents lysis of cells. Na$_2$EDTA may be used instead. After adding EDTA the temperature is adjusted to 37°C. More than 99% of the cells become spherical within 10 minutes and can be observed by phase microscopy.

III.   Appearance of Protoplasts

Samples for electron microscope examination are fixed at 4°C in 3% glutaraldehyde in 0.1 M sodium cacodylate buffer, pH 7, postfixed in 1% osmium tetroxide in 30 mM Veronal buffer containing 0.5 mM CaCl$_2$, pH 6.1, dehydrated in graded acetone, and embedded in Spurr's (1969) resin.
The cell envelope of fixed and sectioned cells is shown in Fig. 1. The

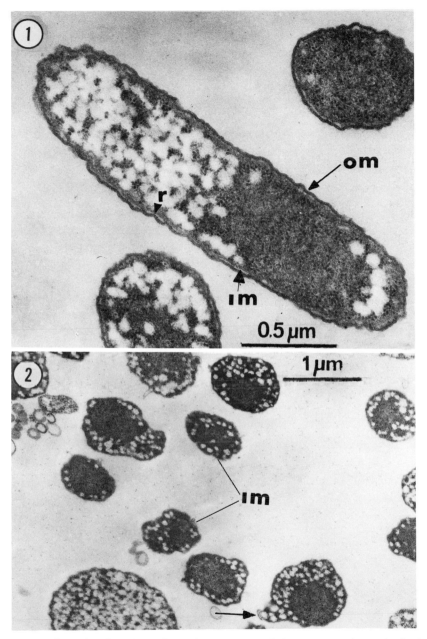

FIG. 1. Control cells. The cell envelope consists of the outer membrane (om), the peptidoglycan layer (r), and the inner or cell membrane (im).

FIG. 2. Protoplasts of *E. coli*. The cell surface is limited by the inner membrane (im); the arrow indicates a wall fragment.

features illustrated are the outer membrane, the peptidoglycan layer, and the inner membrane. A consistent feature of the strain depicted here, but not of other strains, is the presence of small, clear areas within the cytoplasm. This feature is observed both before and after protoplasts are formed and is unique to this strain. A field of cells following lysozyme–EDTA treatment is presented in Fig. 2. In the protoplasts illustrated here the cell wall has been completely removed from 95% of the cells, forming small outer membrane fragments (Fig. 3, inset a). The separation of such fragments is illustrated in Fig. 3, inset b. Here one fragment is partially attached to

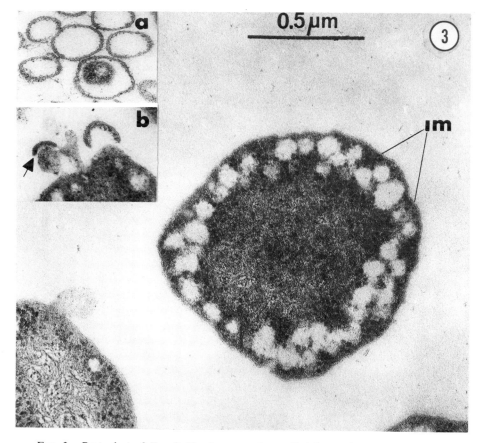

FIG. 3.   Protoplast of *E. coli*. The inner membrane (im) forms the outer surface of the cell. Outer membrane fragments that have separated from the cell are shown in inset a. An outer membrane fragment in the process of separating from the cell is indicated by the arrow in inset b. The marker bar also applies to the insets.

the cell (arrow), whereas the other has separated from it. When the cell surface is observed at higher magnification, as in Fig. 3, the inner membrane can be seen to form the outermost surface boundary, thus confirming cell wall removal. Contaminating cell wall fragments appear on some cells (Fig. 2, arrow). Protoplasts such as these can be centrifuged and resuspended without extensive lysis. Thus one may reasonably expect that membranes can be prepared from such cells and used for further biochemical or structural analysis.

IV. Discussion

Since it has been reported in the literature that gram-negative organisms yield spheroplasts, it is important to consider why this procedure yields protoplasts. Three mechanisms by which this could occur have already been suggested (Weiss, 1976). They are: (1) the enzyme/substrate ratio may be optimal; (2) the constant temperature or ionic conditions may ensure more complete cell wall lysis; and (3) the mild osmotic changes during dilution with EDTA may open up the cell wall.

A consideration of the above proposed mechanisms has led us to suggest ways to adapt this method to organisms other than *E. coli* ML30. The modifications we consider most significant are presented in the following discussion.

A. Lysozyme Concentration

It is well established that lysozyme treatment brings about spheroplast formation in *E. coli* (Kaback, 1971). Concentrations well below the amount used here are apparently sufficient (Birdsell and Cota-Robles, 1967) to peel back the outer membrane with appropriate osmotic treatment. However, it is critically important to have a concentration below that which aggregates the bacteria. While some variation can be tolerated, both minimum (< 20 μg/ml) and excess (< 500 μg/ml) concentrations should be avoided, since they inhibit the efficiency of protoplast formation. When applying this technique to other strains of *E. coli*, it may be necessary to determine the optimum concentration of lysozyme experimentally; for any given organism this concentration may vary with growth conditions. As an example, an organism may respond well to 100 μg/ml in log-phase growth, but require 250 μg/ml under starvation conditions. However, it should be noted that the conditions described here have been used successfully with several ML strains, and it is expected that this procedure will work with other strains such as K12 and B.

B. Temperature and Ionic Strength

It is much less likely that changes in the constant-temperature regime or ionic strength can improve this method when difficulty with protoplast formation is encountered. In this regard however, a general principle is that, when the concentration of lysozyme is increased, the ionic strength of the buffer should be decreased.

C. Osmotic Conditions

The removal of the cell wall depends not only on enzyme activity, but also on the access of the enzyme to the substrate. Since lysozyme is sufficiently active over a wide range of concentrations (Birdsell and Cota-Robles, 1967; Kaback, 1971; Osborn et al., 1972), it seems reasonable to postulate that the osmotic stress attained with the present method allows more complete access of lysozyme to the substrate, perhaps as a result of the egress of loosely bound enzymes. If it is assumed that the release of such enzymes takes place through a break in the outer membrane, the osmotic imbalance would force out the enzymes, creating one or many discontinuities in this layer. Thus, during the general application of this method to gram-negative organisms, it may be advantageous to increase the osmotic stress. Fortunately, a simple approach to determine the experimental requirements is available; it includes increased dilution which may be introduced when adding EDTA to the cells. The direct effect of this and other modifications is rapidly monitored by phase microscopy. An advantage of this method is that morphological changes as well as adherent outer membrane coils of spheroplasts can be visualized. Following this approach more elaborate methods such as electron microscopy and chemical analysis will be required to demonstrate cell wall removal.

V. Summary and Conclusion

Cells of *E. coli* ML30 have been converted into protoplasts using lysozyme–EDTA. This technique has several possible applications for this and other sensitive strains. One of these is the separation of protoplasts from the outer membrane fragments by differential centrifugation. This should allow physiological studies or further manipulation that hopefully will increase the utility of *E. coli* in studies on the structure and function of bacterial membranes.

ACKNOWLEDGMENT

The work described here was supported by a grant from the American Cancer Society (PF-876).

REFERENCES

Birdsell, D. C., and Cota-Robles, E. H. (1967). *J. Bacteriol.* **93**, 427–437.
Fiil, A., and Branton, D. (1969). *J. Bacteriol.* **98**, 1320–1327.
Kaback, H. R. (1971). *In* "Methods in Enzymology" (W. B. Jakoby, ed.), Vol. 22, pp. 84–120. Academic Press, New York.
Osborn, M. J., Gander, J. E., Parisi, E., and Carson, J. (1972). *J. Biol. Chem.* **247**, 3962–3972.
Spurr, A. R. (1969). *J. Ultrastruct. Res.* **26**, 31–43.
Weiss, R. L. (1976). *J. Bacteriol.* **128**, 668–670.

Chapter 13

Fusion of Bacterial Protoplasts

PIERRE SCHAEFFER AND ROLLIN D. HOTCHKISS

Institute of Microbiology, University of Paris-Sud, Orsay, France, and
Rockefeller University, New York, New York

I. Background

The basic information concerning the bacterial cell wall, its chemical nature, the various agents (enzymes and antibiotics) used for removing it, and the properties of the protoplasts and spheroplasts produced by its removal can be found in various books and reviews (Salton, 1964; Rogers and Perkins, 1968; Guze, 1968; Ghuysen *et al.*, 1966), reflecting the early work of Dienes, Weibull, McQuillen, and others.

The plasma membrane being subject to external osmotic influences, protoplasts assume a spheroidal form in a suitable osmotic range such as 0.2–0.5 osM total osmolality. Newly produced protoplasts do not usually divide in liquid media, although they can do so on suitable solid media, leading to "L-form" colonies, and then regenerate to the bacterial form under proper conditions (Clive and Landman, 1970; Wyrick and Rogers, 1973; Castro-Costa and Landman, 1977). We use the term "regeneration" here for the total reestablishment of characteristic bacterial morphology and growth habit, which may include processes other than cell wall biosynthesis. Regeneration may proceed directly from one protoplast to one bacterium or pass through multiunit stages (Hadlaczky et al., 1976), but there is little question that under suitable conditions single protoplasts, which are easily separable bodies, can ultimately give rise to viable bacterial clones.

Bacterial protoplasts, being in some respects comparable to animal cells, have been recently investigated in regard to their ability to undergo fusion. This process has in fact been observed in the light microscope, and followed by time-lapse photography by Stähelin (1954), working with protoplasts of *Bacillus anthracis*. The progeny and indeed the viability of the observed fused protoplasts within a single culture were not investigated. Lederberg and St. Clair (1958) looked unsuccessfully for the appearance of prototrophic bacteria produced as a result of possible fusion between spheroplast populations derived from two polyauxotrophic mutants of F⁻ strains of *Escherichia coli* K12. Essentially similar searches undertaken with protoplasts from *Bacillus* species, submitted to a fusogenic treatment, were successful (Fodor and Alföldi, 1976; Schaeffer et al., 1976) and are the basis of this article.

II.   Rationale for a Fusion Experiment

The experiment is conceived as a cross between two potentially complementing polyauxotrophic parental strains, carried out with their protoplasts in hypertonic medium and eventually yielding prototrophic bacteria. Thus it involves in principle the following steps: (1) precultivation of bacteria, (2) transfer of cells to hypertonic medium, (3) conversion to protoplasts, (4) mixing of parental protoplasts, (5) provision of conditions for fusion, (6) regeneration of bacterial forms on solid hypertonic growth medium, and (7) any selection process required for recovering the presumably rare fused cells and their progeny.

Gram-positive organisms appear to be the material of choice, inasmuch as they are devoid of an outer membrane, a possible obstacle to fusion of the cytoplasmic membranes. Nutritional deficiencies, expected to be recessive, are suitable genetic markers for detecting complementation in heterozygotes. Introducing two such deficiencies into each parental strain makes unlikely the appearance of prototrophs through reverse mutation. Using more than two deficiencies permits some to be reserved as unselected markers.

Lysozyme is chosen as the most efficient agent for making protoplasts. The exhaustiveness with which it is permitted to act is experimentally determined, since cell wall residues may be required as primers in step 6.

Spontaneous fusion has been observed with various kinds of naked cells or protoplasts, but fusion frequency can be increased about 100-fold when various fusogenic agents are applied. Among the latter, polyethylene glycol (PEG) seems to be a very active agent (Kao and Michayluk, 1974; Pontecorvo, 1975) and was originally chosen for the bacterial work. Since then the range of its application has been extended among plant protoplasts (Kao, 1975), yeasts (Sypyczki and Ferenczy, 1977), and filamentous fungi (Anné and Peberdy, 1976).

With eukaryotic cells, fusion is first cellular, or cytoplasmic, leading to heterokaryons. Nuclear fusion may subsequently occur at mitosis, leading to rare viable cells with hybrid nuclei. Only cell fusion is conceivable with bacteria. Diploid cells are the primary fusion products, which may or may not be stable during subsequent growth.

III. Procedures for Bacterial Fusion

A. Bacterial Strains

Successful protoplast fusion with genetically marked bacteria has so far been reported only with polyauxotrophic mutants derived from *Bacillus subtilis* strain 168, or from an asporogenous subline of *Bacillus megaterium* strain KM. In addition, in the former species, a mutation for rifamycin resistance introduced into both parental strains has made it possible practically to eliminate any risk of contamination.

There are presently no contraindications to the use of any other grampositive bacterial species.

B. Media and Buffers

The compositions of the media outlined below are not believed to be critical and in general need some modification for other bacterial strains. For fusion, *B. subtilis* is generally grown in (Difco) nutrient broth usually supplemented with low concentrations of magnesium, calcium, and iron (NB). "Regeneration medium" consists of the rich, hypertonic serum—

agar medium of Wyrick and Rogers (1973), supplemented with the six growth factors required by the parental cells (each at 25 μg/ml) and with DNase and rifamycin (Lepetit Laboratories, Milan, Italy), both at 5 μg/ml. Selection media for *B. subtilis* are based upon the minimal medium of Spizizen (1958), supplemented with DNase (5 μg/ml) and rifamycin (1 μg/ml) and with various combinations of two growth factors (one required by each parental strain, thus leaving one unselected marker for each).

For *B. megaterium*, growth, protoplasting, fusion, regeneration, and selection are carried out in variations of the hypertonic medium (HM) (sucrose–salts–ammonium–glucose; Fodor *et al.*, 1975).

The sucrose–magnesium–maleate (SMM) buffer described by Wyrick and Rogers (1973) is used to prepare stock solutions of lysozyme (4 mg/ml) and of PEG. Generally for the latter, a Merck product of MW 6000 is used to prepare the 40% (w/v stock solution, which is supplemented shortly before use with DNase (at 5 μg/ml from a freshly prepared solution of a purified commercial product). SMM supplemented with DNase is referred to as SMMD.

C. Conducting Fusion Experiments

1. *BACILLUS SUBTILIS* (Schaeffer *et al.*, 1976)

Two rifamycin-resistant triply auxotrophic strains are grown separately at 37°C in NB to an absorbance (A_{570}) of 0.4 to 0.8. Samples containing 6.0 total units of absorbance are centrifuged at 6000 g at 4–15°C for 8 minutes. The cells are resuspended [adjusted to A_{570} = 2.0, or ca. 4 × 10^8 colony-forming units (cfu) per milliliter] in 3 ml of SMMD. Lysozyme is added (200 μg/ml final concentration), and the suspensions in shallow layers are shaken very gently in a waterbath at 37 or 42°C for 30 minutes (at 10 minutes transformation of the rods into spheres is complete). Ordinary nutrient agar plates seeded with 0.10 ml of these suspensions usually remain sterile, indicating that less than 5 × 10^{-8} of the cfu are resistant to osmotic shock.

Samples of 1 ml from each suspension are mixed and centrifuged, along with parental controls (2 ml). Each pellet is resuspended in 0.2 ml of SMMD. To limit exposure to PEG, each of the three protoplast suspensions in succession is carried through to plating. To one suspension, 1.8 ml of a chilled 40% PEG solution in SMMD is added, and the suspension is immediately homogenized by shaking. After a 1-minute exposure to PEG (which agglutinates the protoplasts), serial 3- or 10-fold dilutions in SMMD are made (which dissociate most of the aggregates), of which 0.05-ml samples are plated on regeneration medium. After 48 hours at 37°C, colonies of regenerated bacteria are counted, and plates showing growth are replica-plated (Lederberg and Lederberg, 1952) onto various selection plates. The prototrophic, or partially prototrophic, colonies are counted after 2–3 days of incubation at 37°C. The frequencies of regenerated bacteria and of prototrophs derived from them are calculated from all favorable dilutions available. Since both may vary with the number of protoplasts initially plated, the accepted frequencies are those obtained from two successive dilutions. Based on the original cells, 1–3% of regenerants and $1–3 \times 10^{-4}$ prototrophs have repeatedly been observed. Ways of increasing the yields continue to be investigated (see Section IV).

Under the conditions described the primary prototrophic, or partially prototrophic, colonies obtained on selection media contain stable haploid recombinants. The time is still to be determined within which the heterozygotic first products of fusion segregate to give these or other haploid progeny.

2. *BACILLUS MEGATERIUM* (Fodor and Alföldi, 1976)

Doubly auxotrophic bacterial strains are grown to 5×10^7 cfu/ml at 30°C in HM supplemented with the required growth factors. The cells are centrifuged at 4000 g and 4°C for 20 minutes and then resuspended at a $10 \times$ concentration in HM lacking phosphate, glucose, and supplements. Lysozyme is added to two parental cultures at 100 μg/ml, and aliquots from these cultures immediately mixed in equivalent amounts. After 30 minutes at 30°C without shaking, each of the three completely protoplasted cultures is centrifuged as above and resuspended in the same solution without lysozyme at an additional $10 \times$ concentration. To the suspension is added 10 volumes of a 40% (w/v) solution of PEG (Fluka Company, MW 6000), resulting in visible agglutination. The culture is immediately diluted in HM. Samples of 0.1 ml are added to 2.0 ml soft agar (HM containing 0.4% agar at 40°C) and immediately poured onto HM solidified with 1% agar (Difco) with or without growth factors, depending upon whether regenerants or prototrophs are to be counted.

Prototrophic colonies are obtained on these plates in 3–5 days at 30°C at a frequency ranging from 10^{-5} to 5×10^{-3} per original protoplast. All these prototrophic primary colonies can be subcultured in minimal liquid medium; most contain a proportion of prototrophs after one subculture, and about one-half show the presence of auxotrophs. The prototroph-containing subcultures often still contain prototrophs after further passage; some at least are unstable, apparently giving rise to stable auxotrophic progeny showing various parental and recombinant phenotypes.

IV. Explored Variations and Alternatives

Steps 1 to 7 outlined in Section II are subject to modification and condensation. Precultivation in hypertonic medium (steps 1 and 2), as described for *B. megaterium*, seems to ensure more efficient protoplasting for *B. subtilis*, according to M. Gabor and R. D. Hotchkiss (unpublished). These workers also found evidence that, as an environment for steps 3 to 5, and dilution, SMMD containing 2% bovine albumin fraction V (Sigma) increases bacterial regeneration to as much as 35% of the total cells initially involved and also improves fusion.

In the fusion step 5, PEG preparations of MW 200, 400, 1000, or 20,000, as well as 6000, have given very similar numbers of prototrophs, but for the best effect all require concentrations near 40% (Cami, 1975). Other hydrophilic polymers (dextran, polyvinylpyrolidone, polypropylene glycol, Ficoll) at 40% and concanavalin A (10 μg/ml), even though all except the first agglutinate protoplasts, do not increase the frequency of fusion above control values without PEG.

Exposure to PEG should be brief (about 1 minute); at 4 or 20°C the yields are better than at 37°C (Cami, 1975). Some commercial lots of PEG have excessive toxicity for protoplasts and should be avoided. Stepwise dilution of protoplasts from the high PEG concentration as suggested for animal cell fusion (Pontecorvo *et al.*, 1977) appears not to offer any advantage in yield over the procedure given (A. M. Lhéritier and P. Schaeffer, unpublished).

Calcium salts, frequently recommended for the fusion of plant protoplasts (Keller and Melchers, 1973; Kao, 1975), have not proven generally important for bacterial fusion, although Fodor and Alföldi (1976) describe fusion enhancement by nascent precipitation of calcium phosphate as useful in the absence of PEG.

Landman and co-workers (recently summarized in Castro-Costa and Landman, 1977) and Wyrick and Rogers (1973) investigated the regeneration of bacilli from *B. subtilis* protoplasts, and Hadlaczky *et al.* (1976) did the same for *B. megaterium*. Their findings have been taken into account in the procedures given here. The reversion frequency of *B. subtilis* protoplasts was increased by adding purified bacterial cell wall preparations, or even autoclaved *E. coli* bacteria, to the regeneration system (Clive and Landman, 1970). This finding has been applied by us as follows. Overnight broth cultures of *E. coli* are autoclaved 20 minutes at 110°C, centrifuged, and washed twice in distilled water. The cells, resuspended at a 20-fold concentration in water and preserved in small samples at −20°C, are plated along with equal volumes of protoplast suspension when the latter are placed upon regeneration medium. This modification has given a 3- to 10-fold increase in the yield of regenerated bacteria and prototrophic recombinants, occasionally as high as 1 prototroph from 300 parental protoplasts (A. M. Lhéritier, C. Lévi, and P. Schaeffer, unpublished).

V.   Criteria for Demonstration of Fusion

A.   The Significance of Prototrophic Growth

Having a drug resistance factor in both parents aids in excluding accidental contaminants. When the parental bacteria are multiple auxotrophs, it can be demonstrated that there is very little likelihood of prototrophic growth resulting from back-mutations in a parent.

The observation that two parents have cooperated to produce selective growth, nevertheless, is presumptive evidence of protoplast fusion only if certain more limited possibilities are excluded. Fusion should result not only in the sharing of gene products, complementation, and the sharing or recombination of genomes, but also in the mingling of essentially the total cytoplasm. In bacteria some of these additional implications of fusion can be tested.

B.   Fusion versus Cross-Feeding

If prototrophic growth results from mutual cross-feeding of parental clones in close association, then subcloning should eliminate the effect and give back the nonrecombinant parents. This was not the case in the systems described here. Furthermore, coplating of mixed bacterial parents (in the presence of DNase for *B. subtilis*) has not given evidence of this type of mixed growth in selective media (see also Fodor and Alföldi, 1976).

C.   Fusion versus Genetic Transfer

The semblance of fusion might be produced by transformation, DNA transfer from one parent to another. This is eliminated (demonstrably) by the presence of DNase (Schaeffer *et al.*, 1976) during the entire experiment. Transformation appears to be impossible in *B. megaterium* (Fodor and Alföldi, 1976) and is reported not to occur with fully protoplasted *B. subtilis* (Tichy and Landman, 1969). Mixed parental protoplasts in the absence of PEG constitute another "control," usually yielding very few fusion products.

When the genetic map of the organism is adequately known, the frequency of various recombinant progeny obtained can serve as evidence of a fusion process. Among the products from combining two *B. subtilis* triple auxotrophs, the frequencies of various recombinations distributed around the chromosomes were measured (Schaeffer *et al.*, 1976). Progeny requiring six or four recombination events came to as much as 16 or 27% of those requiring only two (i.e., one double crossover), although the latter were only a small fraction (ca. 10^{-4}) of the total number of protoplasts involved. This result can be taken as evidence that, when the parental genetic material succeeded in recombining at all, whole chromosomes, and not only fragments of them, were available to interact frequently throughout their whole length.

Conjugation and plasmid transfer have not been reported in *Bacillus* species. The *B. subtilis* used were all derived from strain 168. In recent unpublished work, C. Lévi, C. Rivas, and P. Schaeffer prepared from each of two strains virtually "isogenic" triple auxotrophs and found them to be still able to redistribute markers at the usual frequency when treated as described. Unlike the finding of directionality in *E. coli* K12 conjugation (Hayes, 1952), the killing of either parent by streptomycin has the same effect upon the outcome. When the other parent is streptomycin-resistant, fusion occurs with the killed parent and produces prototrophs normally. If both parents are streptomycin-sensitive and one is killed, no prototrophs are produced, perhaps a sign that, in the presumptive fused biparental protoplast, bound streptomycin or some product of its action is able to destroy the viability of the whole complex. If this is correct, then it supplies some evidence that elements other than chromosomal ones, presumably coming from the cytoplasm of the poisoned cell, are transferred to the intact cell.

D.   Cytoplasmic Mixing as a Sign of Fusion

Actual evidence of mixing of cytoplasms depends on having identifying markers for cytoplasm. One such marker is the phage repressor of a lysogenic *B. subtilis* (θ105) parent fused with a phage-sensitive partner (Schaeffer *et al.*, 1976). Mere dilution of the repressor by cytoplasmic fusion should

---

FIG. 1.   A culture of *B. subtilis* incubated at $42°C$ is harvested 3.5 hours after the end of growth (when prespores are formed), protoplasted and PEG-treated as described in the text, fixed with glutaraldehyde, embedded in Epon, sectioned, and stained with lead citrate. Two protoplasts, each carrying one prespore, at an early step in their fusion (A) and after fusion is completed (B) are shown. Prespores, at an early stage IV of their evolution, are surrounded with two membrane layers (om, outer membrane; im, inner membrane). The wall of the enclosed prespore (cw) is already formed, but no cortex or spore coats are visible as yet. (Electron micrograph by Dr. Claude Frehel, Dept. of Molecular Biology, Institut Pasteur, Paris.)

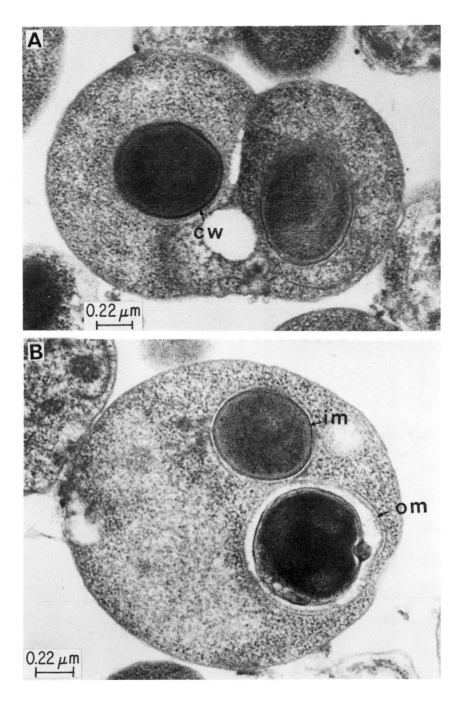

have little effect on the yields of prototrophic recombinants, and this is what was observed, no matter which parent was lysogenic. By contrast, mere chromosome transfer should have resulted in phage induction, drastically reducing the survival of recombinants.

Excellent evidence for cytoplasmic mixing comes from the fusion of protoplasts made while cells are in the process of sporulation (one cell contains no more than one prespore). The fusion of these should produce large protoplasts containing two prespores or more, and such protoplasts are observed, as illustrated in Fig. 1 (unpublished work in collaboration with C. Frehel).

Acknowledgments

The authors acknowledge the advice and help contributed by Dr. M. Gabor in the preparation of the manuscript for this chapter. Financial support was received (P.S.) from CNRS Contract LA 136, and (R.H) from NSF Grant PCM 76-21863.

References

Anné, J., and Peberdy, G. F. (1976). *J. Gen. Microbiol.* **92**, 413.
Cami, B. (1975). University Thesis, University of Paris-Sud, Orsay.
Castro-Costa, M. R., and Landman, O. E. (1977). *J. Bacteriol.* **129**, 678.
Clive, D., and Landman, O. E. (1970). *J. Gen. Microbiol.* **61**, 233.
Fodor, K., and Alföldi, L. (1976). *Proc. Natl. Acad. Sci. U.S.A.* **73**, 2147.
Fodor, K., Hadlaczky, G., and Alföldi, L. (1975). *J. Bacteriol.* **121**, 390.
Ghuysen, J. M., Tipper, D. J., and Strominger, J. L. (1966). *In* "Methods in Enzymology" (E. F. Neufeld and V. Ginsburg, eds.), Vol. 8, p. 685. Academic Press, New York.
Guze, L. B., ed. (1968). "Microbial Protoplasts, Spheroplasts and L-Forms." Williams and Wilkins, Baltimore, Maryland.
Hadlaczky, G., Fodor, K., and Alföldi, L. (1976). *J. Bacteriol.* **125**, 1172.
Hayes, W. (1952). *Nature (London)* **169**, 118.
Kao, K. N. (1975). *In* "Plant Tissue Culture Methods" (O. L. Gamborg and L. R. Wetter, eds.), pp. 22–27. Nat. Res. Counc. Can., Saskatoon, Saskatchewan.
Kao, K. N., and Michayluk, M. R. (1974). *Planta* **115**, 355.
Keller, W. A., and Melchers, G. (1973). *Z. Naturforsch., Teil C* **28**, 737.
Lederberg, J., and Lederberg, E. M. (1952). *J. Bacteriol.* **63**, 399.
Lederberg, J., and St. Clair, J. (1958). *J. Bacteriol.* **75**, 143.
Pontecorvo, G. (1975). *Somatic Cell Genet.* **1**, 397.
Pontecorvo, G., Riddle, P. N., and Hales, A. (1977). *Nature (London)* **265**, 257.
Rogers, H. J., and Perkins, H. R. (1968). "Cell Walls and Membranes." Spon, London.
Salton, M. R. J. (1964). "The Bacterial Cell Wall." Elsevier, Amsterdam.
Schaeffer, P., Cami, B., and Hotchkiss, R. D. (1976). *Proc. Natl. Acad. Sci. U.S.A.* **73**, 2151.
Spizizen, J. (1958). *Proc. Natl. Acad. Sci. U.S.A.* **44**, 1072.
Stähelin, H. (1954). *Schweiz. Z. Allg. Pathol. Bakteriol.* **17**, 296.
Sypyczki, M., and Ferenczy, L. (1977). *Mol. Gen. Genet.* **151**, 77.
Tichy, P., and Landman, O. E. (1969). *J. Bacteriol.* **97**, 42.
Wyrick, P. B., and Rogers, H. J. (1973). *J. Bacteriol.* **116**, 456.

Chapter 14

Preparation of Protoplasts of Plant Cells

CRAIG R. LANDGREN[1]

Department of Biology,
University of Oregon,
Eugene, Oregon

I. Introduction

The first report of the deliberate isolation of protoplasts from higher plant cells came from von Klercker (1892). In this and other early experiments, the plasmolysis of small pieces of tissue was followed by a purely physical disruption of the cell wall to effect release of the protoplasts. More recently these methods have been used (Prat and Roland, 1970; Prat, 1973) to isolate protoplasts free of the influences of the wall-degrading enzymes used in enzymic protoplast isolation techniques. Mechanical isolation generally produces a small number of protoplasts which must be separated from tissue fragments.

In 1960, Cocking reported the isolation of protoplasts of tomato cells using exogenous enzymes to digest the cell walls in plasmolyzed root tips. The pressure of the cover glass on these enzyme-treated tissues apparently

---

[1] *Present address:* Biology Department, Middlebury College, Middlebury, Vermont.

provided a physical disruption of the partially digested cell walls. The need for physical stress in the release of protoplasts is clear in the earliest report of a really large-scale isolation of protoplasts (Gregory and Cocking, 1965), in which teasing of the tissues with tools prepared from carding cloth preceded protoplast release.

With improvements in commercially available enzymes and with modifications in the types of tissues used as cell sources for the isolation of protoplasts, it has been possible to decrease the physical force required for protoplast release. The turgor pressure of the partially plasmolyzed cells or the force of gravity is sufficient to effect protoplast release from enzyme-treated suspension cultures (Schenk and Hildebrandt, 1969) and leaf mesophyll (Power and Cocking, 1970; Banks and Evans, 1976), although even in these systems the stresses of gentle agitation are usually applied (Kao *et al.*, 1970; Takebe *et al.*, 1968).

Many of the early reports of enzymic isolation of protoplasts are found in *Protoplastes et Fusion de Cellules Somatiques Végétales* (Tempé, 1973). The literature on the isolation and culture of protoplasts has been extensively reviewed (Cocking, 1972; Gamborg and Miller, 1973; Eriksson *et al.*, 1974; Eriksson, 1977). Cocking and Evans (1973) have reviewed the literature with an emphasis on protoplast production, while Ruesink (1971) has focused on all the techniques related to protoplast isolation and culture and Takebe (1975) has stressed the use of protoplasts in the study of plant viruses. Constabel (1975) has described generalized techniques for the isolation and culture of higher-plant protoplasts from mesophyll and suspension culture cells. There has also been considerable discussion of the potential uses of protoplasts in plant breeding and in the study of higher-plant genetics (Nickell and Torrey, 1969; Carlson, 1973; Nickell, 1973; Chaleff and Carlson, 1974; Bottino, 1975; Holl, 1975).

II. Protoplast Isolation Procedures

A. Enzymic Protoplast Isolation with Substantial Mechanical Wall Disruption—Pea Root Cortical Cell Protoplasts

One hundred fifty pea seeds (*Pisum sativum* "Little Marvel") were sorted, and those which showed cracks or abrasions were discarded. The remaining seeds were placed in a 250-ml beaker containing 200 ml of diluted commercial bleach (final concentration, 3% sodium hypochlorite). The solution was stirred continuously for 5 minutes and then briefly at 5-minute intervals. After 30 minutes, peas which were floating were removed with a sterile pair of forceps. The bleach was decanted slowly. Sterile distilled water was added and decanted in three 200-ml portions. The seeds were then poured into three 10-cm glass petri plates each of which contained 15 ml of sterile distilled water. The seeds were imbibed in the dark, at 25°C, for 8 hours. The imbibed seeds were transferred with forceps to germination plates containing 15 ml of 0.6% agar in distilled water. Ten peas were placed on each plate. The seeds were germinated in the dark for 48 hours.

Roots were measured, and those which were 1.5–2.5 cm in length were excised near the cotyledons. A cutting tool was used to remove the 10–11 mm region of each of these roots. Explants which were 1.4–1.7 mm in diameter were placed on a punch (Fig. 1), and the central cylinder and part of the inner cortex were removed.

The tissues remaining, called cortical explants, were placed in 5 ml of PS2M medium (Table I) on a 6-cm Nuncware petri plate (Vanguard International, Inc., Red Bank, N. J.). After 30 minutes in this medium, they were transferred to 5 ml of PS2M containing 6% Cellulase "onozuka" (Lot 223011) or 1% Cellulysin (Lot 502106) and 2% Macerozyme (Lot 40064) or 0.2% Macerase (Lot 501661). The enzyme incubation was for 17 hours in the dark at 25°C. The enzyme solution was prepared $\frac{1}{2}$ hour prior to use and was shaken for 5–10 minutes. Immediately before use, it was filter-sterilized through a Swinnex-25, 0.45-μm Millipore filter onto a 6-cm Nuncware petri plate.

After enzyme incubation, the explants were transferred using stainless-steel watchmaker's forceps to 10 ml of PS2M on a 6-cm Nuncware plate. After 30 minutes in this wash, they were transferred to a second identical

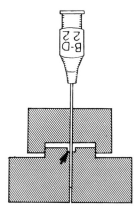

FIG. 1. A longitudinal section of the punch used to remove the central cylinder from root explants. The explant holder (shaded) is turned from two Teflon blocks. The chamber in which the explants are placed (arrow) is 1 mm deep and 1.5 mm in diameter. The plunger is made from a No. 22 B-D needle (approximately 0.7 mm in diameter).

TABLE I

PROTOPLAST ISOLATION AND CULTURE MEDIA[a]

Nutrient	CPW salts	Medium	
		PS2M	F5.9
Inorganic salts			
$CaCl_2 \cdot 2H_2O$	1480	440	850
NH_4NO_3	—	1650	412
KNO_3	101.	1900	525
KH_2PO_4	27.2	170	355
$MgSO_4 \cdot 7H_2O$	246	370	740
KI	0.16	0.83	0.5
$CoCl_2 \cdot 6H_2O$	—	0.025	0.125
H_3BO_3	—	6.2	3.1
$Na_2MoO_4 \cdot 2H_2O$	—	0.25	0.125
$MnSO_4 \cdot H_2O$	—	16.9	11.2
$CuSO_4 \cdot 5H_2O$	0.025	0.025	0.025
$ZnSO_4 \cdot 7H_2O$	—	8.6	4.3
$FeSO_4 \cdot 7H_2O$	—	27.8	13.9
Na_2EDTA	—	37.3	18.6
Vitamins			
Thiamine HCl	—	0.1	1
Nicotinic acid	—	0.5	5
Pyridoxine	—	—	0.5
Folic acid	—	—	0.5
Biotin	—	—	0.05

Nutrient	Medium	
	PS2M	F5.9
Hormones		
Indoleacetic acid	0.175	1
2,4-Dichlorophenoxy-acetic acid	1.1	—
Naphthalene acetic acid	—	2
Kinetin	2	—
6-Benzylaminopurine	—	1
Sugars and sugar alcohols		
Myoinositol	100	100
Sucrose	20 gm/liter	30 gm/liter
Mannitol	109 gm/liter	90 gm/liter
Reduced nitrogen sources		
Glycine	75	1
L-Glutamic acid	147	—
L-Aspartic acid	133	—
L-Arginine	174	—
L-Asparagine	132	—
Urea	60	—
Agar	—	0.6%
pH[b]	5.5	5.8

[a] Concentrations are in milligrams per liter unless otherwise noted.
[b] The pH is not adjusted in CPW salts.

wash for 30 minutes more. Following the second wash, explants were transferred to each of six 50-μl drops of PS2M on a 6-cm Nuncware plate. The explants were teased, using stainless-steel dissecting needles, until maximum protoplast release was attained without a large amount of debris, as observed under a dissecting microscope. The needles were then used to remove the large pieces of tissue from each of the drops. These procedures were repeated until all the explants had been teased. The plates were left undisturbed for 10–20 minutes, during which time the protoplasts settled to the bottom of the drops. A pasteur pipet, pulled to a 200-μm point, was then used to collect all the protoplasts from a single plate. The pooled protoplasts were placed in the center of the plate. After settling a second time, they were collected and distributed in the center of 6-cm Nuncware plates (in two to six drops of 50 μl each containing 300 to 400 protoplasts). Twenty 10-μl drops of medium were placed around the edge of each plate to provide a humid atmosphere.

These techniques have been used to produce small populations (8×10^3 to 10×10^3) of protoplasts in each experiment, from a very limited region of the cortical tissue in pea roots. The protoplasts regenerate cell walls and are mitotically active (Landgren, 1976a,b). Thirty to forty percent of the protoplasts undergo at least one division during the first 10 days in culture, and colonies of 4 to 10 cells have been observed in these cultures after 2 weeks. The techniques for explant production were modified from the work of Libbenga and Torrey (1973).

B. Enzymic Protoplast Isolation with Minimal Mechanical Wall Disruption—White Tobacco Leaf Protoplasts

Seeds of *Nicotiana tabacum* L. "Turkish Samsun" carrying a chloroplast defect (the progeny of plants derived from seeds kindly supplied by S. G. Wildman) were germinated in a layer of vermiculite above a layer of sand in plastic or wooden flats. Seedlings were transplanted successively to 2-, 4-, 6-, and 8-inch pots containing a soil mix (1 part loam, 1 part sand, and 1 part peat, supplemented per cubic foot with 4 gm KNO_3, 4 gm K_2SO_4, 42 gm superphosphate, 126 gm dolomite lime, and 42 gm calcium carbonate lime). When the plants had reached 30–50 cm in height, the mature, non-senescent leaves were suitable for protoplast production. For a single experiment, five to six leaves with distinctive white segments were harvested and placed in a plastic container on moistened paper towels until they could be used (less than 30 minutes). These leaves were generally 7–12 cm wide and 10–15 cm in length.

The leaves were carefully studied in transmitted light, and segments showing even faint green patchiness were marked with indelible ink so that they would not be used. The following procedures were carried out in a laminar-flow hood. Leaves were dipped for a few seconds in 70% ethanol and then submerged in diluted commercial bleach (final concentration, 0.6% sodium hypochlorite) in a sterile 1500-ml beaker. After exposure to the hypochlorite solution for 20 minutes, the bleach was decanted and replaced with at least five 500-ml washes of sterile distilled water. The last wash was decanted.

The leaves were removed one by one and placed on a sterile glass plate. The white segments were sliced with a tool made from 10 double-edged, stainless-steel razor blades mounted on two brass bolts and separated from one another by pairs of 1-mm-thick brass washers. The strips produced in this manner were cut away from the leaf using a single-edged razor blade and floated with the lower epidermis down on 15 ml of 9% mannitol dissolved in CPW salts (Table I) on a 10-cm glass petri plate. After 1 hour in this solution, the strips were transferred to a solution of 1% Cellulysin (Lot 502106) and 0.1% Macerase (Lot 501661) in 9% mannitol–CPW salts. This enzyme incubation was for 16 hours in the dark at 28°C. The pH of the enzyme solution was adjusted to 5.5.

During the enzyme incubation almost all the mesophyll cells were released as protoplasts. Five milliliters of the enzyme solution was transferred to each of three 10-ml screw-cap centrifuge tubes. The petri plate used for enzyme incubation was washed with 5 ml of 9% mannitol in CPW salts, and this solution was placed in a fourth centrifuge tube. Five milliliters of 16% sucrose dissolved in CPW salts was added to each of the tubes, and they were centrifuged at 20 g for 5 minutes in a swing-bucket head on a clinical centrifuge. The protoplasts floated to the top. The fluid and pellet beneath the protoplasts were removed using a 9-inch pasteur pipet. The protoplasts from all the tubes were pooled and mixed with 6 ml of 16% sucrose in CPW salts. After centrifugation at 20 g for 3 minutes, this solution was removed from beneath the floating layer of protoplasts. Another 6-ml portion of 16% sucrose in CPW salts was added to the protoplasts. The mixture was centrifuged at 20 g for 3 minutes, and the sucrose solution was removed from beneath the floating protoplasts. Following this second wash with 16% sucrose in CPW salts, the protoplasts were mixed with 6 ml of F5.9 (without agar, Table I) and centrifuged (20 g, 3 minutes). When this wash was removed from beneath the protoplasts, they were ready to be plated in F5.9 (with agar) at a density of about 10^5 protoplasts/ml.

These techniques have been used to isolate large populations of white tobacco leaf protoplasts. The isolated protoplasts are mitotically active, with 40–60% dividing during the first 10 days in culture. Colonies of 40 to 60 cells were formed after 3 weeks in culture. These procedures were modified from those of Banks and Evans (1976).

III. Discussion

The techniques used for the isolation of higher-plant protoplasts require plasmolysis of the tissues from which the protoplasts will be isolated. In both mechanical and enzymic protoplast isolation systems, this plasmolysis provides stability to the otherwise fragile protoplasts (Fig. 2). Extensive plasmolysis (0.6–1.0 M) is required in mechanical isolation systems to provide space between the cell wall and the protoplast for passage of the cutting tool (Tribe, 1955; Prat and Roland, 1970). In enzymic isolation systems, lower levels of plasmolysis (0.3–0.6 M) are usually effective. Osmotic concentrations nearer 0.6 M are required in tissues with limited intracellular spaces, apparently to provide a pathway for diffusion of the wall-degrading

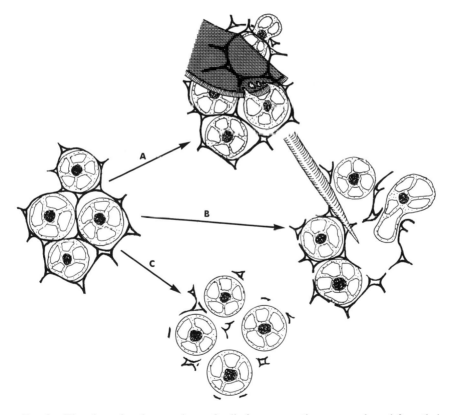

FIG. 2. The plasmolyzed protoplasts of cells from many tissues are released from their cell walls by slicing or chopping (A), or by teasing the tissues after partial digestion of the cell walls (B). The plasmolyzed protoplasts of cells of suspension cultures and of leaf mesophyll are released after wall digestion with little or no mechanical wall disruption (C).

enzymes into the tissue. These tissues also require greater physical stress to effect the release of protoplasts. In the pea root system, insufficient plasmolysis prevented even the softening of cell walls, although the concentration of wall-degrading enzymes was above that necessary for protoplast release at higher osmotic levels.

The most commonly used osmotic agents are mannitol and sorbitol, employed separately or in combination. Metabolically active sugars (Cocking, 1960; Prat and Roland, 1970) and salts (Ruesink, 1973) are occasionally used for their osmotic effects. The choice of osmoticum may have little effect on the actual isolation of the protoplasts. It can, however, have a marked effect on their survival or activity following isolation (Landgren, 1976a). Plasmolyzing agents may be dissolved in distilled water, solutions of nutritive salts, or complex culture media. The components of the medium used for plasmolysis may, however, have some as yet undetermined effect on the survival of the isolated protoplasts. This survival may be increased by preplasmolysis of the tissues to be used for protoplast isolation in less complex media containing neither wall-degrading enzymes nor plant growth hormones (Grambow et al., 1972; Constabel, 1975). This could eliminate osmotically induced pinocytotic uptake of these substances.

There is little limitation on the type of tissue which can be used for protoplast production (see, for example, Otsuki and Takebe, 1969; Schenk and Hildebrandt, 1969; also reviews). Occasionally the number of healthy protoplasts which can be isolated from a tissue is reduced by the effects of cell walls which resist degradation (Taiz and Jones, 1971; Wallin et al., 1977), by excessively fragile plasmalemma (Cocking, 1962), or by excessive cytoplasmic starch (A. J. Crowder and C. R. Landgren, unpublished). In the event of such restrictions, protoplast yields can often be increased by manipulation of the isolation procedures.

The enzymes used for protoplast isolation have been discussed in some detail (Ruesink, 1971; Cocking, 1972; Eriksson, 1977). It should be noted that Cellulase "onozuka" and Macerozyme (All Japan Biochemical Co., Nishinomiya, Japan) are also marketed as Cellulysin and Macerase (Calbiochem, LaJolla, Calif.).

The importance of removing the enzymes from the medium used for protoplast wall regeneration and division has been demonstrated in the pea root system (Landgren, 1976a). In the tobacco system, preliminary results indicate that concentrations of Cellulysin as low as 0.005% cause a 70–80% inhibition of mitotic activity. In systems which require considerable physical force to effect protoplast release, the enzymes are most easily removed prior to the actual protoplast isolation as in the pea root system described.

When protoplasts are released directly into the enzyme incubation medium, they must be isolated from the enzymes by flotation or sedimentation. Buoyant protoplasts can be floated on culture medium, solutions supplemented with sucrose, or media containing agents such as Ficoll (Eriksson and Jonasson, 1969; Gosch et al., 1975). This floatation may occur under natural gravitational forces (Eriksson and Jonasson, 1969) or under the influence of centrifugation.

Protoplasts which are more dense than their medium, or which cannot be made to float on sucrose or Ficoll, can be cleaned by sedimentation. Forces from 50 g (Keller and Melchers, 1973) to 100 g (Cocking and Evans, 1973; Constabel et al., 1973) are commonly used, although we have found that green tobacco protoplasts form a pellet after 3 minutes at 20 g. Longer centrifugation times and higher speeds often result in displacement of the cytoplasm within the protoplast. Starchy protoplasts seem especially susceptible to disruption of their transvacuolar cytoplasmic strands during these cleaning procedures, but the added density of the starch often allows isolation of these protoplasts at the lowest centrifugal forces.

One of the most gentle procedures used to remove enzymes from protoplast suspensions involves floatation of the protoplasts above a filter (8- to 10-μm pore size) (Eriksson and Jonasson, 1969; Kao et al., 1971). Washing medium can be added to the suspended protoplasts as the enzyme-contaminated medium flows through the filter. Because of the mixing that occurs as a result of adding the washing solution, this system works regardless of the density of the protoplasts.

IV. Conclusions

The number of different protoplast isolation systems indicates that a pragmatic approach is essential to the isolation of viable populations of protoplasts from any given tissue. The healthiest protoplasts will be produced in systems which have appropriate osmotic conditions, relatively short enzyme incubation times at the lowest effective level of wall-degrading enzymes, and the least damaging but most effective techniques to remove exogenous enzymes from the medium in which the protoplasts are cultured.

ACKNOWLEDGMENTS

I would like to express my appreciation to Dr. H. T. Bonnett for his informative comments on the manuscript for this chapter, and to Dr. J. G. Torrey in whose laboratory I first began my work on protoplasts.

REFERENCES

Banks, M. S., and Evans, P. K. (1976). Plant Sci. Lett. 7, 409–416.
Bottino, P. J. (1975). Radiat. Bot. 15, 1–16.

Carlson, P. S. (1973). *Proc. Natl. Acad. Sci. U.S.A.* **70**, 598–602.

Chaleff, R. S., and Carlson, P. S. (1974). *Annu. Rev. Genet.* **8**, 267–278.

Cocking, E. C. (1960). *Nature (London)* **187**, 962–963.

Cocking, E. C. (1962). *Nature (London)* **193**, 998–999.

Cocking, E. C. (1972). *Annu. Rev. Plant Physiol.* **23**, 29–50.

Cocking, E. C., and Evans, P. K. (1973). *In* "Plant Tissue and Cell Culture" (H. E. Street, ed.), pp. 100–120. Univ. of California Press, Berkeley.

Constabel, F. (1975). *In* "Plant Tissue Culture Methods" (O. L. Gamborg and L. R. Wetter, eds.), pp. 11–21. Natl. Res. Counc. Can. Saskatoon, Saskatchewan.

Constabel, F., Kirkpatrick, J. W., and Gamborg, O. L. (1973). *Can. J. Bot.* **51**, 2105–2106.

Eriksson, T. (1977). *In* "Proceedings in Life Sciences" (E. Bartz, ed.). Springer-Verlag, Berlin and New York (in press).

Eriksson, T., and Jonasson, K. (1969). *Planta* **89**, 85–89.

Eriksson, T., Bonnett, H., Glimelius, K., and Wallin, A. (1974). *In* "Tissue Culture and Plant Science" (H. E. Street, ed.), pp. 213–231. Academic Press, New York.

Gamborg, O. L., and Miller, R. A. (1973). *Can. J. Bot.* **51**, 1795–1799.

Gosch, G., Bajaj, Y. P. S., and Reinert, J. (1975). *Protoplasma* **85**, 327–336.

Grambow, H. J., Kao, K. N., Miller, R. A., and Gamborg, O. L. (1972). *Planta* **103**, 348–355.

Gregory, D. W., and Cocking, E. C. (1965). *J. Cell Biol.* **24**, 143–146.

Holl, F. B. (1975). *Can. J. Genet. Cytol.* **17**, 517–524.

Kao, K. N., Keller, W. A., and Miller, R. A. (1970). *Exp. Cell Res.* **62**, 338–340.

Kao, K. N., Gamborg, O. L., Keller, W. A., and Miller, R. A. (1971). *Nature (London), New Biol.* **232**, 124.

Keller, W. A., and Melchers, G. (1973). *Z. Naturforsch., Teil C* **28**, 737–741.

Landgren, C. R. (1976a). *Protoplasma* **87**, 49–69.

Landgren, C. R. (1976b). *Am. J. Bot.* **63**, 473–480.

Libbenga, K. R., and Torrey, J. G. (1973). *Am. J. Bot.* **60**, 293–299.

Nickell, L. G. (1973). *Hawaii, Plant. Rec.* **58**, 293–314.

Nickell, L. G., and Torrey, J. G. (1969). *Science* **166**, 1068–1070.

Otsuki, Y., and Takebe, I. (1969). *Plant Cell Physiol.* **10**, 917–921.

Power, J. B., and Cocking, E. C. (1970). *J. Exp. Bot.* **21**, 64–70.

Prat, R. (1973). *J. Microsc. (Paris)* **18**, 65–86.

Prat, R., and Roland, J. C. (1970). *C. R. Hebd. Seances Acad. Sci., Ser. D* **271**, 1862–1865.

Ruesink, A. W. (1971). *In* "Methods in Enzymology" (A. San Pietro, ed.), Vol. 23, pp. 197–209. Academic Press, New York.

Ruesink, A. W. (1973). *Colloq. Int. C.N.R.S.* **212**, 41–49.

Schenk, R. U., and Hildebrandt, A. C. (1969). *Crop Sci.* **9**, 629–631.

Taiz, L., and Jones, R. L. (1971). *Planta* **101**, 95–100.

Takebe, I. (1975). *Annu. Rev. Phytopathol.* **13**, 105–125.

Takebe, I., Otsuki, Y., and Aoki, S. (1968). *Plant Cell Physiol.* **9**, 115–124.

Tempé, J. (ed.) (1973). *Colloq. Int. C.N.R.S.* **212**, 1–549.

Tribe, H. T. (1955). *Ann. Bot. (London)* **19**, 351–368.

von Klercker, J. (1892). *Oefvers. Vetenskapsakad. Foerh., Stockholm* **49**, 463–475.

Wallin, A., Glimelius, K., and Eriksson, T. (1977). *Physiol. Plant.* **40**, 307–311.

Chapter 15

The Cultivation of Animal Cells and Production of Viruses in Serum-Free Systems

LEONARD KEAY

Center for Basic Cancer Research,
Washington University School of Medicine,
St. Louis, Missouri

I. Introduction—The Need for Serum-Free Systems

The perfect tissue culture medium will have the following properties. It should be totally chemically defined and free of serum or any other undefined biological material such as peptones. It should in addition not introduce any contamination such as viruses or mycoplasmas, which is another reason for it to be serum-free and also possibly autoclavable, since this removes any microorganism contamination. In addition, the medium should be commercially available, preferably as a dry powder formulation at low cost. Cells should grow in it without any period of adaptation, and the medium should support the growth of many cell types. The cells should grow to confluency or high cell density in suspension without medium change or replenishment and without any period of adaptation or selection of particular cells. There should be no changes in any of the properties of the cells such as morphology, karyotype, virus susceptibility, cell surface antigens, etc., or at least if any such changes do occur, they should be reversible so that a new cell type is not selected.

How close are we to this perfect medium for cell growth? Unfortunately, not too close at this time. The development of chemically defined media for cell growth was one of the initial investigations as the field of cell and tissue culture developed and is still a subject of considerable concern.

Basically, there are three reasons for the search for chemically defined media which support cell growth: the general desirability of a totally defined growth environment for research investigations and as an aid in the reproducibility of results; economic considerations, especially for large-scale animal cell and virus production, e.g., vaccine production; and finally as a solution to the problem of mycoplasmal and viral contamination.

As pointed out by Higuchi (1973) in his extensive review, the earliest attempt to develop a chemically defined media was made by White (1946) who used a combination of salts, glucose, amino acids, vitamins, and hormones to maintain chick embryo cells in culture, although little proliferation was observed. Morgan *et al.* (1950) developed the still widely used medium 199 containing 68 defined compounds which supported the survival of chick embryo cells but not their proliferation. Most of the earlier work on cell culture media was carried out with tissue explants and primary cells, and the media developed did not support their growth unless supplemented with serum.

Subsequently, much work has been carried out successfully with established cell lines such as the L cell and the HeLa cell, a fact which has been criticized by Ham (1974b) on the basis that these cells are so far removed from the normal diploid cell that there is a large difference in nutritional requirements.

Most earlier studies on cell or tissue cultures used media based on biological fluids such as serum, plasma clots, or ultrafiltrates of embryo extracts. Eagle carried out the most extensive nutritional studies on the newly established mouse L-cell and human HeLa cell lines to find the compounds required as a supplement to dialyzed serum for growth of the cells. The results showed that certain amino acids (the 13 essential amino acids), glucose, certain vitamins, and inorganic ions were required for cell growth, and these components made up the original minimum essential medium (MEM) (Eagle, 1955). Various other studies were carried out leading to formulations of other chemically defined media for the growth of several cell lines (Healy *et al.*, 1954; Swim and Parker, 1958), but all required the presence of dialyzed serum. Therefore the main problem has been to determine the role(s) of serum and replace it with suitable substitutes.

The problems involving the use of serum as a supplement to tissue culture media are lack of chemical definition and variability, introduction of contamination, and cost (Taylor, 1974; Fedoroff et al., 1972). Although serum has been the subject of extensive study and analysis (Boone et al., 1972; Fox and Sanford, 1975), there is no doubt that much is unknown about trace constituents such as hormones, minerals, and organic components. There is a considerable variation among batches of the same type of serum in steroid hormone and insulin content (Esber et al., 1973).

Kniazeff found γ-globulins in 40% of the lots of fetal bovine serum examined, which could be the result of infection or vaccination of a supposed heifer (Kniazeff et al., 1967). This variation in composition and physical and biological properties is a problem, because it makes work irreproducible, makes it difficult to establish nutritional requirements, interferes with metabolic and cell surface receptor studies, etc. Studies on the cloning efficiency of human fibroblastoid cells and embryonic lung cells showed a variation in results according to the serum source (Milo et al., 1976). The results were due to variations in the serum levels of progesterone and 17-β-estradiol and their ratio to one other. The effect of various types of mammalian serum on the attachment of mouse mammary epithelial cells, as well as their effect on thymidine incorporation, has recently been examined (Feldman and Wong, 1977). The results suggest that the sera which provide the highest attachment efficiency are not the best stimulants of DNA synthesis and that an inverse relationship exists between cell attachment efficiency and thymidine incorporation for any given type of mammalian serum.

Evans and Anderson (1966) reported that cells from minced mouse embryos or mouse kidneys became neoplastic when grown in NCTC 135 supplemented with horse serum but not when supplementation was with fetal bovine serum. It was found that horse, fetal horse, calf, or bovine serum does not inhibit "spontaneous neoplastic conversion," whereas fetal bovine serum does (Evans et al., 1972). Bump and Reed (1977) suggest that this inhibition of spontaneous transformation may be due to high levels of protein-glutathione mixed disulfides not found in other sera. Lasfargues and Coutinho (1977) found that, when steroid hormones were removed from serum by treatment with activated charcoal, cell lines could be grown from solid tumors in media supplemented with the serum.

The main objection to the use of serum is that it is the major source of contamination of cell cultures. Serum, when collected, has been shown frequently to be contaminated with bacteria, mycoplasmas, viruses, and phages. Extensive bacteriological activity has been found in unfiltered calf serum collected for tissue culture use (Orr et al., 1975). Of 24 lots examined, only one was found to be free of bacteria, phages, and endotoxin. Bacillus and streptococcus were the predominant contaminants. Bacteriophages have been detected in various commercial sera (Merrill et al., 1972), and the Escherichia coli bacteriophage characterized (Geier et al., 1975). Bacteriophage have also been detected in vaccines and are similar in plaque type and morphology to those found in the serum used in cell production (Petricciani et al., 1973).

In addition to the bacteriophage, several bovine viruses have been detected in fetal bovine serum (Kniazeff et al., 1975; Molander et al., 1972). The viruses were present in over 30% of the 51 lots tested, and included bovine diarrhea virus, parainfluenza type-3-like virus, bovine herpes virus 1, bovine enterovirus type 4, and an unidentified cytopathogenic agent. Of the 51 lots, 20 had been pretested by the suppliers and were considered to be free of known viral contaminants. Five of these 20 lots contained endogenous bovine viruses. These bovine viruses may cause infections in other cell lines, as observed with a cultured hamster cell line which became persistently infected with bovine herpes virus 1 (infectious bovine rhinotracheitis virus) (Michalski and Hsiung, 1976).

Contamination of cell cultures by mycoplasmas is a persistent and disturbing problem (Hayflick, 1965; Barile, 1973; MacPherson, 1966; Stanbridge, 1971). Clearly, serum is the main source of mycoplasma contamination (Barile et al., 1971; Barile and Kern, 1971) and elimination of serum from cell culture systems would eliminate the major source of contamination. Mycoplasmas and viruses cannot be removed by filtration and, although it is possible to sterilize serum by irradiation, Matchett et al. (1977) found that nutritional, morphological, and karyological differences were observed when cells were grown on irradiated serum. Chemical sterilization of serum is also possible. Viruses in serum and media have been inactivated by treatment with 1 mM binary ethyleneimine at 37°C (Bahnemann, 1976). The growth-promoting capacity of bovine serum was not impaired, nor did the inactivant affect the antibody activity of guinea pig hyperimmune serum.

Paracetic acid treatment has been claimed to sterilize serum completely (Schweizer et al., 1972), and serum and trypsin which have been sterilized by treatment with β-propiolactone are commercially available (usually on special order). This serum supports cell growth sufficiently well that it is used for the large-scale production of veterinary vaccines.

The final reason for the elimination of serum is economic. Depending on the type and percentage of serum used, and the cost basis for the remainder of the media (powder or presterilized liquid, with or without labor cost considerations), the serum accounts for 50–90% of the cost of conventional serum-supplemented tissue culture media.

II. The Role of Serum

The serum added to otherwise chemically defined medium may play several roles, and these may vary from one cell type to another and also with the growth conditions used. The major roles suggested have been to supply hormones, vitamins, amino acids, trace elements, and other small molecules, and to supply macromolecules.

The role of macromolecules in serum appears to be twofold. The first involves cell attachment and spreading on surfaces. Studies have shown serum proteins to play a role in cell attachment, and an α_1-globulin, fetuin, which is especially high in fetal bovine serum, may be responsible. This role has been disputed, but since cell attachment is highly dependent on the type of cell involved, it is difficult to relate one study to another.

Culp and Buniel (1976) have shown that serum plays a role in the time, extent and stability of cell attachment, and the pseudopodial spread of cells over the substrate surface. These factors in serum are very trypsin-sensitive. Analyses of the substrate-bound serum fractions and substrate-attached material deposited by the cells suggests that cells do not attach directly to the substrate but to specific classes of substrate-adsorbed serum proteins via deposition of specific classes of cell surface proteins or polysaccharides. Rajaraman et al. (1977) found that SV40-transformed W1-38 cells adhered to glass in the absence of serum but were able to spread only in the presence of serum, whereas untransformed W1-38 cells adhered and spread independently of serum proteins. However, in serum-free media, cells detach more readily from surfaces and, as in certain serum-containing systems, pretreatment of the surfaces with basic proteins or polymers aids in cell attachment (McKeehan and Ham, 1976; Keay, 1975b).

The other role of serum macromolecules appears to involve a protective effect and is probably more significant when cells are grown in agitated suspension systems. This protective effect was first reported by Bryant et al. (1961), who observed that mouse L cells adapted to grow in a serum-free medium rapidly lost viability and disintegrated when grown in an agitated suspension unless a macromolecule such as methyl-cellulose was added (see also Bryant, 1966). Kuchler et al. (1960) observed that the addition of serum or methylcellulose was necessary to obtain a single-cell suspension culture of L cells in TC199 plus 0.5% peptone. The addition of polyvinylpyrrolidone or carboxymethyl cellulose improved the growth of three lymphoblastoid cell lines in RPMI 1640 medium when the serum level was reduced to 2% (Mizrahi and Moore, 1970).

The role of serum in supplying the hormones necessary for cell growth is less clear. Lieberman and Ove (1959) demonstrated that 0.04 IU/ml insulin produced a growth-promoting effect close to that of 10% serum with HeLa and appendix cells. Insulin stimulated glucose utilization and uridine incorporation when added at physiological levels (0.1 mU/ml) to human fetal fibroblasts maintained in a serum-free system (Fujimoto and Williams, 1977), but leucine incorporation was stimulated only at higher insulin levels.

Blaker et al. (1971) and Nagle et al. (1963) added insulin for the growth of HeLa cells, and insulin has been added to a wide range of serum-free media for the growth of several different cell types (Higuchi and Robinson, 1973; Keay, 1976; Lasfargues et al., 1973). Temin (1967) observed that insulin can replace the factors in calf serum which are limiting for the growth of chick embryo cells and observed that in insulin-containing media cells transformed with avian sarcoma virus multiply faster than untransformed cells. Dulak and Temin (1973) isolated a polypeptide from serum which supported the growth of chick embryo fibroblasts as indicated by increased uptake of radioactive thymidine, and Pierson and Temin (1972) isolated the same peptide from medium conditioned by the growth of buffalo rat liver cells.

These multiplication-stimulating (MSA) polypeptides stimulate DNA synthesis and growth, are acid-soluble, are heat-stable, and have molecular weights of about 10,000. They share these properties with somatomedin (Van Wyck et al. 1974) and an acid- and ethanol-soluble, nonsuppressible, insulin-like activity (NSILA) found in human serum (Smith and Temin, 1974). It has been shown, however, that in chick embryo fibroblasts MSA receptors differ from insulin receptors (Rechler et al., 1976).

Epidermal growth factor (EGF) was first isolated from mouse submaxillary glands (Cohen, 1962) and has been immunologically detected in mouse serum (Byyny et al., 1974); a similar factor has been found in human urine (Cohen and Carpenter, 1975). It has been shown to be mitogenic in several cells, but cannot by itself substitute for serum in cell culture, although it can be used to supplement low concentrations of serum (Cohen and Carpenter, 1975) and in some cases replace serum when supplemented with hormones and other growth factors (Westermark, 1976).

A fibroblast growth factor (FGF) has been isolated from bovine pituitary glands (Gospodarowicz, 1975) and has a molecular weight of 13,000. Ovarian growth factor was isolated from the same source (Gospodarowicz et al., 1976). FGF has been shown to be mitogenic for several cell lines, but by itself cannot replace the serum requirement of the cells, although the addition of glucocorticoids as well as FGF stimulates DNA synthesis in 3T3 cells to the level observed with serum (Gospodarowicz and Moran, 1974). A small peptide, glycylhistidyllysine was isolated from fresh human serum (Pickart and Thaler, 1973) and shown to increase survival time, cell growth, and DNA synthesis in both normal and transformed cells. The peptide was synthesized and shown to have the same activity as the natural material (Pickart et al., 1973). These serum growth factors are all active at nanogram per milliliter concentrations.

In addition to these low-molecular-weight factors, serum has also been shown to contain a high-molecular-weight (120,000) mitogen for human diploid fibroblasts (Houck and Cheng, 1973).

Another important and probably related phenomenon is the conditioning of media by the cells and density-dependent cell growth. It was realized that the growth of cells at low cell densities, especially growth from an isolated single cell (cloning), represents a particular problem (Ham, 1974a). Generally, media and conditions which are satisfactory at high initial cell densities do not suffice at low cell densities, and special media are needed (Ham, 1963). As observed by Eagle and Piez (1962), although cells may be able to synthesize certain metabolites such as serine, if the cell density is reduced, exogenous serine must be supplied for cell survival.

This problem seems to be more critical for serum-free media. It was observed that an L-cell derivative (NCTC2071) did not proliferate or even survive when the cell density was below 12,000 cells/ml, but above this inoculum level the cells proliferated. The optimum inoculum was 50,000 to 140,000 cells/ml, although the maximum cell population was independent of the cell inoculum (Fioramonte et al., 1958). The role of serum in supplying vitamins, minerals, and other small molecules is less well documented.

Although analysis has shown that a wide variety of small molecules is present in serum and that addition of these molecules to serum-free media increases cell growth to varying extents, it is not clear how important this is in the requirement for serum for cell growth. Although HeLa cells can grow in Eagle's MEM supplemented with serum, it has been shown that, when serum is replaced in chemically defined or other media, a requirement for both biotin and vitamin B_{12} is introduced (Higuchi, 1969; Keay, 1976; Higuchi and Robinson, 1973).

Similar stimulatory effects of vitamin B_{12} and biotin have been observed with mouse L cells (Evans et al., 1956a,b; Waymouth, 1959; Higuchi, 1969). This requirement for biotin and vitamin B_{12} is probably the reason that HeLa and KB cells grow in modified MEM-α (Stewart; 1973; Stammers et al., 1971), but not in MEM, when supplemented with Bacto-peptone (Keay, 1976). Since L-929 cells grow in MEM supplemented with Bacto-peptone (Keay, 1975a), it is possible that Bacto-peptone contains low levels of essential vitamins, that they increase cell growth but are not essential for it (Waymouth, 1965), or that the requirement is satisfied by some other nutrient present in ill-defined additives such as serum or peptones. The role of other vitamins is less clearly defined.

Fatty acids have been demonstrated to be essential for certain cells. Ham (1963) showed that linoleic acid was required for clonal growth of cells in serum-free and albumin-free media, and similar effects have been observed by Harary *et al.* (1967). More recently, Jenkin and Anderson (1970) found that low levels (10–20 μg/ml) of oleic acid stimulated growth of LLC-MK$_2$ monkey kidney cells in an albumin-containing serum-free system, although the fatty acid became growth inhibitory at higher levels. Jenkin *et al.* (1970) observed that the stimulatory effect of mono- and diunsaturated fatty acids was dependent on the position of the double bonds. Higuchi and Robinson (1973) found that a combination of cholesterol, oleic acid, and lecithin was required for the growth of some cells in a chemically defined medium.

It has been observed that chemically defined media capable of supporting either clonal or bulk cell growth, developed 10 or more years ago, are frequently found to be less satisfactory at the present time. This is probably because present-day biochemical reagents are of a much higher degree of purity than they were in the past, and therefore additional nutritional requirements are now being discovered.

Although the major inorganic ion requirements have been known for a long time, requirements for low levels of such elements as zinc, iron for L cells (Thomas and Johnson, 1967; Higuchi, 1970), selenium for W1-38 and Chinese hamster ovary (CHO) cells (McKeehan *et al.*, 1976) have recently been demonstrated. Manganese, molydenum, and vanadium have been shown to improve clonal growth of W1-38 cells at very low concentrations, 10^{-8}–10^{-9} M (McKeehan *et al.*, 1977). Nickel, silicon, and tin have been implicated as essential trace elements in animal nutrition (Underwood, 1971) but have not yet been shown to be required by animal cells in culture. The role of other low-molecular-weight nutrients is less well defined, and the list of additives to media used with and without serum is long and the need for some additives not substantiated.

One additive found to be present at low levels in serum is putrescine, and it was necessary for clonal growth of CHO cells in a medium containing albumin but not fetuin (Ham, 1965). It was found to be secreted into the medium by human fibroblasts (Pohjampelto and Raina, 1972), and the addition of putrescine-stimulated cell growth.

Other additives have been used in place of serum to promote cell attachment. Lieberman and Ove (1959) and Higuchi (1963) found that salmine (protamine sulfate) promoted cell attachment and cell growth, respectively, but its role is still not clear. Polylysine, protamine, polyornithine, and histone have been shown to have a pronounced effect on the clonal growth of W1-38 cells grown in a chemically defined medium, MCDB 102, supplemented with limited amounts of serum protein (McKeehan and Ham, 1976).

Another role of serum is to protect the cells from the action of trypsin or to assist them in recovering from trypsinization. Pumper *et al.* (1971) showed that RH-PD cells, which normally attach firmly to glass, did not do so after treatment with 0.01 μg/ml crystalline trypsin and that this effect persisted until the third passage. However, when the cells were treated with 1 part in 500 of calf serum, they readily attached. At low levels (up to 0.1 μg/ml) trypsin actually stimulated cell growth, but at higher levels it became inhibitory. The protective effect of serum albumin on cells during trypsinization has also been observed by Tomei and Bertram (1977).

III. The Growth of Animal Cells in Serum-Free Systems

It is not surprising that, since the role of serum has been postulated to be that of a supplier of hormones, trace minerals, protective macromolecules, and organic nutrients such as keto acids, purines, pyrimidines, vitamins, etc., attempts to replace serum have been based upon substitutes for the macromolecules and addition of the other components as chemically defined compounds. Waymouth (1974) stressed that the culture medium should be designed to enhance the genotypic and phenotypic stability of the cells. The cultivation of cells in chemically defined and serum-free systems has been reviewed by Higuchi (1973) and by Katsuka and Takaoka (1973), and work preceding their reviews is not discussed in detail here.

There is no simple way of classifying serum-free media. Attempts to classify them on the basis of cell type or monolayer verus suspension culture are not specific enough, and a classification based on the approaches of addition of hormones (Hayashi and Sato, 1976), addition of trace minerals (McKeehan et al., 1977), or low-cost undefined serum substitutes (Keay, 1975a) are invalid, since most formulations are a combination of these classifications and often involve highly elevated amino acid concentrations, etc., as well.

In this report, the serum-free systems are divided into three basic types:

1. Totally chemically defined, that is, containing only compounds of known molecular structure and configuration and supposedly free of contaminating trace minerals, vitamins, hormones, etc.

2. Chemically defined systems containing additives such as serum albumin, salmine (protamine sulfate), etc., which are fairly well characterized but cannot be guaranteed free of these trace components.

3. Media containing low-cost undefined additives from biological sources such as peptones, lactalbumin hydrolysate, and yeast extract.

Even these divisions are not totally satisfactory, for such additives as methylcellulose and polyvinylpyrrolidone (PVP) are not absolutely defined in terms of molecular weight, polymorphism, solubility, etc.

A. Completely Chemically Defined Media

The early chemically defined media were those of Morgan *et al.* (1950), TC 199, Healy *et al.* (1954, 1955), no. 703 and no. 858 (which with slight modification became CMRL 1066), and Evans *et al.* (1956a,b), NCTC 109, which with deletion of the cysteine became NCTC 135. NCTC 109 and NCTC 135 were used for growing several cell lines (Anderson *et al.*, 1967; Evans *et al.*, 1964). C3H mouse embryo cells were isolated by mechanical disintegration and were never exposed to protein, enzymes, or hormones. After the elimination of epithelial cells, a fibroblastic strain was developed which grew in less complex media as long as vitamin B_{12} and inositol were present, and was tumorigenic (Anderson *et al.*, 1967). HeLa, mouse liver, Chinese hamster, and rat parotid cells were all adapted to grow in NCTC 109 by halving the serum level as soon as the cells appeared to be adapted to the new level (Evans *et al.*, 1964). NCTC 109 contained 68 components, but Sanford *et al.* (1963) showed that many of them were not necessary for the growth of L cells.

Price *et al.* (1966) used NCTC 135 medium to carry out nutritional studies on C3H mouse liver, Chinese hamster, HeLa, and BS-C-1 cells. A new, less complex formulation, NCTC 163, has been developed, and cells previously adapted to NCTC 135 without serum supplementation show greater proliferation in this new medium (Price and Evans, 1977). Merchant and Hellman (1962) grew L-M cells in MEM with double the usual strength of amino acids, vitamins, and glutamine in both monolayer and suspension culture (with added methylcellulose). The monolayers were subcultured after harvesting by scraping with a rubber policeman. The L-M cells had previously been adapted to grow in medium 199 supplemented with 0.5% Bacto-peptone and, when transferred to 2X MEM, the generation time increased from 40 to 56 hours (both significantly higher than in serum-containing systems) but there was little change in morphology. Methylcellulose was found to have a protective effect in suspension culture systems as well as in the harvesting procedure.

Stoner and Merchant (1972) examined the amino acid requirements of strain L-M mouse cells grown in 2X MEM and observed that the glutamine was exhausted after 120 hours without a medium change, while the nonessential amino acids were released into the medium during growth and some were reutilized. Glutamine appeared to be the growth-limiting factor but could be largely but not entirely replaced by glutamic acid. Waymouth (1956) developed a serum-free medium containing crystalline bovine albumin and Bacto-peptone (ML 192/2), which supported the growth of L cells, and later (Waymouth, 1959) modified the formulation to produce a totally chemically defined 40-component system (MB 752/1).

Rat glioma cells have been adapted to grow in unsupplemented F-10 medium by progressive reduction in the levels of serum (Fan and Uzman, 1977). In the absence of serum, the cells showed a morphological differentiation (dendrite-like processes), but the karyotcype remained unchanged, and the rate of production of S-100 protein was only slightly above that in the presence of serum. Polyoma transformed hamster cells and SV40 transformed mouse cells have been cultivated in serum-free Waymouth's medium (Bush and Shodell, 1977). The cells grow more slowly in the absence of serum, and the rate of DNA synthesis is reduced.

Nagle et al. (1963) developed a chemically defined medium containing high concentrations of amino acids, vitamins, glucose, pyruvate, salts, insulin, and methylcellulose. The medium was sterilized by filtration (methylcellulose was sterilized by autoclaving) and supported the growth in suspension of L, HeLa, monkey kidney, rat spleen, guinea pig kidney, and cat kidney cells for several months. Nagle (1968) modified this medium to make it autoclavable. The methylcellulose, glutamine (dry powder), and sodium bicarbonate solution were autoclaved separately and added after cooling. The medium supported the growth in suspension culture of L, HeLa, and cat kidney cells. The medium was further modified by increasing the concentration of choline (Nagle, 1969) Then the glutamine was removed entirely from the medium (Nagle and Brown, 1971) and L-cell growth reported. This medium was used by Tomei and Issel (1975) for the growth of baby hamster kidney (BHK) cells for foot-and-mouth disease virus (FMDV) production (see Section V).

Parsa et al. (1970) designed a chemically defined medium which allowed rat pancrease rudiments to differentiate as well as in serum-containing systems, but these were only short-term (9-day) cultures. Birch and Pirt (1970) developed a defined medium containing salts, vitamins, glucose, keto acids, and high levels of amino acids, which supported the growth of L cells up to a concentration of 3.3×10^6 cells/ml. Blaker et al. (1971) used essentially the same medium and found that, provided insulin was added, good growth of HeLa cells was obtained.

Higuchi (1970) developed a chemically defined medium containing amino acids, vitamins, keto acids, salts, glucose, and methylcellulose, which supported the growth of L cells up to 3×10^7 cells/ml when the medium was changed daily. This medium was modified to include hormones (hydrocortisone, thyroxine, and insulin) as well as oleic acid and protamine sulfate (chemically defined?) (Higuchi and Robinson, 1973).

Parsa and Flancbaum (1975) grew embryronic rat liver explants in a chemically defined medium. By the end of the second week the liver cells showed signs of differentiation and were maintained for 16 weeks. Fischer et al. (1977) grew newborn mouse epidermal cells in modified Waymouth's

752/1 medium supplemented with insulin and hydrocortisone. The enriched medium allows the reduction and possible elimination of the serum requirement.

Jackson *et al.* (1970) grew ALB/N rat embryo cells in NCTC 135 for up to 38 months without their becoming neoplastic, whereas when serum was present they transformed into neoplastic cells. Without serum, the cells had a lower mitotic index and were particularly susceptible to damage when subcultured. Waymouth's 752/1 medium with various additives has been used to grow mouse alveologenic lung carcinoma cells (Franks *et al.*, 1976) and chick embryo liver cells (Goodridge and Adelman, 1976). Chick embryo liver cells have also been grown in F-12 medium supplemented with insulin (Granick *et al.*, 1975).

Several media have been devised for the growth of various human cell lines without serum supplementation (Moore and Woods, 1977). RPMI 1603 was developed for human malignant cell lines, while RPMI 1634 stimulates the growth of normal human hematoietic cells. It is clear that at high cell densities a cell line maintains itself by utilizing protein released from dead cells, and therefore cell density is an important factor to be considered in evaluating serum-free media.

Gerschenson *et al.* (1972) adapted a rat liver cell line from growth in F-12 plus 10% fetal bovine serum to F-12 plus insulin. It was found that the insulin rapidly disappeared from the medium as a result of binding to the plastic surface unless albumin was added. There was still some loss of insulin, apparently due to proteolysis. The cells assumed a fibroblastic morphology but returned to an epithelial type when serum was readded to the medium.

Medium MAB 87/3 was introduced as a medium for the initiation of mouse cell cultures in the absence of serum but has proved useful for long-term culture of mouse fetal liver, postnatal mouse liver, arian liver, and mouse pancreatic cells (Waymouth, 1977). It differs from medium MD 705/1 in trace-metal composition and also includes serine, alanine, and asparagine as well as thymidine and insulin.

Takaoka and Katsuka (1971) developed a medium (DM-120) containing 19 amino acids, 10 vitamins, salts, and glucose. Using a technique called Nagisa culture, they obtained several substrains from rat liver parenchymal, rat thymus reticulum, cynomologous monkey kidney, and rat ascites hepatoma cells. The Nagisa culture involves the cultivation of cells without subculturing for a long time (months) in tubes with flattened surfaces. The tubes are incubated on a slant in a stationary culture, and mutant cell strains grow near the air–liquid interface. The substrains have been shown to have a changed karyotype and morphology (Katsuka and Takoaka, 1973). For example, substrain JTC-21-P3 has a modal chromosome number of 62, while the parent LTC-21 had a modal chromosome number of 69. L-929 had a modal chromosome number of 59, and the substrain L-P3 a modal chromosome number of 64.

Recently, Hayashi and Sato (1976) grew rat pituitary tumor (GH_3) cells in F-12 medium supplemented with five hormones [3×10^{-10} M, Triiodo thyronine (T_3), 1×10^{-9} M thyrotropin-releasing hormone (TRH), 50 mU/ml somatomedin, 5 μg/ml transferrin, and 0.5 ng/ml parathyroid hormone (PTH)]. All the hormones were necessary. BHK cells grew in F-12 supplemented with 0.5 μg/ml luteinizing hormone, 10 mU/ml follicle-stimulating hormone, 5 μU/ml ACTH, 5 ng/ml growth hormone, 10 ng/ml prolactin, 5 ng/ml thyroid-stimulating hormone, 10 ng/ml prostaglandin E_1, 5 μg/ml ceruloplasmin and transferrin, 0.2 μg/ml glycylhistidyllysyl acetate, 50 mU/ml somatomedin A, 50 ng/ml insulin and glucagon, 0.5 ng/ml PTH, 0.1 ng/ml calcitonin, 1×10^{-9} M Thyrotropin releasing factor (TRF), 10 ng/ml Somato tropin releasing inhibitory factor (SRIF) and luteinizing hormone-releasing hormone. An equally complex but different combination of hormones promoted the growth of HeLa cells. Sub-culturing was by trypsinization, and a trypsin inhibitor was added to inactivate residual trypsin and prevent cell damage. Hayashi et al. (1978) have shown that blood meal extract provided growth-stimulating factors for rat pituitary tumor cells (GH_3). The extract was found to contain three isoelectric focusing fractions which were active and could be replaced by fibroblast growth factor and somatomedin C. The hormonal requirements for GH_3, HeLa, a mouse melanoma, and M2R cells were compared. All of the cell lines require insulin and transferrin and also a hormone which localizes in the nucleus for each cell line.

Bauer et al. (1976) used a modification of the Higuchi and Robinson (1973) medium for the growth of mouse mammary tumor lines and the production of mouse mammary tumor virus (MMTV). The medium consisted of MEM to which a chemically defined serum substitute had been added. The mouse mammary tumor cells were grown to confluency in serum-containing medium which was then replaced with serum-free medium. The cells were refed every 2 days and subcultured weekly. Although growth slower in the serum-free system, the cell numbers were comparable to those obtained with serum over an 8-day period. Mouse mammary tumor lines Mm 5 mt/c_1, C_3H/4C-1, Balb/cfC_3H, GR, and RIII were all propagated in this serum-free system, and growth, protein production, and morphology were similar to that in serum-containing systems. (Virus production is discussed in Section V.) The C_3H/HC-1 cells could be split only in a 1:2 ratio, while the others could be split at 1:4 and after five or more transfers Mm5mt/c_1 cells could be split at 1:40. A chemically defined medium (ATCC-CRCM30) has been developed at the American Type Culture Collection which has without serum supplementation been able to support the continuous serial subcultivation of cell lines from nine different animal species for up to five years (Macy and Shannon, 1977). A protein-free synthetic medium MCDB301 has been developed for the clonal growth of CHO cell lines (Hamilton and Ham, 1977). The medium is essentially the F-12 formulation with increased levels of calcium chloride and glutamine, reduced level of cysteine, and the inclusion of twenty other inorganic ions. When supplemented with methylcellulose or insulin, MCDB301 also supports the growth of a chinese hamster lung cell line.

Leiter *et al.* (1974) have established monolayer cultures of endocrine pancreatic cells from both fetal and postnatal mice in a chemically defined medium, MPNL65/C. MPNL65/C is a medium of high osmolality containing inorganic salts, amino acids, vitamins, glucose, glutathione, pyruvate, hypoxanthine, thymidine, insulin, and dexamethasone. After the partial separation of fibroblastoid cells by their tendency to settle and attach in a conventional serum-containing medium, refeeding after 36 hours with MPNL65/C selectively allowed the epithelial cells to grow to form monolayers, the pancreatic β cells predominating, which were endocrinologically competent as shown by their ability to synthesize and secrete insulin. However, the levels of insulin secretion declined with time, and the monolayer disrupted between 3 and 4 weeks *in vitro*, so that a permanent line was not established.

Tomei and Bertram (1977) grew a transformed mouse fibroblast (derived from line C3H-10T 1/2 cells) in a chemically defined medium, FCRC-20. The medium has very high levels of amino acids (e.g., 1200 mg per liter each of leucine and lysine) and was supplemented just prior to use with insulin 0.1 IU/ml and methylcellulose. Initially, cells were removed from the surface by rolling sterile 2-mm glass beads across the monolayer. It was found, however, that the cells tended to adhere as large clumps, and growth was observed only at the periphery of the clumps. Treatment with trypsin gave rise to cells that were incapable of division unless protected by albumin prior to trypsinization. The cell phenotype was drastically altered, and a change in the pathology of the fibrosarcoma was observed with cells grown in chemically defined medium. The cells reverted to the transformed phenotype following the addition of albumin or serum, or exposure to trypsin.

Human lymphoblastoid cells, P3HR-1, have been grown in a chemically defined medium, FCRC-1 (Nagle and Brown, 1974). The FCRC-1 medium was basically that of Nagle and Brown (1971), but with increased levels of glutamine (900 mg per liter) and inositol (50 mg per liter). The cells were transferred from FCRC-1 plus 10% fetal bovine serum to unsupplemented FCRC-1. The cell number doubled in 7 days, and it was observed that some of the cells actually adhered to the glass. The cells could only be propagated for several growth cycles, and as yet no lymphoblastoid cell has been continuously propagated in a chemically defined medium. It was determined by fluorescent antibody techniques that the cells grown in FCRC-1 were producing Epstein-Barr virus.

B. Partially Chemically Defined Media

As indicated previously, the distinction between totally and partially chemically defined may be somewhat artificial or arbitrary. Serum fractions such as albumin have been the most frequently used serum substitute (Sanford et al., 1955. Waymouth, 1956). Since it is not certain whether or not serum fractions contain tightly bound vitamins, hormones, or trace minerals, they cannot be certified as chemically defined. The same argument applies to protamine sulfate, although is usually included at much lower concentrations (1–10 mg per liter) than albumin (500–2000 mg per liter). Since additives such as methylcellulose and PVP are not of animal origin, they cannot be considered sources of hormones or vitamins but can have contaminating trace metals.

Fisher et al. (1959) found that HeLa S-3 cells could be cloned in a synthetic medium provided that both an albumin fraction and fetuin were added. Ham (1962) used a similar synthetic medium, F-7, supplemented with 0.5 gm per liter human serum albumin and 1.0 gm per liter fetuin (F-7AF) to clone diploid CHO cells (CHBOC1D1), and the cells were grown in this medium for up to 6 months. Subsequently, F-10 medium was developed and as F-10AF was used to clone CHO and HeLa cells (Ham, 1963).

Medium F-12 was developed for cloning CHO and Chinese hamster lung cells in plastic dishes without any macromolecular additions (Ham, 1965). L cells were cloned in a modification of this, F-12M. These media were designed for cloning, i.e., very low cell populations, and frequent media changes are required if large populations are to be produced. Ham (1974a) has recently reviewed the unique requirements for clonal cell growth. Improved media and culture conditions for clonal growth of normal diploid cells has been reviewed recently by Ham and McKeehan (1978), and the multiplication stimulating functions of serum were classified operationally as "replaceable" (those that can be replaced by modifying the medium or the culture conditions) and "nonreplaceable" (those that we have not yet been able to replace). The nonreplaceable multiplication-promoting activity from fetal bovine serum for human diploid fibroblasts has been separated from fetuin and serum albumin and partially purified.

Holmes and Wolfe (1961) fractioned serum by curtain electrophoresis and added the fractions to a chemically defined media to test their effects on the growth of Chang liver cells. Three different fractions possessed activity: one associated with albumin and promoting survival, attachment, and flattening, as well as cell proliferation; a second associated with the α-globulins, promoting survival and proliferation; and a third fraction which caused cells to aggregate to form floating clumps.

Jenkin et al. (1970) found that certain isomeric cis-octadecenoic acids stimulated the growth of MK_2 (monkey kidney) cells when added to Waymouth's MB752/1 medium. It was found that oleic acid stimulated growth at low concentrations (up to 20 μg/ml) but at higher levels became inhibitory (Jenkin and Anderson, 1970). The further addition of bovine serum albumin (BSA) increased cell growth, and a medium was developed consisting of Waymouth's MB725/1 medium supplemented with fatty-acid-free BSA (2 mg/ml) and oleate (5 μg/ml) (Morrison and Jenkin, 1972). This medium was used to grow L cells in a suspension culture for the production of Chlamydia psittaci.

Initial attempts to grow BHK cells in this serum-free system by adaptation to growth in reduced amounts of serum and then transfer to Waymouth's–BSA medium supplemented with 20 μg/ml oleate were unsuccessful, and cell death occurred (Guskey and Jenkin, 1975). When the cells were adapted to grow in Waymouth's MB 752/1 medium, HEPES buffer, and 2 mg/ml fatty-acid-free BSA without added oleate and then transferred to the same medium containing 10 μg/ml oleate, there was no appreciable loss in cell viability. After a few transfers the growth stimulation by oleate disappeared.

Yamane et al. (1976) developed a medium comprised of MEM, BSA (usually 10 mg/ml), and 16 additional chemically defined components including insulin. Seven rat, mouse, and human tumor cell lines were grown in suspension culture in this system and, in the case of the Yoshida sarcoma cell, cloning was possible and primary cultures were serially transferred 59 times in 1 year in this serum-free system without any change in karyotype or tumorigenicity. It was found that the addition of basic polymers markedly stimulated cell proliferation (Murakami and Yamane, 1976), although when the albumin and fatty acids were omitted the basic polymers had little effect. The highly basic polymers poly-L-arginine and polyethyleneimine were the most effective polymers.

C. Media Containing Low-Cost Complex Additives

Waymouth (1956) found that a crude complex partial hydrolysate of animal protein (peptone) had a tremendous effect on L-cell growth in ML 192/2 medium. When peptone was omitted, growth was reduced by 90%, but when serum albumin was omitted, growth was only reduced by about 25%. Peptones had been used previously, but the need for both amino acids and peptones in combination had not been so thoroughly examined. Morgan's TC199 medium supplemented with 0.5% peptone had been used extensively to maintain animal cell lines, but little further interest was taken in it until the problem of contamination of cell cultures by viruses and mycoplasmas introduced by serum became apparent.

Other bacterial media ingredients had been examined for use in tissue culture media. Ginsberg et al. (1955) used tryptose phosphate broth as a supplement for the growth of HeLa cells in serum-containing medium and found that cell growth was improved in medium containing 25% tryptose phosphate broth. It was also found that virus replication was higher in trypose phosphate broth-supplemented systems, but no studies were carried out in serum-free systems. Mayyasi and Schuurmans (1956) grew L cells in several bacteriological media combined with horse serum. Various concentrations of yeast extract, proteose peptone, and horse serum were examined, but cell growth did not occur when the serum was omitted. This may be because the yeast extract and peptone were a poor source of amino acids. McCoy et al. (1959) used McCoy 5A medium (a chemically defined medium except for 0.5 gm Bacto-peptone) for the growth of Novikoff hepatoma cells, although serum addition was necessary.

Lactalbumin hydrolysate has been incorporated into various media. Neuman and Tytell (1960) produced a medium which was chemically defined except for lactalbumin hydrolyzate and used it for the growth of Walker carcinosarcoma 256, KB, and human heart cells. Higuchi (1963) developed a medium for monolayer culture which was chemically defined except for lactalbumin hydrolysate. This medium supported the growth of HeLa, human bone marrow, human embryonic intestine, L, Chang liver, bovine kidney, swine testes, chick embryo, rat spleen, kitten lung, monkey kidney, guinea pig kidney, and dog kidney cells. A requirement for insulin for HeLa cell growth was observed. Some of the cells grew satisfactorily immediately upon transfer to serum-free media, while others required extensive periods of adaptation. Trypsinization was found to be detrimental to the cells, as was extensive pipetting to harvest monolayer cultures, and it was found that the addition of 0.05% methylcellulose and the use of glass beads to remove cells (Nagle, 1960) gave excellent cell recoveries. Tribble and Higuchi (1963) modified the medium, omitting the monoolein, for the growth of cells in suspension. Several of the cells listed above and some others were grown in suspension, provided that methylcellulose and insulin were added to the medium.

Bershadsky et al. (1976) grew L cells in medium comprized of equal volumes of Eagle's basal medium and 0.5% lactalbumin hydrolysate. Growth in monolayer culture was about the same with or without added serum, but the addition of 10% fetal bovine serum increased the agglutinability by conconavalin A, increased uptake of deoxyglucose, inhibited the ability to metabolize benzpyrene, and increased colony formation in semi-solid media. The cells also became less spread on the substrate, and the density of microvilli on the surface increased. When serum was removed, these effects were reversed. In serum-free systems a partial phenotypic reversal of transformation is observed. Healy and Macmorine (1972) replaced serum with a soybean protein fraction which supported the growth of primary monkey kidney cells in CMRL 1969, a very complex chemically defined medium (Healy et al., 1971). The cells grown in the CMRL-1969–soy protein medium supported the propagation of polio viruses at levels close to those in serum-supplemented systems.

Pumper (1958) grew lung cells from newborn Swiss mice in TC 199 plus 10% horse serum and after 110 passages transferred them by mechanical scraping into serum-free medium, TC 199 plus 1% Bactro-peptone and 1 gm per liter glucose. When trypsin was used to subculture the cells in the serum-free system, no growth occurred. The same mouse lung cells in both media were examined for their ability to support vaccinia virus production (Pumper et al., 1960). The yield of virus measured 5 days after infection was the same in both media. After four passages in cells grown in serum-free medium the virus did not show reduced infectivity in rabbits.

As pointed out by Pumper et al. (1965), totally chemically defined media require extended periods of adaptation and are very complex, containing as many as 70 components. Rabbit heart cells were adapted from growth in TC 199 plus 10% calf serum to growth in TC 199 plus 0.5% autoclaved Bacto-peptone and 2000 mg per liter glucose. It was found that a dialysate of Bacto-peptone substituted quite well for the crude peptone as a growth stimulant and that the medium itself could be autoclaved and still support cell growth. The peptone dialysate contained various amino acids, but the amino acids did not support cell growth when added to the medium.

The use of peptones and peptone dialysate as serum substitutes has recently been reviewed, with full details provided for the preparation of the solutions as well as their effect on cell karyotypes (Taylor and Parshad, 1977).

Several commercially available peptones were examined for their ability to support the growth of two L-cell derivatives (NCTC 2071 and L-M) and rabbit heart cells (Taylor et al., 1972). Bacto-peptone and peptic peptone were most effective for rabbit heart cells, several peptones supported the growth of L-M cells fairly well, but none supported the growth of NCTC 2071 as well as the basal NCTC 135 medium.

An L-929 derivative NCTC 2071, an LLC-MK$_2$ derivative MK$_2$-BPD, and W1-38 cells were grown in a plastic vessel with a large surface area developed by House et al. (1972) and available commercially as the Dyna-

Cell (Taylor and Evans, 1975). The NCTC 2071 cells were grown in NCTC 109 or 135 medium, the MK_2-BPD cells in MEM–Bacto-peptone, and the W1-38 cells in a serum-supplemented medium. It was found that an adequate inoculum, monitoring of glucose concentration to assess the need for medium replacement, and harvesting of cells with warm pancreatin–EDTA solution were critical for successful cell production.

Holmes (1959) used a modified medium 858 (Healy *et al.*, 1955) to which was added 1 gm per liter Bacto-peptone to grow Chang liver cells. He found that autoclaving did not reduce the activity of the Bacto-peptone. The Bacto-peptone was separated into a series of fractions by curtain electrophoresis, and two peaks of growth-promoting activity were found.

Taylor *et al.* (1972) separated peptone dialysate on a Sephadex G-10 column and examined the peaks for growth-promoting activity. Two peaks, G-10(I) and G-10(II), accounted for 70 and 15% of the starting material, and these two peaks contained the active growth-promoting components. Amino acid analysis of two biologically active fractions isolated from peptone dialysate has shown them to be glycine-rich (Taylor *et al.*, 1974). The molecular weights were both less than 2000, and the glycine contents were about 30 and 10%, respectively, in the two fractions A series of glycine-containing peptides was examined for growth promotion of rabbit heart cells in NCTC 135, but no stimulation was observed unless peptone dialysate was added.

Hsueh and Moskowitz (1973a,b) recently isolated a glycine-rich peptide (MW 700) from whole peptone and reported that this component supported excellent growth of L cells, but the evidence that the growth factor is a peptide is based upon color tests on chromatograms. 6,8-Dihydroxypurine has been isolated from Bacto-peptone (Yamane and Murakami, 1973) and shown to be a stimulator for the growth of Chinese hamster don cells. A mixture of 6,8-dihydroxypurine, inosine, hypoxanthine, uridine, uracil, guanosine, and thymidine were almost as growth-promoting as Bacto-peptone itself. However, 6,8-dihydroxypurine neither supplants the peptone requirement of rabbit heart cells nor amplifies growth in medium containing a growth-limiting amount of peptone dialysate (Taylor and Parshad, 1977). Fisher and Welch (1957) found that the effect of peptone on the growth of L-5178 mouse leukemia cells was reduced by the addition of citravorum factor, indicating that the peptone functions as a source of folic acid or related compounds.

Lasfargues *et al.* (1973) devised a serum substitute which supported the growth of mammary tumor cells. The serum substitute SSK was a solution of 10% Bacto-peptone, 1% dextrose, 5% lactalbumin hydrolysate, and 3.5% PVP-K90. It was autoclaved at 10 psi for 20 minutes and used at a 10% level to supplement RPMI-1640, insulin also being added. Human and mouse mammary tumor cell lines have been shown to proliferate in this medium, but not normal cells which rapidly lost viability in the absence of serum. The formulation was modified (SSK7) as follows: 5% Bacto-peptone, 0.5% dextrose, 3.5% PVP-K90, 2.5% lactalbumin hydrolysate, and 0.5% Yeastolate. When this formulation was used to supplement RPMI-1640 (with insulin added), normal cells were maintained over a 2-month period.

The approach taken in this laboratory was to find an autoclavable medium based on ingredients available commercially, which could lead to a medium marketed in dry powder form. The starting point was the simplest medium available in dry powder form, MEM, and a low-cost autoclavable additive, Bacto-peptone. L cells were chosen as the first cell line to be studied. Two sublines were used, the L-929 clone of Eagle and L-60TM cells [a suspension culture subline derived from L-929 (Mak and Till, 1963) and generally referred to as L-S in this laboratory.]

In the first experiments (Keay, 1975a) several serum substitutes based on Bacto-peptone were examined as supplements to MEM. Bacto-peptone, Bacto-peptone plus PVP, Bacto-peptone plus PVP plus lactalbumin hydrolysate (BP-LH-PVP), and the last-mentioned mixture with added Yeastolate and glucose (SSK-7) all supported growth of L-929 cells in plastic flasks. The cell counts after 8 days were the same as when 10% fetal bovine serum was the supplement. In 8 days 7.5×10^5 cells multiplied to about 3×10^7 cells, a 40-fold increase. The growth of cells in the absence of serum was, however, different. As the cells multiplied, a rather nodular web of cells formed instead of a uniform monolayer.

Various peptones and protine hydrolysates were examined for their ability to support the growth of L-929 cells in MEM. It was found that some peptones supported growth as well as Bacto-peptone, but protein hydrolysates did not. In fact, when the hydrolysates were added together with the peptone, the growth-stimulating effect of the peptone was reduced. This is probably because the hydrolysates contain free amino acids which alter the balance of amino acids in the medium.

It was observed that the cells had a tendency to detach from the plastic surface and proliferate in floating clumps in stationary suspension. This has advantages in that cells can be subcultured by centrifugation and resuspension rather than trypsinization, and can be scaled up in suspension culture systems such as glass spinners. L-929 cells were then grown in glass spinner bottles in MEM supplemented with 0.5% Bacto-peptone. In stationary systems the addition of high-molecular-weight compounds, PVP, methyl-cellulose, etc., had little effect on cell growth, and the level of Bacto-peptone used (0.25–1.0%) had no effect on cell numbers either. The addition of insulin (0.45 IU/ml) had no effect on cell growth.

Results similar to those obtained with L-929 cells were obtained with

L-S cells. L-S cells previously grown in MEM plus 0.5% Bacto-peptone in stationary culture were used to inoculate a spinner using Joklik-modified MEM supplemented with 0.5% Bacto-peptone. In 3 days the cell density rose from 3×10^5 cells/ml to 9×10^5 cells/ml and, after dilution fourfold, it rose back to 9.0×10^5 cells/ml in 4 more days.

Attention was then turned to BHK cells, and initial experiments were carried out with several media and serum substitutes. The results can be summarized as follows: (1) F-12, RPM1 1640, or MEM-α [without ribosides or deoxyribosides (Stewart, 1973)] supported cell growth when supplemented with Bacto-peptone better than MEM, even when supplemented with nonessential amino acids; (2) the cells grew as very nodular clumps on the plastic surface; (3) the addition of insulin had no effect on cell growth.

It was found that, if a polyalkybenzenesulfonate detergent, Darvan 2, was added at a concentration of 300 μg/ml, the cells detached from the plastic surface and grew in suspension as small clumps of cells. F-12 (Ham, 1965) supplemented with 1% Bacto-peptone and Darvan 2 gave the best growth with the most uniform cell suspension. BHK cells have been grown in this system for 6 months with subculture by either dilution or centrifugation and medium replacement two or three times a week. Some difficulty was encountered in quantitating cell growth because of the cell clumps which were often difficult to break up. The problem of cell–cell aggregation in cell counting has been overcome by the use of 0.15% trypsin–0.02% EDTA to wash and disperse the cells (W. G. Taylor, personal communication). Trypsin–EDTA was also used as a diluent for the trypan blue, and the cell suspension to be counted was kept on ice prior to counting. The effect of insulin and of Darvan on BHK cell growth was examined in spinners. In the presence of Darvan the cell count rose from 2×10^5 cells/ml to 1.0×10^6 cells/ml in 4 days. The addition of insulin had no effect, but omission of Darvan completely cut off cell growth, or possibly the cells stuck to the walls of the glass spinner.

Several other cell lines were similarly examined (Keay, 1976). The cell lines were African green monkey kidney (BSC-1) CHO, Balb/c 3T3, HeLa, and KB cells, and the most successful medium was MEM-α (without ribosides or deoxyribosides) supplemented with 1% Bacto-peptone, although supplemented MEM gave fairly good growth for BSC-1 and Balb/c 3T3 cells. When BSC-1 cells previously grown in MEM plus 10% fetal bovine serum were subcultured into MEM or MEM-α supplemented with Bacto-peptone with or without lactalbumin hydrolysate, in all cases monolayer cell proliferation occurred. It was noted, however, that lactalbumin hydrolysate reduced the growth-promoting effect of Bacto-peptone, as had been observed with L cells.

When confluent plastic flasks were trypsinized to quantitate cell growth, it was found that the addition of insulin did not increase cell growth, but actually gave slighly lower cell counts. MEM-α gave slightly higher cell counts than MEM. When confluent, cells grown on MEM-α plus 1% Bacto-peptone showed the rather nodular whorls observed when grown in the presence of serum (see Fig. 1). CHO cells were subcultured after mechanical harvesting from a monolayer grown in MEM-α plus 10% fetal bovine serum. The cells were subcultured into MEM, MEM-α, and F-12, supplemented with Bacto-peptone, BP-LH-PVP, or the complex substitute SSK7 (Lasfargues *et al.*, 1973).

It was observed that in all systems the cells grew as grapelike clusters in stationary suspension, provided that insulin was added. When the cells were maintained without insulin, they remained viable for up to 2 weeks, but in the presence of insulin the viability fell rapidly when the cell density became high. Little difference could be detected between MEM and MEM-α when supplemented with 1% Bacto-peptone and 0.11 IU/ml insulin. From an inoculum of 5×10^4 cells/ml, 9.3×10^5 cells/ml were produced in 7 days in MEM and 8.1×10^5 cells/ml in MEM-α. The CHO cells were maintained in these serum-free systems for several months by subculturing two or three times weekly by centrifugation. When finally subcultured into MEM-α plus serum, they attached to the plastic surfaces and regained their original morphology.

Balb/c 3T3 cells were grown in Dulbecco's modified Eagle's medium

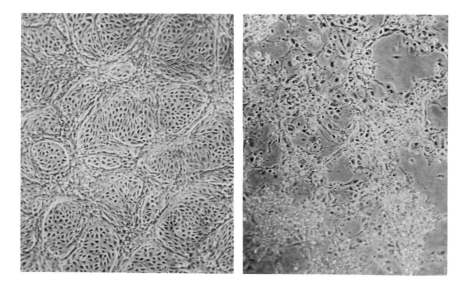

FIG. 1. BS-C-1 cells growing in MEM-α plus 1% Bacto-peptone plus 0.46 IU/ml insulin. Left: Confluent culture showing whorls; right: cytopathic effects after infection with SV40.

(DME) supplemented with 10% fetal bovine serum and then harvested by trypsinization. They were washed twice with phosphate-buffered saline by centrifugation and transferred to MEM-α or DME, supplemented with Bacto-peptone, BP-LH-PVP, or SSK7, insulin being added to all systems at a concentration of 0.46 IU/ml. The cells were refed after 3 days. In the MEM-α systems the cells were attached and confluent, while in the DME systems they were partly attached and partly in suspension. The cells in MEM-α appeared to be healthier and were subcultured. It was found that, when cells grown in MEM-α plus Bacto-peptone and insulin were subcultured into new plastic flasks, they attached and grew extensively as a monolayer, sometimes to confluency but, when subcultured back into a used flask, did not attach but grew in stationary suspension (see Fig. 2). When a clump of cells was transferred to an unused flask, the clump attached and an outgrowth of a monolayer occurred (see Fig. 2). Accurate cell counting was difficult because of the size of the cell clumps which were difficult to break up, but attempts were madt to determine quantitatively the effect of insulin and the growth in MEM versus that in MEM-α. It was clear that insulin (0.11 IU/ml) had a pronounced effect on cell growth, and the growth in MEM was greater than in MEM-α. The cells were then grown in a glass spinner in MEM plus 1% Bacto peptone and 0.35% PVP-360 and insulin (0.23 IU/ml). In 4 days the cell density rose from 1×10^5

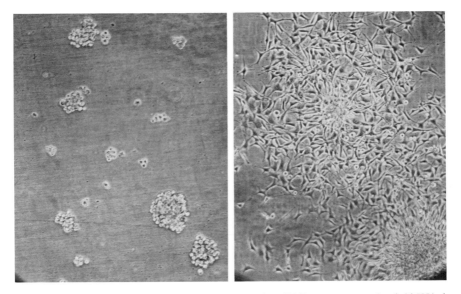

FIG. 2.   Balb/c 3T3 cells growing in MEM-α plus 1% Bacto-peptone plus 0.46 IU/ml insulin. Left: Cells growing in stationary suspension; right, monolayer outgrowth from floating clumps transferred to fresh plastic flasks.

cells/ml to 4.8×10^5 cells/ml. When transferred back to DME plus serum, the cells returned to their normal contact-inhibited monolayer morphology.

When HeLa cells were set up in plastic flasks with MEM or MEM-α supplemented with Bacto-peptone, BP-LH-PVP, or the complex substitute SSK7 (Lasfargues *et al.*, 1973), plus insulin, rapid cell proliferation occurred. The cells did not attach to the plastic surface but grew in stationary suspension. The complex substitutes BH-LH-PVP and SSK7 appeared to cause a more rapid loss in viability and variation in cell size, and further experiments were carried out with Bacto-peptone alone as the serum substitute.

It was found that insulin had an effect on cell growth at levels as low as 0.01 IU/ml and that MEM-α was necessary because it contained vitamin B_{12} and biotin which have been shown to be required for the growth of HeLa cells (Tribble and Higuchi, 1963; Higuchi, 1963; Blaker et al., 1971). In a spinner with MEM-α, Bacto-peptone, PVP-360, and insulin, an inoculum of 2×10^5 cells/ml grew to 1.4×10^6 cells/ml in 5 days.

Results with KB cells were very similar to those obtained with HeLa cells. MEM-α gave better growth than MEM, insulin was necessary, Bacto-peptone or the more complex substitutes all stimulated growth, and the cells grew in stationary suspension instead of the monolayer observed when grown in serum-containing systems (Fig. 3). The requirement for MEM-α was again shown to be due to its vitamin B_{12} and biotin content.

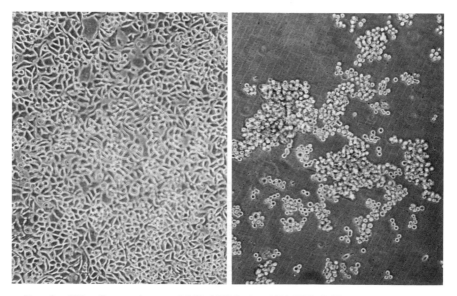

FIG. 3.   KB cells growing on Joklik MEM plus 10% FBS (left) and MEM-α plus 1% Bacto-peptone plus 0.11 IU/ml insulin (right).

After various cell lines were grown in systems comprized of simple chemically defined media, supplemented with an autoclaved Bacto-peptone solution and in some cases insulin, attention was turned to a completely autoclavable system. Autoclavable systems have been developed by Nagle and Brown (1971) and Yamane et al. (1968), although that of the latter required the addition of both serum and glutamine. Yamine et al. modified MEM by the addition of succinate to buffer the pH at 4.5 during autoclaving, as it was determined that the growth-supporting power of autoclaved media was highly dependent on the pH during autoclaving.

Attempts were made to combine an autoclavable MEM formulation with Bacto-peptone for the growth of L cells (Keay, 1977). Studies on the effect of autoclaving conditions indicated that the Bacto-peptone solution could be autoclaved at 15 psi for 30 minutes without loss of growth-promoting activity, but MEM autoclaved at pH 7.0 rapidly loses its activity and growth promotion is not restored by the addition of glutamine. Further experiments showed that unneutralized regular MEM to which 0.5% Bacto-peptone had been added supported the growth of L cells after autoclaving at 15 psi for 20 minutes. The succinate-modified MEM seems to retain its growth-promoting properties only if the Bacto-peptone is autoclaved separately and glutamine is added after autoclaving. Apparently the lower pH due to the succinate inactivates both the glutamine and some active component in the Bacto-peptone.

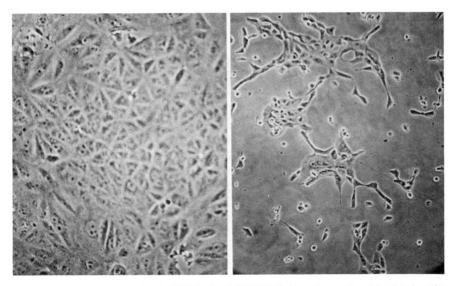

FIG. 4. Vero cells growing in MEM plus 10% FBS (left) and autoclaved MEM plus 1% Bacto-peptone (right).

Using a medium prepared by autoclaving MEM (minus $NaHCO_3$) plus
0.5% Bacto-peptone and 0.1% methylcellulose at 15 psi for 15 minutes,
cooling, and neutralizing with $NaHCO_3$, L-S cells were grown in glass
spinners. Starting from a small spinner, a 1-liter spinner was set up
at 2.5×10^5 cells/ml, which in 4 days reached 1.5×10^6 cells/ml and was
diluted to give a 3-liter spinner which reached 1.0×10^6 cells/ml in 4 days.
Several other peptones were found to support the growth of L cells.

The autoclaved MEM–bacto-peptone medium was then used in attempts
to grow Chang liver and Vero cells. Although both cells appeared to grow in
the MEM–Bacto-peptone system, the Chang cells detached from the
plastic and appeared not to multiply in suspension. Vero cells did not
detach as readily and grew to confluency in plastic flasks (see Fig. 4).
Insulin did not stimulate cell growth. Vero cells were grown in the autoclaved MEM–Bacto-peptone system for about 6 weeks and finally transferred to glass roller bottles where they grew almost to confluency before
detaching from the surface.

Low cost additives will not always totally replace serum as a supplement.
Lactalbumin hydrolysate supplementation increases growth of SV40 transformed 3T3 cells tenfold, but an obligatory level (0.15%) of calf serum is
necessary (Young, 1976). The stimulatory factors in lactalbumin hydrolysate
have molecular weights 1500 to 4000 and do not correspond to any known
growth factors in serum. Similarly, Mizrahi (1977) has found that a soluble
peptone Primatone RL can only partially replace the serum requirement for
BHK cells and a lymphocyte cell line from a patient with multiple myeloma.
The compositions of various serum-free media are shown in Table I,
which illustrates the wide range of ingredients and concentrations
used.

IV.   The Growth of Insect Cells in Serum-Free Systems

The subject of nonmammalian cell growth has been much less studied,
but some research has been carried out on the growth of insect cells in
serum-free systems. Originally, insect cells were cultivated in defined media
supplemented with insect hemolymph, an expensive item which is difficult
to procure on a large scale. The interest in large-scale production of insect
cells is related to the potential use of viruses for insect control (Vaughn,
1976). Since the use of primary cultures of insect cells is not feasible on a
large scale, the development of insect cell lines was a necessary prerequisite
for insect virus production. Ignoff and Hink (1971) list valid cell lines from 16

insect species, 5 of which are important agricultural pests and 4 of which are mosquitoes which are vectors of human diseases.

The use of nuclear polyhedrosis virus (NPV) to control insects was first demonstrated by Trager (1935). One of the first requirements for practical production of insect viruses was the production of low-cost media. Initially, insect hemolymph was replaced by serum, albumin fractions, or chicken egg ultrafiltrate (Yunker et al., 1967), but subsequently serum-free media were developed, often using low-cost undefined additives such as peptones, lactalbumin hydrolysate, and yeast extract. The most successful so far is the BML-TC/7A medium of Gardiner and Stockdale (1975), which is a simple medium containing lactalbumin hydrolysate and tryptose broth. Hink et al. (1977) modified Grace's medium with Yeastolate, lactalbumin hydrolysate, Bacto-tryptose, and serum proteins and grew TN-368 cabbage looper cells to a density of over 2×10^6 cells/ml. The cells were then adapted to a medium without the serum proteins.

Chen and Levi-Montalcini (1969) developed a chemically defined synthetic medium for the growth of cockroach embryonic nervous tissue. The medium was a composite of Eagle's basal medium and Schneider's Drosophila medium (Schneider, 1964). Seshan (1976) modified the medium to overcome the problem of salt precipitates. Marks et al. (1967) used an M-14 medium which contained a serum substitute containing PVP and synthetic peptides for the growth of cockroach tissue in vitro. A review of invertebrate cell lines and media for their propagation has recently been published (Hink, 1976).

NPV have been produced in several insect cell lines, and highly susceptible and productive cells have been developed (Grace, 1967). Insect cell lines have been shown to grow in suspension, with methylcellulose and Darvan 2 added to reduce cell clumping and mechanical damage, and high cell densities obtained in suspension culture equipment, so that the ultimate objective may eventually be achieved.

V. The Production of Viruses in Serum-Free Systems

It appears that serum is not required for virus production in nonproliferating cell systems. Eastern equine encephalomyelitis virus was prepared in chick embryo cells in both monolayer and suspension systems (White et al., 1971). Medium supplemented with 5% fetal bovine serum was used for monolayer cultures in order to allow monolayer formation and

TABLE I

Composition of Some Serum-Free Media[a]

Component	NCTC 135 Evans et al. (1956b)	CMRL 1066 Parker (1961)	Birch and Pirt (1970)	Nagle and Brown (1971)	DM-120 Takaoka and Katsuta (1971)	Parsa et al. (1970)	Higuchi and Robinson (1973)	Yamane et al. (1976)	Guskey and Jenkin (1976)	MCDB 104 McKeehan et al. (1977)	FCRC-20 (Tomei) and Bertram (1977)
Amino acids											
Alanine	31.48	25.0	90	400	400	32				8.909	800
α-Aminobutyric acid	5.51										
Arginine						6					
Arginine HCl	31.16	70.0	450	100	100	100	32	126	75	210.7	110
Asparagine				300		10	150				
Asparagine–H_2O	9.19									15.01	
Aspartic acid	9.91	30.0			25	40				13.31	
Cysteine						50					
Cysteine HCl		260.0		75	80		22		90	8.78	75
Cystine	10.49	20.0	75			15		24	15		
Glutamic acid	8.26	75.0	150 (Na salt)	150	150	150			150	14.71	300
Glutamine	136.7	100.0	100		100	400	198	584	350	365.3	900
Glycine	13.51	50.0	15		15	50		7.5	50	7.507	
Histidine					30	180					
Histidine HCl			99	60			63	42	150	20.97	30
Histidine HCl–H_2O	26.65	20.0									
Hydroxyproline	4.09	10.0				10					
Isoleucine	18.04	20.0	180	150	150	35	33	52.5	25	3.936	75
Leucine	20.44	60.0	180	300	400	80	26	52.4	50	13.32	1200
Lysine					100	200					
Lysine HCl	38.43	70.0	175	300		100	28	72.5	240	36.54	1200
Methionine	4.44	15.0	30	60		20	15	15	50	4.476	60
Ornithine HCl	9.41					50					
Phenylalanine	16.53	25.0	70	120	80	90	33	32	50	4.956	120
Proline	6.13	40.0		300	12	40	115		50	34.53	300
Serine	10.75	25.0	20	300	80	4	105	10.5		10.51	300
Taurine	4.18										
Threonine	18.93	30.0	100	135	100	100	12	48	75	11.91	135
Tryptophan	17.50	10.0	20	60	40	80	6	10	40	2.042	60
Tyrosine	16.44	40.0	70	120	50	50	46	36	40	5.436	240

Compound											
Valine	150	11.72	65	46	35	100	85	150	150	25.0	25.00
Carbohydrates and related compounds											
Ascorbic acid			17.5			50	40			50.0	50.0
Sodium acetate											50.0
Sodium acetate–3H$_2$O										83.0	
Ethyl alcohol										16.0	40.0
D-Glucosamine HCl					360						3.85
Glucose	3000	720.64	5000	3000	1800	3000	1000	3000	3000	2000	1000
Gluconolactone					178	2					
Sodium glucuronate						2					
Sodium glucuronate–H$_2$O											1.8
D-Glucuronolactone					2.0	2					1.8
Inositol										4.2	
Isonositol	10	18.02		14	2.0			2.0	1	0.050	0.125
Myoinositol–2H$_2$O									80.0		
α-Ketoglutanate			1.0						220		
Sodium pyruvate	110	110	111.0		110	110		110			
Sodium mucate						23					
Nucleotide precursors											
Adenine		1.351				10				10.0	10.0
Deoxyadenosine				1		10				10.0	10.0
Deoxycytidine				0.03							
Deoxycytidine HCl						10				10.0	10.0
Deoxyguanosine						10			10.0		
Hypoxanthine			25			1					0.1
5-Methylcytosine											
5-Methyldeoxy-cytidine										0.1	
Thymidine				1		10				10.0	10.0
Uridine triphosphate										1.0	
6,8-Dihydroxypurine				0.3							
Cofactors and vitamins											
p-Aminobenzoic acid						0.125					0.125
Biotin	1	0.0727	0.02		0.002	0.05	0.002	1	0.1	0.010	0.025
Calciferol						0.25					0.250
Choline					17	150		50			
Choline chloride	1	13.96	250		10.0	10.0	250	50	20.0	0.50	1.25

TABLE I (continued)

COMPOSITION OF SOME SERUM-FREE MEDIA[a]

Component	NCTC 135 Evans et al. (1956b)	CMRL 1066 Parker (1961)	Birch and Pirt (1970)	Nagle and Brown (1971)	DM-120 Takaoka and Katsuta (1971)	Parsa et al. (1970)	Higuchi and Robinson (1973)	Yamane et al. (1976)	Guskey and Jenkin (1976)	MCDB 104 McKeehan et al. (1977)	FCRC-20 (Tomei) and Bertram (1977)
Cocarboxylase (TPP)	1.0	1.0				1					
Coenzyme A	2.5	2.5				3					
DPN	7.0	7.0				30					
FAD	1.0	1.0				1					
Folic acid	0.025	0.010	1.0	1	0.01	1.0	1.0	1.0	0.4		1
Folinic acid								1.0		0.0006	
Leucovorin Ca											
Lipoic acid						0.0002				0.002	
Menadione	0.025					0.25					
Nicotinix acid	0.0625	0.025									
Nicotin amide	0.0625	0.025	1.2	1	5.0	1	0.1	1.0	1.0	6.105	1
Pantothenic acid					1.0						
Calcium pantothenate	0.025	0.010	1.2	2		10	0.2	1.0	1.0	0.238	20
Pyridoxal HCl	0.0625	0.025		1		10	0.05	1.0	1.0	0.1129	1
Riboflavin	0.025		0.2	0.1	1.0	10	0.10	0.1	1.0		0.1
Thiamine					10.0						
Thiamine HCl	0.025	0.010	2.0	1			0.20	1.0	10.0	0.3373	1
α-Tocopherol phosphate Na₂	0.025					0.25					
TPN	1.0					1					
Vitamin A	0.25	1.0				0.25					
Vitamin B₁₂	10.0		1.0	0.002	0.005	1		1.0	0.2	0.1355	0.020
Lipids and detergents											
Cholesterol		0.20					0.002				
Lecithin						0.08	5×10^{-6} M				
Methyl oleate						15					
Oleic acid							0.5		10		
Tween-80	12.5	5.0				12.5					
Other additives											
Albumin							10,000		2000		
Glutathione	10.0	10.0									
Hydrocortisone						30	10^{-8}M				
Insulin (IU/liter)							64	23			100

198

	1	2	3	4	5	6	7	8	9	10
Methylcellulose			1000	500			600			200
PVP			1000							
Protamine sulfate (salmine)										
Putrescine–2HCl						2.0	0.1			
L-Thyroxine						2.5×10^{-9} M			0.00016	
Salts										
NaCl	6800	6799	5900	7400	8000	7400	7400	5000	5845	7400
KCl	400	400	400	400	200	300	400	150	223.6	400
CaCl$_2$	200	200			264	144				
CaCl$_2$ · 2H$_2$O			1	265			120	120	183	130
CaCl$_2$ · 4H$_2$O										
MgSO$_4$	97.7						97.7			
MgSO$_4$ · 7H$_2$O		200	200	275	154			200	246.38	275
MgCl$_2$ · 6H$_2$O					160	275	275			
Na$_2$HPO$_4$								200		
Na$_2$HPO$_4$ · 7H$_2$O								240	804.3	
Na$_2$HPO$_4$ · 12H$_2$O					15			300		
KH$_2$PO$_4$	140	140	140	300	190	100	100			150
NaH$_2$PO$_4$ · H$_2$O					17	40	140	80		
NaHCO$_3$	2200	2200	2550	1000	2920	1140	2200	2240	1040	
NaOH										
CuSO$_4$ · 5H$_2$O			0.5		0.004				0.00025	
FeSO$_4$ · 7H$_2$O			1			4.8	0.8		1.39	
FeNH$_4$(SO$_4$)$_2$ · 12H$_2$O										
FeNH$_4$ · citrate				3						
MnSO$_4$ · H$_2$O									0.00324	3
MnCl$_2$ · 4H$_2$O			0.1							
(NH$_4$)$_6$ · Mo$_7$O$_{24}$ · 4H$_2$O									0.00123	
ZnSO$_4$ · 7H$_2$O			1	0.3	0.28		0.029		0.1438	0.15
NiCl$_2$ · 6H$_2$O									0.000119	
H$_2$SeO$_3$									0.00387	
Na$_2$SiO$_3$ · 9H$_2$O									0.1421	
SnCl$_2$ · 2H$_2$O									0.000113	
NH$_4$VO$_3$									0.000585	
Indicators and Antibiotics										
Phenol red	20.0	20.0	10	10	10	10	10	10	1.242	10
Penicillin (IU/liter)		200,000								
Streptomycin sulfate		200								
EDTA							10		10^{-5} M	
HEPES	2980	2200	2550		2920	1140	5750		11,915	

[a]Concentrations are in milligrams per liter unless otherwise noted.

199

cell proliferation, but in suspension culture virus preparation the cells were suspended in MEM with an additional 1% glucose and 4 mM glutamine and immediately infected, and high titers of virus were obtained.

It was observed that, when chick embryo cells were infected with Semliki Forest virus, the yield of virus produced was highly dependent on the presence of glucose but much less dependent on the presence of amino acids, vitamins, or serum (Zwartouw and Algar, 1968). Similar results were observed with the production of Sindbis virus in chick embryo cell suspensions (Keay and Schlesinger, 1974). Taylor-Robinson et al. (1961) infected embryonic rabbit kidney cells with polio virus and found that the only requirements for virus production were the presence of glucose and oxygen. Ito et al. (1974) found that glutamine was required for the production of Sendai virus in BHK cells. Without glutamine, little neuraminidase or hemagglutinating activity is produced. Probably, if serum has any effect, it is where cell growth continues in the presence of serum and therefore increases the virus yield. Serum may also give an apparent increase in virus yield if the presence of serum increases virus stability, as in the prevention of virus adsorption on the walls of the vessel of in the inhibition of cellular proteases.

Venezuelan equine encephalomyelitis virus has been produced in L cells, HeLa cells, and cat kidney cells grown in suspension culture in serum-free and chemically defined media (Tribble et al., 1971, Hearne et al., 1971). The media used were the lactalbumin hydrolysate medium (LAH) of Higuchi (1963) with the ommission of monoolein, and the defined medium (D) of Nagle et al. (1963). Suspended-cell cultures were incubated in 100-ml serum bottles on a gyrotary shaker at 35°C and inoculated with virus. After adsorption for 45 minutes, unadsorbed virus was removed by centrifugation and samples were removed at 24-hour intervals for assay. In all three cell lines it was observed that, with the addition of 10% calf serum at the time of infection, the maximum production of virus was delayed by 48–96 hours compared with virus growth in LAH medium. The virus titer was higher when serum was added, probably because of the higher cell growth which occurred. Apparently serum is not a prerequisite for the replication of virus but is a stimulus for virus proliferation, as observed with Rift Valley fever virus (Walker et al., 1969).

Swaney (1976) observed that seven strains of FMDV formed larger plaques on monlayers of a fetal pig kidney cell line in the absence of serum than in its presence under a tragocouth overlay.

Tomei and Issel (1975) grew BHK-S cells in an autoclavable serum-free medium, GFAD (Nagle and Brown, 1971), in flasks on a gyrotary shaker where they rapidly proliferated to form loosely associated clusters of about 10 to 50 cells. The cells did not, however, survive in spinners for more than a few days. The cells were then collected by centrifugation and resuspended in fresh medium prior to infection with each of the seven major types of FMDV (from both animal lesions and tissue culture passage). After 12 hours the virus was harvested, assayed, and subsequently inactivated prior to use in monovalent and trivalent vaccines which were tested in guinea pigs and steers for immunogenicity.

The peak infectivity of virus produced was observed to be the same with cells grown in suspension in serum-supplemented media as in chemically defined media, but higher cytopathogenic effects (as estimated by loss in cell viability) were observed in the chemically defined medium. The cells grown in chemically defined medium were susceptible to the seven major types of FMDV, and the vaccines produced from three of the strains produced neutralizing antibodies in both guinea pigs and steers, and when challenged none of the vaccinated animals gave a febrile response.

Recently, H. G. Bahnemann (private communication) grew another strain of BHK cells in a filter-sterilized F-12 nutrient medium supplemented with 1% Bacto-peptone, 2000 mg per liter glucose, 0.0015 M CaCl$_2$, 0.001 M MgCl$_2$, 2 mM glutamine, and MEM vitamins. These BHK cells have been grown in volumes up to 1 liter, seeding at 5 × 10^5 cells/ml and attaining cell densities of 2.5 × 10^6 to 4.0 × 10^6 cells/ml in 92 hours. The cells were then infected with FMDV either after resuspension in fresh medium or directly in the growth medium, both methods giving about the same virus yield. When the cells were infected in the growth medium, it was supplemented with MEM vitamins and 1000 mg per liter glucose and the pH adjusted with NaHCO$_3$ before virus infection.

It was observed, however, that FMDV produced in BHK cells grown in suspension is antigenically different and therefore may not be of great practical value. This antigenic difference has been described by Cowan et al. (1974) and Meloen (1976).

It appears that the cellular membranes of suspension culture cells are different from those of monolayer culture cells, with accompanying changes in the cellular receptor sites, hence adsorption and replication of the virus. Brugh (1975a) showed that the susceptibility of bovine and swine kidney cells to FMDV is inversely proportional to the concentration of serum. Transfer of cells to serum-free medium increased the titer of virus two- to five-fold, but only after 24–72 hours of serum deprivation. The replication of FMDV is accelerated in serum-deprived cells, apparently as a result of a more efficient release of virus from infected cells (Brugh, 1975b).

It was shown recently that BHK cells grown in suspension culture on the surface of DEAE-Sephadex beads in a serum-containing system demonstrate the FMDV susceptibility characteristics of BHK cells grown in Roux bottles and not those of cells grown in free suspension culture

(Spier and Whiteside, 1976). It appears therefore that the change in virus susceptibility is due to the change in physical growth conditions and not the change to serum-free media. If serum-free media can be used with a DEAE-Sephadex bead suspension culture system, the problem of changes in susceptibility to different virus strains may be overcome.

Fine *et al.* (1976) found that several factors influenced MMTV production in Mn5mt/C_1 cells. Media providing good cell growth were not necessarily optimal for virus expression. In systems supplemented with serum it was found that MMTV expression was highest in media with the highest glucose content. The addition of insulin alone had little effect on MMTV expression (by reverse transcriptase assay), but when added together with dexamethasone it enhanced virus expression 19- to 45-fold, dexamethasone only increasing it 10- to 30-fold. In a chemically defined system (Bauer *et al.,* 1976) insulin produced an 8-fold increase, dexamethasone a 100-fold increase, and a combination of insulin and dexamethasone a 150-fold increase.

Recently, Nagle and Fine (1977) showed that insulin, hydrocortisone, and dexamethasone increase MMTV production per cell, and maximum production occurs when all are present.

Cell growth and MMTV expression increased when thyroxine, asparagine, proline, and serine were omitted, and on this basis a modified medium was produced which gave about a 50% increase in MMTV expression as well as an increase in maximal cell numbers at confluency. It was subsequently shown (Nagle *et al.*, 1977) that MMTV virus is maximal when ferric ammonium sulfate and zinc sulfate are present at 12 mg per liter and 7 mg per liter, respectively, and oleic acid is present (0.5 mg per liter) but cholesterol and lecithin are omitted. High- and low-producer clones of Mm5mt/c_1 in the chemically defined medium continued to produce equivalent levels of MMTV with no concommitant increase in the expression of murine leukemia virus.

The production of viruses in media containing low-cost Bacto-peptone serum substitutes has also been studied. BHK cells grown in plastic flasks in serum-free media for 2 days were infected with Sindbis virus (Keay, 1975a). Similar flasks of BHK cells grown in stationary suspension in F-12–Bacto-peptone and Darvan were infected, except that the contents of one flask were centrifuged to remove the Darvan. The yield of Sindbis virus was approximately the same in the Bacto-peptone systems as in the serum-containing system. The yield of virus in the flask containing the Darvan was much lower, as the detergent appears to interfere with virus adsorption (Keay and Schlesinger, 1974).

The production of SV40 in BSC-1 cells grown in MEM-α plus Bacto-peptone and insulin was demonstrated by Keay (1976). Plastic flasks of BSC-1 cells grown to confluency in the serum-free medium were infected with SV40 and, when cytopathic effects were observed (7–8 days, see Fig. 1), the virus was isolated from the cells and media by a combination of freeze-thawing, deoxycholate treatment, cushioning onto CsCl, and finally isolation by equilibrium ultracentrifugation.

Adenovirus-2 was produced in HeLa and KB cells grown in MEM-α, Bacto-peptone, PVP-360, and insulin (Keay, 1976). The cells were grown in suspension, centrifuged, and resuspended to a concentration of 6×10^6 cells/ml, infected at an multiplicity of infection (MOI) of 20 PFU/ml and after 1 hour diluted to 3×10^5 cells/ml and incubated at 37°C for 28.5 hours. The cells were collected and ruptured by freeze-thawing, sonication, and homogenization with a fluorocarbon. The virus was cushioned on CsCl and then spun to equilibrium in a CsCl gradient and the virus band collected. The yield and titer of the virus was the same as when the cells were grown and infected in serum-containing systems.

VI.   Problems Associated with the Use of Serum-Free Media and Its Current Position in Cell Culture

The problems have been mentioned in various parts of the preceding sections but are summarized here. Many different chemically defined and serum-free media have been devised which support the growth of various cell lines, but no one medium appears to support the growth of a wide range of cell lines and so far, with the possible exception of the recent formulations of McKeehan et al. (1977), none of the media support the growth of primary, normal, or diploid cells.

The problems of cell attachment remain, many cells not attaching to surfaces at all, or attaching initially but detaching as the surface ages or the cell proliferates so that suspension cultures are the only form of growth. Although this has the advantage of ease of scale-up of the suspension culture, it makes the problem of refeeding more difficult, and with serum-free media frequent refeeding appears to be necessary.

In serum-free media, where cells do attach, subculture by trypsinization is often not possible because of the sensitivity of the cells and the lack of trypsin inactivation normally occurring with the addition of serum. Approaches have been the addition of albumin (Tomei and Bertram, 1977), the use of lower trypsin concentrations and multiple washings with saline to remove trypsin, or mechanical harvesting with glass beads or by scraping.

In suspension cultures and to a lesser extent in monolayer cultures, the need for protective macromolecules such as methylcellulose, PVP, etc., is necessary, but their precise function is not understood.

Often the cells have to be adapted to growth in serum-free media over a period of time and by gradual reduction in the level of serum. It may be that certain cells are being selected, since the morphology changes as does the karyotype in some cases. Although some cells have been shown to grow in serum-free media without extensive adaptation and revert back to their original morphology when serum is added, the possibilities of adaptation and selection are a problem.

Although the use of serum-free media has the potential of eliminating contamination problems, autoclavability is the surest way to solve this particular problem. However, autoclaving on a large scale poses engineering problems, as the ability to heat rapidly and cool a large volume of liquid in such a way as to sterilize it without destroying the nutrients would require elaborate and expensive equipment.

The production of viruses in serum-free or chemically defined systems appears to present fewer problems than the growth of cells. In general it seems that the yield of virus is dependent largely on the number of healthy cells which are infected. The antigenic specificity of some viruses may change, but it may be due to growth in suspension versus monolayer culture rather than to the omission of serum.

The commercial production of viruses for human vaccine production does not make use of established cell lines but is restricted to primary cells (with the exception of W1-38), although veterinary vaccines are being developed using cell lines such as BHK-21. Therefore the serum-free systems for the growth of cells (and the production of viruses) developed so far probably have little or no commercial significance. The production of viruses in nonproliferating primary cell suspension may be of more value in some cases, since serum apparently is not required, provided that virus production is rapid and that large-scale purification of the virus can be carried out.

In summary, the need for serum-free or chemically defined media is clearly established and many successful formulations have been published, but in practice these media are not being used to any great extent, and the problems involved in their use are well documented. It is clearly a situation where the old saying, "Caveat emptor," applies.

ACKNOWLEDGMENT

The comments and suggestions of Drs. L. D. Tomei, W. G. Taylor, and H. G. Bahnemann are gratefully acknowledged.

REFERENCES

Anderson, W. F., Price, F. M., Jackson, J. L., Dunn, T. B., and Evans, V. J. (1967). *J. Natl. Cancer Inst.* **38**, 169–183.

Bahnemann, H. G. (1976). *J. Clinical Microbiol.* **3**, 209–210.

Barile, M. (1973). *In* "Contamination in Tissue Culture" (J. Fogh, ed.), pp. 131–172. Academic Press, New York.

Barile, M. F., and Kern, J. (1971). *Proc. Soc. Exp. Biol. Med.* **138**, 432–437.

Barile, M. F., Hopps, H. E., Grabowski, M. W., Rigg, D. B., and DelGuidice, R. A. (1971). *Ann. N.Y. Acad. Sci.* **225**, 251–264.

Bauer, R. F., Arthur, L. O., and Fine, D. L. (1976). *In Vitro* **12**, 558–563.

Bershadsky, A. D., Gelfand, V. I., Guchstein, V. J., Vasiliev, J. M., and Gelfand, J. M. (1976). *Int. J. Cancer* **18**, 83–92.

Birch, J. R., and Pirt, S. J. (1970). *J. Cell Sci.* **7**, 661–670.

Blaker, G. J., Birch, J. R., and Pirt, S. J. (1971). *J. Cell. Sci.* **9**, 529–537.

Boone, C. W., Mantel, N., Caruso, T. D., Kazam, E., and Stevenson, R. E. (1972). *In Vitro* **7**, 174–189.

Brugh, M. (1975a). *Zbl. Vet. Med.* **22**, 285–294.

Brugh, M. (1975b). *Zbl. Vet. Med.* **22**, 295–302.

Bryant, J. C., Evans, V. J., Schilling, E. L., and Earle, W. R. (1961). *J. Natl. Cancer Inst.* **26**, 239–252.

Bryant, J. C., (1966). *Ann. N.Y. Acad. Sc.* **139**, 143–161.

Bump, E. A., and Reed, D. J. (1977). *In Vitro* **13**, 115–119.

Bush, H., and Shodell, M. (1977). *J. Cell Physiol.* **90**, 573–583.

Ryyny, R. L., Orth, D. N., Cohen S., and Doyne, E. S. (1974). *Endocrinology* **95**, 776–782.

Chen, J. S., and Levi-Montalcini, R. (1969). *Science,* **166**, 631–632.

Cohen, S (1962). *J. Biol. Chem.* **237**, 1555 1562.

Cohen, S., and Carpenter, G. (1975). *Proc. Natl. Acad. Sci. U.S.A.* **72**, 1317–1721.

Cowan, K. M., Erol, N., and Whiteland, A. P. (1974). *Bull. Off. Int. Epizoot.* **81**, 1271–1298.

Culp, L. A., and Buniel, J. F. (1976). *J. Cell Physiol.* **88**, 89–106.

Dulak, N. C., and Temin, H. M. (1973). *J. Cell Physiol.* **81**, 161–170.

Eagle, H. (1955). *Science* **122**, 43–46.

Eagle, H., and Piez, N. (1962). *J. Exp. Med.* **116**, 29 43.

Esber, H. J., Payne, I. J., and Bodgen, A. F. (1973). *J. Natl. Cancer Inst.* **50**, 559–562.

Evans, V. J., and Anderson, W. F. (1966). *J. Natl. Cancer Inst.* **37**, 247–249.

Evans, V. J., Bryant, J. C., Fioramonte, M. C., McQuilkin, W. J., Sanford, K. K., and Earle, W. R. (1956a). *Cancer Res.* **16**, 77–86.

Evans, V. J., Bryant, J. C., McQuilkin, W. T., Fioramonte, M. C., Sanford, K. K., Westfall, B. B., and Earle, W. R. (1956b). *Cancer Res.* **16**, 87–94.

Evans, V. J., Bryant, J. C., Kerr, H. A., and Schilling, E. L. (1964). *Exp. Cell Res.* **36**, 439–474.

Evans, V. J., Price, F. M., Sanford, K. K., Kerr, H. A., and Handleman, S. L. (1972). *J. Natl. Cancer Inst.* **49**, 505–511.

Fan, K., and Uzman, B. G. (1977). *Exp. Cell Res.* **106**, 397–401.

Federoff, S., Evans, V. J., Hopps, H. E., Sanford, K. K., and Boone, C. W. (1972). *In Vitro* **7**, 161–167.

Feldman, M. K., and Wong, D. L. (1977). *In Vitro* **13**, 275–279.

Fine, D. L., Arthur, L. O., and Young, L. J. T. (1976). *In Vitro* **12**, 693–701.

Fioramonte, M. C., Evans, V. J., and Earle, W. R. (1958). *J. Natl. Cancer Inst.* **21**, 579–583.

Fischer, S. M., Berry, D. L., and Slaga, T. J. (1977). *In Vitro* **13**, 186.

Fisher, G. A., and Welch, A. D. (1957). *Science* **126**, 1018–1019.

Fisher, H. W., Puck, T. T., and Sato, G. (1959). *J. Exp. Med.* **109**, 649–660.
Fox, C. H., and Sanford, K. K. (1975). *Tissue Cult. Assoc. Man.* **1**, 233–237.
Franks, L. M., Carbonell, A. W., Hemmings, V. J., and Riddle, P. N. (1976). *Cancer Res.* **36**, 1049–1055.
Fujimoto, W. Y., and Williams, R. H. (1977). *In Vitro* **13**, 268–274.
Gardiner, G. R., and Stockdale, H. (1975). *J. Invertebr. Pathol.* **25**, 363–370.
Geier, M. R., Attallah, A. R. M., and Merrill, C. R. (1975). *In Vitro* **11**, 55–58.
Gerschenson, L. E., Okigaki, T., Anderson, M., Molson, J., and Davidson, M. B. (1972). *Exp. Cell Res.* **71**, 49–58.
Ginsberg, H. S., Gold, E., and Jordan, W. S. (1955). *Proc. Soc. Exp. Biol. Med.* **89**, 66–71.
Goodridge, A. G., and Adelman, T. G. (1976). *J. Biol. Chem.* **251**, 3027–3032.
Gospodarowicz, D. (1975). *J. Biol. Chem.* **250**, 2515–2520.
Gospodarowicz, D., and Moran, J. S. (1974). *Proc. Natl. Acad. Sci. U.S.A.* **71**, 4584–4588.
Gospodarowicz, D., Moran J. S., Braun, D., and Birdwell, C. (1976). *Proc. Natl. Acad. Sci. U.S.A.* **73**, 4120–4124.
Grace, T. D. C. (1967). *In Vitro* **3**, 104–117.
Granick, S., Sinclair, P., Sassa, S., and Grieninger, G. (1975). *J. Biol. Chem.* **250**, 9215–9225.
Guskey, L. E., and Jenkin, H. M. (1975). *Appl. Microbiol.* **30**, 433–438.
Guskey, L. E., and Jenkin, H. M. (1976). *Proc. Soc. Exp. Biol. Med.* **155**, 221–224.
Ham. R. G. (1962). *Exp. Cell Res.* **28**, 489–500.
Ham, R. G. (1963). *Exp. Cell Res.* **29**, 515–526.
Ham, R. G. (1965). *Proc. Natl. Acad. Sci. U.S.A.* **53**, 288–293.
Ham, R. G. (1974a). *J. Natl. Cancer Inst.* **53**, 1459–1463.
Ham. R. G. (1974b). *In Vitro* **10**, 119–129.
Ham, R. G., and McKeehan, W. L. (1978). *In Vitro* **14**, 11–22.
Hamilton, W. G., and Ham, R. G. (1977). *In Vitro* **13**, 537–547.
Harary, I., Gerschenson, L. E., Haggerty, D. L., Desmond, W., and Mead, J. F. (1967). *Lipid Metab. Tissue Cult. Cells, Symp., 1966* Wistar Inst. Symp. Monogr. No. 6, pp. 17–30.
Hayashi, I., and Sato, G. H. (1976). *Nature (London)* **259**, 132–134.
Hayashi, I., Larner, J., and Sato, G. (1978). *In Vitro,* **14**, 23–32.
Hayflick, L. (1965). *Tex. Rep. Biol. Med.* **23**, Suppl. 1, 285–303.
Healy, G. M., and Macmorine, H. G. (1972). *Prog. Immunobiol. Stand.* **45**, 202—208.
Healy, G. M., Fisher, D. C., and Parker, R. C. (1954). *Can. J. Biochem. Physiol.* **32**, 327–337.
Healy, G. M., Fisher, D. C., and Parker, R. C. (1955). *Proc. Soc. Exp. Biol. Med.* **89**, 71–77.
Healy, G. M., Teleki, S., Seefried, A. V., Walton, M. J., and Macmorine, H. G., (1971). *Appl. Microbiol.* **21**, 1–5.
Hearn, H. J., Tribble, H. R., Nagle, S. C., and Bowersox, D. C. (1971). *Appl. Microbiol.* **21**, 342–345.
Higuchi, K. (1963). *J. Infect. Dis.* **112**, 213–220.
Higuchi, K. (1969). *Fed. Proc.,* **29**, 627.
Higuchi, K. (1970). *J. Cell. Physiol.* **75**, 65–72.
Higuchi, K. (1973). *Adv. Appl. Microbiol.* **16**, 111–136.
Higuchi, K., and Robinson, R. C. (1973). *In Vitro* **9**, 114–121.
Hink, W. F. (1976). *In* "Invertebrate Tissue Culture" (K. Maramarosch, ed.), pp. 319–369. Academic Press, New York.
Hink, W. G., Strauss, E. M., and Lynn, D. E. (1977). *In Vitro* **13**, 177.
Holmes, R. (1959). *J. Biophys. Biochem. Cytol.* **6**, 535–536.
Holmes, R., and Wolfe, S. W. (1961). *J. Biophys. Biochem. Cytol.* **10**, 389–401.
Houck, J. C., and Cheng, R. F. (1973). *J. Cell Physiol.* **81**, 257–270.

House, H. Shearer, M., and Maroudas, N. G. (1972). *Exp. Cell. Res.* **71**, 293–296.
Hsueh, H. W., and Moskowitz, M. (1973a). *Exp. Cell Res.* **77**, 376–382.
Hsueh, H. W., and Moskowitz, M. (1973b). *Exp. Cell Res.* **77**, 383–390.
Ignoff, C. M., and Hink, W. F. (1971). *In* "Microbial Control of Insects and Mites" (H. D. Burgess and N. W. Hussey, eds.), pp. 541–580. Academic Press, New York.
Ito, Y., Kumura, Y., and Kunii, A. (1974). *J. Virol.* **13**, 557–566.
Jackson, J. L., Sanford, K. K., and Dunn, T. B. (1970). *J. Natl. Cancer Inst.* **45**, 11–23.
Jenkin, H. M., and Anderson, L. E. (1970). *Exp. Cell Res.* **59**, 6–10.
Jenkin, H. M., Anderson, L. E., Holman, R. T., Ismail, I. A., and Gunstone, F. D. (1970). *Exp. Cell Res.* **59**, 1–5.
Katsuka, H., and Takaoka, T. (1973). *Methods Cell Biol.* **6**, 1–42.
Keay, L. (1975a). *Biotechnol. Bioeng.* **17**, 745–764.
Keay, L. (1975b). *Tissue Cult. Assoc. Man.* **1**, 177–180.
Keay. L. (1976). *Biotechnol. Bioeng.* **18**, 363–382.
Keay, L. (1977). *Biotechnol. Bioeng.* **19**, 399–411.
Keay, L., and Schlesinger, S. (1974). *Biotechnol. Bioeng.* **16**, 1025–1044.
Kniazeff, A. J., Rimer, V., and Gaeta, L. (1967). *Nature (London)* **214**, 805–806.
Kniazeff, A. J., Wopsch 11, L. J., Hopps, H. E., and Morris, C. S. (1975). *In Vitro* **11**, 400–403.
Kuchler, R. J., Marlowe, M. L., and Merchant, D. J. (1960). *Exp. Cell Res.* **20**, 428–437.
Lasfargues, E. Y., and Coutinho, W. G. (1977). *In Vitro* **13**, 204.
Lasfargues, E. Y., Coutinho, W. G., Lasfargues, J. C., and Moore, D. H. (1973). *In Vitro* **8**, 494–500.
Lehrer, E. H., Coleman, D. L., and Waymouth, C. (1974). *In Vitro* **9**, 421–433.
Lieberman, I., and Ove, P. (1959). *J. Biol. Chem.* **234**, 2754–2758.
McCoy, T. A., Maxwell, M., and Kruse, P. F. (1959). *Proc. Soc. Exp. Biol. Med.* **100**, 115–118.
McKeehan, W. L., and Ham, R. G. (1976). *J. Cell Biol.* **71**, 727–734.
McKeehan, W. L., Hamilton, W. G., and Ham, R. G. (1976). *Proc. Natl. Acad. Sci. U.S.A.* **73**, 2023–2027.
McKeehan, W. L., McKeehan, K. A., Hammond, S. L., and Ham, R. G. (1977). *In Vitro* **13**, 399–416.
MacPherson, I. (1966). *J. Cell Sci.* **1**, 145–168.
Macy, M. I., and Shannon, J. E. (1977). *Tissue Culture Association Manual*, **3**, 617–621.
Mak, S., and Till. J. E. (1963). *Radiat. Res.* **20**, 600–618.
Marks, E. P., Reinecke, J. P., and Caldwell, J. M. (1967). *In Vitro* **3**, 85–92.
Matchett, A., Wessman, S. J., Andrew, B. B., and Mason, P. S. (1977). *In Vitro* **13**, 153.
Mayyasi, S. A., and Schuurmans, D. M. (1956). *Proc. Soc. Exp. Biol. Med.* **93**, 207–210.
Meloen, R. H. (1976). *Arch. Virol.* **51**, 299–306.
Merchant, D. J., and Hellman, K. B. (1962). *Proc. Soc. Exp. Biol. Med.* **110**, 194–198.
Merrill, C. R., Friedman, T. B., Attallah, A. F., Geier, M. R., Krell, K., and Yarkin, R. (1972). *In Vitro* **12**, 682–686.
Mickalski, F. J., and Hsiung, G. D. (1976). *In Vitro* **12**, 682–686.
Milo, G. E., Malarkey, W. B., Powell, J. E., Blakeslee, J. R., and Yohn, D. S. (1976). *In Vitro* **12**, 23–30.
Mizrahi, A. (1977). *Biotechnol. Bioeng.*, **19**, 1557–1560.
Mizrahi, A., and Moore, G. E. (1970). *Appl. Microbiol.* **19**, 906–910.
Molander, C. W., Kniazeff, A. J., Boone, C. W., Paley, A., and Imagawa, D. T. (1972). *In Vitro* **7**, 168–173.

Moore, G. E., and Woods, L. K. (1977). *Tissue Cult. Assoc. Man.* **3**, 503–509.

Morgan, J. T., Morton, H. J., and Parker, R. C. (1950). *Proc. Soc. Exp. Biol. Med.* **73**, 1–8.

Morrison, S. J., and Jenkin, H. M. (1972). *In Vitro* **8**, 94–100.

Murakami, O., and Yamane, I. (1976). *Cell Struct. Funct.* **1**, 285–290.

Nagle, S. C. (1960). *Bacteriol. Proc.* p. 192.

Nagle, S. C. (1968). *Appl. Microbiol.* **16**, 53–55.

Nagle, S. C. (1969). *Appl. Microbiol.* **17**, 318–319.

Nagle, S. C., and Brown, B. L. (1971). *J. Cell. Physiol.* **77**, 259–264.

Nagle, S. C., and Brown, B. L. (1974). *Appl. Microbiol.* **28**, 518–520.

Nagle, S. C., and Fine, D. L. (1977). *In Vitro* (in press).

Nagle, S. C., Tribble, H. R., Anderson, R. E., and Gary, N. D. (1963). *Proc. Soc. Exp. Biol. Med.* **112**, 340–344.

Nagle, S. C., Fine, D. L., and Kmetz, J. P. (1977). *In Vitro* **13**, 173.

Neumann, R. E., and Tytell, A. A. (1960). *Proc. Soc. Exp. Biol. Med.* **104**, 252–256.

Orr, H. C., Sibinovic, K. H., Probst, P. G., Hochstein, H. D., and Littlejohn, D. C. (1975). *In Vitro* **11**, 230–234.

Parker, R. C. (1961). "Methods of Tissue Culture," 3rd ed., pp. 74–75, Harper, New York.

Parsa, I., and Flancbaum, L. (1975). *Dev. Biol.* **46**, 120–131.

Parsa, I., Marsh, W. H., and Fitzgold, P. T. (1970). *Exp. Cell Res.* **59**, 171–175.

Petricciani, J. C., Chu, F. C., and Johnson, J. B. (1973). *Proc. Soc. Exp. Biol. Med.* **144**, 789–792.

Pickart, L., and Thaler, M. M. (1973). *Nature New Biol.* **243**, 85–87.

Pickart, L., Thayer, L., and Thaler, M. M. (1973). *Biochem. Biophys. Res. Commun.* **54**, 562–566.

Pierson, R. W., and Tewin, H. M. (1972). *J. Cell. Physiol.* **79**, 319–330.

Pohjampelto, P., and Raina, A. (1972). *Nature (London), New Biol.* **235**, 247–249.

Price, F. M., and Evans, V. J. (1977). *Tissue Cult. Assoc. Man.* **3**, 497–501.

Price, F. M., Kerr, R. A., Andersen, W. F., Bryant, J. C., and Evans, V. J. (1966). *J. Natl. Cancer Inst.* **37**, 601–618.

Pumper, R. W. (1958). *Science* **128**, 363–364.

Pumper, R. W., Alfred, L. J., and Sackett, D. L. (1960). *Nature (London)* **158**, 123–124.

Pumper, R. W., Yamashiroya, H. M., and Molander, L. T. (1965). *Nature (London)* **207**, 662–663.

Pumper, R. W., Fagan, P., and Taylor, W. G. (1971). *In Vitro* **6**, 266–268.

Rajaraman, R., MacSween, J. M., and Fox, R. A. (1977). *In Vitro* **13**, 205.

Rechler, M. M., Podskalny, J. M., and Nissley, S. P. (1976). *Nature (London)* **259**, 134–136.

Sanford, K. K., Westfall, B. B., Fioramonte, M. C., McQuilkin, W. T., Bryant, J. C., Peppers, E. V., and Earle, W. R. (1955). *J. Natl. Cancer Inst.* **16**, 789–802.

Sanford, K. K., Dupree, L. T., and Covalesky, A. B. (1963). *Exp. Cell Res.* **31**, 345–375.

Schneider, I. (1964). *J. Exp. Zool.*, **156**, 91–104.

Schweizer, H., Sprossig, M., Mucke, H., and Wutzler, P. (1972). *Nature New Biology*, **240**, 61–62.

Seshan, K. R. (1976). *Tissue Cult. Assoc. Man.*, **2**, 319–322.

Smith, G. L., and Temin, H. M. (1974). *J. Cell. Physiol.* **84**, 181–192.

Spier, R. E., and Whiteside, J. P. (1976). *Biotechnol. Bioeng.* **18**, 659–667.

Stammers, C. P., Elicieri, G. L., and Green, H. (1971). *Nature (London), New Biol.* **230**, 52–53.

Stanbridge, E. (1971). *Bacteriol. Rev.* **35**, 206–227.

Stewart, C. C. (1973). *J. Reticuloendothel. Soc.* **14**, 332–349.

Stoner, G. D., and Merchant, D. J. (1972). *In Vitro* **7**, 330–343.

Swaney, L. M. (1976). *Am. J. Vet. Res.* **37**, 1319–1323.

Swim, H. E., and Parker, R. C. (1958). *Can. J. Biochem. Physiol.* **36**, 861–868.
Takaoka, T., and Katsuka, H. (1971). *Exp. Cell Res.* **67**, 295–304.
Taylor, W. G. (1974). *J. Natl. Cancer Inst.* **53**, 1449–1457.
Taylor, W. G., and Evans, V. J. (1975). *Biotechnol. Bioeng.* **17**, 1847–1851.
Taylor, W. G., and Parshad, R. (1977). *Methods Cell Biol.* **15**, 421–434.
Taylor, W. G., Dworkin, R. A., Pumper, R. W., and Evans, V. J. (1972). *Exp. Cell Res.* **74**, 275–279.
Taylor, W. G., Evans, V. J., and Pumper, R. W. (1974). *In Vitro* **8**, 278–285.
Taylor-Robinson, D., Zwartouw, H. T., and Westwood, J. C. N. (1961). *Br. J. Exp. Pathol.* **42**, 317–323.
Temin, H. M. (1967). *J. Cell. Physiol.* **69**, 377–384.
Thomas, J. A., and Johnson, M. J. (1967). *J. Natl. Cancer Inst.* **39**, 337–345.
Tomei, L. D., and Bertram, J. S. (1978). *Cancer. Res.* **38**, 444–451.
Tomei, L. D., and Issel, C. J. (1975). *Biotechnol. Bioeng.* **17**, 765–778.
Trager, W. (1935). *J. Exp. Med.* **61**, 501–514.
Tribble, H. R., and Higuchi, K. (1963). *J. Infect. Dis.* **112**, 221–225.
Tribble, H. R., Hearn, H. J., and Nagle, S. C. (1971). *J. Gen. Virol.* **10**, 231–236.
Underwood, E. J. (1971). "Trace Elements in Human and Animal Nutrition," 3rd ed. Academic Press, New York.
Van Wyck, J. J., Underwood, L. E., Hintz, R. L., Clemmons, D. R., Voina, S. J., and Weaver, R. P. (1974). *Recent Prog. Horm. Res.* **30,** 259–295.
Vaughn, J. L. (1976). *In* "Invertebrate Tissue Culture" (K. Maramarosch, ed.), pp. 295–303. Academic Press, New York.
Walker, J. S., Carter, R. C., Klein, F., Snowdon, S. E., and Lincoln, R. E. (1969). *Appl. Microbiol.* **17**, 658–663.
Waymouth, C. (1956). *J. Natl. Cancer Inst.* **17**, 315–325.
Waymouth, C. (1959). *J. Natl. Cancer Inst.* **22**, 1003–1017.
Waymouth, C. (1965). In "Cells and Tissues In Culture" (E. N. Wilmer, ed.), Academic Press, New York, Vol. 1, 99–141.
Waymouth, C. (1974). *J. Natl. Cancer Inst.* **53**, 1443–1448.
Waymouth, C. (1977). *Tissue Cult. Assoc. Man.* **3**, 521–525.
Westermark, B. (1976). *Biochem. Biophys. Res. Commun.* **69**, 304–310.
White, A., Berman, S., and Lowenthal, J. P. (1971). *Appl. Microbiol.* **22**, 909–913.
White, P. R. (1946). *Growth* **10**, 231–239.
Yamane, I., and Murakami, O. (1973). *J. Cell. Physiol.* **81**, 281–284.
Yamane, I., Matsuya, C., and Jimbo, K. (1968). *Proc. Soc. Exp. Biol. Med.* **127**, 335–336.
Yamane, I., Murakami, O., and Kato, M. (1976). *Cell Struct. Funct.* **1**, 279–284.
Yunker, C. E., Vaugh, J. L., and Cory, J. (1967). *Science* **155**, 1565–1566.
Zwartouw, H. T., and Algar, D. J. (1968). *J. Gen. Virol.* **2**, 243–250.

Chapter 16

A Rapid-Mixing Technique to Measure Transport in Suspended Animal Cells: Applications to Nucleoside Transport in Novikoff Rat Hepatoma Cells

ROBERT M. WOHLHUETER, RICHARD MARZ,[1]
JON C. GRAFF,[2] AND PETER G. W. PLAGEMANN

*Department of Microbiology, University of Minnesota Medical School,
Minneapolis, Minnesota*

[1] Current affiliation: Institute of Molecular Biology, Austrian Academy of Sciences, Salzburg, Austria.
[2] Current affiliation: Department of Microbiology and Immunology, Duke University Medical Center, Durham, N.C.

I.   Introduction

The uptake of nutrients into animal cells consists, in most instances, of two processes operating in tandem: transport of the nutrient across the cell membrane, followed by intracellular metabolism. (Throughout the text we use "transport" to denote the transfer of a substance across the cell membrane as mediated by a saturable, selective carrier. "Uptake" or "incorporation" denotes the appearance of radioactivity, derived from an exogenous, labeled substrate, within the cell, or some fraction of it.) For purposes of studying metabolism the membrane barrier is readily shed, but investigators of transport are forced to retain a compartmented system, often the intact cell surrounded by an isotonic medium. We depend therefore on techniques which permit an operational separation of transport and metabolism. This separation can be achieved by genetic, chemical, or kinetic manipulation, or a combination thereof. The selection of mutants unable to metabolize a given substrate, the choice of chemical analogs of natural substrates, and the attainment of appropriate time scales for measurement all yield useful information about the transport system per se.

Transport scientists are becoming increasingly aware that a key element in the successful assessment of the transport process is the rapidity with which the two compartments—cells and medium—can be separated from one another. Several techniques have been developed to cope with the rapid rate encountered in transport studies.

Early on, Sen and Widdas (1962) used light scattering to detect volume changes in erythrocytes elicited by the net movement of (osmotically active) hexoses across the membrane. Light scattering has been measured also by Sha'afi *et al.* (1967) in conjunction with a stopped-flow mixing apparatus which permits monitoring to commence within a few milliseconds of mixing. The use of an auxiliary enzyme to provide an electric signal was introduced by Taverna and Langdon (1973). Erythrocyte ghosts were loaded with glucose oxidase, and the oxygen consumption due to inflowing D-glucose was continuously monitored with an oxygen electrode. Such methods avoid the whole problem of having to separate cells and medium; they lack, however, versatility with respect to the range of substrates and/or concentrations which can be employed.

The efflux of substrate from erythrocytes preloaded with radioactive substrate has been followed at short intervals (about 10 seconds) by sampling the medium through Millipore filters (Mawe and Hempling, 1965; Lassen, 1967). In general, efflux measurements preclude the use of metabolized substrates and, where the appearance of substrate in the medium is followed, are dependent on the use of comparatively large total intracellular volumes. The reverse protocol, namely, following the uptake of radioactive substrates into cells by rapidly filtering off the medium and analyzing for cell-associated radioactivity, has proven unsatisfactory in our hands. Either radioactivity in the extracellular space was relatively high and variable or, with thorough rinsing, intracellular radioactivity was leached from the cells.

Attached cells constitute an immobilized transport compartment and lend themselves to another approach. Thus Berlin and co-workers (Hawkins and Berlin, 1969; see also a review by Berlin and Oliver, 1975) allowed rabbit polymorphonuclear leukocytes to attach as monolayers on glass coverslips. Such monolayers can be immersed in substrate solutions and then rapidly rinsed. Their utility is limited by the relatively small intracellular volume available with such a monolayer (typically 0.7 μl).

Several potent inhibitors of transport processes are known and suggest another strategy, that of rapidly quenching transmembrane flux, followed by centrifugal separation of cells from the medium. Cass and Patterson (1972), for instance, used a nucleoside transport inhibitor—hydroxynitrobenzyl-thioguanosine—in this capacity to demonstrate the efflux of uridine from Ehrlich ascites cells at 5-second intervals, while the addition of phloretin plus $HgCl_2$ to erythrocytes effectively quenches hexose transport (Hankin and Stein, 1972). Quenching methods are useful where an inhibitor has been demonstrated to act with the requisite potency and speed on the transport system under investigation. A technique for the rapid addition of inhibitor at short time intervals is still required by quenching methods, though the necessity for rapid separation of cells from the medium is of course obviated.

Centrifugation of mitochondria through inert oil layers has long been applied (Werkheiser and Bartley, 1957) to the study of the distribution of substrates across the membrane of these subcellular particles, and the sedimentation of chloroplasts through silicone oil–glycerol mixtures has been visualized stroboscopically by Portis and McCarty (1976). By centrifuging erythrocytes into dibutylphthalate, Oliver and Paterson (1971) observed that uridine uptake ceased within 1 minute. And recently Strauss et al. (1976) studied adenosine and thymidine transport in lymphocytes using a silicone oil system similar to the one we employ. Centrifugation through oil seems to accomplish the separation of cells and medium sufficiently rapidly to warrant its use in conjunction with rapid-mixing techniques. This conjunction of rapid mixing and centrifugation through oil is the basis of our approach of transport measurements.

The methodology we describe here has been used extensively over the past 2 years in our laboratory to investigate the transport of nucleosides, nucleic acid bases, and hexoses into several mammalian cell lines grown in suspension culture. It combines rapid mixing by means of a dual-syringe device with centrifugation through silicone oil and requires only a modest investment in equipment.

Rapid kinetic techniques provide one with the capability of assessing the kinetics of transport systems in animal cells, but the interpretation of such data in terms of the molecular basis of transport is still uncertain. Kinetic equations have been derived for a simple carrier model (Eilam and Stein, 1974; Segel, 1975) and for more complicated models (Eilam, 1975). The derivation and nomenclature of Eilam and Stein make clear several distinctions between simple enzyme kinetics and the kinetics of transport. Their derivation defines resistance coefficients, which correspond to the time for a round trip of the carrier loaded, unloaded, or loaded one way only with substrate. The apparent Michaelis constant of a zero-trans experiment K_m^{zt} (where "zt" stands for zero-trans), for example, encompasses terms for carrier mobility as well as for the carrier–substrate association constant K.

Sections II,E and IV,C treat the questions of how different experimental protocols can be employed with the mixing-syringe technology to measure the various parameters for simple carrier-mediated transport and what sort of information can be extracted from following the complete time course of uptake of a nonmetabolized substrate.

II. Apparatus and Procedures

The method of rapid sampling we employ consists of mixing constant aliquots of substrate solution and cell suspension by means of a dual-syringe apparatus, dispensing the effluent from the mixing chamber directly into a series of centrifuge tubes according to an appropriate time schedule, and separating the cells from the incubation mixture by centrifugation through a

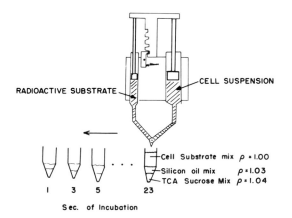

FIG. 1. Schematic view of sampling procedure described in the text. The longest-time sample is dispensed first; the countdown ends when the centrifuge is started at zero seconds. The trichloroacetic acid phase (TCA–sucrose mix) is optional where termination of cellular metabolism is necessary.

denser-than-water oil phase. Operated manually, this method provides 12 discrete samplings at intervals as short as 1 second. Figure 1 depicts the method schematically.

Two crucially important advantages are realized by the oil centrifugation technique. The first is the rapid separation of cells from the medium, commensurate with the rapid rates of substrate influx. The second is the avoidance of a washing step which, thanks to equally rapid substrate efflux, runs the risk of washing out intracellular substrate not anchored by phosphorylation, etc. Even without a wash, extracellular fluid is so efficiently stripped off cells by their passage through the oil layer that, with cultured Novikoff cells, it comprises only about 12% of the total fluid space in the cell pellet (see Section IV,A).

A. Dual-Syringe Apparatus

Our apparatus has as its antecedents the dual-syringe devices used for stopped-flow enzyme kinetics (Gibson, 1969). It is easily made by a machinist from standard components and can accommodate a variety of syringe sizes

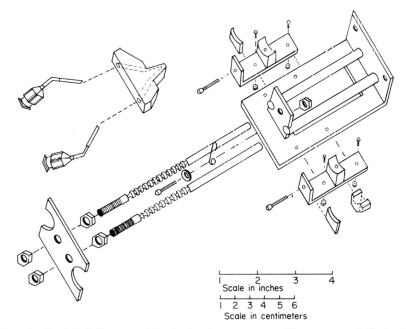

Scale in inches

Scale in centimeters

FIG. 2. Exploded diagram of the dual-syringe apparatus. The apparatus is designed to accommodate plastic syringes (Monoject, Sherwood Medical Industries, Deland, Fla.) 6-ml or smaller, with the use of adaptors. The notched slides are interchangeable and removable. The apparatus is used in conjunction with an Eppendorf Model 3200/30 centrifuge.

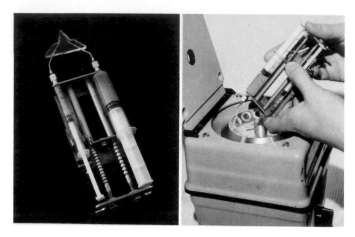

F̴ɪɢ. 3. Photographs of the dual-syringe apparatus and its use in conjunction with a microcentrifuge.

and volume programs. An exploded view of the apparatus is given in Fig. 2, and its use illustrated in Fig. 3.

An essential feature is the 13 circumferential grooves evenly spaced along a guide rod of the push bar (see Fig. 2). These grooves engage a ratchet spring, such that 12 equal linear displacements of the syringe plungers are attained by advancing the bar from one groove to the next. The fixed linear displacement is converted to a desired volume by choosing a syringe of appropriate internal diameter. Thus the actual delivery volumes can be tailored to experimental needs, limited by the total volume which can be spun in capless tubes in the (fixed-angle) rotor without spillage. We routinely use 1.5-ml centrifuge tubes which can accommodate about 600 μl. A 0.138-inch spacing of the ratchet grooves provides 448 μl (S.D. = 5 μl for 12 ratchet intervals) from a 6-ml disposable plastic syringe (cells) plus 61.2 μl (S.D. = 3.2 μl) from a 1-ml syringe (substrate).

Mixing is accomplished with a Y-bored (1/32 inch ϕ) mixing block of Plexiglas, with virtually no dead volume after confluence of the two streams. Initially both syringes are filled with the push bar drawn past the first ratchet groove, and the bar is then returned to the first groove to fill the mixing block just prior to beginning the timed run.

B. Centrifuge and Oil System

We use an Eppendorf Model 3200/30 microcentrifuge (Brinkmann Instruments, Westbury, N.Y.) with a 12-place, fixed-angle rotor. (This model has now been superseded by Model 1542 which performs similarly but is of less convenient design.) This machine accelerates to its maximum speed,

15,000 rpm (= 12,000 g), in 5 seconds and can be started by closing its lid. The caps of standard 1.5-ml polyethylene centrifuge tubes (Brinkmann Instruments, or less expensive imitations) are detached before use, as time does not permit capping during a transport run.

Dow Corning 550 silicone fluid (density 1.07 gm/ml, viscosity 125 cs at 25° C) has proven advantageous as the basic constituent of the oil phase. Eighty-four parts by weight of this oil is mixed with 16 parts of light mineral oil (density 0.844 gm/ml) to a final density of 1.034 gm/ml at 24° C. This oil mixture underlies the aqueous cell suspension but is readily penetrated by cells. (The apparent density of Novikoff rat hepatoma cells is about 1.07 gm/ml.)

Our formulation of the oil system was arrived at empirically after testing several varieties and mixtures with less success. Aside from the obvious factors of density and viscosity which affect cell sedimentation, other more subtle factors may contribute to the suitability of a given formulation. For example, the sedimentation of Novikoff cells with some oils tested was not as clean in glass as in polyethylene tubes.

If total incorporation of radioactivity into cells is to be measured, it suffices to add 100 μl of oil mixture to the centrifuge tubes; cells are simply pelleted below the oil and set aside until the workup (see Section II,D). If, in addition to separating cells from the suspension medium, it is necessary to stop cellular metabolism, an aqueous layer (50 μl) of 0.2 M trichloroacetic acid in 10% (w/v) sucrose (density 1.04 gm/ml) can be added under the oil layer. In this case we increase the oil layer to 250 μl (to prevent accidental mixing of acid and cell suspension upon syringing the cell suspension into the tube) and decrease the total delivery into the tubes to 405 μl by using a 0.110-inch ratchet spacing. Even so, the presence to a pH indicator in the suspending medium helps to detect an occasional mishap.

C.   Time Schedules

The time during which transport occurs is defined by the interval between delivery of the cell–substrate mixture from the mixing block and sedimentation of the cells into the oil phase. The lag between starting the centrifuge and effective sedimentation is considered in Section IV,B; it is about 2 seconds.

One dispenses the longest-time sample first, works backward, and starts the centrifuge soon after the last (shortest-time) sample has been dispensed. Scheduling is of course dictated by the transport system under consideration and by the experimental strategy. Take for example, the zero-trans influx of thymidine into a thymidine kinase-less subline of Novikoff rat hepatoma cells (Plagemann *et al.*, 1976). In these cells thymidine is not metabolized, and the equilibrium between extra- and intracellular thymidine is mediated by a saturable, nonconcentrative transport system. At exogenous concentrations of thymidine below the K_m^{zt} (i.e., $< 85\,\mu M$), influx is sufficiently rapid at 24°C to attain, intracellularly, 50% of the equilibrium level in 6–10 seconds (see Fig. 4; also Wohlhueter *et al.*, 1976). The range of time in which true initial rates are obtainable is impractically short. Therefore we follow the complete time course to equilibrium and calculate initial rates by fitting the influx data to an integrated rate equation (see Section IV,C). Equilibration is virtually complete by 60 seconds, so that scheduling aims to spread the 12 samples over this time.

Following the time course of a transport process can also yield other in-

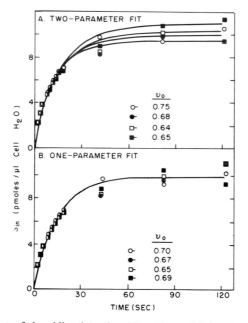

FIG. 4.   Transport of thymidine into thymidine kinase-deficient Novikoff rat hepatoma cells. 83.1 μM [³H]thymidine (10 μCi/ml) in a 1-ml syringe was mixed with a suspension of 2.5×10^7 cells/ml basal medium 42 in a 6-ml syringe (final concentration of thymidine 10 μM) and pelleted through oil. Total radioactivity in the pellet was measured. Four identical runs were carried out over a 2.5-hour period (open circles, 30 minutes; solid circles, 50 minutes; open squares, 120 minutes; solid squares, 150 minutes). Cell-associated radioactivity was corrected for 2.0 μl extracellular space/pellet, and the intracellular concentration was calculated from a cellular water space of 15.6 μl per pellet. (A) The results were fit to Eq. (1), whereby both k and $S_{in,\infty}$ were treated as parameters to be fit. (B) The same data were fit to Eq. (1), whereby $S_{in,\infty}$ was fixed at S_{out}. The ambient temperature was 24°C. In both cases, initial velocity v_0 is calculated as $kS_{in,\infty}$ in units of picomoles per microliter of cell water per second.

formation of interest: on the capacity of the transport system to generate a concentration gradient and on the intracellular space available to a given substrate compared to intracellular water space.

Whatever the schedule selected, we pace syringe operation with a standard metronome which provides from 40 to 208 beats per minute. Fixed time intervals are useful, and we manage 1.5-second intervals (i.e., every fourth beat at 160 beats per minute) comfortably and 1.0-second intervals with effort. Alternatively, and especially when full time courses are to be observed, an arithmetic progression of sampling times is advantageous. At 80 beats per minute, for instance, counting 12 beats between tubes 12 and 11, 11 between 11 and 10, etc., and 1 between tube 1 and centrifuge start yields a time sequence of 0.75, 2.25, . . ., 58.5 seconds.

If the duration of an experiment is to exceed more than a few minutes, the settling of unstirred cells may become a problem. Settling of cells in the syringe can be minimized by including a small air bubble and inverting the syringe between samplings; settling in the centrifuge tubes is a more intractable problem. In such experiments we generally take the short-term samples by means of the syringe, and the long-term samples by conventional pipetting from stirred reaction vessels.

In a typical transport run (to measure the zero-trans influx of 1 μM thymidine into thymidine kinase–less Novikoff cells, for instance), we set up the syringe apparatus with a 6-ml syringe containing a suspension of about 2×10^7 cells/ml basal medium, a 1-ml syringe containing 8.31 μM [*methyl-*^3H]thymidine (10 μCi/ml), and a ratchet slide with 0.138-inch spacing. This setting yields mixtures of 448 μl of cell suspension plus 61.2 μl of substrate solution. These mixtures are prepared in 1.5-second intervals (every fourth beat with a metronome setting of 160 beats per minute) and dispensed in sequence into 12 centrifuge tubes containing 100 μl of oil mixture each. After centrifugation for about 10 seconds, the samples are transferred to racks (inexpensively made to any desired size from stock aluminium shelving) to await processing as described in Section II, D.

Experiments are carried out at ambient temperature which, if need be, is thermostatically controlled.

D.   Sample Workup

1.   TOTAL INCORPORATION

When total isotope incorporation into cells is to be measured, as in the above example, a two-phase system (oil under cell suspension) suffices. The cell-free aqueous layer is aspirated, and the tube is filled with water, gently so as not to perturb the oil-covered cell pellet. The rinse water is then aspirated along with most of the oil, and the pellet is analyzed for radioactivity. Rinsing removes > 99.99% of radioactivity not associated with the pellet.

Generally, we add 200 μl of 0.5 M trichloroacetic acid to each pellet and immediately mix on a vortex-type mixer. (Immediate mixing is essential for a finely dispersed precipitate.) The tubes are then incubated at 70° C for 30 minutes, after which the tube, cap, and contents are transferred to a scintillation vial containing a dioxane-based scintillator solution. Alternatively, the pellet may be dissolved in a tissue solubilizer (e.g., NCS, Amersham/Searle, Arlington Heights, Ill.) and counted in a toluene-based scintillation solution. This method is more costly than trichloroacetic acid digestion and suffers from a progressive color quenching if scintillation counting is delayed.

Figure 4 demonstrates the rapidity and reproducibility of results obtained using this methodology to study thymidine transport into thymidine kinase–less cells. Repeated runs were performed over a period of 2.5 hours with the same cell suspension, and the data were fit to an integrated rate equation by two approaches, as described in Section IV,C.

2. Fractionation of Intracellular Radioactivity

When monitoring the uptake of metabolized substrates, it may be desirable to fractionate intracellular radioactivity into individual metabolites or classes of metabolites. Metabolism is brought to a halt by centrifuging into a trichloroacetic acid phase underlying the oil. Tubes in which cells have been centrifuged into acid are stored on ice, rinsed as described above, and extracted with 1 ml of cold diethyl ether, which removes oil residues and most of the trichloroacetic acid. Subsequent fractionations will depend on experimental objectives. With wild-type Novikoff cells and [^{3}H]thymidine as substrate we have fractionated the acid phase into free nucleoside, nucleotides, and DNA as follows (at 0 to 4° C). After ether extraction, 5 μl of a carrier solution is added (either 1 M inorganic phosphate or 10 mM dTTP) to the residue, followed by 500 μl of a solution composed of 0.1 M LaCl$_3$ (Gallard-Schlesinger Corporation, Carle Place, N.Y.), 50 mM triethanolamine, and 20 mg Hyflo Super-Cel (Fisher Scientific Company, Chicago, Ill. per milliliter, pH 7.8 (adjusted with HCl). The mixture is allowed to stand $\geq$ 15 minutes and then spun in the Eppendorf centrifuge. An aliquot of the supernate is analyzed for radioactivity (free thymidine); the remaining supernate is aspirated, and the pellet washed once with the LaCl$_3$ solution minus Hyflo Super-Cel. The washed pellet is extracted with 500 μl of cold (0° C), 0.1 M trichloroacetic acid, which dissolves the lanthanum–nucleotide precipitate, and an aliquot of supernate is analyzed for radioactivity (nucleotides). The remaining supernate is aspirated, and the pellet is washed with 0.1 M trichloroacetic acid. The washed pellet is hydrolyzed in 200 μl 0.5 M trichloroacetic acid at 70°C for 30 minutes, and the tube

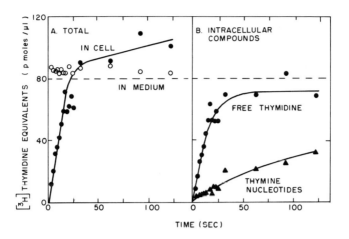

FIG. 5.   Time courses of thymidine incorporation into intracellular free thymidine and thymine nucleotide pools by wild-type Novikoff hepatoma cells. The incorporation of [³H]thymidine (final concentration 80 μM, 1.2 μCi/ml) was followed in a suspension of 6.3 × 10⁷ Novikoff hepatoma cells per milliliter of basal medium containing D-glucose, at 24°C. Cells were centrifuged through oil into 0.2 M trichloroacetic acid, and the cell-associated radioactivity fractionated into thymidine, thymine nucleotides, and DNA (see text). (A) The sum of isotope in all three fractions as a function of time (solid circles), and that in the supernatant medium (open circles). (B) Isotope incorporation into intracellular thymidine (solid circles) and into thymine nucleotides (triangles). Isotope incorporation into DNA at 24°C was negligible at this coordinate scale and is not illustrated.

and contents transferred to a scintillation vial for counting (DNA). The results of such an experiment are shown in Fig. 5 (and discussed in Section IV,C).

E.   Experimental Protocols

The examples cited so far have been for zero-trans influx. In this protocol radioactively labeled substrate (in the smaller syringe) is mixed with a cell suspension in which the concentration of substrate in extra- and intracellular space is (or is assumed to be) zero, and the net uptake of label into the cells is measured.

The dual-syringe apparatus also lends itself to two further experimental approaches, equilibrium exchange and infinite-cis, as diagramed in Fig. 6. Both types of experiments are applicable of course only with nonmetabolizing systems. In equilibrium exchange, cells are preincubated with unlabeled substrate for a time sufficient to ensure complete equilibration of substrate between intra- and extracellular space. This mixture of cells and substrate is then loaded into the larger syringe and mixed with the same substrate at the

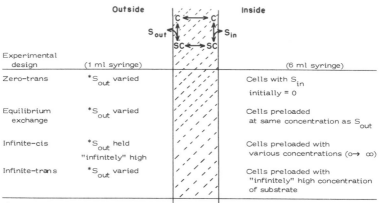

F IG. 6.   The simple carrier model in various experimental configurations. Nomenclature is that of Eilam and Stein (1974), who have developed rate equations for the model depicted. With the dual-syringe apparatus radioactive substrate is added to the "outside" space from one syringe; cells are added from the other syringe in which the concentration of nonradioactive substrate in the "inside" space is varied by preloading the cells as indicated.

same concentration, but radiolabeled, loaded into the smaller syringe. The rate of appearance of radioactivity in the sedimented cells is equivalent to the rate of change in specific radioactivity in the intracellular substrate pool and involves no net flux of substrate.

For an infinite-cis experiment cells are preincubated with various discrete concentrations of nonlabeled substrate until equilibrium is attained. This cell suspension is then loaded into the larger syringe and mixed with radioactive substrate from the smaller syringe at an "infinite" concentration (i.e., 10 to 100 times the K_m for equilibrium exchange K_m^{ee}). Measured is the influx of labeled substrate into the cell against a known intracellular concentration. It should be noticed that in this case there is a net flux of substrate across the membrane; it is, however, not this net flux but rather the undirectional influx of radioisotope that is measured. The kinetic equations applicable to this case are different from those derived by Sen and Widdas (1962) or Eilam and Stein (1974) for infinite-cis experiments in which net flux is measured. The appropriate equation is presented in Section IV, C. It should also be appreciated that an integrated rate equation is not available for the infinite-cis experiment, and that the kinetic equation discussed in Section IV,C pertains to the time period in which the infinite substrate concentration remains effectively constant. For these reasons, initial rates must be measured directly.

An infinite-trans protocol (and, by extension, a countertransport protocol) is not practicable with the dual-syringe methodology, because substrate remaining outside the cells after preincubation with a high concentration of nonradioactive substrate dilutes the radioactivity of the added, labeled substrate. Such an experiment can, however, be conducted in the following manner (Marz *et al.*, 1978). Cells are preloaded to equilibrium with an infinite concentration of unlabeled substrate. The equilibrated cells are collected by centrifugation and washed rapidly in cold (0°C) basal medium by resuspension and centrifugation. The cells are then resuspended to the original density (1–2 × 10⁷ cells/ml) in medium containing varying concentrations of labeled substrate. At various times of incubation, duplicate 0.5-ml samples of the suspension are layered over an oil layer and centrifuged in an Eppendorf centrifuge. Time points as short as 18 seconds can be taken using one centrifuge, and shorter time points are possible with the alternate use of two centrifuges. In infinite-trans or countertransport experiments the labeled substrate accumulates transiently within the cell against a concentration gradient (see Plagemann *et al.*, 1976), and influx is constant with time for 60 seconds or more. Thus initial rates of entry of labeled substrate can be readily estimated from the initial time points, even though these are at much longer intervals compared to those attained with the other protocols compatible with the dual-syringe technique.

III. Cells and Substrates

A. Relationship of Transport to Metabolism

Consider a hypothetical situation in which transport is very rapid and nonconcentrative, in which the transported substrate is phosphorylated more slowly and thus sequestered within the cell, and in which phosphorylation proceeds linearly for relatively long times. In such a situation, the measurement of incorporation of exogenously supplied, isotopically labeled substrate into cells on a leisurely time scale may provide considerable information about cellular metabolism but not about the transport system.

In fact, this hypothetical situation fairly well describes the incorporation of nucleosides, nucleic acid bases, and hexoses into several mammalian cell lines (Wohlhueter *et al.*, 1976). It is the reality of this situation which provides impetus for the use of rapid kinetic techniques and of nonmetabolizable substrates.

B.   Transport without Metabolism

The problem of disentangling transport from metabolism is most easily resolved by choosing substances which are substrates for the transport system under study but are not metabolized by the cells used in the study. Several tactics have been employed to establish this situation, usually involving either the choice of nonmetabolized analogs of natural substrates or the manipulation of cells so as to put a metabolic pathway normally present out of commission.

The synthetic compounds 3-O-methyl-D-glucose (reviewed by Plagemann and Richey, 1974) and γ-aminoisobutyric acid (Christensen, 1975) have been widely applied to study hexose and amino acid transport, respectively, and illustrate the use of nonmetabolized analogs. However, no generally nonmetabolized nucleoside or base analogs have been described.

Other normally metabolized substrates have been rendered nonmetabolizable by genetic manipulation. The selection of Novikoff cells for resistance to bromodeoxyuridine results in sublines devoid of thymidine kinase and unable to alter thymidine metabolically. Such a cell line has been exploited to study thymidine transport (Plagemann *et al.*, 1976). Likewise, deoxycytidine-deficient Yoshida sarcoma cells (Mulder and Harrap, 1975) and hypoxanthine-guanine phosphoribosyltransferase–deficient Novikoff hepatoma cells (Zylka and Plagemann, 1975) are metabolically inert toward deoxycytidine–cytosine arabinoside and hypoxanthine-guanine, respectively, and lend themselves to studies of the transport of these compounds.

Another approach is aimed at the functional impairment of enzymes. Christopher (1977) inhibited galactose kinase of NIL hamster kidney cells by treatment with N-ethylmaleimide and thus rendered galactose a nonmetabolized substrate for the hexose transport system of these cells. Our laboratory has used ATP depletion as a means of preventing the phosphorylation (or phosphoribosylation) of substrates by kinases (or phosphoribosyltransferases). Incubation of Novikoff cells for 10 minutes in 5 mM KCN plus 5 mM iodoacetate in a basal medium lacking D-glucose reduces the ATP pool by 99% (Plagemann *et al.*, 1976). Although such cells are somewhat fragile and deteriorate after about 2 hours, the thymidine transport system characterized in them resembles closely that in thymidine kinase–deficient sublines (Plagemann *et al.*, 1976; Marz *et al.*, 1977).

Quinlan and Hochstadt (1976) advocate the use of membrane vesicles prepared from mammalian cells as a means of segregating transport from metabolism. The concept is attractive, but conventional techniques for preparing vesicles do not guarantee their being devoid of cytoplasmic enzymes and one must carefully assess the contribution of residual enzyme activities.

It should be emphasized that the use of nonmetabolizing systems does not obviate the need for rapid kinetic techniques for measuring transport rates. On the contrary, it is precisely such systems that have most clearly identified this need. Furthermore, it is not always sufficient to incapacitate a single enzyme. Adenosine, for example, is rapidly deaminated even in cells lacking adenosine kinase, or depleted of ATP. Mammalian cells generally lack a kinase for guanosine and inosine, but both substrates are rapidly degraded by nucleoside phosphorylase. Uridine is also phosphorolyzed even in uridine kinase–deficient mutants. In such situations rapid sampling techniques may be validly applied to estimate initial rates of incorporation, as

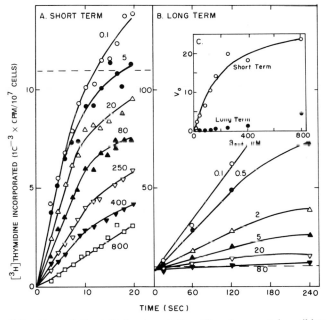

FIG. 7. Estimation of the initial rates of thymidine transport by wild-type Novikoff hepatoma cells (A), long-term incorporation values (B), and Michaelis–Menten analyses of data (C). Samples of a suspension of cells and solutions of $[^3H]$thymidine were mixed as described in the text (Section II,C) to a final cell density of 1×10^7 cells/ml, and the indicated final concentrations of thymidine (micromolar; 840,000 cpm/ml in each case). After incubation at 24°C, the cells were collected by centrifugation through oil, and the cell pellets were analyzed for radioactivity (see Section II,D). Initial rates v_0 were estimated from the linear portions of the short-term (A) and long-term (B) incorporation curves. The broken lines indicate the intracellular radioactivity expected at equilibrium with that in the extracellular medium. Kinetic parameters for the short-term velocities (C) are $K_m = 192\ \mu M$ and $V_{max} = 30$ pmoles/μl cell water per second.

demonstrated for thymidine transport into wild-type cells (Fig. 7). But the burden of proof lies with the would-be investigator of transport that the results of his or her investigations reflect indeed transport and not metabolism (cf. Figs. 5 and 7).

C. Cell Density and the Sensitivity of Transport Measurements

Obviously the methodology described here requires cells grown in (or brought into) suspension. A further limitation is imposed if a nonconcentrative transport system is studied, namely, the number of cells needed to determine transport rates accurately. The intracellular volume of 10^6 Novikoff cells, for example, comprises about 1.2 μl. At a final isotopic concentration of 1 μCi/ml, the maximum ($t \rightarrow \infty$) radioactivity appearing in 10^6 pelleted cells is about 600 cpm. We generally use 10^7 cells per sample, but this requires, for a single run of 12 time points, the expenditure of 12×10^7 cells. We have found, however, that thymidine influx (in picomoles per microliter of cell water per second) into thymidine kinase–deficient Novikoff cells is constant over a density range from 5×10^6 to 2×10^8 cells/ml (see Fig. 8A).

IV. Evaluation of Data

A. Estimation of Intra- and Extracellular Spaces

In order to quantitate intracellular pools of substrate, the contribution of substrate in the extracellular space to the total radioactivity in the pellet must be evaluated. Where an unambiguous, linear initial rate is obtained, extrapolation to zero time may suffice. Analogously, if a decelerating uptake curve is to be fit to a nonlinear equation, it is possible to treat the zero-time "background" as an additional parameter to be fit. For an independent estimation of extracellular space, however, we measure $[carboxyl\text{-}^{14}C]$ carboxylinulin space, and simultaneously, total ^3H_2O space.

A solution containing 5 μCi of $[^{14}C]$inulin and 10 μCi ^3H_2O per milliliter of balanced salt solution is mixed with the cell suspension in the same proportion as delivered by the syringes (normally 218 μl of inulin–H_2O solution to 1600 μl of cell suspension). Triplicate samples (normally 509 μl) are removed to centrifuge tubes after about 1 minute. Centrifugation through oil and a sample workup are carried out as for transport samples. Tritium and carbon-14 are counted with an internally standardized, dual-channel scintillation spectrometer, and the data are evaluated assuming that $[^{14}C]$inulin is distributed evenly throughout the extracellular space

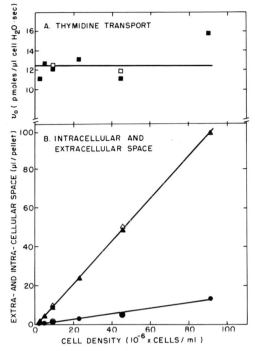

FIG. 8. Zero-trans initial velocities of thymidine transport in thymidine kinase deficient Novikoff hepatoma cells (A) and extra- and intracellular water space (B) as a function of cell density. [*Carboxyl*-14C] carboxylinulin space (circles), 3H2O space (triangles), and zero-trans influx of thymidine (squares; final concentration 320 μM) were measured as described in Sections IV,A and II,C, respectively. Five dilutions of the same batch of cells were assayed in triplicate. The open symbols refer to measurements repeated approximately 2 hours after the first set. The average initial velocity v_0 was 12.5 pmoles/μl cell water per second (S.E. = 0.5). The slope of the line for intracellular water corresponds to 1.09 μl/106 cells, and that for extracellular space to 0.146 μl/106 cells (11.8% of total pellet space).

and 3H2O throughout the total water space in the cell pellet; the intracellular space is calculated by difference. Figure 8B shows that both spaces are directly proportional to cell number, in accordance with virtually no detectable radioactivity left in the centrifuge tubes in the absence of cells. From Table I it is apparent that the intracellular space and the extracellular space in a cell pellet vary with cell type and, to some extent, within the same cell line in different experiments. Intracellular volume depends on the growth stage of the cells, while extracellular space depends presumably on the surface area per cell. In any case, from raw uptake data a background of radioactivity is subtracted equivalent to that of substrate radioactivity contained in the extracellular space.

The inulin–water procedure also gives an independent estimate of intra-

TABLE I

KINETIC PROPERTIES OF THYMIDINE TRANSPORT IN VARIOUS

CELL LINES TREATED WITH KCN AND IODOACETATE[a]

Cell line	K_m^{zt}	V_{max}^{zt}	Intracellular space	Extracellular space
	(μM)	(pmoles/μl cell water per second)	(μl/10^7 cells)	(% of total pellet water)
Novikoff hepatoma, N1S1-67 (TK$^-$)[b]	70–140	10–24	10–15	8–20
Novikoff hepatoma, N1S1-67 (wild type)	85	27.	13	12
Chinese hamster ovary, suspension culture	35	3.5	13	26
Mouse leukemia, P388	43	5.6	5–7	22
Mouse fibroblasts, L (wild type)	44	4.0	15	10

[a]Zero-trans influx of $[^3H]$thymidine was measured over a concentration range of 10–800 μM in various cell lines treated with KCN and iodoacetate to deplete them of ATP. Uptake data were fit to Eq. (1), and initial velocities computed as the slope of the curve at zero time. The kinetic parameters K^{zt} and V_{max}^{zt} were obtained by fitting initial velocities to the Michaelis–Menten equation. Methods for ATP depletion, influx measurements, and extra- and intracellular volume estimations are described in the text.

[b]Control cell line not reated with KCN plus iodoacetate; data are the range of values from several experiments.

cellular water space. In the case of thymidine uptake into thymidine kinase–deficient or ATP-depleted Novikoff cells, the intracellular thymidine space at equilibrium agrees (generally within 10–15%) with the intracellular water space. On this agreement we base our conclusion that thymidine transport is nonconcentrative. The converse logic, i.e., that greater-than-equilibrium amounts of substrate within the cell imply a concentrative transport system, however, is not necessarily sound. We offer two examples in which it is unsound. (1) Prednisolone is taken up into Novikoff cells to an extent about twice that expected for equilibrium in cellular water (Plagemann and Erbe, 1976). The intracellular radioactivity is recoverable as prednisolone. It is the solubility of the lipophilic substrate in the lipids of the cell and its binding to cytoplasmic receptors which account for the manifest concentrative uptake. (2) 8-Azaguanine loses a proton with p$K_a =$ 6.5 (Parks and Agarawal, 1975); the undissociated form enters the cell, whereas the anionic form is not taken up. Equilibrium across the cell membrane is rapidly attained for the undissociated form but, if the extracellular pH is lower than the intracellular pH, the ratio of dissociated to undissociated is higher in the cell, giving rise to a greater-than-equilibrium concentration of radioactive azaguanine in the cell (Plagemann, 1978).

When a substrate is metabolized, intracellular concentrations of radioactivity derived from the substrate may well exceed that in the medium (see, for example, Figs. 5 and 7). In such a case, if intracellular pools of isotope are to be expressed on a concentration basis, an independent estimate of cellular water space is essential.

B. Estimation of Effective Zero Time

There is a finite time between switching on the centrifuge and the cells becoming effectively out of contact with the medium. This time is certainly not negligible with respect to the rates of nucleoside transport into Novikoff cells, and its estimation is probably one of the largest sources of error in measuring these rates. We have arrived at a figure of 2 seconds from the following considerations.

At initial concentrations of thymidine well below its transport K_m, the entry of thymidine into cells follows first-order kinetics, and its approach to equilibrium is described by a standard, integrated form of the first-order rate equation:

$$S_{in,t} = S_{in,\infty} (1 - e^{-kt}) \tag{1}$$

where k = first-order rate constant, $S_{in,t}$ = intracellular cocentration at time t, and $S_{in,\infty}$ is its asymptotic value. A graph of eq. (1) passes through the origin. To shift the time axis by a constant amount t_0, Eq. (1) can be rewritten as:

$$S_{in,t} = S_{in,\infty} (1 - e^{-k(t+t_0)}) \tag{2}$$

Data from several experiments were fit to Eq. (2) by a nonlinear, least-squares method (see Section IV, C). Best-fitting values for t_0 cluster around -2 seconds. We have adopted this value as the effective zero time, and it seems in reasonable accord with the 5-second acceleration time of the Eppendorf centrifuge. Our nominal sampling time is thus incremented by 2 seconds for purposes of calculating uptake rates.

C.   Analysis of Transport Data in Terms of Theoretical Models

1.   Direct Estimation of Initial Velocities

With suitably slow uptake processes, and necessarily with substrates subject to metabolism, initial rates of entry can be obtained from graphical analysis or linear regression. This is illustrated in Fig. 7 for thymidine uptake into wild-type Novikoff cells. Figure 7A shows the actual initial rates of entry of [³H]thymidine as measured by the rapid-sampling technique. Graphical estimation of initial slopes, plotted against exogenous thymidine concentration (Fig. 7C) indicates a K_m (192 μM) and a V_{max} (30 pmoles/μl cell water per second) which are ascribable to the transport system (see Fig. 9A). It is obvious, however, that at most substrate concentrations incorporation is linear for not more than 5–8 seconds (Fig. 7A).

Figure 7B illustrates the pitfalls inherent in choosing inappropriately long time schedules when characterizing a transport system. Long-term points from the same experiment (namely, those at 19.5, 60, 120, and 240 seconds) are plotted separately. One is tempted to regard the slopes depicted as initial slopes, an error which leads to very different conclusions with respect to kinetic parameters (Fig. 7C).

The reasons for this are elucidated by the experiment shown in Fig. 5B, where the intracellular radioactivity is fractionated into free thymidine and thymine nucleotides. By 50 seconds the thymidine pool is in a steady-state; the increase in total radioactivity from 1 minute onward measures the rate of phosphorylation of thymidine, and the apparent, high zero time points in Fig. 7B reflect the rapid intracellular accumulation of free thymidine to equilibrium levels.

For very rapid transport with which metabolism does not interfere, it is advantageous to apply nonlinear equations. For the simple, carrier-mediated transport model depicted in Fig. 6 (operating in the absence of metabolism) Eilam and Stein (1974) developed both initial velocity and integrated rate equations corresponding to many of the experimental protocols described above (Section II,E). The application of these equations to zero-trans, equilibrium exchange, and infinite-cis—the experimental protocols practicable with the dual-syringe technology—permit estimation of true initial velocities where the linear portion of the curve is impractically short. A comparison of the kinetic parameters measured in each of these protocols with one another can, theoretically at least, also yield valuable information about the molecular properties of the transport system. Application of these rate equations to each of the experimental configurations is discussed in turn.

2. ZERO-TRANS INFLUX

Zero-trans initial velocity is a Michaelis function of substrate concentration, where the apparent V_{max}^{zt} and K_m^{zt} are functions of the rates of movement of the loaded and unloaded carrier, and of the carrier–substrate association constant K. The various rate and association constants can all be extracted from analysis of the time courses of uptake at a series of initial concentrations, and Eilam and Stein (1974) present a graphical method for doing this. The graphical method relies on a logarithmic transformation in which the relative errors become large as the intracellular substrate concentration S_{in} approaches the extracellular concentration S_{out}, and proper weighting becomes difficult.

Empirically, we find that the time courses of zero-trans influx of thymidine and 3-O-methylglucose into Novikoff cells is well described by a simple, integrated, first-order rate equation [See Eq. (1)]. For nonconcentrative transport $S_{in, \infty}$ should equal S_{out}, which is assumed to be constant with time. (The assumption is reasonable at a density of 2×10^7 Novikoff cells per milliliter, which corresponds to an intracellular volume equal to about 2.5% of the medium volume.) However, k is a pseudo-first-order constant, in that it is in fact a function of S_{out}.

The aptness of Eq. (1) in describing thymidine transport is illustrated in Fig. 4 (see also Wohlhueter et al., 1976), and we rationalize it as follows. The integrated rate equation of Eilam and Stein (1974) can be written formally as

$$S_{in,t} = S_{in, \infty} \left\{ 1 - \exp\left[- \frac{t + f_2(S_{in,t})}{f_1(S_{out})} \right] \right\} \tag{3}$$

Here $f_1(S_{out})$ is a quadratic function of S_{out}, and $f_2(S_{in,t})$ is a linear function of $S_{in,t}$, not segregated to the left side of the equation. $f_2(S_{in,t}) = 0$ when $t = 0$ and becomes constant $[= f_2(S_{out})]$, and thus negligible, as $t \to \infty$. If one neglects it at all values of t, Eq. (3) reduces to Eq. (1) with $k = 1/f_1 (S_{out})$. At any rate, to the extent to which Eq. (1) describes adequately the experimental points, its first derivative at $t = 0$, namely, $kS_{in,\infty}$, may be taken as the initial slope of the curve v_0, and it is for this purpose that Eq. (1) is very useful.

Plots of v_0 versus S_{out} are hyperbolic and allow calculation of the zero-trans kinetic constants K_m^{zt} and V_{max}^{zt}. Zero-trans kinetic parameters for thymidine transport in Novikoff cells are illustrated in Fig. 9A, and those for several suspension culture lines are summarized in Table I. v_0 and V_{max}^{zt}, computed as outlined here, have the dimensions concentration per time.

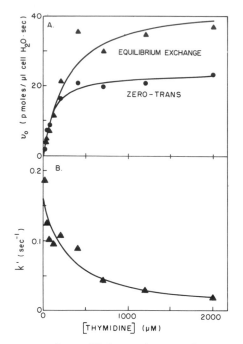

FIG. 9.   Kinetic parameters for equilibrium exchange and zero-trans influx of thymidine in Novikoff hepatoma cells. The initial velocities of entry and of equilibrium exchange of [³H]thymidine at various concentrations (4.1 or 8.2 μCi/ml) into thymidine kinase–deficient Novikoff cells were measured as described in the text and computed by fitting uptake data to Eq. (1). The ambient temperature was 24°C. In (A) the initial velocities of entry (circles; $k' \times S_{in,\infty}$) and equilibrium exchange (triangles; $k' \times$ [thymidine]) are plotted against [thymidine]. For entry the least-squares regression line is for $v_0 = V_{max}^{zt} S/K_m^{zt} + S)$, where $V_{max}^{zt} = 24$ pmoles/μl cell water per second and $K_m^{zt} = 106$ μM. For equilibrium exchange $V_{max}^{ee} = 43$ pmoles/μl cell water per second and $K_m^{ee} = 223$ μM. In (B) the pseudo-first-order rate constant k' is plotted directly against [thymidine]. The regression line is for $k' = V^{ee}/K_m^{ee} + S)$, where $V^{ee} = 37.1$ pmoles/μl cell water per second and $K_m^{ee} = 244$ μM.

They are thus contingent on cell volume and can be converted to the conventional dimensions of permeation (moles per area per time) only if the relationship between cell volume and surface area is known or assumed.

Data can be fit to Eq. (1) by iterative nonlinear, least-squares procedures. *Statistical Package for the Social Sciences* (Nie *et al.*, 1975) and *Biomedical Computer Programs* (Dixon, 1975) provide appropriate computation programs on a subscription basis, and the authors can make available comparable programs in FORTRAN or Hewlett-Packard's HPL language (Dietrich and Rathmann, 1975). The latter, a generalized, nonlinear least-squares program, has been used with a Hewlett-Packard model 9825 desk-top computer for all the theoretical curves illustrated here.

3.   EQUILIBRIUM EXCHANGE

For equilibrium exchange, too, an integrated rate equation has been solved for the model in Fig. 6 (Eilam and Stein, 1974). It can be rearranged in the same form as Eq. (1):

$$N_{in,t} = N_{in,\infty} \left[1 \left(- \exp - \frac{t V_{max}^{ee}}{K_m^{ee} + S_{out}} \right) \right]$$

where $N_{in,t}$ = intracellular radioactivity per sample at time t and is proportional to the specific radioactivity of the intracellular substrate. $N_{in,\infty} = N_{out}$ = radioactivity per equivalent volume of medium, and K_m^{ee} and V_{max}^{ee} are the Michaelis–Menten constants for equilibrium exchange. Letting $V_{max}^{ee}/(K_m^{ee} + S_{out}) = k$ reduces Eq. (4) formally to Eq. (1). Replots of k versus S_{out}, fit by a nonlinear, least-squares procedure, allow evaluation of K_m^{ee} and V_{max}^{ee} (see Fig. 9B). Alternatively, the product kS_{out} expresses the unidirectional flux in dimensions of moles per liter per time unit and can be plotted against S_{out} $V_{max}^{ee}/(K_m^{ee} + S_{out})$ to yield a familiar Michaelis–Menten plot (Fig. 9A). These alternative methods of fitting data do not necessarily give identical results for K_m^{ee} and V_{max}^{ee}, for in essence each weights the data differently.

For thymidine we find that V_{max}^{ee} is about two to three times greater than V_{max}^{zt} (see Fig. 9A). In terms of the simple carrier model, $V_{max}^{ee} > V_{max}^{zt}$ implies that the loaded carrier moves faster than the unloaded carrier (Eilam and Stein, 1974).

4.   INFINITE-CIS

The rate equation for infinite-cis experiments, as developed by Sen and Widdas (1962) and Eilam and Stein (1974), is based on a measurement of net flux. Utilization of radiolabeled substrates, as in the protocol outlined in Section II,E, results in one's measuring the unidirectional flux of the isotope and requires a different equation.

Starting from a general equation for unidirectional flux (Eq. 16 of Eilam and Stein, 1974) and introducing the simplifying assumptions that $S_{out} \gg K$ (the carrier–substrate dissociation constant) and is essentially constant, we obtain for the initial velocity of influx v_{influx}^{ic} (inward, as the present methodology measures, where "ic" stands for infinite cis):

$$v_{influx}^{ic} = \frac{K + S_{in}}{(K/V_{efflux}^{zt}) + (S_{in}/V_{max}^{ee})} \tag{5}$$

where V^{zt}_{efflux} = maximum velocity of zero-trans efflux which, if the system is symmetrical, equals V^{zt}_{max} of influx.

This equation is not Michaelian in form, so that K_m^{ic} is not defined, and moreover, as V^{zt}_{efflux} approaches V^{ee}_{max}, the equation degenerates into an identity. If, however, V^{zt}_{efflux} and V^{ee}_{max} are sufficiently different, the equation takes on some useful attributes. A plot of V^{ic}_{influx} versus S_{in}, for example, is hyperbolic with a horizontal asymptote = V^{ee}_{max} (the maximal velocity of equilibrium exchange), with Y intercept = V^{zt}_{efflux} and X intercept = $-K$.

Thus, theoretically, an infinite-cis series with varying concentrations of substrate preloaded in the cells can directly yield a value for the substrate–carrier dissociation constant, as well as values for the maximum velocities of zero-trans exit and equilibrium exchange.

As a practical method for determination of the carrier–substrate dissociation constant, the infinite-cis protocol (detailed in Section II,E), however, has not been satisfactory in our hands. Because S_{out} is infinitely high, the relative rates of entry are slow, but linear for a considerable time. Thus initial velocities can be determined by graphical analysis, but the radioactivity in the extracellular space is high in comparison to that in the intracellular space. The unfavorable "signal-to-noise ratio" results in imprecise estimations of initial velocities and consequently in unsatisfactory fits to Eq. (5).

The condition in which $V^{ee}_{max} > V^{zt}_{max}$ has been observed also for uridine in Ehrlich ascites cells and termed "accelerated exchange diffusion" by Cass and Paterson (1972), although these investigators did not apply the appropriate equation for unidirectional, infinite-cis flux. As noted above, this inequality implies that the loaded carrier moves faster than the unloaded carrier. It is just this condition which is most succinctly demonstrated by an infinite-cis experiment.

V.   Conclusions

The transport of various compounds across mammalian cell membranes is frequently found to occur with a rapidity which necessitates collecting data at intervals of a few seconds. By means of a dual-syringe device, suspended cells can be mixed nearly instantaneously with radioactively labeled substrate and separated from the substrate again within seconds by centrifugation into silicone oil.

These techniques have been employed to measure the transport of nonmetabolized substrates, as well as to distinguish kinetically transport from metabolism, where the latter occurs. We have used these techniques mostly to study the transport of nucleosides and hexoses into Novikoff rat hepatoma cells, but also with other cell lines and classes of substrates, and we believe that they are of general utility in measuring the uptake of any. substrate which rapidly enters (or exits) cells in suspension.

Depending on the cell–substrate system under investigation, initial transport velocities may be either measured directly or calculated from the time course with which equilibrium across the membrane is attained.

With nonmetabolizing systems, the dual-syringe apparatus is adaptable to a variety of experimental protocols—zero-trans, equilibrium exchange, and infinite-cis—which in combination make possible a thorough kinetic characterization of a transport system. With cell–substrate systems in which the substrate is metabolized, the methodology can be extended to probe the relationship of transport to metabolism—to follow, for example, the establishment of steady-state, intracellular pools of substrate and the rate of conversion of substrate *in situ* to metabolic products.

The syringe apparatus itself is simple and easily constructed; the necessary auxiliary equipment is relatively inexpensive; and the computational software, if needed, is available upon request.

ACKNOWLEDGMENTS

We are indebted to John Erbe of the Department of Microbiology and John Connett of the Biometry Department for their indispensible help with computer programing, to Willi Rudi for masterfully machining the syringe apparatus, to Heather Kutzler for technical drawings of that apparatus, and to Cheryl Thull for her help in preparation of the manuscript. Special thanks goes to Franklyn Prendergast for his constructive criticism of the manuscript. This work was supported by Public Health Service Research Grant CA 16228 and Training Grant CA 09138 (R.M.) from The National Cancer Institute, NIH.

REFERENCES

Berlin, R. D., and Oliver, J. M. (1975). *Int. Rev. Cytol.* **42**, 287–336.
Cass, C. E., and Paterson, A. R. P. (1972). *J. Biol. Chem.* **247**, 3314–3320.
Christensen, H. N. (1975). "Biological Transport," 2nd ed. Benjamin, Reading, Massachusetts.
Christopher, C. W. (1977). *J. Supramol. Struct.* **6**, 485–494.
Christopher, C. W. (1977). *Fed. Proc., Fed. Am. Soc. Exp. Biol.* **36**, 692.
Dietrich, O. W., and Rathmann, O. S. (1975). "Keyboard" Vol. 7, pp. 4–6. (Hewlett-Packard, Inc Loveland, Colorado).
Dixon, W. J. (1975). "Biomedical Computer Programs," pp. 541–572. Univ. of California Press, Berkeley.
Eilam, Y. (1975). *Biochem. Biophys. Acta* **401**, 349–363.
Eilam, Y., and Stein, W. D. (1974). *Methods Membr. Biol.* **2**, 283–354.
Gibson, Q. H. (1969). *In* "Methods in Enzymology" (K. Kustin, ed.), Vol. 16, pp. 187–249. Academic Press, New York.

Hankin, B. L., and Stein, W. D. (1972). *Biochim. Biophys. Acta* **288**, 127–136.

Hawkins, R. A., and Berlin, R. D. (1969). *Biochim. Biophys. Acta* **173**, 324–337.

Lassen, U. V. (1967). *Biochim. Biophys. Acta* **135**, 146–154.

Marz, R., Wohlhueter, R. M., and Plagemann, P. G. W. (1977). *J. Supramol. Struct.* **63**, 433–440.

Marz, R., Wohlhueter, R. M., and Plagemann, P. G. W. (1978). Submitted for publication.

Mawe, R. C., and Hempling, H. G. (1965). *J. Cell. Comp. Physiol.* **66**, 95–104.

Mulder, J. H., and Harrap, K. R. (1975). *Eur. J. Cancer* **11**, 373–379.

Nie, N. H., Hull, C. H., Jenkins, J. G., Steinbrenner, K., and Bent, D. H. (1975). "Statistical Package for the Social Sciences," 2nd ed. McGraw-Hill, New York.

Oliver, J. M., and Paterson, A. R. P. (1971). *Can. J. Biochem.* **49**, 262–270.

Parks, R. E., Jr., and Agarwal, K. C. (1975). *Handb. Exp. Pharmacol.* **38**, Part 2, 458–467.

Plagemann, P. G. W., and Erbe, J. (1976). *Biochem. Pharmacol.* **25**, 1489–1494.

Plagemann, P. G. W., and Richey, D. P. (1974). *Biochim. Biophys. Acta* **344**, 263–305.

Plagemann, P. G. W., Marz, R., and Erbe, J. (1976). *J. Cell. Physiol.* **89**, 1–18.

Plagemann, P. G. W. (1978). In preparation.

Portis, A. R., and McCarty, R. E. (1976). *J. Biol. Chem.* **251**, 1610–1617.

Quinlan, D. C., and Hochstadt, J. (1976). *J. Biol. Chem.* **251**, 344–354.

Segel, I. H. (1975). "Enzyme Kinetics." Wiley, New York.

Sen, A. K., and Widdas, W. F. (1962). *J. Physiol. (London)* **160**, 392–416.

Sha'afi, R. I., Rich, G. T., Sidel, V. W., Bossert, W., and Solomon, A. K. (1967). *J. Gen. Physiol.* **50**, 1377–1399.

Strauss, P. R., Sheehan, J. M., and Kashket, E. R. (1976). *J. Exp. Med.* **144**, 1009–1021.

Taverna, R. D., and Langdon, R. G. (1973). *Biochim. Biophys. Acta* **298**, 412–421.

Werkheiser, W. C., and Bartley, W. (1957). *Biochem. J.* **66**, 79–91.

Wohlhueter, R. M., Marz, R., Graff, J. C., and Plagemann, P. G. W. (1976). *J. Cell. Physiol.* **89**, 605–612.

Zylka, J. M., and Plagemann, P. G. W. (1975). *J. Biol. Chem.* **250**, 5756–5767.

Chapter 17

Cloning of Normal and Transformed Cells on Plastic Films

LAWRENCE E. ALLRED

Department of Molecular, Cellular and Developmental Biology,
University of Colorado,
Boulder, Colorado

I. Introduction

Whenever possible, experiments performed with mammalian cells in culture should begin with homogeneous, isogenic populations of cells. Because variant cells arise within any population of cultured cells, especially transformed cells, populations should be examined for homogeneity periodically and pure populations reestablished. This is particularly necessary for experiments in which morphology is examined, since morphological variants arise frequently.

Ideally, a large population of cells should be grown from a single cell. The population should be tested for homogeneity, and a sufficient quantity should be placed in frozen storage to supply fresh cells at frequent intervals throughout an entire series of experiments. Whenever it becomes necessary to replenish the frozen stocks, pure clones of cells should be directly compared with cells taken from original stocks. Then any clone identified as having the proper characteristics can be cultured, tested for homogeneity, and frozen.

Pure populations of cells are obtained by isolating single cells which are cultured to become large populations. This is the process of *cloning*. If the rate of variant formation for the cells is nominal, satisfactory homogeneity will be maintained as they proliferate. Here we use the term "colony" for a group of cells growing in a cluster and reserve the use of the word "clone" only for colonies that grow from a single progenitor cell.

II.   Standard Cloning Procedures

In practice, clones are obtained in two ways. The most commonly used technique is called *dilution culture* (Ham, 1973). A few cells are innoculated into culture vessels such that the resulting clones are widely separated on the growth surface (typically no more than one clone per 2 cm² of surface). The cultures are allowed to incubate undisturbed for about 10 generation times or to about 1000 cells. A simple examination of each clone by light microscopy allows approximations of generation time (clone size), migratory ability, cell morphology, and clonal growth pattern to be made. Selected clones are then isolated with cloning rings. Cloning rings are glass or stainless-steel cylinders, 6 mm in diameter and 8 mm in length, which are sealed down to the substrate with sterile silicone stopcock grease placed on the end of the cylinder. Standing a cloning ring on end over a clone isolates the clone from the rest of the culture vessel. The clone is then harvested by trypsinization.

This procedure is the most practical first step when a population has considerable heterogeneity and reduced viability, such as after mutagenic or selection procedures. The main disadvantage of this procedure is that there is never absolute certainty that a clone is uncontaminated by other cells.

Contamination occurs in several ways. Clumps of two or more cells can grow into colonies that are indistinguishable from one-cell clones. Even in sparse cultures, pairs of clones may fortuitously grow too close together to allow separate harvesting. Cells are susceptible to becoming dislodged from the substrate during cell division. Such cells can infiltrate other clones or form secondary clones. This increases the clonal density in the vessel. This is aggravated by even the slightest disturbance of the culture vessel, especially in the case of transformed cells which typically adhere poorly to the substrate. It is very important that incubator shelves with growing clones remain untouched during the growth period.

Cloning rings introduce other complications. Cells may be physically dislodged during the placement of the cloning ring. The silicone grease may cover cells as the ring is pressed down to form a seal. This grease adds small globules of material to the medium in which the cells are harvested and ultimately to the medium in which they are cultured.

Another problem with the technique of dilution culture is the inefficiency of clonal growth. The innoculation number in any vessel must be low, and the volume of medium must be generous since the cultures cannot be fed with fresh medium during the proliferation period. These conditions can cause many types of cells to have lower efficiencies of survival and clonal growth. Desirable cells may not survive. [See McKeehan and Ham (1976) for improvements in clonal growth conditions.]

In summary, this type of cloning culture is a useful method for selecting clones but is unreliable as a method for isolating pure clones.

The only procedure that ensures that a clone will be of single-cell parentage is to isolate a single cell and allow a clone to grow from it. One simple method of single-cell isolation is to innoculate a few cells into a culture vessel with medium and many small, sterile fragments of cover-slip glass (Martin, 1973). After the cells attach, fragments of glass are found that have only one cell attached. Each of these is transferred to a small tissue culture vessel such as a well in a 96-well, disposable microtray (Linbro or Falcon). These cells are cultured in small volumes of medium. Since the cells can be refed at any time, conditioned medium can be used at the beginning to improve the chances of survival for individual cells. However, glass coverslip fragments leave portions of the vessel surface uncovered and allow cells to slip between them. The preparation, sterilization, and handling of glass fragments is tedious and time-consuming, and the work increases as the size of the fragments decreases.

An alternative procedure is to place cloning rings around individual cells soon after they have attached and thereby isolate them from other clones forming in the same vessel.

III. Cloning on Plastic Films

Either method of clonal isolation can be improved by growing cells on a substrate of plastic film rather than the glass or rigid plastic of a tissue culture vessel. The cell or clone to be isolated is removed from the vessel by cutting out a section of the plastic film and transferring the cut piece to a new, small vessel. The cut piece can be washed in saline solution to remove any contaminating cells which may have begun to attach after the culture had been disturbed by removal of it or sister cultures from the incubator shelf. Careful use of the scalpel blade allows highly selective isolation of clones or even portions of clones. If it is desired to isolate single cells, this procedure increases the efficiency of isolation, since any part of the total substrate area can be removed. Furthermore, the plastic film technique allows cells to attach to the substrate and to maintain this attachment uninterrupted by the intrusion of trypsin or other harsh treatment.

Preparation

Since one advantage of this system is saving time, no special washing technique or coating technique was used. The films were cut into 5-cm squares which would fit conveniently on 60-mm petri plates (which we refer to here as P60s). The squares of film were soaked overnight in distilled water, followed by rinsing three to four times with distilled water. While they were still wet, they were placed on dry P60 plates. Capillary attraction by the water between the film and the floor of the P60 plate was sufficient to adhere the film to the bottom of the plate. If this adhesion is broken, some of the lighter plastic films will float to the surface of the medium, but this was not found to be a problem if plates were not vigorously disturbed. For larger pieces of film that cover the entire bottom of the plates this should be less of a problem. For small cut pieces with attached cells, the film may be allowed to float on top of the medium, cell side down. However, after several days of soaking, Saran wrap becomes denser than water, and this aleviates the problem of floating film.

The plates and plastic film were sterilized by one exposure to ultraviolet light for a period of 1 hour. The apparatus consisted of a General Electric, 25-W germicidal lamp 25 cm above the floor of a box (35 × 60 cm) lined with aluminium foil. It was not necessary to irradiate both sides of the plastic films separately.

IV.   Growth on Plastic Films

Chinese hamster ovary (CHO) cells and WI-38 diploid human fetal cells were subcultured by trypsinization (0.05%). CHO cells were diluted into Ham's F-12 medium (Gibco) and WI-38 cells into BME (Gibco). Both media were supplemented with 10% fetal calf serum (Kansas City Biological). This was pipetted onto the P60 plates with the squares of plastic film. Tissue culture–grade P60 plates were used but are not necessary. The cultures were inspected regularly during the week after inoculation, and the cells were scored for attachment, spreading, and growth on the film. An internal control was provided by scoring the growth of cells on the uncovered surface of the P60 plates (Figure 1).

Several brand-name food wraps were tested and a few speciality films (Table 1); Saran wrap, Reynolds' Brown-In-Bag, yellow Deccofilm, and Transilwrap polyester worked well for CHO cells, but only Transilwrap polyester and yellow Deccofilm served well for WI-38 cells. Water color-treated acetate film and Mylar film with a matte finish on one side were found to be toxic.

The general observation was that WI-38 cells could attach and elongate on films that CHO cells could not attach to. However, WI-38 cells failed to spread well laterally on these films, and this was correlated with the failure of the WI-38 cells to grow. CHO cells did not require attachment to the substrate to proliferate. They grew as well on Reynolds' Brown-In-Bag and Saran wrap as on tissue culture ware, but sometimes a few days of incubation were required before they became well spread. This may be due to the soaking of the plastic film or perhaps to a conditioning effect of the cells. This was not a problem with Transilwrap polyester or Deccofilm.

A. Improvements

Small pieces of lightweight film float on top of liquid medium. If the tissue culture vessel is quite small, the capillary attraction of a water layer under the film may be sufficient to attach a small piece to the bottom of the vessel. However, the piece may be allowed to float cell side down on top of the medium if this is a problem.

If microorganism contamination is noticed in an early stage, the plastic film can be removed from the vessel, washed in sterile saline with high concentrations of antibacterial or antifungal agents, and placed in a new vessel with a nominal, prophylactic concentration of the anticontaminant.

For cells with adeherence difficulties, coating the films with basic polymers such as protamine sulfate or poly-D-lysine may improve adherence and cloning efficiency. This has been examined by McKeehan and Ham (1976). Prolonged soaking of the films may be useful for increasing attachment and has been found to cause Saran wrap to become denser than water after several weeks.

B. Other Uses

While these observations suggest that plastic films are useful for cloning transformed or normal cells, they may be useful for other purposes also.

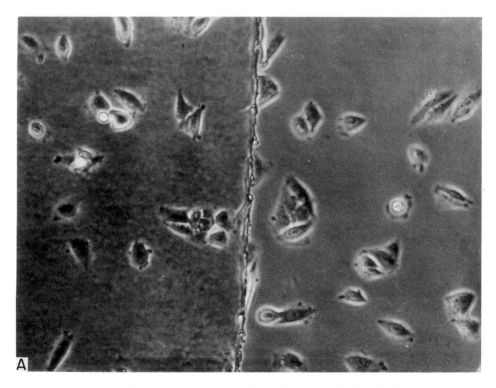

FIG. 1. CHO cells are shown growing both on Saran wrap (left) and adjacent Corning tissue culture plastic (right) at low (A) and high (B) density. There is no general significance to the greater cell density on the Saran wrap in (B). This is merely a fortuitous local inhomogeneity resulting from the inoculation of a small number of cells onto the tissue culture plate.

Firket (1973) showed that polyester sheeting can be advantageously used for *in situ* Epon embedding of cells, irradiation of cells with heavy particles, and as a low-"noise" substrate for cells being examined by scanning electron microscopy.

With flexible plastic films several aliquot samples of cells growing in monolayer can be cut from a single population. It becomes simple to divide one group of cells into several samples to be tested redundantly or to be tested by several different types of analyses. This is much more difficult with glass coverslips. Also, wherever one wishes to cover the entire bottom of any vessel, plastic films can be cut to fit.

The observations of adhesiveness to the various plastic films are paradoxical. At early times after inoculation of the cells, WI-38 normal cells adhere better than CHO transformed cells. This is consistent with the

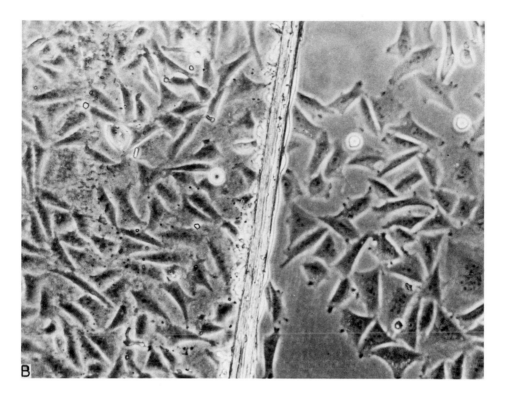

discovery that the LETS protein (Hynes and Bye, 1974), found on normal but not transformed cells, is an adhesive protein (Yamada *et al.*, 1976; Ali *et al.*, 1977). After a few days, however, the transformed cells are found to adhere to substrate that normal cells still do not fully spread upon. Apparently this acquired superiority of adhesiveness of the transformed cells is caused by the CHO cells which continue to proliferate despite the lack of anchorage to the substrate.

Another question raised by these adherence observations is, How do the WI-38 cells elongate without proper lateral spreading? Ivanova *et al.* (1976) and Bragina *et al.* (1976) showed that, in the normal sequence of spreading events, cells spread first radially and then, after 3 hours, begin to elongate and acquire bipolarity. The preliminary observation presented here of the light microscopy of WI-38 cells on poorly adherent substrates is that they can elongate but are not seen to have significant radial or lateral spreading. Since these cells apparently do not increase in number, elongation alone must be insufficient for growth of these anchorage-dependent cells. Obviously anchorage dependence is a more complex phenomenon

TABLE I
GROWTH CHARACTERISTICS OF CELLS ON PLASTIC FILMS

Plastic film[a]	CHO cells				WI-38 cells			
	Attachment	Spreading	Growth	Growth on plate	Attachment	Spreading	Growth	Growth on plate
1[b]	+	+	++	++[c]	+	+	+	++
2	-	-	+	++	+	+	-	++
3	+	-	+	++	+	+	-	++
4[b]	+	+	+	++	+	+	-	++
5	+	-	+	++	+	+	-	++
6	+	-	+	++	+	+	-	++
7	+	-	++	++[c]	-	-	-	++
8	-	-	-	Toxic	-	-	-	Toxic
9	+	+	++	++[c]	+	+	++	++[c]
10	-	-	-	Toxic	-	-	-	Toxic
11	-	-	-	++	-	-	-	++
12	+	+	++	++[c]	+	+	++	++[c]
13[b,d]	+	+	++	++[c]	+	+	++	++[c]
14	-	-	-	+	-	-	-	+

[a] (1) Acetate (0.88 mil) "berry cap" sheets, Zellerbach, Inc., Denver, Colorado; (2) Goodyear Choice-mx Vitafilm, shrink wrap for meat packaging; (3) Topco clear plastic wrap; (4) Reynolds' Brown-In-Bag; (5) Gladwrap (polyethylene), Union Carbide Corporation; (6) Handiwrap, Dow Chemical Company; (7) Saran wrap, Dow Chemical Company; (8) acetate (3 mil) treated for water color, Transilwrap West, Inc., South San Francisco, Calif.; (9) polyester (5 mil) Transilwrap West, Inc.; (10) Mylar film (matte-smooth finish) Transilwrap West, Inc.; (11) Mylar film (matte-matte finish) Transilwrap West, Inc.; (12) Deccofilm (yellow) Amscan, Inc., Harrison, N.Y.; (13) Deccofilm (pink) Amscan, Inc.; (14) Cellophane (clear) Bienfang Paper Company, Metuchen, N.J.

[b] More dense than medium.

[c] Supports growth equivalent to that in tissue culture ware, determined after 1 week.

[d] Color dissolves in medium.

than is currently understood. Perhaps the various adhesive characteristics of different plastic films may be useful in studying anchorage-dependent growth control in detail (Paranjpe *et al.*, 1975).

ACKNOWLEDGMENT

The work reported here was supported by National Institutes of Health Grant No. AG-003 10-04 to Keith R. Porter and by American Cancer Society Grant No. VC-193 to David M. Prescott. I wish to thank Dr. Gretchen Stein for her helpful discussion and criticisms.

REFERENCES

Ali, I. U., Moutner, V., Lanza, R., and Hynes, R. O. (1977) *Cell* **11**, 115–126.
Bragina, E. E., Vasiliev, J. M., and Gelfand, I. M. (1976). *Exp. Cell Res.* **97**, 241–248.
Firket, H. (1973). *In* "Tissue Culture: Methods and Applications" (P. F. Kruse, Jr. and M. K. Patterson, eds.), p. 378. Academic Press, New York.
Ham, R. G. (1973). *In* "Tissue Culture: Methods and Applications" (P. F. Kruse, Jr. and M. K. Patterson, eds.), p. 254. Academic Press, New York.
Hynes, R. O., and Bye, J. M. (1974). *Cell* **3**, 113–120.
Ivanova, O. Y., Margolis, L. B., Vasiliev, J. M., and Gelfand, I. M. (1976). *Exp. Cell Res.* **101**, 207–219.
McKeehan, W. L., and Ham, R. G. (1976). *J. Cell Biol.* **71**, 727–734.
Martin, G. M. (1973). *In* "Tissue Culture: Methods and Applications" (P. F. Kruse, Jr. and M. K. Patterson, eds.), p. 264. Academic Press, New York.
Paranjpe, M. S., Boone, C. W., and del Ande Eaton, S. (1975). *Exp. Cell Res.* **93**, 508–512.
Yamada, K. M., Yamada, S. S., and Pastan, I. (1976). *Proc. Natl. Acad. Sci. U.S.A.* **73**, 1217–1221.

Chapter 18

A Simple Replica-Plating and Cloning Procedure for Mammalian Cells Using Nylon Cloth

L. KARIG HOHMANN

Department of Experimental Biology
Roswell Park Memorial Institute
Buffalo, New York

I. Introduction

Mutants of mammalian tissue culture cells have become a valuable and irreplaceable resource in genetic studies. They have found use in the elucidation of biochemical pathways, regulatory mechanisms, and other cellular processes, and as markers for genetic mapping. In addition, the mutant characteristic can provide a selection procedure that can be used, for example, in hybrid complementation of auxotrophs to eliminate parental cells from the culture. The majority of mutants generated in laboratories fall into the following categories: (1) auxotrophic mutants obtained by the selective killing of prototrophic cells in minimal growth medium (Kao and Puck, 1974), (2) drug-resistant mutants obtained as a result of resistance to usual killing by a cell poison (Thompson and Baker, 1973), (3) temperature-sensitive mutants obtained by the selective killing of temperature-resistant cells at a non-permissive temperature (Basilico and Meiss, 1974; Naha, 1974), and (4) cell surface mutants obtained as the result of a selectable variation in a surface-related property, e.g., adhesiveness (Atherly *et al.*, 1977, Pouysségur and Pastan, 1976). In all these methods, it is possible to eliminate the nonmutant phenotype without sacrificing the mutant phenotype.

This leaves a large number of very interesting mutants inaccessible, because the procedure that would indicate their presence is also lethal to the cell. For example, Brenner *et al.* (1976) have listed 46 classes of enzymes that have the potential for use in colorimetric assays in the cell, but the assays are lethal to the cell. The sib selection technique (Marin, 1969) is one way to circumvent this problem. A more direct solution has been the development of replica-plating procedures for tissue culture cells similar to those that have been used to great advantage with bacteria (Lederberg and Lederberg, 1952) and fungi (Roberts, 1959). There are three general methods in use.

The first method (Goldsby and Zipser, 1969; Suzuki and Horikawa, 1973) uses 96-well plates and a syringe replicator capable of sampling the 96 wells simultaneously and dispensing cells into additional 96-well plates. The cells in each well represent a cloned population. The procedure requires three steps: detachment of the cells from the well surface, dispersement into single cells, and finally transfer to additional plates by the replicator. The efficiency of replication can be as high as 100%, and six replica copies can be made.

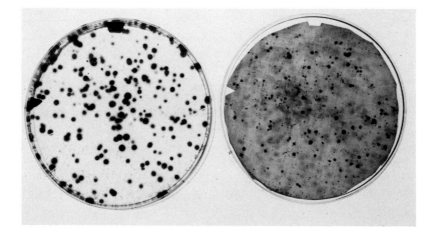

FIG. 1. Example of actual replica plating. CHO cells were plated on a 60-mm plate and grown in F-12 medium supplemented with 10% fetal calf and 5% calf serum for 6 days. The Nitex was in contact with the cells on the plate for 7 hours. After several additional days in culture, the cells on both the Nitex and the plate were fixed in 10% Carnoy's fixative and stained with crystal violet. The mottled appearance of the Nitex is from the stain. With the orientation marks aligned, the replication of colonies onto the Nitex can be seen. Also, it is apparent that colonies replicate onto the Nitex to different degrees (L. Karig Hohmann and B. J. Barnhart, unpublished data).

The second method (Kuroki, 1975) is a direct adaptation of velveteen replica plating for bacteria (Lederberg and Lederberg, 1952) to tissue culture cells. The cells must be capable of growth on agar. After the colonies have reached an appropriate size, velveteen is used in the same way as with bacteria to transfer cells to other agar plates. Efficiencies of 95% are reported for this method.

The third method (Stamato and Karig Hohmann, 1975; Karig Hohmann and Barnhart, 1975), to be described here, utilizes the ability of tissue culture cells to attach to and grow on nylon cloth. The cells on the nylon cloth become the replicate copy (Fig. 1). A monofilament nylon cloth, 10-μm Nitex (HD3-10, Tetko, Inc., Elmsford, N.Y.), has been used for this purpose, but other nylon materials without special finishes can presumably be used. In addition, small pieces of Nitex can be used to clone colonies without the need for trypsinization.

II. Replica Plating Using Nitex Nylon Cloth

A. General Method

The Nitex cloth is prepared as follows (Stamato and Karig Hohmann, 1975). Circles of Nitex are cut to fit a 60-mm culture plate and notched twice for orientation purposes. The circles are washed in hot water with an appropriate tissue culture detergent, rinsed extensively with distilled, deionized water, and dried. They are placed between blotting-paper spacers in a glass petri dish and autoclaved 15 minutes at 15 pounds of pressure.

The cells to be replicated are seeded at low density (100 to 300 cells) onto the surface of 60-mm tissue culture plates, and colonies are allowed to grow to the appropriate size for replication. Since different proportions of cells from each colony attach to the Nitex, it may take several days for all the colonies to reach an adequate size for analysis. It is best to use the smallest colony size possible without sacrificing replicating efficiency. This minimizes the possibility of overgrowth of the master plate while waiting for the Nitex copy to grow. Alternatively, the master can be kept at a lower temperature to slow cell growth.

The medium is removed from the plate to be replicated, and the plate is rinsed with fresh medium. The sterile Nitex circle is completely wetted by immersion in medium on a separate culture plate (Karig Hohmann and Barnhart, 1975) and then laid on the plate to be replicated. The position of the notches in the Nitex cloth are marked on the plate for orientation. The plate is then incubated for the appropriate time in a humidified CO_2 incubator. Finally, the Nitex is carefully pulled off, minimizing sliding motion, and placed cell side up on another culture plate. Fresh medium is added to both plates, and the colonies are allowed to grow.

The selection procedure can be performed on either the Nitex or the master culture plate. With the orientation marks aligned, a position on one copy can be correlated with that on the other and, if an interesting colony is found, it can be cloned from the untreated plate. Cells are difficult to see on the Nitex but can be visualized with various vital stains. Colonies can be cut out of the Nitex with sterile scissors or cloned in the usual manner from the master plate.

Different cells require different conditions for optimal replication. Stamato and Karig Hohmann (1975) demonstrated that colony size and time of contact between nylon and culture plate affect the efficiency of replica plating. Plating efficiencies may also be a factor (Karig Hohmann and Barnhart, 1975). As shown in Fig. 2, with Chinese hamster ovary (CHO) cells 90% replication is achieved with 5-day colonies (Karig Hohmann and Barnhart, 1975, also unpublished observations). Similar results were obtained with the CHO-Kl subclone of CHO cells (Stamato and Karig Hohmann, 1975). Eleven-day-old colonies of ALR-1 BsAg-9 cells, a hybrid of RAG (mouse) cells and a human fibroblast, can replicate with up to 93% efficiency with a 1/2-hour contact time but are rapidly killed if the contact time increases (L. Karig Hohmann, C. E. Wright, and T. B. Shows, unpublished observations).

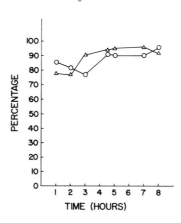

FIG. 2. Nitex replication of 6-day-old colonies of CHO cells as a function of time of contact of Nitex with the colonies. Triangles, Plating efficiency of master plate; circles, efficiency of replication of cells remaining on master plate (L. Karig Hohmann and B. J. Barnhart, unpublished data).

B. Applications

Nitex replica plating is a simple, rapid method useful for screening large numbers of cells for a single characteristic. It is less tedious than the 96-well replicator method, but in general only one good copy of the plate can be obtained. Since the cells to be replicated grow on standard culture plates, the need for agar growth conditions is unnecessary. It has been used to obtain CHO variants lacking glucose-6-phosphate dehydrogenase (Stamato and Karig Hohmann, 1975; Stamato, 1975) and lactate dehydrogenase (Stamato, 1975). It should be useful in the isolation of other enzyme variants assayed colorimetrically. The isolation of cell lines deficient in DNA repair mechanisms should be facilitated by this method (Kuroki and Miyashita, 1977). In addition, it should have application in the isolation of temperature-sensitive and auxotrophic mutants. One copy can be used to assay for the mutant, while the other copy remains unexposed to the selective agent and is used to isolate the mutant for future growth.

III. Clone Isolation Using Nitex

Because of their ability to attach to and grow on Nitex, cells can be cloned (Karig Hohmann and Barnhart, 1977) using small pieces, a few millimeters square. These pieces are rinsed with ethanol and water, and dried. They are then placed between two sheets of blotting paper in a glass petri dish and autoclaved 15 minutes at 15 pounds of pressure.

The cells to be cloned are plated at low density (less than 100 cells) on a 100-mm plate. Colonies of the appropriate size for replication are allowed to form. The clones to be isolated are located under an inverted microscope, and their position marked. The medium on the plate is removed, and the plate rinsed with fresh medium. A square of Nitex, wetted with medium, is placed over each clone and left in place for the appropriate period of time. Times that give optimal whole-plate replication are also optimal for cloning. The Nitex is then carefully removed and placed cell side down in one well of a 24-well culture plate containing medium. As the cells on the floating Nitex divide, some are shed, fall to the bottom of the plate, attach, and grow (Fig. 3). After the plate is seeded, the Nitex can be used to seed additional plates.

Suspected temperature-sensitive mutants of CHO cells were cloned by this method (Karig Hohmann and Barnhart, 1977). Cells were grown in F-12 medium supplemented with 15% fetal calf serum to the 50- to 100-cell stage. Of 56 clones selected, 55 were successfully isolated with a 3- to 5-hour colony-

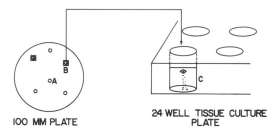

FIG. 3. Cloning with Nitex. (A) Colony on tissue culture plate. (B) Colony on a plate covered with Nitex. (C) One well of a 24-well tissue culture plate. After the small piece of Nitex is in place over the clone for the necessary time, it is placed cell side down in one well of the 24-well tissue culture plate. As cells grow on the Nitex, they float to the bottom of the well and establish a colony.

to-Nitex contact time. Three plates were seeded successively with each clone. Seven-day colonies of LM/TK$^-$ cells grown in Dulbecco's modified Eagle's medium were cloned with Nitex. The contact time was 2 hours. Of 20 clones selected for isolation, cells from 15 clones attached to the Nitex, leaving some cells on the original plate. Two clones were totally lifted, and three did not attach to the Nitex at all but remained on the culture dish, available for further attempts. Colonies of ALR-1 BsAg-9 were cloned with a 1/2-hour contact time (L. Karig Hohmann, C. E. Wright, and T. B. Shows, unpublished observations).

This method of cloning has several advantages over the standard metal ring isolation method (Ham and Puck, 1962). Mechanically, it is simpler. Each Nitex piece is used once. The method does not require trypsinization or pipetting. The problems of trypsin leakage from the rings and ring cleanup are avoided. Cells naturally attach to Nitex. The detachment of the cells to seed the culture plate is probably due to the lessened attachment of the cells to Nitex during mitosis and the force of gravity. Also, since only part of a colony attaches to the Nitex, a section is left on the master plate for other clonings or second attempts.

IV.   Conclusion

A method of replica plating has been described here which is based on a cell's ability to attach to Nitex nylon cloth. Two other methods have also been described in the literature. The Nitex method is mechanically simpler than the use of 96-well plates and a replicator; however, only one copy of a culture plate can be made rather than several. Because this procedure is simpler, many more cells can be rapidly screened for a given trait. Also, the colonies to be replicated grow on standard culture plates, eliminating the need for special agar growth conditions. Cells that attach to Nitex can also be cloned in a way that is again simpler than those described previously. This method has been used to isolate mutants deficient in glucose-6-phosphate dehydrogenase and lactate dehydrogenase from CHO cells and to clone temperature-sensitive CHO cells. Other cells have also shown the ability to be replicated and cloned.

ACKNOWLEDGMENTS

The studies with T. D. Stamato were supported by FDA Contract No. 73-213 from the National Center for Toxicological Research and American Cancer Society Grant VC-81E to the Eleanor Roosevelt Institute for Cancer Research. The author was supported by a NIH biochemical training grant (GM 00550). The work with B. J. Barnhart was partially supported by Energy Research and Development Agency and a NCI fellowship, 1F22 CA 01945-01, awarded to the author.

REFERENCES

Atherly, A. G., Barnhart, B. J., and Kraemer, P. M. (1977), *J. Cell. Physiol.* **90**, 375–386.
Basilico, C., and Meiss, H. K. (1974). *Methods Cell Biol.* **8**, 1–22.
Brenner, M., Dimond, R. L., and Loomis, W. F. (1976). *Methods Cell Biol.* **14**, 187–194.
Goldsby, R. A., and Zipser, E. (1969). *Exp. Cell Res.* **54**, 271–275.
Ham, R. G., and Puck, T. T. (1962). *In* "Methods in Enzymology" S. P. Colowick and N. O. Kaplan, eds.), Vol. 5, pp. 90–119.
Kao, F.-T., and Puck, T. T. (1974). *Methods Cell Biol.* **8**, 23–39.
Karig Hohmann, L., and Barnhart, B. J. (1975). *J. Cell Biol.* **67**, 176a.
Karig Hohmann, L., and Barnhart, B. J., (1977). In preparation.
Kuroki, T. (1975). *Methods Cell Biol.* **9**, 157–178.
Kuroki, T., and Miyashita, S. Y. (1977). *J. Cell. Physiol.* **90**, 79–90.
Lederberg, J., and Lederberg, E. M. (1952). *J. Bacteriol.* **63**, 399–406.
Marin, G. (1969). *Exp. Cell Res.* **57**, 29–36.
Naha, P. M. (1974). *Methods Cell Biol.* **8**, 41–46.
Pouysségur, J. M., and Pastan, I. (1976). *Proc. Natl. Acad. Sci. U.S.A.* **73**, 544–548.
Roberts, C. F. (1959). *J. Gen. Microbiol.* **20**, 540–548.
Stamato, T. D. (1975). *J. Cell Biol.* **67**, 416a.
Stamato, T. D., and Karig Hohmann, L. (1975). *Cytogenet. Cell Genet.* **15**, 372–379.
Suzuki, F., and Horikawa, M. (1973). *Methods Cell Biol.* **6**, 127–142.
Thompson, L. H., and Baker, R. M. (1973). *Methods Cell Biol.* **6**, 209–281.

NOTE ADDED IN PROOF   T. D. Stamato and C. A. Waldren [*Somatic Cell Genetics* 3 (1977), 431–440] have combined Nitex replica plating with a selective detachment technique to isolate UV-sensitive variants of CHO-KI Cells.

Chapter 19

Human Minisegregant Cells

R. T. JOHNSON, A. M. MULLINGER, AND C. S. DOWNES

Department of Zoology, University of Cambridge, Cambridge, England

I. Introduction

Fusion between whole cells followed by segregation of phenotypic markers is a powerful genetic tool and has permitted the assignment of numerous traits to chromosomes and regions of chromosomes (see Ruddle, 1972).

*This section was written by R. T. Johnson, R. H. J. Brown, and M. J. Berry.

255

However, the stability of chromosomes in hybrids and the ways in which they segregate are poorly understood. Though some successful efforts have been made to direct chromosome loss in hybrids (Pontecorvo, 1971), there is as yet no general means for controlling the chromosome constitution of these cells. By reconstruction between whole cells and cells of reduced DNA content it is possible to obtain hybrids with at least a bias toward the desired chromosome balance. Cell reconstruction therefore, holds out considerable promise as a general approach to the analysis of gene expression and localization in somatic cells.

The production of fragmented cells with reduced DNA content has been reported by three groups: (1) Ege and Ringertz (1974) produce multimicronucleate rodent cells by prolonged exposure to Colcemid (Levan, 1954; Stubblefield, 1964) and then induce enucleation by cytochalasin B. They thus obtain microcells with reduced DNA content which may be as low as one chromosome equivalent. Fournier and Ruddle (1977) use such microcells to complement enzyme deficiency and to generate proliferating hybrids. Unlike rodent cells which respond to Colcemid by producing scattered anaphase clusters of chromosomes, human cells are generally blocked irreversibly in a C-type mitosis (Brinkley et al., 1967), which is probably the reason why this method has not been extended to human cells. (2) Johnson et al. (1975) produce minisegregant human cells from mitotic precursors by inducing an abnormal type of division (extrusion subdivision). In this process chromosome segregation and cytokinesis are severely disturbed, and a single mitotic cell gives rise to many daughters, some of which contain small amounts of DNA and reduced numbers of chromosomes. Minisegregants have been successfully isolated according to size and DNA content, and used to complement enzyme deficiency and to generate proliferating hybrids (Johnson et al., 1975; Schor et al., 1975; Mullinger and Johnson, 1976; Tourian et al., 1978). (3) Cremer et al. (1976) obtain a low yield of fragmented human diploid cells by prolonged exposure to Colcemid.

The aims of this chapter, which includes previously unpublished material, are: (1) to describe methods for the production of human minisegregant cells from mitotic precursors; (2) to assess the nature of the perturbation induced in the mitotic cell, and the roles of substances that either inhibit or promote this aberrant division; (3) to describe the fractionation of minisegregant cells and some of their properties; (4) to assess their competence to transfer information by means of cell fusion.

II. Extrusion Subdivision and the Production of Minisegregant Cells

A. The Standard Method for Inducing Extrusion

Human minisegregant cells are produced from mitotic HeLa cells which are themselves usually obtained as follows. HeLa cells with a generation time of 22 hours are routinely grown in suspension culture in Eagle's minimal essential medium (MEM) (Eagle, 1959) supplemented with nonessential amino acids, sodium pyruvate, antibiotics, and 5% fetal calf serum (MEMFC). Partial synchrony is obtained by first administering 2.5 mM thymidine to a suspension culture at a density of 2×10^5 cells/ml for 20 hours

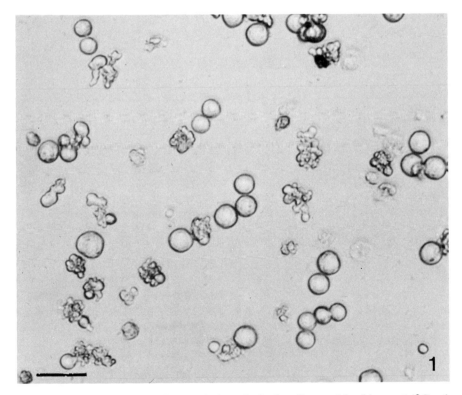

FIG. 1. Photomicrograph of a population of mitotic cells stored for 6 hours at 4°C and then plated out in HBSS for 3 hours at 37°C. While some cells remain normal, others have extruded. Bar represents 25 μm. Reproduced from Johnson et al. (1975) by courtesy of the Royal Society of London.

to arrest cells either in S phase or at the G_1–S boundary. Cells are released from the thymidine arrest by centrifugation and resuspension in MEMFC; 4 hours later they are placed in a 5.4-atm (5.066 × 10⁵ N/m²) nitrous oxide mitotic block for 9 hours (Rao, 1968). For this last block the suspension of cells is transferred to plastic dishes, at a density of 6 × 10⁶ cells per 150-mm dish. In nitrous oxide cells progress asynchronously into mitosis, are arrested at pseudometaphase (Brinkley and Rao, 1973), round up, and usually detach from the surface; routinely more than 90% of the cells detach in a 9-hour period. More than 95% of these floating cells are mitotic, as determined by inspection of cell monolayers prepared by means of a cytocentrifuge (Shandon Instruments Ltd.). Details of the pressure vessels employed for nitrous oxide treatment and designed for use with large tissue culture plates are given in the Appendix. Immediate incubation of nitrous oxide-arrested mitotic cells in MEMFC at 37°C results in synchronous progress into interphase within 1.5 hours (Rao, 1968; Rao and Johnson, 1972a). When the arrested cells are stored at 4°C for 4–12 hours immediately after release of the mitotic block and are subsequently incubated at 37°C in MEMFC, division is delayed for up to several hours, and many of the cells display highly abnormal patterns of division. The most extreme form of behavior, which we call extrusion subdivision, results in the formation of clusters of small cells (minisegregants) resembling bunches of grapes (BOGs), as shown in Fig. 1.

B. Topographic Changes in Extrusion

The topographic changes accompanying extrusion subdivision can be followed by time-lapse photography and scanning electron microscopy (Mullinger and Johnson, 1976). Such studies show that extrusion is a highly variable process but that in general the cell surface gives rise to multiple protrusions which enlarge at the expense of the parent and develop into clusters of small daughter cells as illustrated schematically in Fig. 2.

1. TIME-LAPSE ANALYSIS

A typical example of extrusion subdivision is shown in the time-lapse series in Fig. 3 and is described in detail in the legend. At the start of incubation the cold-stored mitotic cell was spherical, and this shape was maintained for 70 minutes. Extrusion began with the formation of small blebs which appeared during a 2-minute period and were distributed nonuniformly over the visible cell surface. During the next few minutes many blebs grew in size and changed shape, some forming elongated structures ("fingers") (Fig. 3A); individual blebs developed into fingers within 40–90

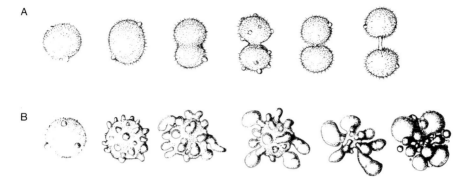

FIG. 2. Schematic drawings of sequential stages of two dividing HeLa cells to highlight the progressive differences in the surface topology between cleavage division (A) and extrusion subdivision (B). Reproduced from Mullinger and Johnson (1976) by courtesy of the Company of Biologists Ltd.

seconds. As extrusion continued, the protrusions enlarged, coalesced, and changed shape before developing into more-or-less spherical minisegregants. The time taken to reach this stage (BOG) from the appearance of the first bleb was about 10 minutes. During a subsequent 3 hours of observation the appearance of the BOG continued to change, but more slowly, and some minisegregants fused or divided further.

Most extruding cells divide into BOGs in a broadly similar manner, although there are considerable differences in details such as the size, shape, and number of surface protrusions, the kinetics of their production, and the extent of change in shape. One example with more regularly shaped protrusions giving rise to a BOG with fewer minisegregants is shown in Fig. 4. Figure 5 shows another cell in which contractile waves pass along a larger protrusion "balloon" from tip to base at a rate of 15–20 μm per minute. Finally, the continuing instability of the surface of extruding cells and the possible confusion that this can cause in attempts to quantitate the phenomenon are illustrated in Fig. 6, which shows the reversal of an extruded cell to an apparently normal spherical body. A glossary of terms used in describing extrusion is provided in Fig. 7. Further examples and analyses of time-lapse sequences are included in a previous article (Mullinger and Johnson, 1976).

2. SCANNING ELECTRON MICROSCOPY

For examination by scanning electron microscopy cells were plated out onto polylysine-coated coverslips (PL coverslips; Mazia *et al.*, 1975). Mitotic and extruding HeLa cells adhere to polylysine, though not to glass or plastic; but the shapes of extruding HeLa cells on PL coverslips do not differ greatly,

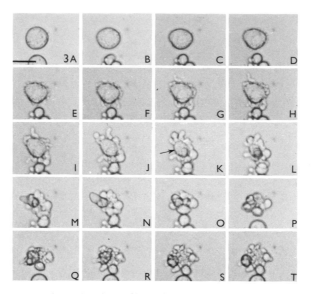

Figs. 3–6. Frames from time-lapse films of cold-stored mitotic HeLa cells incubated at 37°C in various media. Figures 3, 4, and 6: 10-hour storage at 4°C and incubation in BSSH at pH 8.0, including 1 µg/ml Colcemid. Figure 5: 5-hour storage at 4°C and incubation in the same medium with 2 mM dithiothreitol but without Colcemid. Bar represents 20 µm. Reproduced from Mullinger and Johnson (1976) by courtesy of the Company of Biologists Ltd.

Fig. 3. Typical extrusion subdivision into a BOG. Blebs and fingers are formed, and the BOG, once produced, is further subdivided. Total time of sequence, 196 minutes 24 seconds. (A–D) Cell bulges and blebs appear. (E) Further development of blebs. (F–G) Some blebs elongate into fingers. (H–I) asynchronous and varied development of surface protrusions; some blebs remain, while others elongate into fingers. (J) Protrusions enlarge, and some coalesce; the parent cell is smaller. (K–L) Protrusions become spherical; the parent cell is much smaller; by this stage one area of the cell surface (arrow) has shown no apparent surface activity. (M) Typical BOG composed of a tight cluster of minisegregants; the parent cell is obscured. (N–T) Further development of BOG; rearrangement of cluster; some minisegregants are subdivided. A, 74 minutes 16 seconds from start; A–B, 33 minutes 4 seconds; B–C, 10 minutes 8 seconds; C–D, 1 minute 12 seconds; D–E, 1 minute 12 seconds; E–F, 32 seconds; F–G, 40 seconds; G–H, 24 seconds; H–I, 1 minute 4 seconds; I–J, 1 minute 20 seconds; J–K, 1 minute 12 seconds; K–L, 40 seconds; L–M, 56 seconds; M–N, 1 minute 4 seconds; N–O, 5 minutes 4 seconds; O–P, 35 minutes 36 seconds; P–Q, 24 minutes 16 seconds; Q–R, 33 minutes 52 seconds; R–S, 27 minutes 36 seconds; S–T, 16 minutes 32 seconds.

as far as can be seen by light microscopy, from those seen in nonadherent extrusion. PL coverslips are prepared by cleaning in mild detergent, rinsing in distilled water, dipping in 0.01% poly-L-lysine hydrobromide (Sigma, Grade 1B, MW 70,000 and over), washing thoroughly in distilled water, and air-drying. Cold-stored mitotic cells, added in a drop of medium to PL

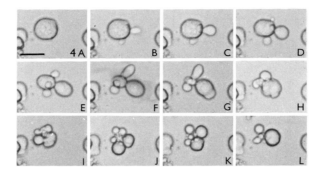

FIG. 4. Extrusion subdivision into a BOG, by means of a few regularly shaped protrusions. Total time of sequence, 164 minutes 32 seconds. (A–B) two protrusions appear. (C–D) First two protrusions enlarge; third and fourth protrusions appear. (E–F) Protrusions enlarge further, particularly the first and fourth; the parent cell simultaneously becomes smaller. (G–J,) Contraction waves appear on largest protrusion (balloon) which is then withdrawn partially and subdivided; the parent cell is much smaller and finally is no longer visible; BOG formed. (K–L) Further development of BOG; rearrangement of cluster and, finally, fusion of some minisegregants. A, 143 minutes 4 seconds from start; A–B, 1 minute 52 seconds; B–C, 1 minute 4 seconds; C–D, 1 minute 44 seconds; D–E, 1 minute 4 seconds; E–F, 40 seconds; F–G, 24 seconds; G–H, 6 minutes 0 seconds; H–I, 13 minutes 20 seconds; I–J, 22 minutes 8 seconds; J–K, 35 minutes 52 seconds; K–L, 80 minutes 24 seconds.

coverslips, settle within 5 minutes and adhere firmly; they can then be incubated to allow extrusion to occur, and subsequently fixed and processed for scanning electron microscopy *in situ* by methods described by Mullinger and Johnson (1976).

Scanning electron micrographs of extruding cells can be correlated with

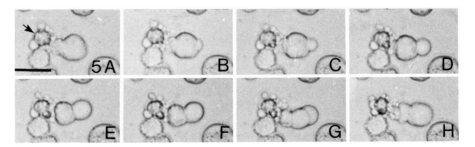

FIG. 5. Contractile waves on an extruded balloon. Total time of sequence, 1 minute 56 seconds. (A) Extruding cell (arrow) with many small blebs and one large balloon. (B–F) Contraction wave passes along balloon from tip to base. (G–H) Second contraction wave appears at tip of balloon and passes back. A, 95 minutes 56 seconds from start; A–B, 8 seconds; B–C, 8 seconds; C–D, 12 seconds; D–E, 16 seconds; E–F, 8 seconds; F–G, 12 seconds; G–H, 52 seconds.

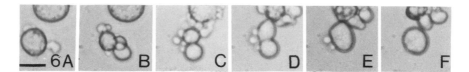

FIG. 6. Reconstitution of parent cell after extrusion subdivision into a small number of minisegregants. Total time of sequence, 68 minutes 16 seconds. (A–B) Blebs appear, enlarge, and become approximately spherical; the parent cell becomes smaller. (C) BOG with about six minisegregants. (D–F) Further development of BOG; minisegregants reconstituted into a single cell. No further surface activity was observed after (F). A, 78 minutes 56 seconds from start; (A–B) 6 minutes 56 seconds; B–C, 36 minutes 16 seconds. C–D, 16 minutes 0 seconds; D–E, 2 minutes 48 seconds; E–F, 6 minutes 16 seconds.

stages seen in time-lapse photography and show more detail. Blebs, fingers, balloons, and BOGs are recognizable (Figs. 8–12). During extrusion the abundant microvilli of the mitotic HeLa surface are progressively lost. Both protrusions and the fully formed BOG have a smooth surface. We do not know whether or not the topographic changes at the fine-structural level seen during extrusion indicate changes in the surface area and/or composition of the cortex (for discussion, see Mullinger and Johnson, 1976).

C.   Internal Events in Extrusion

1.   LIGHT MICROSCOPY

The redistribution of DNA during extrusion subdivision is demonstrated by the study of BOGs formed from mitotic cells prelabeled with [³H]

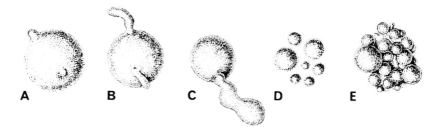

FIG. 7. Schematic representation of terms used to describe the process of extrusion subdivision. (A) *Bleb*: Generally small, localized protrusion approximately 0.5–2 μm in diameter but sometimes larger, e.g., Fig. 3E. (B) *Finger*: Elongated, localized protrusion, approximately 7–10 μm in length, e.g., Fig. 3H. (C) *Balloon*: Large protrusion of localized origin, approximately 10–25 μm in length, e.g, Fig. 5. (D) Minisegregants: Products of extrusion subdivision of mitotic cell, variable in size (1–10 μm) and with or without DNA. (E) *BOG*: "Bunch of grapes", group of attached minisegregants derived from a single parent cell by extrusion subdivision, e.g., Figs. 3N–T and 4J–K. Reproduced from Mullinger and Johnson (1976) by courtesy of the Company of Biologists Ltd.

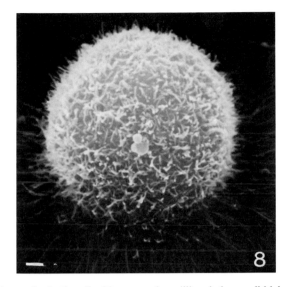

FIG. 8. Cold-stored mitotic cell with many microvilli and also small blebs.

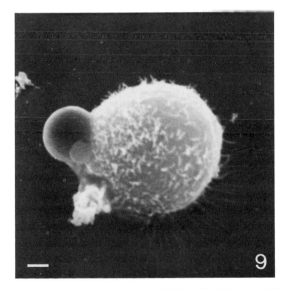

FIGS. 8–12. Scanning electron micrographs of HeLa cells. Nitrous oxide-arrested mitotic cells were stored at 4°C for 11 hours, plated out on PL-coverslips, and incubated at 37°C in BSSH, pH 7.2, and fixed 2 hours (Figs. 9 and 10) or 3 hours (Figs. 8, 11, and 12) after the start of incubation. Bar represents 2 μm. Reproduced from Mullinger and Johnson (1976) by courtesy of the Company of Biologists Ltd.

FIG. 9. Early stage of extrusion; development of smooth-surfaced blebs.

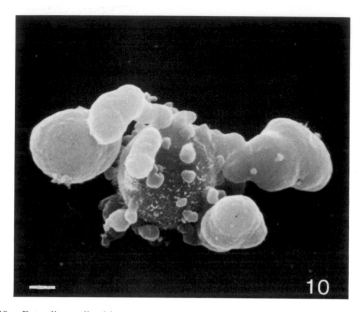

Fig. 10. Extruding cell with many protrusions of highly variable size and shape. The surfaces of the protrusions are generally smooth. One half of the parent cell surface bears microvilli, while the other half is smooth.

Fig. 11. Cluster of typical minisegregants, or BOG. This stage probably corresponds to Fig. 3T.

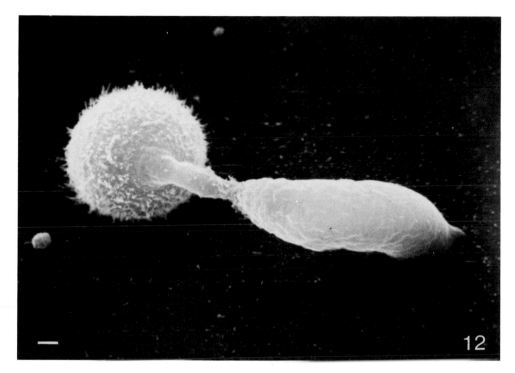

FIG. 12.   Extruding cell with a large balloon.

thymidine. Autoradiographs of sectioned material (Fig. 13) clearly demonstrate the presence of DNA in some minisegregants.

Further details of chromosome segregation are shown by Feulgen staining of material adhering to PL glass slides, fixed at intervals in neutral, buffered formaldehyde for 24 hours (Lillie, 1965) and stained according to standard procedures (Pearse, 1968). This method demonstrates that, as with topographic events, the redistribution of DNA is both asynchronous and variable within a population, and the following description applies to the majority of cells stored in mitotic arrest at 4°C for 10–12 hours and incubated in MEMFC. At the end of the period of cold storage chromosomes of nitrous oxide-arrested cells appear similar to those of normal metaphase cells, but during incubation at 37°C the chromosomes in an increasing proportion of cells condense further into irregularly shaped aggregates which stain more intensely (Fig. 14A–C). Further chromosome aggregation often gives rise to compact, more-or-less spherical bodies which stain uniformly with an intense, magenta color; we call these bodies "dense nuclei" (Schor *et al.*, 1975) to distinguish them from "normal nuclei" which are

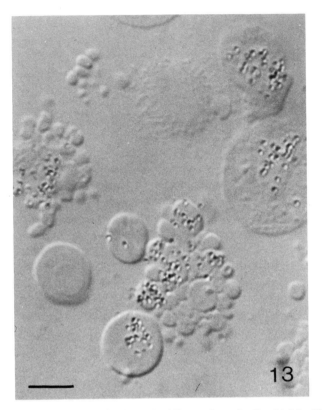

FIG. 13.   Autoradiograph of a 1-μm Araldite section of cells which had been prelabeled with [³H]thymidine before mitotic arrest and incubated in HBSS, pH 7.2. Silver grains over minisegregants indicate the presence of DNA. Normarski differential-interference contrast. Bar represents 5 μm. Reproduced from Johnson *et al.* (1975) by courtesy of the Royal Society of London.

similar in appearance to those of interphase cells and stain nonuniformly with a paler magenta color.

Cells showing early extrusion activity often contain a single "nucleus" (diameter when spherical about 3.5–7 μm), although more than one such body or aggregated chromosomes have been seen in cells at this stage. BOGs usually have more than one "dense nucleus," suggesting that fragmentation occurs during the later stages of extrusion subdivision (Fig. 14D–F). There is considerable variation in the number of "dense nuclei" in different BOGs and in their distribution among the minisegregants (Fig. 14G–I). Analysis of the "dense nuclei" in BOGs shows: (1) that the number varies from 1 to more than 30, (2) that their diameter varies from less than 0.2 μm to 7 μm, (3) that they may be concentrated in one or two minisegregants of a BOG or

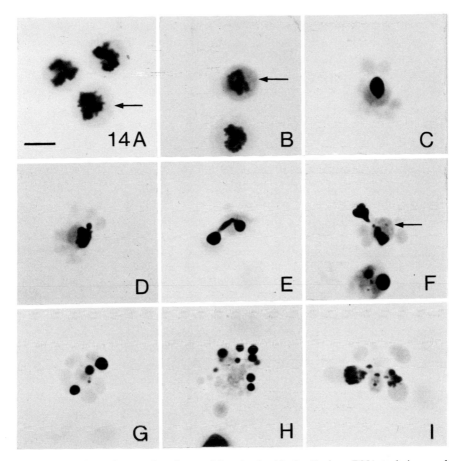

Fig. 14. Photomicrographs of material stained with the Feulgen DNA technique and counterstained with fast green. Nitrous oxide-arrested HeLa cells were stored at 4°C for 9 hours, plated out on PL slides, incubated in MEMFC at 37°C, and fixed 1–1.5 hours later. (A–C) Progressive stages in chromosome aggregation. (C–F) Early stages of extrusion in cells with "dense nuclei". (D–F) Extruding cells with "dense nuclei" which are possibly fragmenting. (G–H) BOGs with "dense nuclei." (I) BOG with "dense nuclei" and partially condensed chromosomes. Bar represents 10 μm.

more widely distributed, and (4) that individual minisegregants may contain 1 or more or none.

Other types of chromatin distribution are also found in extruding cells: BOGs and minisegregants with "normal nuclei," both with and without additional "dense nuclei," have been seen.

2. Transmission Electron Microscopy

The variation in internal features seen by light microscopy is also observed at the fine-structural level; the following description applies to cells after 10–12 hours of cold storage and incubation in MEMFC. The surface membranes of BOGs and minisegregants (but not internal membranes) are less stable than those of normal cells, as judged by their imperfect preservation.

Stages of chromosome aggregation into "dense nuclei" observed in Feulgen-stained material are also seen at the fine-structural level (Figs. 15 and 16), and there is no nuclear envelope around either the aggregating chromosomes or the "dense nuclei" (Fig. 17). The presence of DNA in "dense nuclei" is clearly demonstrated in autoradiographs of minisegregants formed from mitotic cells prelabeled with [³H]thymidine (Fig. 18).

At the stage of early aggregation, chromosomes lie in a region largely free of other organelles, whereas the more peripheral cytoplasm contains mitochondria, vesicles, free ribosomes, extensive smooth endoplasmic reticulum, and annulate lamellae, the last-mentioned often arranged in multiple concentric layers (Fig. 15). By the time chromosome aggregation is complete and/or extrusion has begun, this distribution has often changed. Initial blebs and fingers are relatively devoid of larger organelles which often become concentrated in one region of the cell (Fig. 19). As extrusion continues, more organelles may enter the protrusions, and minisegregants of the final BOG are highly variable in composition.

In other cells, extrusion is accompanied by different internal events and, for example, "normal nuclei" with nuclear envelopes are found in some minisegregants (Johnson et al., 1975; Fig. 20). Further details of fine-structural features of extruding cells will be presented elsewhere.

D. Modification of Extrusion

Manipulation of extrusion behavior has two main purposes. The first is to obtain the greatest number of extruding cells, and therefore minisegregant cells, as rapidly as possible. The second is to shed light on the nature of the processes responsible for the cortical instability of extrusion. It is also

---

Figs. 15–20. Transmission electron micrographs of sections of nitrous oxide-arrested HeLa cells which had been stored at 4°C for 9 hours and incubated in MEMFC for 1.5–2.5 hours at 37°C. After gentle centrifugation cells were fixed in HEPES- or phosphate-buffered glutaraldehyde, postfixed in osmium tetroxide, and stained with lead citrate and uranyl acetate by methods similar to those described by Johnson et al. (1975). Bar represents 1 μm.

Fig. 15. Cell with aggregated chromosomes (c) in central cytoplasm and mitochondria and concentric layers of endoplasmic reticulum and annulate lamellae in peripheral regions.

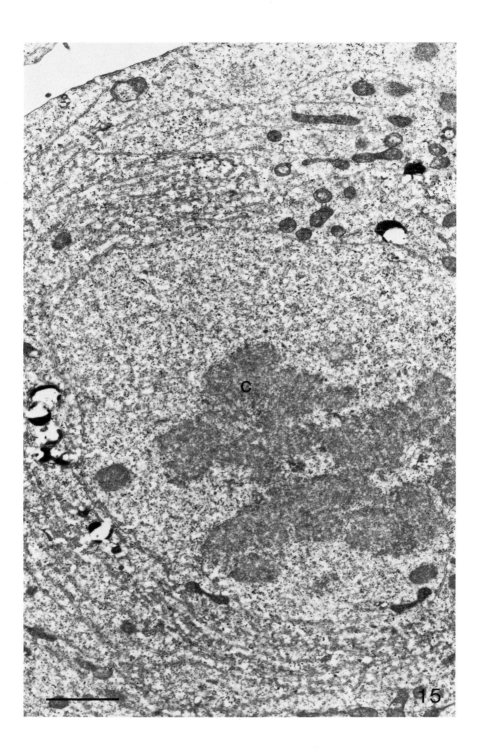

15

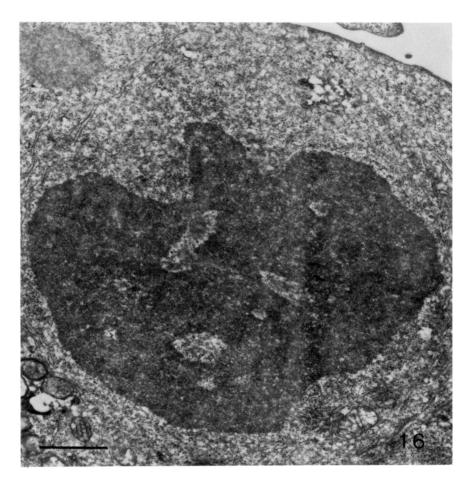

FIG. 16.   Late stage of chromosome aggregation into "dense nucleus".

possible that variations in the way extrusion is promoted may result in different patterns of chromosome segregation into minisegregants.

Extrusion activity can be promoted by a variety of treatments. The mitotic cells used in these studies are in all cases, except where otherwise specified, obtained by nitrous oxide arrest, although essentially similar results are obtained when either Colcemid-arrested or shake-off (selective detachment) mitotic cells are used. Among the promoters of extrusion are cold storage, Colcemid and other mitostatic agents, high pH, thiol-reducing agents, and trypsin (Johnson *et al.*, 1975; Mullinger and Johnson, 1976; Section D).

Further examination of the effects of these agents requires quantitation of

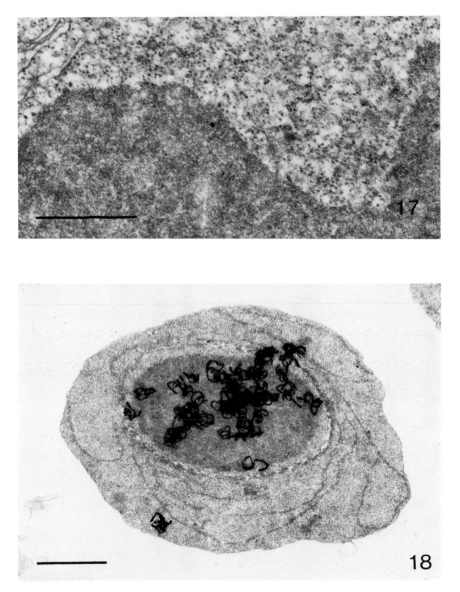

FIG. 17. Enlargement of part of Fig. 16, showing the absence of membrane around the "dense nucleus".

FIG. 18. Autoradiograph of minisegregant obtained by extrusion subdivision of a, [³H]thymidine-prelabeled mitotic cell. Silver grains are concentrated over the "dense nucleus". Ilford L4 emulsion, developed in Microdol-X.

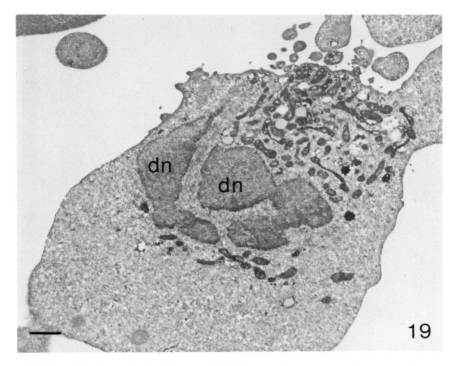

FIG 19.   Extruding cell with "dense nuclei" (dn), mitochondria, and other organelles concentrated in one region of the cytoplasm.

extrusion activity. This can be carried out in several ways, the most accurate of which is time-lapse sequence analysis. Although such a method suffers from the disadvantages of small sample numbers and the time required for satisfactory analysis, it permits a precise measure of the heterogeneity of both the nature and the kinetics of behavior and is perhaps the only way of unequivocally demonstrating unanticipated behavior such as reversal (Fig. 6). It also allows one to distinguish between normal cells that are the product of successful mitotic division, and cells that have remained inactive throughout; in cases in which a large proportion of cells divide normally, ignorance of the extent of division causes great uncertainty in estimation of the proportion of cells extruding. Another, less accurate, method of analysis is to fix populations of extruding cells attached to PL coverslips at different

FIG. 20.   Extruding cell with membrane-bound "normal nuclei" (n) in both parent cell (p) and extrusion (e).

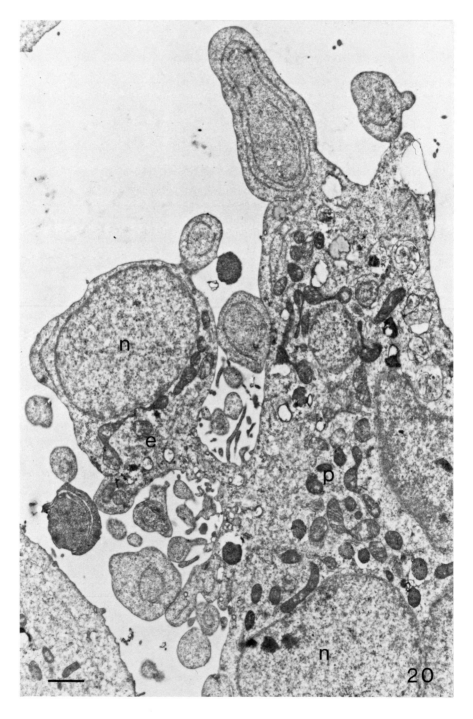

Key

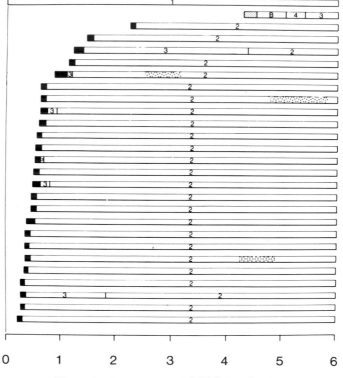

■ Cleavage
▨ Extrusion
▨ Long phase with active balloon
▨ Bubbling
⊡ 1 cell
⊡ 2 daughters
⊡ 3 daughters
⊡ 4 daughters
⊡ Bunch of grapes (more than 4 daughters)
⊡ Incomplete division into 3 daughters
⊡ Incomplete division into 4 daughters

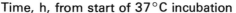

Time, h, from start of 37°C incubation

FIGS. 21 AND 22 (facing page).   Diagrams showing the kinetics of division of individual cells in populations of mitotic HeLa cells. Figure 21: Cells incubated at 37°C in MEMFC immediately after the release of a nitrous oxide block. Figure 22: Cells stored at 4°C for 10 hours after release from nitrous oxide and then incubated at 37°C in BSSH, pH 8.0, with 0.1 μg/ml Colcemid. Reproduced from Mullinger and Johnson (1976) by courtesy of the Company of Biologists Ltd.

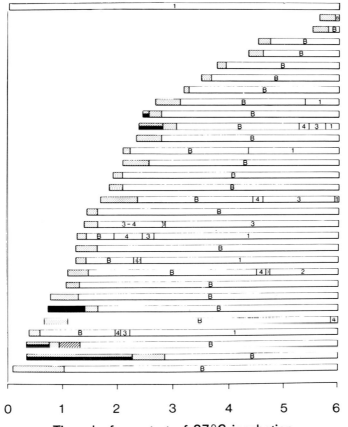

Time, h, from start of 37°C incubation

times after incubation at 37°C. The populations can then be scored at leisure. Spot counts on living material can also be used as a means of assessing the progress of extrusion activity within a population. Because of the effect of temperature on extrusion, such counts must be made at 37°C.

Analysis of the effects of different agents on extrusion is also complicated: (1) by the relative asynchrony of the onset of extrusion compared with cleavage (Figs. 21 and 22); (2) by the interaction or potentiation of the effects of different agents, resulting in either a different proportion of extruding cells or a faster rate of extrusion; and (3) by the variety of behavior, ranging from full-scale extrusion to pure cleavage, and even mixtures of these two processes (Mullinger and Johnson, 1976).

1. Cold Storage and Mitostatic Agents

For all experiments involving cold storage, mitotic cells suspended in MEMFC are transferred from the nitrous oxide chamber to bottles and placed in beakers containing precooled water at 4°C. After the desired period of cold storage aliquots of the cells are removed and, if necessary, resuspended in appropriate media before incubation at 37°C. Table I indicates the potentiating effect of cold storage on extrusion behavior of cells incubated in a variety of different media. In the absence of cold storage, Hanks' basal salt solution (HBSS, Hanks and Wallace, 1949) promotes the greatest amount of extrusion while, after cold storage, there is relatively little difference between the various media. Although the proportion of cells that finally extrude shows considerable variation from one experiment to another, there is good correlation between the duration of cold storage and the proportion of cells that extrude rather than cleave (Table II).

An example of the potentiating effect of two different treatments on extrusion is seen in the relation between length of cold storage and length of nitrous oxide arrest (Fig. 23). The synchronization schedule was modified for this experiment, since the procedure described in Section II,A produces cells which have been arrested in mitosis for different periods of time, up to a maximum of 9 hours. HeLa cells were plated out in plastic dishes with fresh MEMFC after a 20-hour exposure to 2.5 mM thymidine and, 4 hours later, placed in nitrous oxide in MEMFC buffered with 20 mM N-2-hydroxyethylpiperazine-N'-2-ethanesulfonic acid (HEPES), with the pH adjusted

TABLE I

EFFECTS OF DIFFERENT MEDIA ON EXTRUSION[a,b]

Medium	Percent of extruded cells after various times of incubation at 37°C		
	0.75 hours	1.5 hours	3.5 hours
MEMFC	3 (1)	21 (5)	34 (5)
MEM	3 (1)	31 (2)	52 (5)
HBSS	4 (4)	25 (3)	48 (12)
HBSS plus 5% fetal calf serum	1 (1)	10 (2)	43 (7)
HBSS without glucose	0.5 (0.5)	15 (1)	38 (4)

[a] Reproduced from Johnson et al (1975) by courtesy of the Royal Society of London.

[b] Mitotic cells were collected from nitrous oxide arrest, cooled at 4°C, and stored for 4 hours before incubation in various media at 37°C, all at pH 7.4. Five hundred cells were scored for each sample at the times noted, using unfixed material. The figures in parentheses refer to the scores obtained from parallel samples which were not stored at 4°C before incubation.

TABLE II

RELATIONSHIP BETWEEN EXTRUSION AND COLD STORAGE[a,b]

Time of storage at 4°C (hours)	Incubation medium	Percent of cells showing extrusion activity	Total number of cells observed
0	MEMFC, pH 7.2	0	27
6	MEMFC, pH 7.2	28	18
8	MEMFC plus 20 mM HEPES, pH 7.2	34	18
10	MEMFC, pH 7.2	73	15
6	BSSH, pH 8.0	42	17
8	BSSH, pH 8.0	80	10
10	BSSH plus 0.1 μg/ml Colcemid, pH 8.0	84	31

[a]Abridged from Mullinger and Johnson (1976), reproduced by courtesy of the Company of Biologists Ltd.

[b]Data obtained from the analysis of time-lapse sequences of nitrous oxide-arrested HeLa cells stored at 4°C and incubated in various media at 37°C for 4 hours.

to 7.2 with sodium hydroxide. After 1 hour cells already in mitosis were collected, spun down, and resuspended in MEMFC–HEPES, pH 7.2, at a concentration of 2×10^5 cells/ml, and either returned to nitrous oxide for further periods of time or used immediately. To prepare mitotic cells without nitrous oxide treatment, a similar schedule was used, except that a 1-hour normal incubation was substituted for the 1-hour nitrous oxide arrest; cells in mitosis were collected by shake-off and resuspended in appropriate media. Figure 23 shows that prolongation of cold storage above a threshold value of about 1 hour increases the proportion of extruding cells. Similarly, prolongation of time previously spent in nitrous oxide arrest (up to 8 hours) also leads to a greater number of extruding cells, provided there has also been cold storage. However, very prolonged exposure of mitotic cells to either nitrous oxide or Colcemid is alone sufficient to induce extrusion (Table III), a result which suggests that the fragmentation of human diploid cells described by Cremer et al. (1976) is essentially an extrusion phenomenon. It is noticeable that the combined effects of nitrous oxide and Colcemid are not additive but resemble the effects of Colcemid alone. This may be explained by the possibility that Colcemid blocks mitosis at an earlier stage than nitrous oxide (Brinkley and Rao, 1973).

2. pH OF INCUBATION MEDIUM

Alkaline medium can by itself induce a low level of extrusion in nitrous oxide-arrested mitotic cells; after 4 hours of cold storage the extent of extrusion is further increased by incubation in medium of high rather than low

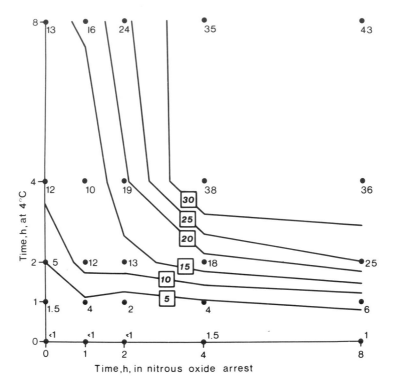

FIG. 23. Diagram showing the combined effects on extrusion activity of the length of time spent in nitrous oxide arrest and in cold storage. Mitotic HeLa cells were incubated at 37°C on PL coverslips in BSSH, pH 7.6, after various periods in nitrous oxide and at 4°C, as described in the text. After 4 hours, samples were fixed and the percentage of extruding cells scored. In the diagram these percentages (figures beside points) are positioned in relation to the time spent in nitrous oxide (x axis) and at 4°C (y axis). The contour lines with boxed percentage values are interpolated between these experimental points to join together estimated points having the same percentage of extrusion activity.

pH, though the lag period before extrusion is initiated is longer at higher pH, as in the case of normal mitotic cleavage (Table IV).

3. TRYPSIN

Nitrous oxide-arrested HeLa cells treated with 0.25% trypsin (Gibco, Inc.) in normal saline extrude when subsequently incubated in various media. In experiments involving trypsin, which eventually detaches cells from poly-L-lysine, cells were either trypsinized first and then placed on PL coverslips which already bore drops of HBSS buffered with 20 mM HEPES (BSSH) containing 0.2% soybean trypsin inhibitor, or trypsinized after adhesion to coverslips coated with poly-D-lysine hydrobromide (Sigma, Grade 1B, MW

TABLE III

EFFECT OF PROLONGED MITOTIC ARREST ON EXTRUSION[a]

Mitostatic treatment	Percent of extruded cells after various times of arrest at 37°C		
	12 hours	24 hours	36 hours
Nitrous oxide	1	5	34
Nitrous oxide plus 0.1 μg/ml Colcemid	0.4	11	14
Colcemid, 0.1 μg/ml	2	13	13

[a]HeLa cells were synchronized at the start of S phase by a standard double thymidine block (Rao and Johnson, 1972a) and exposed to mitostatic agents in MEMFC plus 20 mM HEPES, pH 7.4, 6 hours after release of the second thymidine block. Samples were fixed at intervals and scored. Over 200 cells were scored for each sample.

70,000 and over) which is not affected by trypsin. Table V presents the results of one such trypsin experiment performed with nitrous oxide-arrested mitotic cells. All operations were carried out at 37°C, and the proportion of extruding cells in the population was found to be related to the duration of the trypsin treatment. In this context it is of interest to note that neuraminidase does not potentiate extrusion activity.

TABLE IV

EFFECT OF pH ON EXTRUSION[a,b]

pH of HBSS	Percent of extruded cells after various times of incubation	
	1.5 hours	3.5 hours
6.8	3 (12)	4 (16)
7.2	3 (10)	4 (28)
7.6	5 (14)	11 (34)
8.0	1 (5)	18 (45)
8.4	4 (7)	22 (60)

[a]Reproduced from Johnson et al. (1975) by courtesy of the Royal Society of London.
[b]Mitotic HeLa cells plated out in HBSS immediately after release from nitrous oxide arrest and incubated at 37°C. The figures in parentheses represent the proportion of extruding cells in parallel samples stored at 4°C for 4 hours before incubation at 37°C. Five hundred cells were scored for each sample at the times noted, using unfixed material.

TABLE V

EFFECT OF TRYPSIN ON EXTRUSION[a]

Length of 0.25% trypsin treatment (minutes)	Percent extruding and dividing at various times after release from trypsin treatment and incubation in HBSS, pH 8.0					
	1 hour		2 hours		3 hours	
	Extruding	Dividing	Extruding	Dividing	Extruding	Dividing
5	—	26	1	53	6	42
10	—	61	4	46	5	53
20	—	37	11	36	14	47
30	1	51	1	68	14	24

[a]Nitrous oxide-arrested HeLa cells were washed thoroughly twice in excess warm (37°C) normal saline before trypsin treatment. They were then incubated at 37°C in BSSH buffered to pH 8.0 containing 0.2% soybean trypsin inhibitor. Cells were fixed at the times indicated, and over 200 cells were scored for each sample.

4. STATE OF CELL THIOLS

Agents affecting extrusion through thiol groups fall into two classes.

a. Thiol-Reducing Agents. We first observed considerable extrusion in HeLa cells as a response of mitotic cells to 2 mM dithiothreitol (DTT). At this concentration and without cold storage, up to 5% of mitotic cells extrude during a 4-hour period. DTT also significantly increases the proportion of cells extruding after a period of cold storage. The potentiation by DTT and other thiol-reducing agents including dithioerythritol, sodium thioglycollate, 2-mercaptoethanol, reduced glutathione, and cysteine (each at 2 mM), is between three- and sixfold at pH 7.4. DTT also substantially increases the number of extruding cells through a wide pH range, the enhancing effect being greatest at low pH (Johnson *et al.*, 1975).

b. Thiol-Blocking Agents. Mitosis in several systems is known to be blocked by agents that form stable adducts of thiol groups (e.g., Okazaki *et al.*, 1973). Similarly, extrusion appears to require some thiol groups, since thiol-blocking agents, such as N-ethylmaleimide and sulfonic acid diazonium chloride, block extrusion. Both these agents are toxic, however, killing cold-stored HeLa cells within 10 minutes even at 1 μM. Mercurial agents such as the cell penetrant p-chloromercuribenzene (PCMB) and p-chloromercuriphenyl sulfonate (PCMPS), which is relatively nonpenetrant (Vansteveninck *et al.*, 1965), are less rapidly lethal; brief exposures to 100 μM leave HeLa cells alive after 1 hour. Both PCMB and PCMPS block extrusion in cold-stored HeLa cells when given either immediately after cold storage or after poststorage incubation at 37°C of up to 45 minutes. How-

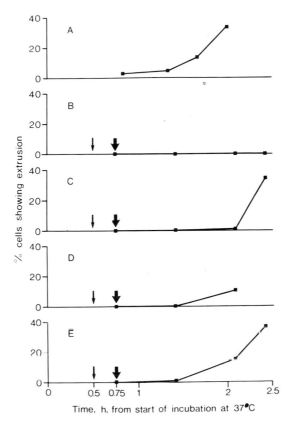

FIG. 24. Effects of PCMPS on extrusion. HeLa cells, synchronized in mitosis by a thymidine–nitrous oxide block, were stored in MEMFC at 4°C for 8 hours and plated out onto PL coverslips; several coverslips were transferred to each of five dishes containing BSSH, pH 8.0, and incubated at 37°C. The cells in dish A were allowed to extrude without interruption. After a 30-minute incubation, PCMPS dissolved in BSSH, pH 8.0, was added to dishes B through E to a final concentration of 100 μM (B and C) or 10 μM (D and E). After 15 minutes of PCMPS treatment, coverslips from dishes B and D were rinsed in BSSH, pH 8.0, and transferred to fresh BSSH, pH 8.0; coverslips from dishes C and E were rinsed in BSSH, pH 8.0, and transferred to 1 mM DTT in BSSH, pH 8.0. All manipulations were carried out at 37°C. At various times, coverslips were removed from the dishes, fixed in warm, buffered formaldehyde, and later scored for the presence of extruding cells. The beginning and the end of PCMPS treatment are marked by light and heavy arrows, respectively.

ever, cells that have already extruded do not revert to normal when treated with mercurials. The effect of PCMB cannot be reversed by thiol-reducing agents, but it is possible to reverse the effect of the nonpenetrant PCMPS by DTT (Fig. 24). These data strongly indicate that the cell cortex is the likely site of the thiol groups implicated in extrusion activity.

E.   Extrusion Activity in Other Cell Types

Extrusion can be readily induced in the following heteroploid human cell strains: HeLa, KB, HEp-2, and D98/AH$_2$. Under appropriate conditions diploid human fibroblasts can also be induced to extrude (Mullinger and Johnson, 1976). For diploid fibroblasts we found that mitotic cells arrested by conventional nitrous oxide pressure treatment extrude most readily after 12-hour storage at 4°C followed by incubation in BSSH, pH 8.0, at 37°C in the presence of 2 mM DTT; though the proportion of cells forming BOGs is much lower than for heteroploid lines, some diploid cells clearly give rise to free minisegregants (Fig. 25A), as do HeLa cells. The topography of extruding diploid cells is also similar to that seen in HeLa cells (Fig. 25B).

Extrusion activity resulting in the production of minisegregants can also be induced in mouse TA3 and Chinese hamster ovary (CHO-K1) cells by the following procedure. Mitotic cells are accumulated in the presence of Colcemid (0.05 μg/ml) for 4 hours and stored for 14 hours at 4°C in growth medium. The cells are then resuspended either in MEM without serum or in BSSH, each buffered to pH 7.4 with 20 mM HEPES and containing 8 mM DTT and Colcemid at 0.25 μg/ml. Within 1 hour of incubation at 37°C up to 20% of CHO-K1 cells, and slightly fewer TA3 cells, extrude to form classic BOGs. Gentle agitation by pipetting releases many of the minisegregants, and these commonly contain DNA as judged by Feulgen staining.

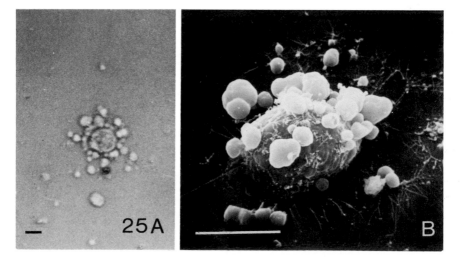

FIG. 25.   Extrusion of human diploid fibroblasts. (A) Photomicrograph of spontaneously detaching minisegregants. (B) Scanning electron micrograph of BOG. Bar represents 10 μm. Reproduced from Mullinger and Johnson (1976) by courtesy of the Company of Biologists Ltd.

F. Extrusion in Interphase HeLa Cells

It is also possible to induce a form of extrusion in interphase HeLa by the same treatments as are effective in mitotic cells; similar "zeiotic blebbing" was reported earlier (Dornfeld and Owczarzak, 1958; Rose, 1966). However, extrusion in interphase cells shows certain differences when compared with that in cells derived from originally mitotic populations. The surface protrusions tend to be smaller, and the parent interphase cell is often still identifiable in the most extruded cells (Fig. 26), so that fully developed BOGs are rare (less than 4% of all cells). In some cells "dense nuclei" may be formed, apparently by condensation and fragmentation of the interphase nucleus (Fig. 27A and B), whereas in others the nucleus may remain uncondensed.

Interphase extrusion is promoted by cold storage, prolonged trypsiniza-

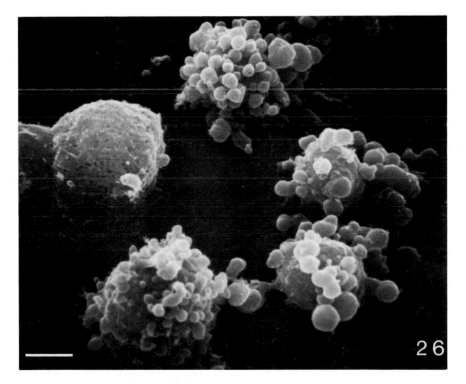

Fɪɢ. 26. Scanning electron micrograph of extruding interphase HeLa cells. Nitrous oxide-arrested cells were allowed to progress into interphase by incubation at 37°C in MEMFC for 4 hours, stored at 4°C for 8 hours, plated out on PL coverslips, incubated at 37°C in BSSH, pH 8.0, with 0.25 μg/ml Colcemid for 2 hours, fixed, and processed for scanning electron microscopy by methods described by Mullinger and Johnson (1976). Bar represents 5 μm.

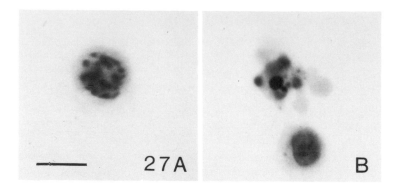

FIG. 27. Photomicrographs of extruding interphase cells. Synchronized S-phase cells were obtained by a standard double thymidine block (Rao and Johnson, 1972a), followed by incubation in MEMFC for 3 hours; after storage at 4°C for 8 hours cells were plated out on PL-slides and incubated at 37°C in BSSH, pH 8.0, for 1–2 hours, fixed in formaldehyde, Feulgen-stained, and counterstained in fast green. (A) Condensing interphase nucleus. (B) Extruding cell with "dense nuclei". Bar represents 10 μm.

tion, and the presence of Colcemid in the incubation medium, although none of these agents promotes extrusion on the scale seen in parallel experiments with mitotic cells. Interphase extrusion is blocked by PCMPS, this effect being reversed by DTT.

G.   The Basis of Extrusion Subdivision

There is a striking correspondence between the factors that induce extrusion and some known to affect the cellular systems of microtubules (MTs) and microfilaments (MFs) which are involved in cell movement and normal mitotic division.

The degree of polymerization of MTs is affected by, *inter alia*: (1) Temperature: MTs and spindles dissociate within 15 minutes after cooling to 4°C (Inoué, 1952; Brinkley and Cartwright, 1975). (2) Redox state of thiols: Tubulin contains about eight SH groups (Lee *et al.*, 1973), and chemical blockage, reduction, or oxidation of these groups prevents further polymerization and depolymerizes existing MTs (Mazia, 1958; Stephens *et al.*, 1966; Mellon and Rebhun, 1976). (3) pH: MTs *in vitro*, isolated spindles, and spindles in intact cells are dissolved at alkaline pH (Went, 1959; Cande *et al.*, 1974). (4) Mitostatic agents: Colchicine and Colcemid cause MTs to depolymerize. Nitrous oxide does not, but leads to disorganization of MTs in the spindle (Brinkley and Rao, 1973). (5) Trypsin: This compound leads to the disorganization and loss of MTs (Lucky *et al.*, 1975; Osborn and Weber, 1976).

The stability and functioning of MFs is known to be affected by, *inter alia*: (1) Temperature: The actin–troponin–actinin components extracted from cytoplasm form a potentially contractile gel which dissolves on cooling to 4°C (Stossel and Hartwig, 1976; Pollard, 1976). (2) Redox state of thiols: The actomyosin ATPase is dependent on free thiol groups, and thiol-reducing agents accelerate the reformation of cytoplasmic gels on warming from 4°C (Stossel and Hartwig, 1976). (3) Trypsin: This compound leads to the breakdown of the complex three-dimensional meshwork of MFs associated with the cell membrane (Lazarides, 1976).

This information suggests therefore that a disturbance in the organization of MTs and/or MFs plays an important role in extrusion subdivision. We do not imply that these are the only cellular systems disturbed, nor that the effects of the factors listed above on MTs or MFs are necessarily direct. (They may act, for instance, by changing intracellular Ca^{2+} levels, e.g., Weber and Murray, 1973; Morimoto and Harrington, 1974; Harris, 1975.) A fuller discussion of possible mechanisms underlying extrusion is given in Mullinger and Johnson (1976). It suffices to say here that any explanation should take into account: (1) the occurrence of extrusion in interphase cells and the lack of inhibition by mitostatic agents, suggesting that extrusion is not likely to be a variation of a specifically mitotic process; (2) the rare spontaneous occurrence of BOGs in cultures of HeLa cells (Mullinger and Johnson, 1976), and (3) the similarities between extrusion subdivision and the naturally occurring phenomenon of apoptosis, described by Kerr *et al.* (1972) and Bird *et al.* (1976) in solid tumors, and as a possible pathway of programmed cell disintegration during embryogenesis. Extrusion may be an *in vitro* homolog of apoptosis.

III. Isolation and Properties of HeLa Minisegregants

A. Separation of Minisegregant Cells according to Size

Since extruding cells are fragile and liable to clumping, the following method for the production of minisegregant HeLa cells has been chosen for its relative gentleness. Nitrous oxide-arrested cells in MEMFC are cooled to 4°C and stored thus for 9–12 hours. The cells are then resuspended and plated out into 150-mm plastic dishes at 5 × 10⁶ cells per dish and placed in a humidified carbon dioxide incubator at 37°C. Within 2 hours extrusion activity can be detected, and usually within 4 hours up to 60% of the cells form BOGs. Minisegregants can be readily separated on the basis of cell diameter by sedimentation under gravity through a 1–2% Ficoll-buffered

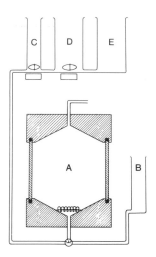

Fig. 28.   Apparatus for separating minisegregant cells. The sedimentation chamber (A in diagram, 14.5 cm in diameter, 17 cm high, approximately 1800-ml capacity) consists of a cylindrical plastic (Rohm and Haas Plexiglas) midpiece fitted into two Plexiglas blocks which form the lower and upper portions of the chamber. Rubber O-rings ensure water-tight connections, and the entire chamber is bolted together by four metal rods. Liquid enters the chamber from below through a perforated plastic disk covered with glass beads which reduce the turbulence and increase the resolution of the gradient. Reservoirs B and C are connected to the chamber by means of Teflon tubing via a three-way valve. Screw clamps are located between all the compartments to control liquid flow. The contents of reservoirs C and D are mixed by means of magnetic stirrers. The tubing connecting A with both B and C is first primed with medium (MEMFC), and 50 ml is also introduced into A. Reservoirs C, D, and E are filled with, respectively, 0.33% (150 ml), 1% (800 ml), and 2% (800 ml) Ficoll in MEMFC. The cell suspension containing the minisegregants is then added to B and allowed to flow slowly into A. Fresh medium is used to clear the connecting tube of cell suspension. The linear step gradient is formed by allowing the contents of reservoirs C, D, and E to flow into the sedimentation chamber. The flow rate is adjusted by a screw clamp between A and C to approximately 4 ml per minute for the first 150 ml and then to 15 ml per minute for the rest. The cells are allowed to sediment for the desired time, and then fractions are collected by replacing the contents of reservoirs C, D, and E with 1 M sucrose which is allowed to flow into the chamber. The contents of the gradient are thus forced through the top exit and are collected in a series of 50-ml tubes. Reproduced from Schor *et al.* (1975) by courtesy of the Company of Biologists Ltd.

step density gradient. All operations are carried out at room temperature. The apparatus used (modified from Miller and Phillips, 1969, and Denman and Pelton, 1973) is shown in Fig. 28. The extruding population of cells must be *gently* pipetted to break up aggregates and to liberate as many minisegregants as possible. This step is monitored by phase-contrast microscopy. Material derived from between 4 and 5×10^7 mitotic cells is concentrated by low-speed centrifugation, gently resuspended in approximately 30 ml of MEMFC, and added to the gradient apparatus. The Ficoll gradient is introduced below the cells at a slow rate (4 ml per minute) until the meniscus reaches the vertical walls of the chamber. The cells now appear as a narrow band near the top of the gradient. The remainder of the gradient is then introduced at a faster rate (approximately 15 ml per minute) until the meniscus reaches the top of the cylindrical section. At this point the cells are allowed to sediment through the gradient for up to 5 hours. The main function of the gradient is to prevent mixing of the different layers of cells by convection during loading, unloading, and separation. It is likely that density changes in the gradient have little effect on the sedimentation rate of nucleated cells, although this is probably not true for very small cells without chromatin. Miller and Phillips (1969) clearly demonstrated that, despite slight differences in density, cells could be adequately separated on the basis of size, the sedimentation velocity depending primarily on the cell radius. For this reason in particular, it is critical to handle the extruding mitotic cells gently during the breakup of the minisegregant aggregates so as to produce the maximum number of spherical cells and the minimum amount of irregular debris. Spherical cells might be expected to exhibit uniform hydrodynamic behavior in the gradient; broken cells and clumped aggregates do not appear to sediment according to size on these gradients and can clearly introduce errors into subsequent analysis of the minisegregants.

A buffered step gradient (Miller and Phillips, 1969) was chosen because greater numbers of cells can be loaded in a small volume without promoting streaming artifacts. We generally use Ficoll gradients made up in growth medium containing 5% serum, but adequate gradients can also be made in saline G (Kao and Puck, 1974) supplemented with 5% fetal calf serum.

After the period of sedimentation 1 M sucrose is introduced from below, thus forcing the contents of the gradient out through the top exit tube (see legend for Fig. 28 for details) at a rate of approximately 10 ml per minute. Up to 30 fractions (50 ml each) can be routinely collected in this manner. After low-speed centrifugation, the supernatant is removed from each fraction and replaced with complete MEMFC for subsequent analysis. The cells and minisegregants are stable at 4°C for at least 12 hours. All operations can be carried out under sterile conditions if required, and to this end we routinely run gradients and collect fractions in a laminar air flow cabinet.

The initial mixture of cells placed on the gradient is extremely heterogeneous, consisting of mono- and multinucleate cells of various sizes, including the minisegregant cells. The initial mixture also contains some cells in which extrusion division is incomplete. By means of the gradient it is possible to resolve the mixture of cells into a number of fractions each enriched in minisegregants of a particular size range. A typical initial mixture and various fractions are shown in Fig. 29. Scanning micrographs of the surface

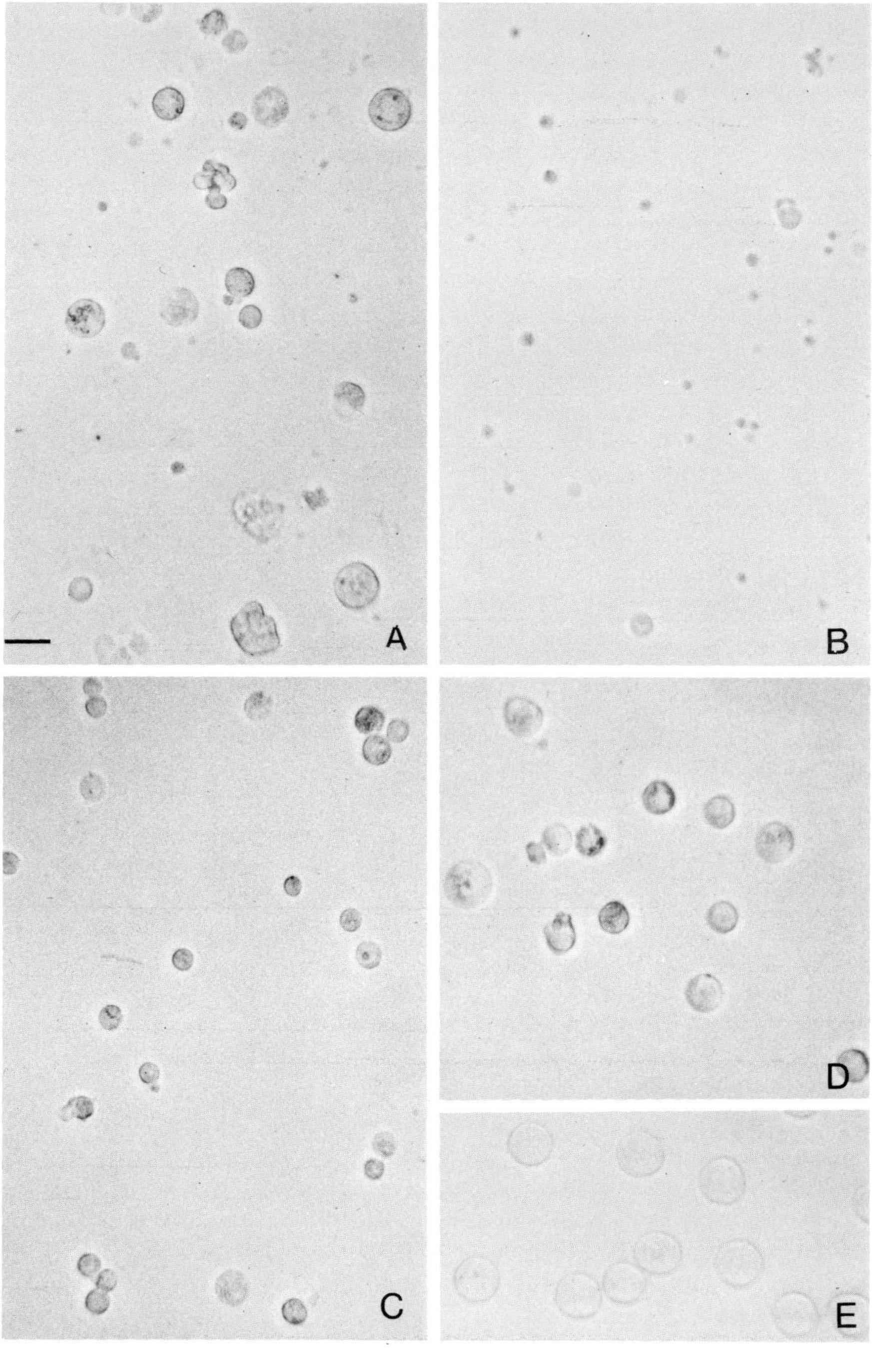
A
B
C
D
E

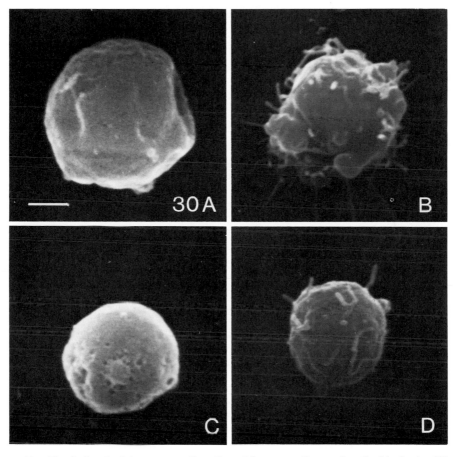

FIG. 30. Isolated minisegregant cells collected from a gradient as described in Section III and prepared for scanning electron microscopy as described in Mullinger and Johnson (1976). (A and B) Fixed immediately after collection from the gradient. (C and D) Fixed after incubation on PL coverslips in MEMFC for 24 hours at 37°C. Bar represents 1 μm.

of isolated minisegregants from the gradient both before and after incubation are shown in Fig. 30. Cells in the initial mixture usually vary in size between about 1 and 30 μm (measured on living material). In Fig. 31 the size distribution of minisegregants has been analyzed for the first 18 fractions

FIG. 29. Photomicrographs of the initial mixture layered on the gradient (A), and of isolated gradient fractions 1–2 (B), fractions 9–10 (C), fractions 19–20 (D), and HeLa cells arrested in mitosis (E). Bar represents 20 μm. Reproduced from Schor *et al.* (1975) by courtesy of the Company of Biologists Ltd.

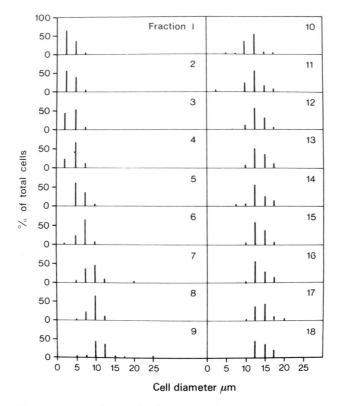

FIG. 31. Histograms showing the distribution of cells with different diameters in the first 18 fractions of a gradient separation. Cell diameter was determined on a minimum of 100 cells with a calibrated eyepiece micrometer.

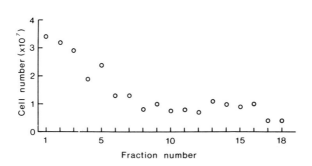

FIG. 32. The number of minisegregants in each of the first 18 fractions of a gradient separation. Cell counts were made using a hemocytometer.

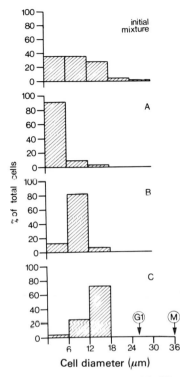

FIG. 33. Histograms showing the distribution of cells with different diameters in the initial mixture of cells layered on the gradient and in several pooled fractions. (A) From gradient fractions 1–8. (B) From gradient fractions 9–16. (C) From gradient fractions 17–24. Arrows indicate the mean diameters of G_1 (G1) and mitotic (M) HeLa cells. Reproduced from Schor *et al.* (1975) by courtesy of the Company of Biologists Ltd.

of a gradient. The number of minisegregants from each gradient fraction declines rapidly between fractions 1 and 7, thereafter remaining constant (Fig. 32). The size distribution of minisegregants from another gradient is shown in Fig. 33. Here various gradient fractions have been pooled to obtain several larger samples with clear differences in mean cell size. Trypan blue dye exclusion tests on both the initial mixture of cells and the individual gradient fractions reveal that between 95 and 100% of the cells routinely exclude the dye and can be judged viable by this criterion.

B.   DNA Content of Minisegregant Cells

1.   PRESENCE OF DNA

The distribution of DNA in cells of the heterogeneous initial mixture and in the various fractions can be assessed by Feulgen staining (Fig. 34). In a

TABLE VI
PRESENCE OF DNA IN MINISEGREGANTS[a,b]

Sample	Percent of Feulgen-positive cells
Initial	58
Fractions, 1–3 pooled	15
Fractions 4–5 pooled	29
Fractions 11–12 pooled	84
Fraction 14	78
Fraction 20	81

[a]Abridged from Schor *et al* (1975) by courtesy of the Company of Biologists Ltd.

[b]Minisegregants from the initial mixture of extruding cells placed on the gradient, and various minisegregant fractions isolated according to size by means of a standard Ficoll gradient. Each sample was fixed and Feulgen-stained. Three hundred cells were scored for each sample.

representative experiment 42% of the cells in the initial mixture were Feulgen-negative cytoplasts (Table VI). The remaining 58% of the cells each contained one or more Feulgen-positive areas, but these showed considerable variation in size and appearance among different cells and even within one cell, as expected from studies of whole extruding cells. Both "dense nuclei" and "normal" nuclei were sometimes seen in the same minisegregant but, more commonly, when a cell contained more than one Feulgen-positive body they were all of the same type. The percentage of cells with any detectable Feulgen-positive area increased from 15% in pooled fractions 1 through 3 to 81% in fraction 20. An analysis (Schor *et al.*, 1975) showed that cells with two or more "normal nuclei" were more common in fractions of

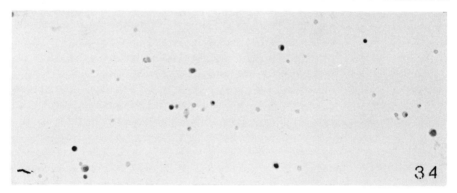

FIG. 34.   Isolated minisegregant cells collected from a gradient, as described in Section III. Fractions 4 and 5 were pooled, fixed and stained for Feulgen DNA, and counterstained with fast green. Bar represents 10 μm. Reproduced from Schor *et al.* (1975) by courtesy of the Company of Biologists Ltd.

higher number, and those with more than four "normal nuclei" were observed only in the initial mixture, presumably because they came to lie in regions of the gradient below fraction 20. Cells with a single "dense nucleus" were particularly common in the fractions obtained from the top of the gradient.

2.   DNA Quantitation

Microdensitometry of Feulgen-stained preparations (Schor *et al.* 1975) permits a quantitative estimate of the DNA content of minisegregant cells from different regions of the gradient. An analysis of this type is illustrated in Fig. 35. The Feulgen values for the initial mixture varied between 0.01 and 4C compared with a population of HeLa cells in which an average chromosome would have a value of 0.03C (data recalculated from DuPraw, 1970). There was a relatively uniform distribution of values across this range. Values of less than 0.01C almost certainly lie beyond the limits of detection, and it is probable that there were many minisegregants with extremely small quantities of DNA which could perhaps only be detected by genetic techniques.

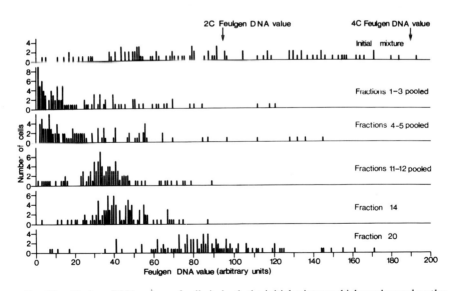

Fig. 35.   Feulgen DNA values of cells in both the initial mixture which was layered on the gradient and in isolated fractions. Each sample included both mononucleate and multinucleate cells and also nuclei of both normal and dense types (see text). One hundred cells with Feulgen-positive staining were scored for each sample. The mean DNA Feulgen values with standard deviations are, in arbitrary units: initial mixture, 84.7 ± 48.9; fractions 1–3 pooled, 25.8 ± 27.2; fractions 4–5 pooled, 31.6 ± 32.1; fractions 11–12 pooled, 35.0 ± 16.1; fraction 14, 42.6 ± 14.4; and fraction 20, 85.7 ± 30.0. Arrows indicate the 2C and 4C Feulgen values for HeLa cells. Reproduced from Schor *et al* (1975) by courtesy of the Company of Biologists Ltd.

Minisegregants in the isolated fractions had a narrower range of Feulgen values than those in the initial cell mixture, and the mean DNA value increased from lower to higher fraction number. It should be noted that the wide range of Feulgen staining intensities found in the minisegregants makes it difficult to give precise DNA values for the data presented in Fig. 35, since there is likely to be a considerable Feulgen artifact, particularly for the "dense nuclei" (Garcia, 1970). The range of Feulgen DNA values found in the minisegregant cells does, however, resemble the range measured by Ege and Ringertz (1974) in rodent microcells.

C. The Chromosomes of Minisegregant Cells

Although knowledge of the amount of DNA in minisegregant cells is important, we would also like to know the number and the nature of the chromosomes in these fragments. Since we have never observed cell division in isolated minisegregant cells from any position on the gradient, details of the chromosomes cannot be obtained using conventional techniques; we have therefore employed the method of premature chromosome condensation to analyze the minisegregant karyotype. When mitotic and interphase cells are fused, one common consequence is the induction of prematurely condensed chromosomes (PCCs) from the interphase nucleus. The precise morphology of the PCCs is related to the position of the cell in interphase at the time of fusion. Thus G_1 PCCs consist of single chromatids, G_2 PCCs of similar bivalent elements, and S PCCs of a heterogeneous mixture of condensed and uncondensed regions (Johnson and Rao, 1970; Rao and Johnson, 1974). There is good evidence that PCCs are the interphase counterparts of mitotic chromosomes with the same chromosome number and banding patterns (Johnson et al., 1970; Unakul et al., 1973; Waldren and Johnson, 1974; Röhme, 1974). Analysis of PCCs therefore provides an adequate and unique means of chromosome analysis in differentiated cells that never divide, and also in minisegregant cells (Johnson et al., 1970; Schor et al., 1975).

PCCs obtained from HeLa minisegregant cells by fusion with HeLa mitotic cells are shown in Fig. 36. Minisegregant cells are readily fused by means of Sendai virus (for details of fusion techniques and the induction of PCCs,

FIG. 36. Chromosome spreads containing PCCs induced from minisegregant cells after fusion with mitotic HeLa cells. (A) Minisegregant PCCs with about 30 condensed G_1 elements. (B) PCCs with about 10 attenuated G_1 elements. (C) Seven condensed PCCs, probably bivalent. (D) Three extended and damaged bivalent PCCs and a separate group of damaged, condensed G_1 PCCs. (E) Highly fragmented PCCs. (F) Autoradiograph of G_1 PCCs derived from a mitotic cell which had been prelabeled with [^{3}H]thymidine. Bar represents 5 μm. Reproduced from Schor et al. (1975) by courtesy of the Company of Biologists Ltd.

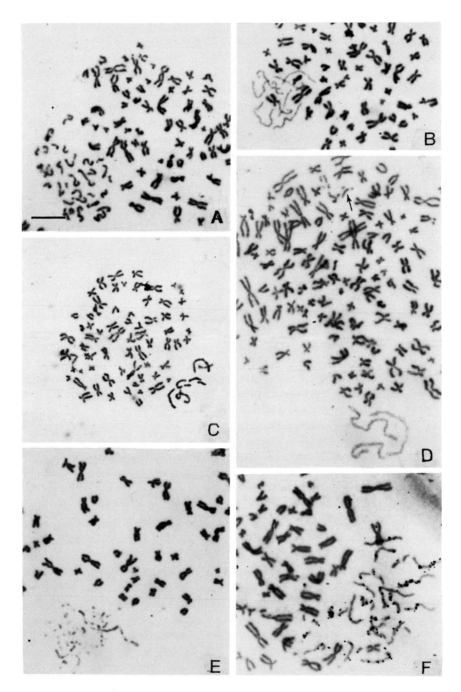

see Rao and Johnson, 1972a), and the PCCs can be counted. The probability of multiple minisegregant–single mitotic cell fusions was reduced by adjusting the ratio of mitotic to minisegregant cells in the fusion mixture to approximately 4:1. An analysis of minisegregant PCCs is presented in Fig. 37 which shows the distribution of chromosome numbers in three pooled fractions, the cell size distributions of which are shown in Fig. 33. These results show that a good correlation exists between chromosome number and minisegregant diameter, the smaller cells generally containing fewer than the larger. Minisegregants from the top of the gradient (pooled fraction A) contain a greatly reduced chromosome complement. However, although Feulgen values suggest that many of the smallest cells should contain a single chromosome, only a few such fused cells have been observed.

PCCs not only permit the quantitation of chromosome number in minisegregants but also provide information about the nature and integrity of the chromosomes as a function of cell size. Chromosome spreads most commonly contain single clusters of minisegregant PCCs which are either mono- or bivalent, i.e., derived from separated or nonseparated mitotic chromosomes, respectively. The bivalent PCCs are often attenuated, indicating a considerable degree of decondensation from their initial mitotic condition (Fig. 36D). The G_1 PCCs also vary in their attenuation (Fig. 36A and B). Some PCCs show various degrees of damage, frequently in the form of localized breaks, despiralization, or fragmentation (Fig. 36D and E). Damaged PCCs are most common in the smallest cells and bivalent PCCs are more likely to be damaged than monovalent chromosomes. There is a greater proportion of monovalent PCCs in the fraction containing the largest cells; apparently the larger the minisegregant, the more normal were both the division of its parent cell and the segregation of sister chromatids before division.

D. Synthesis of Macromolecules by Isolated Minisegregants

The synthesis of DNA, RNA, and protein by isolated minisegregants of different sizes has been established by autoradiography (Fig. 38). Table VII shows the results obtained when pooled minisegregant fractions from different regions of a gradient were incubated in the presence of radioactive precursors of DNA, RNA, and protein. Considerably fewer of the smaller minisegregants are capable of synthesizing RNA and DNA compared with the larger, although incorporation of leucine was similar in both. Protein synthesis by isolated cell fragments has also been described by Cremer *et al.* (1976), whose results are essentially similar to those described here. It remains possible that the cells from each minisegregant fraction capable of synthesizing DNA and RNA (6% of the smaller fraction, average chromo-

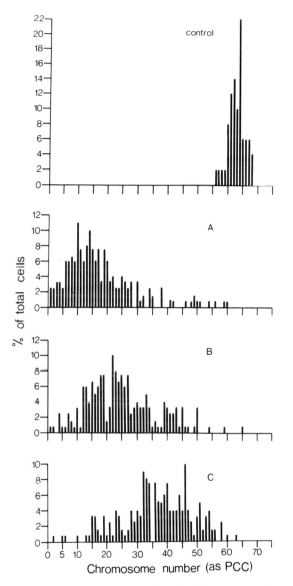

FIG. 37. Histograms showing the distribution of minisegregant cells containing different numbers of PCCs in various pooled fractions. (A) From gradient fractions 1–8. (B) From gradient fractions 9–16. (C) From gradient fractions 17–24. These data are based on the same isolation as that in Fig. 33. Each pooled fraction was fused with mitotic HeLa cells to induce PCCs. Chromosome preparations were made, and the number of PCCs in at least 200 PCC clusters was counted for each fusion sample. The mean chromosome numbers with standard deviations in the pooled fractions are: A, 17.3 ± 11.7; B, 26.0 ± 12.9; and C, 35.7 ± 12.2. Few spreads containing PCCs were observed in control mitotic × mitotic fusions. These were invariably of standard G_2 morphology and displayed a tight distribution around 64, the modal chromosome number for this strain of HeLa cell. Reproduced from Schor *et al.* (1975) by courtesy of the Company of Biologists Ltd.

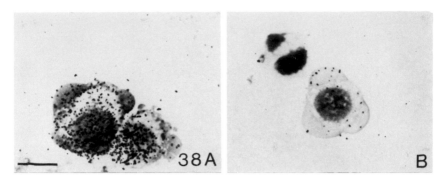

FIG. 38. Incorporation of [³H]uridine (A) and [³H]leucine (B) into TCA-insoluble products in minisegregant cells from pooled fraction B in Fig. 33 labeled for 3 hours immediately after gradient separation. Cytocentrifuge preparations. Bar represents 10 μm. Reproduced from Schor et al. (1975) by courtesy of the Company of Biologists Ltd.

some number 17, and 34% of the larger fraction, average chromosome number 26) correspond to the larger end of the size range in each population. Further studies are needed to determine both the chromosomal constitution of the cells capable of DNA synthesis and also the extent to which DNA replication proceeds in cells with different karyotypes.

A more detailed analysis of the incorporation of [³H]uridine and [³H]leucine by isolated minisegregant populations is shown in Fig. 39, in which

TABLE VII

MACROMOLECULAR SYNTHESIS IN ISOLATED MINISEGREGANTS[a,b]

Synthesis	Length of exposure to radioactive precursor (hours)	Pooled fraction[c]	Nucleated cells labeled (%)
DNA	24	A	6 (236)
		B	34 (202)
RNA	3	A	14 (176)
		B	52 (341)
Protein	3	A	47 (321)
		B	57 (251)

[a]Reproduced from Schor et al. (1975) by courtesy of the Company of Biologists Ltd.
[b]For labeling, cells were placed in plastic dishes; after the appropriate period they were trypsinized to remove adhering cells, and a cytocentrifuge preparation made of the pooled floating and attached cells. The numbers in parentheses denote the total number of cells scored for each sample.
[c]Pooled fractions A and B (mean chromosome numbers 17 and 26 and pooled gradient fractions 1–8 and 9–16, respectively; see Figs. 33 and 37) were incubated in the presence of [³H]thymidine (5 μCi/ml), [³H]uridine (5 μCi/ml), or [³H]leucine (5 μCi/ml).

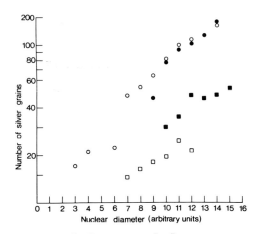

FIG. 39. The incorporation of [³H]leucine and [³H]uridine by isolated minisegregants as a function of nuclear diameter. Both the number of silver grains and the nuclear diameter in the fixed material were determined for at least 100 cells from each sample. Open circles, [³H]uridine incorporation in pooled fraction A (gradient fractions 1–8); solid circles, [³H]uridine incorporation in pooled fraction B (gradient fractions 9–16); open squares, [³H]leucine incorporation in pooled fraction A; solid squares, [³H]leucine incorporation in pooled fraction B. Reproduced from Schor *et al.* (1975) by courtesy of the Company of Biologists Ltd.

the average number of silver grains has been plotted as a function of the nuclear diameter. There is an approximately linear relationship between the amount of precursor incorporated per cell and the nuclear diameter.

E. Conclusion

These studies indicate that human cells with much reduced DNA and chromosome content can be produced by simple techniques and isolated according to size. These cells are capable of limited macromolecular synthesis and, in addition, are readily fused by means of Sendai virus; because of this they are potential vectors for gene transfer.

IV. The Fusion of Minisegregant and Other Cell Fragments and the Transfer of Genetic Information

A. Early Postfusion Hybrids

Human minisegregant cells with greatly reduced chromosome numbers are readily fused with whole cells, a point that emerges from the studies of PCCs. For minisegregants to be useful in gene complementation and hybrid

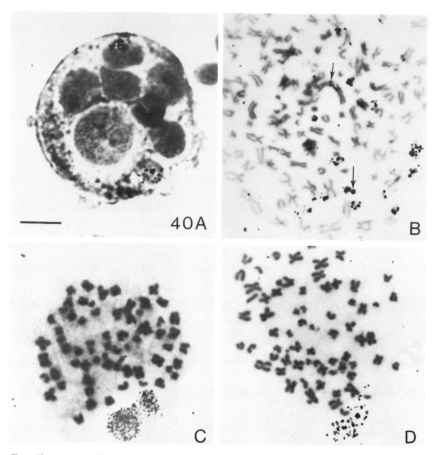

Fig. 40. Autoradiographs showing products of fusion between minisegregant cells labeled with [³H]thymidine and random HeLa cells fixed during the first and second cycles after fusion. (A) Cytocentrifuge preparation of multinucleate cell, from the first interphase after fusion, containing one minisegregant nucleus (identified by silver grains) and seven HeLa nuclei. (B) Hybrid metaphase chromosome spread, from the second mitosis after fusion, showing heavily labeled chromosomes of minisegregant origin and also a more lightly labeled chromosome presumably arising from chromosome rearrangements within the fused cell. (C) Metaphase chromosome spread associated with two small, prelabeled minisegregant nuclei. (D) Metaphase chromosome spread with adjacent region of radioactivity presumably associated with highly decondensed minisegregant chromatin. Bar represents 10 μm. Reproduced from Schor *et al.* (1975) by courtesy of the Company of Biologists Ltd.

production, it is necessary that their DNA become stably integrated and expressed in a foreign environment. We have followed the early fate of minisegregant DNA from HeLa cells previously labeled with [³H]thymidine in cells during early cell cycles after fusion with unlabeled HeLa cells. Several hours after fusion, heterokaryons can be detected with one or more highly labeled micronuclei (Fig. 40A). When chromosome preparations are made from cells collected during the first and second mitosis after fusion, several contain labeled micronuclei or chromosomes of minisegregant origin (Fig. 40B–D). In some cases several of the chromosomes show uniform labeling, while others have local concentrations of radioactivity (Fig. 40B). These observations clearly demonstrate that minisegregant chromosomes can be incorporated into a single mitotic figure within 24 hours of fusion.

B.   HeLa Minisegregant—Mouse A₉ Hybrids

We have recently described (Tourian et al., 1978) a series of proliferating mouse–human hybrid cells obtained by the fusion of HeLa minisegregants and mouse A_9 cells deficient in hypoxanthine phosphoribosyltransferase (HPRT). The minisegregants used for the two series of hybridizations were on both occasions obtained from fraction 4 from the top of standard Ficoll gradients. Feulgen staining of the minisegregants from fraction 4 from each gradient revealed that 20 and 26% of the cells, respectively, contained DNA. Previously, quantitative Feulgen densitometry of pooled fractions 4 and 5 showed that 97% of the cells containing DNA had a HeLa value, of less than 2C, and over 80% a value much less than 1C (Schor et al., 1975). Phase-contrast microscopy of the minisegregants from fraction 4 before fusion did not reveal any cells with a diameter greater than 12 μm.

For cell fusion, 2×10^6 minisegregants were mixed with 2.5×10^6 A₉ cells in HBSS with 250 hemagglutinating units of ultraviolet-inactivated Sendai virus and, after incubation for 30 minutes at 37°C, 4×10^5 cells were plated in plastic culture dishes in Dulbecco's modified MEM supplemented with 10% fetal calf serum. After 24 hours the medium was replaced and the cells selected under standard HAT conditions (Littlefield, 1964). Colonies were picked after 3 weeks, using cloning rings. Two hybrids were obtained from the first hybridization, of which one has survived, and over 50 from the second.

To assess the capacity of fractionated minisegregants to proliferate, and to determine whether or not they were contaminated by whole cells, half of the population of cells from fraction 4 was incubated in parallel with each fusion mixture. In these and numerous other control experiments no cells from this position in the gradient were found to survive and proliferate. The hybrids have been maintained for 2 years in HAT selection and contain up to seven human chromosomes. Back-selection of the hybrids in the presence of 20 μg/ml 6-thioguanine resulted in cells which lost the human X chromosome and also the capacity to express HPRT. The other human chromosomes were usually retained in these revertants.

The stability of HeLa minisegregant–A_9 hybrids in the absence of selection for the retention of HPRT varies both among the different hybrids and also for each hybrid as a function of the length of time out of HAT selection. One hybrid in particular consistently differed in segregation behavior from the rest; for up to approximately 15 days out of HAT selection, counterselection in the presence of 6-thioguanine did not reveal any colonies, and thereafter the rate of segregation rose dramatically. All other hybrids revealed essentially similar patterns of segregation, increasing smoothly with time out of HAT selection. The rates of segregation in these hybrids vary from an average of 4×10^{-6} per cell per generation during the first 6 days of culture under nonselective conditions, to 1×10^{-4} after 14 days and 5×10^{-4} after 21 days (Tourian et al., 1978).

Several independently derived hybrid clones were characterized and compared with the parental phenotypes (mouse and human) with respect to HPRT and two other X-linked enzymes, glucose-6-phosphate dehydrogenase and phosphoglycerate kinase. Human isozymes of each of the three enzymes were present in each hybrid examined. Although the specific activity of partially purified HPRT in this series of hybrids is only 10% of that of the HeLa parent, tests to evaluate its activity (e.g., heat denaturation, Michaelis constants for substrate and cofactor) all argue in favor of its being structurally and functionally normal. Further tests have not revealed evidence for enhanced utilization of the product of HPRT activity (inosine monophosphate) which would spuriously affect the estimation of enzyme specific activity (Tourian et al., 1978). It is possible that the low activity of human HPRT in the hybrids is partly related to gene dosage. The HeLa cells used to provide minisegregants are nearly triploid, and there is evidence that HeLa cells contain two genetically active X chromosomes (Czaker, 1973). Each of the hybrids in question contains a single X chromosome, and HPRT activity should therefore be half that of the HeLa parent. In fact, the specific activities are much lower.

C. Survey of Cell Reconstruction Hybrids

Several reconstruction hybrids have been produced from cell fragments and intact cells by various techniques (Table VIII), and some have been examined with respect to the expression, stability, segregation, and regulation of constitutive markers. Of the enzymes studied HPRT has been investigated in the most detail.

1. EXPRESSION OF HPRT

In addition to the hybrids produced by minisegregant–A_9 fusions, described above, other hybrids with HPRT of low specific activity have been produced by reconstruction. In hybrids formed after the transfer of isolated human metaphase chromosomes from heteroploid HPRT[+] cells strains to HPRT[−] mouse or Chinese hamster cells, the specific activity of human HPRT was always found to be much lower (2–30%) than that of the human parental cell, and also variable from one hybrid to another (Willecke and Ruddle, 1975; Wullems *et al.*, 1975). In contrast, the transfer of metaphase chromosomes from near-diploid HPRT[+] to HPRT[−] Chinese hamster cells results in hybrids with uniform HPRT activity identical to that of the HPRT[+] donor parent (Wullems *et al.*, 1975, 1976a). Low activity of otherwise normal constitutive enzymes in interspecific cell hybrids may therefore reflect problems experienced by regulatory elements in recognizing heterospecific signals. Alternatively, the possibly abnormal regulation of enzyme activities in heteroploid cells may not be readily transferable to hybrids, particularly if partial genomes are introduced first. In the future it will be necessary to undertake comparative studies to test these alternatives with intra- and interspecific hybrid cells formed after metaphase chromosome transfer, and after rodent or human microcell and minisegregant fusion. Additionally, it will be important to examine the regulation of enzyme activity in such hybrids when donor cell fragments or chromosomes are derived from diploid and heteroploid sources and, finally, to compare results obtained when whole cells are fused and before segregation has taken place.

2. HYBRID STABILITY AND HPRT PHENOTYPE

Hybrids formed by cell reconstruction techniques show variable stability in the absence of positive selection. With one exception (Wullems *et al.*, 1976b) it has not been possible to detect whole or partial donated chromosomes in hybrids obtained after metaphase chromosome transfer. Although some of these hybrids show what amounts to permanent integration of the HPRT gene (Burch and McBride, 1975; Wullems *et al.*, 1976a), most of them show rapid loss of the donated phenotype in the absence of positive selection (McBride and Ozer, 1973; Willecke and Ruddle, 1975; Wullems *et al.*, 1975, 1976a,b). Only in the case in which a donated chromosome has been found has it been possible also to detect the expression of other X-linked human loci in the hybrids (Wullems *et al.*, 1976b).

The rates of segregation observed in minisegregant–A_9 hybrids overlap those of whole-cell Chinese hamster interspecies hybrids (Chasin, 1972) and are generally much lower than those found in the rapidly segregating interspecific chromosome transfer hybrids (Willecke and Ruddle, 1975).

304

TABLE VIII

CELL FRAGMENTATION AND RECONSTRUCTION EXPERIMENTS

Cell lines used	Method of fragmentation[a]	Terminology of fragments and fusion products	Fusion partners	Proliferating hybrids, selection system, and characteristic selected for	Reference
Experiments involving anucleate fragments					
Avian erythrocytes, mouse L cells	CB	Anucleate cells	Avian erythrocytes × enucleated L cells	−	Ladda and Estensen (1970)
Avian erythrocytes, mouse macrophages, mouse L cells	CB	Anucleate cells, heterokaryons, hybrids	Avian erythrocytes × enucleated mouse macrophages and L cells	+	Poste and Reeve (1972)
Avian erythrocytes, mouse L and A_9 cells	CB	Anucleate cells reconstituted cells	Avian erythrocytes × enucleated L and A_9 cells	−	Ege et al. (1975)
Mouse macrophages, mouse L cells, human HEp-2 cells	CB	Anucleate cells, hybrids, heterokaryons	Enucleated macrophages × L cells; Enucleated L cells × HEp-2 cells	+, trypsin, selecting for macrophage surface component	Poste and Reeve (1972)
African green monkey kidney (AGMK) CV-1 cells, SV40-transformed mouse mKSBu 100 cells	CB	Enucleated cells, fused cells	Enucleated AGMK cells × mouse cells	+	Croce and Koprowski (1973)
AGMK cells, SV40-transformed mouse 3T3 cells (SV3T3)	CB	Anucleate cells, heterokaryons	Enucleated AGMK cells × SV3T3 cells	+	Poste et al. (1974)

Cell type	Technique	Products	Experiment	Result	Reference
Mouse L cells	CB	Anucleate cells, cybrids	Enucleated L cells resistant to chloramphenicol × L cells resistant to BrdU	+, chloramphenicol–BrdU, selecting for chloramphenicol resistance and thymidine kinase⁻	Bunn et al. (1974)
Human WI-38 cells	CB	Anucleate cells, heteroplasmons	Iodoacetate-killed WI-38 cells × enucleated WI-38 cells	+, selecting for rescue of iodoacetate killing	Wright and Hayflick (1975)
Experiments involving anucleate cells and nucleated cells with much reduced cytoplasm but normal gene complement					
Mouse L cells	CB	Karyoplasts			Wise and Prescott (1973); Ege et al. (1974a)
Rat L6 myoblasts	CB	Minicells (none proliferating)			Lucas et al. (1976)
Mouse L cells	CB	Karyoplasts (up to 1.4% proliferating)			
Mouse L 929 cells	CB	Cytoplasts (-DNA) karyoplasts (+DNA), reconstructed cells	Marked cytoplasts × marked karyoplasts	-(but some dividing cells observed)	Veomett et al. (1974)
Rat L6 myoblasts	CB	Anucleate cells, minicells, cybrids, reconstituted cells	[³H]thymidine-labeled L6 minicells × [³H]-leucine-labeled enucleated L6 cells	—	Ege et al. (1974b)
Rat L6 myoblasts	CB	Anucleate cells, minicells, cybrids, reconstituted cells	Marked, HPRT⁺ L6 minicells × marked, HPRT⁻ enucleated L6 myoblasts	—	Ege and Ringertz (1975)

TABLE VIII (*Continued*)

Cell lines used	Method of fragmentation[7]	Terminology of fragments and fusion products	Fusion partners	Proliferating hybrids, selection system, and characteristic selected for	Reference
Mouse L cells, A₉ cells	CB	Cytoplasts, karyoplasts, hybrids	A₉ karyoplasts (8-azaguanine-resistant) × L-cell cytoplasts (8-azaguanine-sensitive)	+, select with 8-azaguanine for HPRT⁻	Lucas and Kates (1976)
Rat L6 myoblasts, mouse A₉ cells	CB	Minicells, cytoplasms, reconstituted cells	L6 myoblast minicells × A₉ cytoplasms	+, HAT, selecting for HPRT⁺	Krondahl *et al.* (1977)
Experiments involving nucleated cells with reduced DNA content					
Rat L6 myoblasts	Colcemid-induced micronucleation followed by CB-induced enucleation	Microcells, heterokaryons	³H-labeled microcells × L6 myoblasts	—	Ege and Ringertz (1974) Ege *et al.* (1974b)
Mouse A₉, mouse B82, mouse CT11C₁, HeLa, and Chinese hamster E36 cells, mouse embryo fibroblasts	For rodent cells; colcemid-induced micronucleation followed by CB-induced enucleation	Microcells, heterokaryons, hybrids	A₉ microcells (HPRT⁻) × B82 (TK⁻) cells CT11C₁ microcells (HPRT⁺) × Chinese hamster E36 (HPRT⁻) cells mouse fibroblast microcells × HeLa cells	+, HAT, selecting for TK⁺, HPRT⁺ +, HAT, selecting for HPRT +, ouabain, selecting for mouse component	Fournier and Ruddle (1977)

Cell type	Method	Products	Cross	Result	Reference
HeLa cells	Extrusion of cold-arrested mitotic cells	Minisegregants			Johnson et al. (1975), Mullinger and Johnson (1976)
HeLa cells	Extrusion of cold-arrested mitotic cells	Minisegregants, cytoplasts, heterokaryons	HeLa minisegregants × HeLa cells	− (mitotic activity of fused cells observed)	Schor et al. (1975)
HeLa cells, mouse A$_9$ cells	Extrusion of cold-arrested HeLa mitotic cells	Minisegregants, hybrid cells	HeLa minisegregants, (HPRT$^+$) × mouse A$_9$ (HPRT$^-$) cells	+, HAT, selecting for HPRT$^+$	Tourian et al. (1978)
Human primary fibroblasts	Colcemid-induced nuclear and cell fragmentation	Cell fragment (+DNA)			Cremer et al. (1976)

Experiments involving reconstitution of cells by transfer of metaphase chromosomes

Cell type	Method	Products	Cross	Result	Reference
Rat ascites hepatoma AH 130 cells, primary mouse embryonic fibroblasts, Chinese hamster diploid cells	MC		Rat ascites chromosomes × mouse fibroblasts	− (but hybrid metaphase figures observed)	Sekiguchi et al. (1973)
			Chinese hamster chromosomes × Chinese hamster fibroblasts	− (but hybrid metaphase figures observed)	
Chinese hamster V79, mouse A$_9$, and HeLa cells	MC	Chromosomes, hybrids	V79 metaphase chromosomes × A$_9$ (HPRT$^-$) cells	+, HAT, selecting for HPRT$^+$	McBride and Ozer (1973)
			HeLa metaphase chromosomes × A$_9$ (HPRT$^-$) cells	+, HAT, selecting for HPRT$^+$	

TABLE VIII (*Continued*)

Cell lines used	Method of fragmentation[a]	Terminology of fragments and fusion products	Fusion partners	Proliferating hybrids, selection system, and characteristic selected for	Reference
HeLa and mouse A₉ cells	MC	Chromosomes, hybrids	HeLa metaphase chromosomes × A₉ (HPRT⁻) cells	+, HAT, selecting for HPRT⁺	Willecke and Ruddle (1975)
HeLa, human lymphoblastoid, and mouse A₉ cells	MC	Chromosomes, hybrids	HeLa and lymphoblastoid cell metaphase chromosomes × A₉ (HPRT⁻) cells	+, HAT, selecting for HPRT⁺	Burch and McBride (1975)
HeLa and human T cells, Chinese hamster Wg 3-h cells and Don fibroblasts	MC	Chromosomes, hybrids	Metaphase chromosomes from Don (HPRT⁺) fibroblasts X Wg 3-h (HPRT⁻) cells	+, HAT, selecting for HPRT⁺	Wullems *et al.* (1975)
			HeLa or T-cell metaphase chromosomes × Wg 3-h (HPRT⁻) cells	+, HAT, selecting for HPRT⁺	Wullems *et al.* (1975, 1976a)
Chinese hamster (CH) diploid cells, mouse L cells	MC	Chromosomes, hybrids	CH metaphase chromosomes × L cells (thymidine kinase⁻)	+	Sekiguchi *et al.* (1975)
Chinese hamster Wg 3-h–human lymphocyte hybrids, HeLa cells	MC	Chromosomes, hybrids	HeLa metaphase chromosomes (HPRT⁺) X Chinese hamster–human hybrids (HPRT⁻)	+ HAT, selecting for HPRT⁺	Wullems *et al.* (1976b)

[a]CB, Cytochalasin-B enucleation; MC, isolation of metaphase chromosomes.

V. Conclusions

The ability to fragment human cells by a variety of simple procedures and to produce DNA-containing minisegregants makes possible new experiments in somatic cell genetics. These studies are just beginning and can be expected to yield interesting information with respect to gene localization and perhaps regulation. Reconstitution experiments between whole cells and cell fragments will form a useful contrast and complement to whole-cell fusions and should help to simplify analysis of regulatory phenomena, not least by predetermining the direction of chromosome balance in the hybrid. Chromosome segregation normally occurs very slowly in intraspecies crosses (Kao *et al.*, 1969; Rao and Johnson, 1972b; Bengtsson *et al.*, 1975), and rather rapidly and unidirectionally in interspecies crosses (Kao and Puck, 1970). Attempts to direct chromosome loss in hybrids have so far met with limited success (Rao and Johnson, 1972b; Pontecorvo, 1974; Bengtsson *et al.*, 1975). Since minisegregants and microcells can be separated according to size and DNA content (Schor *et al.*, 1975; Fournier and Ruddle, 1977), it should be possible to control the amount of DNA introduced by these agents and therefore to construct the desired hybrid immediately. With increasing knowledge of the extrusion process in mitotic cells and of the factors which influence and promote it, there is hope that the abnormal segregation of chromosomes can be better controlled to provide minisegregants containing specified chromosomes or chromosome groups.

VI. Appendix: The Assembly of a Pressure Chamber for Synchronizing Human Cells

R. T. JOHNSON, R. H. J. BROWN, AND M. J. BERRY*

In 1968 Rao described the synchronization of human tissue culture cells in mitosis by high-pressure (5.4-atm) nitrous oxide treatment. Unlike the more commonly used methods of mitotic accumulation using Colcemid or vinblastine, the nitrous oxide pressure method is fully reversible, thereby permitting large synchronized populations of postmitotic cells to be obtained (Rao and Johnson, 1972a). Nitrous oxide pressure arrest can be used to accumulate mitotic cells from a variety of permanent human heteroploid

*Department of Zoology, University of Cambridge, Cambridge, England.

cell strains (HeLa, HEp-2, KB, D98/AH$_2$), in addition to diploid human cells from secondary tissue cultures (numerous fibroblastic populations). Although commercial pressure chambers are available, they are expensive and are not designed to accept the largest tissue culture plastic petri dishes. We have therefore designed and constructed a pressure chamber that is relatively inexpensive and accepts all sizes of commercially available plastic culture dishes. This makes it possible to collect more than 10^8 mitotic human cells with ease.

The apparatus includes a cylindrical steel chamber, in which the tissue culture plates are stacked, and a top plate with an inserted manifold and valves for gas introduction and evacuation. A pressure gauge is fitted to the manifold, and the top plate is bolted to the chamber. The top plate is connected via the manifold to the gas supply by a flexible pressure hose. In this manner several chambers can be attached compactly to a gas supply.

A diagram of the apparatus is shown in Fig. 41. To save time, as much use as possible is made of commercially available components, the specifications of which are listed in the legend.

The pressure chamber is made up from a length of 6-inch-diameter steel tubing with a wall thickness of $\frac{1}{4}$ inch. It is machined as shown in the figure with registers to locate an inserted baseplate and a top flange. These are both $\frac{1}{2}$ inch thick and are welded in position as shown. The top flange is machined flat after welding, and a standard $\frac{1}{8}$-inch O-ring groove is turned in its upper surface. The top plate, nominally $\frac{1}{2}$ inch thick, is machined all over and is bolted to the flange by eight $\frac{1}{4}$-inch-high tensile socket-head cap screws. A manifold with two side arms fabricated from a brass bar is screwed into the top plate and carries a pressure gauge and inlet and outlet valves.

The need for a long, flexible gas inlet pipe is satisfactorily met by a vehicle brake pipe which routinely operates at a pressure of 500 psi. We use a British Leyland part, and thus the fittings at each end are threaded $\frac{3}{8}$ inch UNF.

To avoid corrosion of the chamber by accidental spillage of medium or by water-saturated gas (the chamber operates at 37°C and often contains a large volume of liquid), the internal surfaces of the chamber are coated with an epoxy resin.

The chamber has been designed to operate at 5.4 atm (80 psi, 5.066×10^5 N/m^2) for periods exceeding 9 hours at 37°C with a considerable margin of safety. The tube and plates have safety factors exceeding 50 to avoid the necessity of highly specialized welding and subsequent normalization. The socket-head screws have a safety factor of about 4. They are used in preference to bolts to avoid the risk of overstressing by the use of large spanners. It is recommended that a test be carried out every 6 months in which the

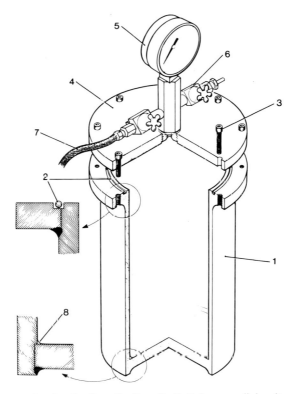

FIG. 41. Pressure chamber for the collection of mitotic human cells by nitrous oxide arrest. (1) Chamber consisting of (a) steel pipe with $\frac{1}{4}$-inch-thick wall, 6-inch inner diameter, 13 inches high to flange; (b) steel baseplate, $\frac{1}{4}$ inch thick, 6 inches in diameter, set into pipe and welded as illustrated in figure detail; (c) steel flange, $\frac{1}{2}$ inch thick, 9 inches in diameter, welded to pipe as illustrated in figure detail. Eight holes are drilled and turned to accept $\frac{1}{4}$-inch BSF screws; $\frac{1}{8}$ inch O-ring groove ($\frac{1}{8}$ inch deep and $7\frac{1}{2}$ inches in diameter) cut into flange. (2) O-ring (No. R4700 from G. Angus & Co., Ltd., Fluid Steel Division, Coast Road, Wallsend, Northumbria, U.K.). (3) $\frac{1}{4}$-inch BSF socket-head screws. (4) Steel top plate, $\frac{1}{2}$ inch thick, 9 inches in diameter. (5) One pressure gauge, 0–200 psi. (6) Two valves, die-cast bodies, threaded $\frac{1}{4}$ inch BSP (7) Brake pipe. (8) Epoxy resin.

chamber is *completely* filled with water and the internal pressure increased to 200 psi.

Tissue culture plates containing cells are stacked in the chamber either in a simple carrier or taped together. The internal dimensions of the chamber permit the largest commercially available plastic culture plates to be used, e.g., 150-mm Integrid dishes (Falcon Plastics, Inc.) or 140-mm dishes (Sterilin, Ltd). With the use of a chamber of the dimensions given here it is possible to stack thirteen 150-mm Integrid dishes. Since each dish is seeded with 6×10^6 HeLa cells and 95% are routinely arrested in mitosis during the nitrous oxide pressure treatment, more than 7×10^7 mitotic HeLa cells can be readily obtained from a single chamber. The degree of mitotic synchrony achieved by the nitrous oxide block depends of course on the length of the cell cycle. Usually populations of cells having a 20- to 24-hour cycle are partially synchronized by means of a 2.5 mM thymidine block before the nitrous oxide treatment (Rao, 1968; Rao and Johnson, 1972a).

Approximately 200 ml of carbon dioxide is added to a chamber of this size before nitrous oxide is admitted, so as to maintain adequate physiological pH during the high-pressure treatment. After the desired period of nitrous oxide exposure the chamber should be evacuated slowly over a 15-minute period and the plates of cells removed. We operate the chamber in a constant-temperature room although, provided the outer surface is treated with a corrosion-resistant material, it can be immersed in a water bath. If a constant-temperature room is used, the nitrous oxide cylinder should be kept outside and the chambers evacuated to the outside for safety.

ADDENDUM

Ege *et al.* (1977), in a review on the preparation of microcells, have used the term "microcell" to describe all subdiploid cell fragments.

ACKNOWLEDGMENTS

We appreciate many valuable contributions from the following collaborators during various stages of the investigation described here: Drs. S. L. Schor, R. J. Skaer, A. Tourian, K. Burg, S. Nicolson, K. Sperling, and M. Samaranayaka. We also thank Mr. R. Northfield, Mr. J. Rodford, and Mrs. E. G. Wilkin for excellent assistance, and Dr. A. R. S. Collins for helpful comments on the manuscript. This work was supported by the Cancer Research Campaign of which R. T. J. is a research fellow.

REFERENCES

Bengtsson, B. D., Nabholz, M., Kennett, R., Bodmer, W. F., Povey, S., and Swallow, D. (1975). *Somatic Cell Genet.* **1**, 41–64.
Bird, C. C., Wyllie, A. H., and Currie, A. R. (1976). *In* "Scientific Foundations of Oncology" (T. Symington, ed.), pp. 52–62. Heinemann, London.
Brinkley, B. R., and Cartwright, J. (1975). *Ann. N.Y. Acad. Sci.* **253**, 428–439.
Brinkley, B. R., and Rao, P. N. (1973). *J. Cell Biol.* **58**, 96–106.
Brinkley, B. R., Stubblefield, E., and Hsu, T. C. (1967). *J. Ultrastruct. Res.* **19**, 1–18.
Bunn, C. L., Wallace, D. C., and Eisenstadt, J. M. (1974). *Proc. Natl. Acad. Sci. U.S.A.* **71**, 1681–1685.
Burch, J. W., and McBride, O. W. (1975). *Proc. Natl. Acad. Sci. U.S.A.* **72**, 1797–1801.
Cande, W. Z., Snyder, J., Smith, D., Summers, K., and McIntosh, J. R. (1974). *Proc. Natl. Acad. Sci. U.S.A.* **71**, 1559–1563.
Chasin, L. A. (1972). *Nature (London), New Biol.* **240**, 50–52.
Cremer, T., Zorn, C., Cremer, C., and Zimmer, J. (1976). *Exp. Cell Res.* **100**, 345–355.
Croce, C. M., and Koprowski, H. (1973). *Virology* **51**, 227–229.
Czaker, R. (1973). *Humangenetik* **19**, 135–144.

Denman, A. M., and Pelton, B. K. (1973). *Methodol. Dev. Biochem.* **2**, 185–199.

Dornfeld, E. J., and Owczarzak, A. (1958). *J. Biophys. Biochem. Cytol.* **4**, 243–250.

DuPraw, E. J. (1970). "DNA and Chromosomes." Holt, New York.

Eagle, H. (1959). *Science* **130**, 432–437.

Ege, T., and Ringertz, N. R. (1974). *Exp. Cell Res.* **87**, 378–382.

Ege, T., and Ringertz, N. R. (1975). *Exp. Cell Res.* **94**, 469–473.

Ege, T., Hamberg, H., Krondahl, U., Ericcson, J., and Ringertz, N. R. (1974a). *Exp. Cell Res.* **87**, 365–377.

Ege, T., Krondahl, U., and Ringertz, N. R. (1974b). *Exp. Cell Res.* **88**, 428–432.

Ege, T., Zeuthen, J., and Ringertz, N. R. (1975). *Somatic Cell Genet.* **1**, 65–80.

Ege, T., Ringertz, N. R., Hamberg, H., and Sidebottom, E. (1977). *Methods Cell Biol.* **15**, 339–357.

Fournier, R. E. K., and Ruddle, F. H. (1977). *Proc. Natl. Acad. Sci. U.S.A.* **74**, 319–323.

Garcia, A. M. (1970). *In* "Introduction to Quantitative Cytochemistry" (G. L. Wied and G. F. Bahr, eds.), Vol. 2, pp. 153–170. Academic Press, New York.

Hanks, J. H., and Wallace, R. E. (1949). *Proc. Soc. Exp. Biol. Med.* **71**, 196–200.

Harris, P. (1975). *Exp. Cell Res.* **94**, 409–425.

Inoué, S. (1952). *Biol. Bull.* **103**, 316.

Johnson, R. T., and Rao, P. N. (1970). *Nature (London)* **226**, 717–722.

Johnson, R. T., Rao, P. N., and Hughes, S. D. (1970). *J. Cell. Physiol.* **76**, 151–158.

Johnson, R. T., Mullinger, A. M., and Skaer, R. J. (1975). *Proc. R. Soc. London Ser. B.* **189**, 591–602.

Kao, F. T., and Puck, T. T. (1970). *Nature (London)* **228**, 329–332.

Kao, F. T., and Puck, T. T. (1971). *Methods Cell Biol.* **8**, 23–39.

Kao, F. T., Johnson, R. T., and Puck, T. T. (1969). *Science* **164**, 312–314.

Kerr, J. F. R., Wyllie, A. H., and Currie, A. R. (1972). *Br. J. Cancer* **26**, 239–257.

Krondahl, U., Bols, N., Ege, T., Linder, S., and Ringertz, N. R. (1977). *Proc. Natl. Acad. Sci. U.S.A.* **74**, 606–609.

Ladda, R. L., and Estensen, R. D. (1970). *Proc. Natl. Acad. Sci. U.S.A.* **67**, 1528–1533.

Lazarides, E. (1976). *J. Cell Biol.* **68**, 202–219.

Lee, J. C., Frigon, R. P., and Timasheff, S. N. (1973). *J. Biol. Chem.* **248**, 7253–7262.

Levan, A. (1954). *Hereditas* **40**, 1–64.

Lillie, R. D. (1965). "Histopathologic Technic and Practical Histochemistry." McGraw-Hill, New York.

Littlefield, J. W. (1964). *Science* **145**, 709–710.

Lucas, J. J., and Kates, J. R. (1976). *Cell* **7**, 397–405.

Lucas, J. J., Szekely, E., and Kates, J. R. (1976). *Cell* **7**, 115–122.

Lucky, A. W., Mahoney, M. J., Barrnett, R. J., and Rosenberg, L. E. (1975). *Exp. Cell Res.* **92**, 383–393.

McBride, O. W., and Ozer, H. L. (1973). *Proc. Natl. Acad. Sci. U.S.A.* **70**, 1258–1262.

Mazia, D. (1958). *Exp. Cell Res.* **14**, 486–494.

Mazia, D., Schatten, G., and Sale, W. (1975). *J. Cell Biol.* **66**, 198–200.

Mellon, M., and Rebhun, L. I. (1976). *In* "Cell Motility" (R. Goldman, T. Pollard, and J. Rosenbaum, eds.), Book C, pp. 1149—1164. Cold Spring Harbor Lab., Cold Spring Harbor, New York.

Miller, R. J., and Phillips, R. A. (1969). *J. Cell. Physiol.* **73**, 191–201.

Morimoto, K., and Harrington, W. F. (1974). *J. Mol. Biol.* **88**, 693–709.

Mullinger, A. M., and Johnson, R. T. (1976). *J. Cell Sci.* **22**, 243–285.

Okazaki, Y., Mabuchi, I., Kimura, I., and Sakai, H. (1973). *Exp. Cell Res.* **82**, 325–334.

Osborn, M., and Weber, K. (1976). *Proc. Natl. Acad. Sci. U.S.A.* **73**, 867–871.

Pearse, A. G. E. (1968). "Histochemistry Theoretical and Applied," Vol. I. Churchill, London.

Pollard, T. D. (1976). *J. Cell Biol.* **68**, 579–601.

Pontecorvo, G. (1971). *Nature (London)* **230**, 367–369.

Pontecorvo, G. (1974). *In* "Somatic Cell Hybridization" (R. L. Davidson and F. de la Cruz, eds.), pp. 65–69. Raven, New York.

Poste, G., and Reeve, P. R. (1972). *Exp. Cell Res.* **73**, 287–294.

Poste, G., Schaeffer, B., Reeve, P. R., and Alexander, D. J. (1974). *Virology* **60**, 85–95.

Rao, P. N. (1968). *Science* **160**, 774–776.

Rao, P. N., and Johnson, R. T. (1972a). *Methods Cell Physiol.* **5**, 75–126.

Rao, P. N., and Johnson, R. T. (1972b). *J. Cell Sci.* **10**, 495–513.

Rao, P. N., and Johnson, R. T. (1974). *Adv. Cell Mol. Biol.* **3**, 135–189.

Röhme, D. (1974). *Hereditas* **76**, 251–258.

Rose, G. G. (1966). *J. R. Microsc. Soc.* **86**, 87–102.

Ruddle, F. H. (1972). *Adv. Hum. Genet.* **3**, 173–235.

Schor, S. L., Johnson, R. T., and Mullinger, A. M. (1975). *J. Cell Sci.* **19**, 281–303.

Sekiguchi, T., Sekiguchi, F., and Yamada, M. (1973). *Exp. Cell Res.* **80**, 223–236.

Sekiguchi, T., Sekiguchi, F., Tachibana, T., Yamada, T., and Yoshida, M. (1975). *Exp. Cell Res.* **94**, 327–338.

Stephens, R. E., Inoué, S., and Clark, J. I. (1966). *Biol. Bull.* **131**, 409.

Stossel, P. T., and Hartwig, J. H. (1976). *J. Cell Biol.* **68**, 602–619.

Stubblefield, E. (1964). *Symp. Int. Soc. Cell Biol.* **3** 223–248.

Tourian, A., Johnson, R. T., Burg, K., Nicolson, S., and Sperling, K. (1978). *J. Cell Sci.* (in press).

Unakul, W., Johnson, R. T., Rao, P. N., and Hsu, T. C. (1973). *Nature (London), New Biol.* **242**, 106–107.

Vansteveninck, J., Weed, I. R., and Rothstein, A. (1965). *J. Gen. Physiol.* **48**, 617–632.

Veomett, G., Prescott, D. M., Shay, J., and Porter, K. R. (1974). *Proc. Natl. Acad. Sci. U.S.A.* **71**, 1999–2002.

Waldren, C. A., and Johnson, R. T. (1974). *Proc. Natl. Acad. Sci. U.S.A.* **71**, 1137–1141.

Weber, A., and Murray, J. M. (1973). *Physiol. Rev.* **53**, 612–673.

Went, H. A. (1959). *J. Biophys. Biochem. Cytol.* **5**, 353–356.

Willecke, K., and Ruddle, F. H. (1975). *Proc. Natl. Acad. Sci. U.S.A.* **72**, 1792–1796.

Wise, G. E., and Prescott, D. M. (1973). *Exp. Cell Res.* **81**, 63–72.

Wright, W. E., and Hayflick, L. (1975). *Proc. Natl. Acad. Sci. U.S.A.* **72**, 1812–1816.

Wullems, G. J., van der Horst, J., and Bootsma, D. (1975). *Somatic Cell Genet.* **1**, 137–152.

Wullems, G. J., van der Horst, J., and Bootsma, D. (1976a). *Somatic Cell Genet.* **2**, 155–164.

Wullems, G. J., van der Horst, J., and Bootsma, D. (1976b). *Somatic Cell Genet.* **2**, 359–371.

Chapter 20

The Use of Tween-80-Permeabilized Mammalian Cells in Studies of Nucleic Acid Metabolism

DANIEL BILLEN AND ANN C. OLSON

The University of Tennessee, Oak Ridge Graduate School of Biomedical Sciences,
Oak Ridge, Tennessee

I. Introduction

A. Usefulness

There are experiments in cell biology in which the availability of cells made permeable to small molecules is useful. Ideally this permeability should be reversible and nonlethal. Such cells offer to the experimenter the possibility of assaying the cellular machinery *in situ*.

Here we describe the use of Tween-80 as an agent that produces a reversible permeability to nucleotides in Chinese hamster ovary (CHO) cells. We have used such cells to study DNA replication and repair following irradiation with ultraviolet light or X rays. One of the main advantages of

315

using permeabilized cells in DNA metabolism studies is that nucleoside triphosphate precursor pools are more easily controlled. Under these conditions a more accurate measure of DNA synthesis using radioactively labeled precursors is possible. Nearest-neighbor analysis of nucleic acid newly synthesized *in situ* can also be performed. In addition, manipulation of the cofactors necessary for the activity of certain enzymes and the use of nucleotide analogs are possible.

B.   Background

In bacteria, toluene treatment has been used to great advantage in studies on DNA metabolism (Moses and Richardson, 1970; Masker and Hanawalt, 1973; Billen and Hellerman, 1974). Ether-treated bacterial cells can synthesize RNA and protein in addition to DNA when provided with appropriate precursors (Vosberg and Hoffmann-Berling, 1971). However, essentially all the bacteria so treated are nonviable. Several techniques have been described for rendering mammalian cells permeable to exogenous nucleoside triphosphates for the purpose of studying DNA metabolism. These include cold shock (Berger and Johnson, 1976) and cold shock in a hypotonic medium (Seki *et al.*, 1975). It is likely that these techniques result in cell death and certainly bring about dramatic changes in cell constituents. Kay (1965) described the use of the detergent Tween-80 to increase the permeability of Ehrlich ascites carcinoma cells to dyes and metabolites without affecting their viability. We have modified and extended the use of this technique to other mammalian cells and to the study of DNA synthesis and repair, as described in this chapter. A brief description of this procedure has been published elsewhere (Billen and Olson, 1976).

II.   Methods

A.   Cell Culture

CHO cells (line CHO-K1, CCL 61, from the American Type Culture Collection) were maintained in nutrient mixture F-12 supplemented with 10% fetal calf serum (both from Grand Island Biological Company) in Falcon plastic tissue culture flasks at 37°C in a humidified atmosphere containing 10% CO_2. The cultures were checked regularly for, and found free of, mycoplasmas.

Cells for experiments were grown as above in larger flasks or in suspension. Monolayer cultures were seeded at 7×10^5 cells per 75-cm² flask and grown as described above for 2 days to a density of 4×10^6 to 8×10^6 cells per flask. They were washed with Hanks' balanced salt solution (Grand Island) and incubated in 0.1% trypsin (1:300, Grand Island) in calcium-free, magnesium-free, phosphate-buffered saline for 2 minutes at room temperature. The trypsin was decanted; the cells were dislodged by a sharp rap and resuspended in 10 ml of F-12 plus serum. Suspension cultures were seeded at 7×10^4 cells/ml and grown for 2 days in spinner flasks (Wheaton), or 125-ml glass culture flasks with magnetic stirring bars, to a density of 4×10^5 to 8×10^5 cells/ml.

Cells were usually prelabeled for 1–2 days with 0.005 μCi/ml of [¹⁴C]thymidine (Schwarz-Mann).

B. Permeabilization

The entire permeabilization procedure was carried out at room temperature. Suspension cells or trypsinized monolayer cells were collected from the medium by centrifugation at 1000 rpm for 5 minutes in 15 or 50 ml plastic centrifuge tubes, washed with 0.25 M sucrose, and finally resuspended in 1% Tween 80 (Sigma Chemical Company) in 0.25 M sucrose at a concentration of 2.0×10^7 to 2.8×10^7 cells/ml. They were immediately (within 5 minutes) assayed for their ability to synthesize DNA, as described in Section II,C.

C. Assay for Capacity to Synthesize DNA

Cells suspended in Tween–sucrose as described above (1.5×10^6 to 2.0×10^6 cells in 50–100 μl) were incubated at 37°C in 0.25 ml (total volume) of reaction mixture containing 60 mM potassium phosphate buffer (KPB), pH 7.4, 13 mM MgSO$_4$, 1.3 mM ATP, 33 μM dATP, 33 μM dCTP, 33 μM dGTP, 20 μM dTTP (all from Sigma), and 10 μCi/ml [$methyl$-³H]dTTP (New England Nuclear Corporation). At 15-minute intervals, 25-μl aliquots were spotted on a 1×2 cm strip of Whatman 3MM chromatographic paper, which had been previously soaked in 10% trichloroacetic acid (TCA), and the strips dropped into cold 10% TCA. At the end of the assay (90 minutes) the strips were washed twice in 10% TCA for 30 minutes, for 15 minutes in ethanol–diethyl ether (3:1), and in diethyl ether. The paper strips were dried and counted in a toluene–Liquifluor (New England Nuclear) cocktail.

III.   Results

A.   Appearance and Viability of Permeabilized Cells

Scanning electron micrographs (see Billen and Olson, 1976) showed that permeabilized cells are essentially identical in appearance to control cells treated with sucrose alone. Autoradiographs (Fig. 1) show that the cells are more or less intact after 60–90 minutes in the reaction mixture and also that the [³H]dTMP has been incorporated into the nuclei. Almost no label is seen over the cytoplasm. Furthermore, the fraction of nuclei labeled in this way is the same (0.56 in one experiment, 0.48 in another) as the fraction labeled when untreated cells are incubated with [³H] thymidine (0.55 and 0.46).

Cells treated with Tween–sucrose for 30 minutes are viable. In plating experiments such cells formed 80–85% as many colonies as cells treated with sucrose alone. However, viability was rapidly lost after transfer to the reaction mixture: After 20 minutes essentially all the cells were dead. We have made no attempt to adjust the composition of the reaction mixture to ensure survival of the cells.

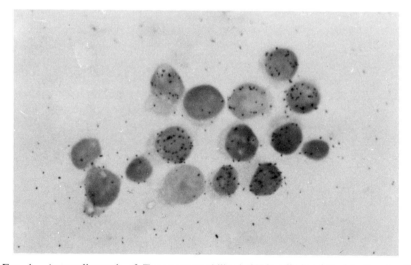

FIG. 1.   Autoradiograph of Tween-permeabilized CHO cells incubated in the standard DNA synthesis reaction mixture containing [³H]dTTP. After 60 minutes of incubation aliquots of the reaction mixture were spread on subbed slides, air-dried, fixed in Carnoy's fixative for 30 minutes, rinsed in ethyl alcohol–water (3:1) for 30 minutes, and dried. The slides were dipped in NTB2 liquid emulsion diluted 1:1 with distilled water, exposed for 4 days, and then processed through D-19 developer, rinse water, and Kodak fixer, all at 17°C. After rinsing in running water they were stained with Harris's hematoxylin and alcoholic eosin. ×500.

B. Variation in the Permeabilization Procedure

When we first began these studies, we permeabilized the cells with a much longer exposure to Tween: 90 minutes at room temperature followed by centrifugation and incubation in fresh Tween–sucrose for another 30 minutes. Our procedure has become progressively shorter, with no reproducible alteration in the subsequent assay results with respect to the level of dTMP incorporation, nucleotide dependence, or shape of the time course. In fact, CHO cells need only be suspended in 0.25 M sucrose and assayed in a reaction mixture containing 0.2–0.5% Tween. The sucrose, however, appears to be necessary. Final resuspension of the cells in Hanks' or F-12 (without serum) medium resulted in only 20–45% of the control activity. This is shown in Table I, where the dependence of subsequent incorporation on the concentration of Tween is also given.

TABLE I

INCORPORATION OF [³H]dTMP UNDER VARIOUS CONDITIONS

Stage of modification	Condition	[³H]dTMP incorporated (% of control)[a]
Permeabilization	1% Tween (control)	100
	0.3% Tween	50
	0.1% Tween	28
	0% Tween	11
	Hanks' solution in place of sucrose cells in F-12 minus serum	21[b]
	0.4% Tween in reaction	45
Reaction	0.1 M Tris instead of KPB	10[b]
	Plus 2 mM DTT	~50
	Minus ATP	10
	Minus dATP	10[b]
	Minus dATP, dCTP, dGTP	<4[b]
	Plus 0.67 mM UTP, CTP, GTP	85
	Plus 0.25 M sucrose	71
	Plus 10% glycerol	40
	Plus 0.2 mM NAD	106
	Plus 0.8 mM NAD	112
	Plus 8 × 10⁻⁶ M araCTP	67
	Plus 8 × 10⁻⁵ M araCTP	23
	Plus 8 × 10⁻⁴ M araCTP	9
	Plus 2 × 10⁻³ M araCTP	7
	Plus 1.1% Tween	80
	Plus 2.3% Tween	66

[a]Values for ³H incorporated were taken as the average of the three or four points in the plateau region of the time course graph.

[b]DTT (2 mM) was included in both control and test reactions.

Cells in Tween–sucrose can be left at room temperature for 1–2 hours with little loss in their capacity to synthesize DNA. However, storage overnight at 4°C results in a loss of about half of this activity. Loss of permeability on return to growth medium is fairly rapid. Cells exposed to Tween for 1 hour, returned to growth medium for 15 minutes, and then assayed showed little more incorporation of dTMP than cells that had not been exposed to Tween (data not shown).

C.   Properties of the Reaction and the Product

1.   PROPERTIES OF THE REACTION

Representative time courses of [³H]dTMP incorporation by CHO cells are shown in Fig. 2. There is often a lag of up to 30 minutes, followed by rapid incorporation for 30–45 minutes. A plateau value is usually reached at 45–60 minutes, which may decline slightly as the incubation is continued. Figure 2B includes data on the ¹⁴C bulk label and shows that there is only slight, if any, degradation of the prelabeled DNA. The data in Fig. 2A verify that cells not exposed to Tween incorporate very little dTMP, while Fig. 2B shows that there is also very little incorporation by Tween-treated cells in the absence of ATP. This information is included in Table I which shows

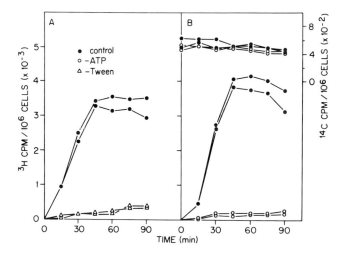

FIG. 2.   Time courses of [³H]dTMP incorporation. Cells were grown in suspension, harvested, permeabilized, and assayed as described in Section II unless otherwise noted. (A) Permeabilization was as follows: the cells, in growth medium, were centrifuged, resuspended in Tween–sucrose (control) or sucrose only, held at room temperature for 30 minutes, centrifuged, resuspended in fresh Tween–sucrose or sucrose alone, and assayed. (B) The cells were prelabeled for 1 day with [¹⁴C] thymidine and assayed with or without ATP.

the relative amount of dTMP incorporated under various conditions. In addition to being dependent on ATP, the reaction is completely dependent on the four deoxyribonucleoside triphosphates. When one was omitted (dATP, Table I), the incorporation of dTMP dropped to 10% of the control values, while when only dTTP was present (with ATP), its incorporation into acid-insoluble material was not detectable. Any of the other three deoxyribonucleoside triphosphates can be used as the labeled precursor, resulting in levels of incorporation comparable to that of dTMP (data not shown).

We have not systematically determined the reaction conditions for maximal dTMP incorporation, although we have looked at the effects of some additions, deletions, and substitutions (see Table I). Dithiothreitol (DTT), originally included in the reaction mixture, was found to be fairly inhibitory, as were sucrose (0.25 M) and glycerol (10%). The addition of Tween to the reaction (beyond the 0.2–0.4% introduced with the cells) resulted in some inhibition of dTMP incorporation. No stimulation of dTMP incorporation resulted from including the other three ribonucleoside triphosphates in the reaction mixture.

NAD was tested because it is the precursor of poly (ADP-ribose), and synthesis of poly (ADP-ribose) has been correlated with stimulation or inhibition of DNA synthesis in various cell types (for reviews, see Honjo and Hayaishi, 1973; Sugimura, 1973). In our system we observed a slight stimulation. Cytosine arabinoside has been shown to inhibit DNA synthesis in a large variety of prokaryotes and eukaryotes. The nucleotide, (cytosine arabinoside triphosphate, inhibits eukaryotic DNA polymerases *in vitro* (Furth and Cohen, 1968) and DNA synthesis in isolated HeLa cell nuclei (Wist *et al.*, 1976) and in mouse L cells permeabilized by cold shock (Berger and Johnson, 1976). In our system also, DNA synthesis was strongly inhibited by cytosine arabinoside triphosphate.

Tween-permeabilized CHO cells also incorporate [³H]UTP into acid-insoluble material in a reaction dependent on the four ribonucleoside triphosphates. In a reaction mixture containing UTP, CTP, and GTP at 150 μM (with 1.3 mM ATP and no dNTPs), twice as many picomoles of UMP were incorporated as picomoles of dTMP in the standard DNA synthesis reaction.

Addition of more of the dNTPs or of ATP after the reaction began to level off did not increase the final level of dTMP incorporation.

2. CHARACTERIZATION OF THE PRODUCT

That the product is indeed DNA is shown by equilibrium sedimentation in CsCl density gradients (Fig. 3). Figure 3A shows the profile of DNA (pre-labeled with [¹⁴C]thymidine) labeled with [³H]dTTP in a standard reaction

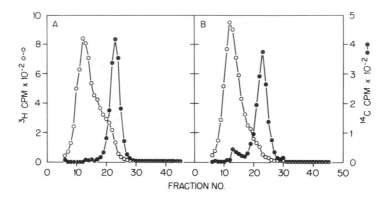

FIG. 3. CsCl density gradient profiles of CHO cell DNA synthesized in the reaction mixture with BrdUTP substituted for dTTP. A suspension culture, prelabeled for 1 day with [14C]thymidine, was split, and 5 μg/ml of BrdU was added to half. Two hours later, cells were harvested from both cultures, permeabilized, and incubated in the standard reaction mixture with BrdUTP in place of TTP. After 60 minutes at 37° C the cells were centrifuged out of the reaction mixture, taken up in 0.2 ml of 0.1% SDS, and frozen. A few days later they were thawed and added to 4.5 ml of CsCl (refractive index 1.402) and centrifuged for 72 hours at 29,000 rpm in an SW 50.1 rotor. Gradients were collected from the bottom onto filter paper strips, washed in 5% TCA and in 95% ethyl alcohol, dried, and counted as described in the text. (A) No preincubation with BrdU. (B) Preincubation with BrdU.

in which BrdUTP was substituted for dTTP. The method of preparation of the lysates for these experiments (described in the legend for Fig. 3) resulted in DNA fragments of double-strand molecular weight ~2 × 10⁶ as judged by sedimentation in neutral sucrose gradients [5–20%, w/v, in 0.5% sodium dodecyl sulfate (SDS), 5 mM Tris, 10 mM NaCl, 1 mM sodium citrate, pH 7.6, modified from Bonura et al., 1975]. (We have recently modified our procedure and obtained much larger DNA. Cells centrifuged out of the reaction mixture are thoroughly dispersed in 1 volume of 0.25 M sucrose; then 4 volumes of 0.5% SDS–1 mM EDTA is added. These lysates are passed through a syringe to shear the DNA and are mixed with CsCl.) The ³H profile has a peak at the position of fully substituted hybrid DNA, with a shoulder on the side toward the unsubstituted ¹⁴C parental DNA. This shoulder of intermediate-density material is almost completely eliminated by incubation of the cells with BrdU before permeabilization and incubation with BrdUTP and [³H]TTP (Fig. 3B). This indicates that the intermediate-density material consists of fragments containing the transition points between dTMP- and BrdUMP-containing regions and that most of the DNA synthesized in the postpermeabilization reaction is an extension of the regions being synthesized before permeabilization.

Neutral and alkaline sucrose gradient centrifugation (data not shown) demonstrated that after permeabilization the DNA synthesized in 60–90 minutes of reaction was large. Using DNA sheared to a double-strand molecular weight of 2×10^6 or 25×10^6 (lysates prepared as in the modified procedure described above) there was little difference between the molecular weights of newly synthesized and parental DNA calculated from profiles in either neutral or alkaline sucrose gradients.

D.   Permeabilization of Human Cell Lines

We have succeeded in permeabilizing two human cell lines, namely, HSBP (a line of normal diploid human fibroblast derived from fetal skin; Regan and Setlow, 1974) and Raji (a human lymphoblast line obtained from a Burkett's lymphoma patient; Pulvertaft, 1965). Exponentially growing HSBP cells required a longer exposure (90 minutes) to Tween-80 than CHO cells, and the DNA-synthesizing activity of the former was quite variable among preparations. Moreover, HSBP cells were more difficult to suspend by trypsinization, and the prolonged exposure to trypsin could temporarily affect their ability to synthesize DNA. The Raji cells were grown in T flasks where they remained unattached to the surface and therefore did not require trypsinization. Exponentially growing Raji cells were made permeable to nucleoside triphosphates by a 60-minute exposure to Tween-80 in 0.25 M sucrose (Fig. 4). One percent Tween-80 was more effective than 0.2% Tween-80 (compare Fig. 4A and B in terms of incorporation of dTMP). The Raji cell line, like the CHO cells, requires ATP for DNA synthesis, since the omission of ATP from the assay mixture resulted in a marked decrease in [³H]dTMP incorporation.

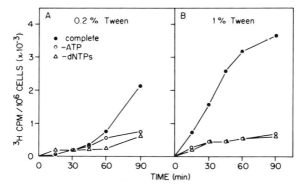

FIG. 4.   Incorporation of [³H]dTMP by Tween-permeabilized Raji cells. A 4-day culture, containing 1×10^7 cells in 15 ml of growth medium, was harvested, washed with Hanks' solution, and resuspended at a concentration of 5×10^6 cells/0.5 ml of 0.2% (A) or 1% (B) Tween in 0.25 M sucrose. They were left at room temperature for 60 minutes and then assayed in the standard DNA synthesis reaction mixture, except for the omission of ATP, or of dCTP, dGTP, and dATP where indicated.

In the study with the Raji cell line it was noted that the morphology of the cells changed during Tween-80 treatment. Phase-microscope observations indicated loss of cell refractility by about 50% of the cells after 90 minutes of incubation in Tween–sucrose. We have not determined cell survival for the Raji line.

IV.   Concluding Remarks

The use of Tween-80 to permeabilize mammalian cells seems to have general applicability. CHO cells are reversibly permeabilized by Tween; this has obvious advantages for studies in which subsequent cell growth is important. Whether other cell types can be reversibly permeabilized probably depends on whether or not appropriate conditions for achieving it are attainable. The possibility of using this technique to introduce enzymes and other proteins into cells has not been explored, but clearly, if such could be done, the usefulness of this permeabilization process would be increased for *in situ* studies of cellular activities.

ACKNOWLEDGMENT

The authors thank Mr. William Gude for carrying out the autoradiography. The research described here was supported in part by the Energy Research and Development Administration through Contract AT-(40-1)-4568.

REFERENCES

Berger, N. A., and Johnson, E. S. (1976). *Biochim. Biophys. Acta* **425**, 1–17.
Billen, D., and Hellerman, G. R. (1974). *Biochim. Biophys. Acta* **361**, 166–175.
Billen, D., and Olson, A. C. (1976). *J. Cell Biol.* **69**, 732–736.
Bonura, T., Town, C. D., Smith, K. C., and Kaplan, H. S. (1975). *Radiat. Res.* **63**, 567–577.
Furth, J. J., and Cohen, S. S. (1968). *Cancer Res.* **28**, 2061–2067.
Honjo, T., and Hayaishi, O. (1973). *Curr. Top. Cell. Regul.* **7**, 87–127.
Kay, E. R. M. (1965). *Cancer Res.* **25**, 764–769.
Masker, W. E., and Hanawalt, P. C. (1973). *Proc. Natl. Acad. Sci. U.S.A.* **70**, 129–133.
Moses, R. E., and Richardson, C. C. (1970). *Proc. Natl. Acad. Sci. U.S.A.* **67**, 674–681.
Pulvertaft, R. J. V. (1965). *J. Clin. Pathol.* **18**, 261–273.
Regan, J. D., and Setlow, R. B. (1974). *Cancer Res.* **34**, 3318–3325.
Seki, S., LeMahieu, M., and Mueller, G. C. (1975). *Biochim. Biophys. Acta* **378**, 333–343.
Sugimura, T. (1973). *Prog. Nucleic Acid Res. Mol. Biol.* **13**, 127–151.
Vosberg, H.-P., and Hoffmann-Berling, H. (1971). *J. Mol. Biol.* **58**, 739–753.
Wist, E., Krokan, H., and Prydz, H. (1976). *Biochemistry* **15**, 3647–3652.

Chapter 2 1

Nucleic Acid Synthesis in Permeabilized Eukaryotic Cells

NATHAN A. BERGER

The Hematology-Oncology Division of The Department of Medicine,
The Jewish Hospital of St. Louis,
Washington University School of Medicine, St. Louis, Missouri

I. Introduction

DNA synthesis is a complex process that requires multiple approaches to the elucidation of its mechanism and control. *In vitro* studies with isolated enzymes have been important in defining the characteristics and kinetics of individual components. The genetic approach with complementation of mutants has been useful in identifying the multitude of enzymes and proteins involved in DNA synthesis, and studies with intact cells have been useful in identifying the physiological changes that affect DNA synthesis. Studies of the replication process in intact cells are complicated by processes which interfere with the transport and phosphorylation of nucleosides, as well as by other manipulations which alter the intracellular pool sizes of deoxynucleosides or deoxynucleotides.

Permeable cell systems were developed in bacteria (Mordoh *et al.*, 1970; Moses and Richardson, 1970; Pisetsky *et al.*, 1972; Scholler *et al.*, 1972) and disrupted or permeable cell systems were developed in eukaryotes (Burgoyne, 1972; Seki *et al.*, 1975; Berger and Johnson, 1976; Berger *et al.*, 1976; Tseng and Goulian, 1975) to allow exogenously supplied deoxynucleoside triphosphates and other reaction components to be supplied directly to the replication complex functioning on its intrinsic DNA template. These techniques allow the replication complex to be assayed under conditions which maintain many of the relations that exist in intact cells. For optimal utility, a permeable cell system should permit exogenously added phosphorylated compounds and other agents of interest to be supplied in precisely controlled concentrations to the enzymes functioning inside the nucleus. The concentrations of these compounds should be independent of transport and pool sizes, and the phosphorylated compounds should be used directly without any intervening breakdown and rephosphorylation. The activity of the DNA replication complex in the permeable cells should vary with the replicative activity of the cells. The DNA synthesized in the permeable cells should be the product of semiconservative, replicative DNA synthesis, and it should occur as extensions of replication forks that were active *in vivo*. The permeable eukaryotic system, described below, meets these criteria and can be used for studies of DNA, RNA, and poly (adenosine diphosphoribose) [poly (ADPR)] synthesis.

II.   Materials

A.   Permeabilizing Buffer

This buffer consists of 0.01 M Tris–HCl, pH 7.8, 0.25 M sucrose, 1 mM EDTA, 30 mM 2-mercaptoethanol, and 4 mM MgCl$_2$. It works well with mouse L cells and Chinese hamster ovary (CHO) cells. The same buffer with the sucrose omitted can be used for HeLa cells, human and mouse lymphocytes, and L1210 cells.

B.   DNA Polymerase Reaction Mix

This mixture consists of 0.1 M HEPES, pH 7.8, 0.02 M MgCl$_2$, 0.21 M NaCl, 15 mM ATP, 0.3 mM dATP, 0.3 mM dCTP, 0.3 mM dGTP, and

0.3 mM dTTP. The radioactive tracer is usually included as [^{3}H]dTTP or [α-^{32}P]dTTP (specific activity 10–50 × 10^3 cpm/nmole). Successful reactions have been performed with dTTP at concentrations as low as 2–5 μM and specific activities of 3 × 10^4 cpm/pmole. The radioactive nucleotides are usually purchased in a solution containing ethanol and are taken to complete dryness under nitrogen before they are combined in a reaction mix. This is imperative, since small amounts of ethanol inhibit the DNA polymerase reaction (Berger and Johnson, 1976).

C.   RNA Polymerase Reaction Mix

The components are identical to those listed for the DNA synthesis reaction mix, except that the ribonucleoside triphosphates, ATP, UTP, GTP, and CTP, are substituted for the dNTPs. [^{14}C]UTP (specific activity 10 × 10^3 cpm/nmole) can be used for the radioactive tracer.

D.   Poly (ADPR) Polymerase Reaction Mix

This mixture consists of 100 mM Tris–HCl, pH 7.8, 120 mM MgCl$_2$, and 1 mM NAD. [^{3}H]-or [^{14}C]NAD (specific activity 10 × 10^3 cpm/nmole) with the radioactive label in the adenine portion of the molecule should be used for the radioactive tracer.

E.   Agents Affecting Nucleic Acid Synthesis

To test the effect of any compound on nucleic acid synthesis, it should be prepared in a buffer with the final pH adjusted to 7.8. Control reaction tubes should contain equivalent amounts of buffer and any other solvents used.

III.   Procedure

The starting material should be a monodisperse suspension of cells. Cells growing in suspension culture are the best source. Tumor cells harvested from an ascites tumor or blood cells prepared on Ficoll-Hypaque gradients have also been used. Mitogen-stimulated lymphocytes have been used after vigorous vortexing to break up cell clumps. A cell count is performed on the starting suspension. From this point until the start of the reactions, the cells are manipulated at 0–4°C to minimize loss of enzyme activity. The cells are collected by centrifugation at 1000 g for 10 minutes at 4°C. The supernatants are decanted, and the sides of the tubes are wiped free of excess medium. The cell pellets are rapidly suspended in ice-cold

permeabilizing buffer by aspiration with plastic pipets. They are vortexed at
an intermediate setting for 20–30 seconds to ensure a monodisperse suspension and then diluted to a concentration of 2×10^6 cells/ml and incubated in
an ice-water bath for 15 minutes. The cells are collected by centrifugation
at 1000 g for 10 minutes at 4°C. The supernatants are decanted, the tube
walls are wiped dry, and the cells are resuspended at 2.5×10^7 cells/ml in ice-
cold permeabilizing buffer by vortexing for 30 seconds. At this point, microscopic examination of the cells reveals that they are in a monodisperse
suspension. They appear swollen but morphologically intact. The cells
should all be permeable as demonstrated by the uptake of trypan blue
(Boyse et al., 1964). Figure 1 is an electron micrograph of L cells before (A)
and after (B) permeabilization, and after a 30-minute incubation at 37°C (C).
The permeabilized cells appear swollen and washed out relative to the untreated control cells. They retain the cytoplasmic and nuclear membrane
throughout the processing and incubation (Berger and Johnson, 1976). As
noted above, L cells and CHO cells can be readily permeabilized by this
technique. HeLa cells, L1210 cells, and human lymphocytes can be permeabilized by the same technique, except that sucrose should be omitted
from the permeabilizing buffer.

The reaction components are combined in an ice-water bath in the following order: 0.1 ml of reaction mix followed by a 0.2-ml aliquot containing
5×10^6 permeabilized cells in permeabilizing buffer. When agents are to be
tested for their effects on DNA synthesis, they are usually added to the
reaction tubes just before the cells. Triton X-100 at a final concentration of
0.05% increases the permeability of the cells, so that macromolecules such
as proteins can gain rapid access to the nucleus. This is accomplished by
adding 10 μl of 1.5% Triton X-100 to the reaction tube just before the cells
are added; i.e., the cells are permeabilized in the usual buffer system, and the
Triton X-100 is added when all the reaction components are combined. As
demonstrated in Fig. 1D, the Triton-treated cells contain swollen nuclei
surrounded by cytoplasm; the nuclear membrane appears intact, while the
cell membrane is disrupted.

Reactions are started by transferring tubes to a 37°C water bath for
measurement of DNA or RNA synthesis, or to a 30°C water bath for
measurement of poly (ADPR) synthesis. After appropriate incubation
periods, the reactions are terminated by rapidly immersing the reaction
tubes in an ice-water bath and adding 5 ml of 10% trichloroacetic acid (TCA)
and 2% $Na_4P_2O_7$. The cell pellets are collected by centrifugation at 2200 g
for 7 minutes at 4°C and then resuspended in 5% TCA and 1% $Na_4P_2O_7$ in
the case of DNA or RNA synthesis, or in 10% TCA and 2% $Na_4P_2O_7$ for
poly (ADPR) synthesis. The TCA precipitates are sonicated for 10 seconds,
centrifuged at 10,000 g for 10 minutes 4°C, and then washed again with 95%

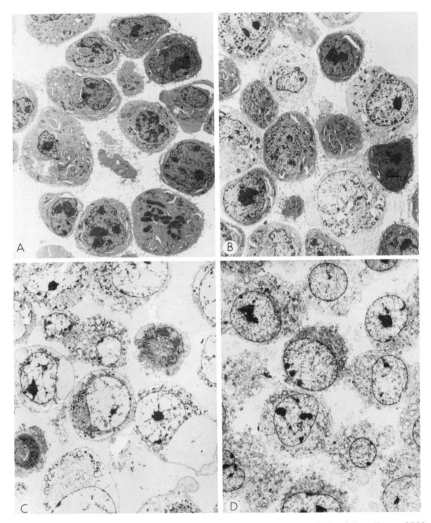

FIG. 1. (A–C) Electron micrographs of untreated and permeabilized L cells. × 2730. (A) Untreated cells direct from spinner culture. (B) After complete permeabilization procedure and a 1-minute incubation at 37°C. (C) Permeabilized cells after a 30-minute incubation at 37°C in DNA synthesis reaction mix. (D). Permeabilized cells incubated in the presence of 0.05% Triton X-100. × 2650.

ethanol and collected by centrifugation. The pellets are solubilized by overnight treatment with 0.5 ml Protosol and then counted in 10 ml of toluene scintillation fluid containing 6 gm PPO and 50 mg POPOP per liter.

A high-specific-activity DNA synthesis mix can be used to prepare newly synthesized DNA for velocity sedimentation analysis. It can also be used to scale down the number of cells required for the reaction. The size distribution of newly synthesized DNA was examined on 5–20% alkaline sucrose gradients in 0.9 M NaCl–10 mM EDTA–0.3 N NaOH (Berger *et al.*, 1977). Cells were permeabilized as above and suspended at 2.5 × 10^7 cells/ml. The reaction mix contained all the components listed above, except that the concentration of [³H]dTTP was 2.2 μM (specific activity 3 × 10^4 cpm/pmole). The reaction contained 25 μl of this high-specific-activity DNA synthesis mix and 50 μl containing 1.25 × 10^6 cells. The reactions were incubated at 37°C for 30 minutes and then stopped by immersing the tubes in an ice-water bath. Aliquots were taken directly from the reaction tubes and transferred to a layer of 0.5 ml of 20 mM EDTA on top of the preformed alkaline sucrose gradients. Then 0.5 ml of 0.6 N NaOH and 100 μl of 20% sodium lauroyl sarcosine were added to the top of each gradient. The cells were allowed to lyse on top of the gradients for 12–18 hours at 4°C. The tubes were centrifuged at 22,500 rpm at 4°C for 6 hours in the SW 25.1 head of a Beckman L2 centrifuge. Then 1.1-ml fractions were collected, and each was precipitated with 10% TCA and 2% Na$_4$P$_2$O$_7$ on glass-fiber filters (Whatman GF/C). They were subjected to two successive washes with 5% TCA and 1% Na$_4$P$_2$O$_7$ and one wash with 95% ethanol. The filters were dried, treated overnight with 0.5 ml of Protosol, and counted in toluene scintillation fluid as described above.

With the high-specific-activity reaction mix, a smaller number of cells can be used to obtain significant incorporation of radioactive nucleotide into DNA. With this small number of cells, the incorporation of radioactivity into macromolecules can also be detected by entrapment on filters. These reactions are stopped by the addition of an excess of 10% TCA and 2% Na$_4$P$_2$O$_7$; the precipitate is sonicated, poured over Whatman GF/C filters, washed twice with 10% TCA and 2% Na$_4$P$_2$O$_7$ and once with ethanol, and then dried and prepared for scintillation counting as described above.

IV.   Characteristics of Nucleic Acid Synthesis

A.   Kinetics of the System

Figure 2 shows the synthesis of DNA, RNA, and poly (ADPR) in the permeable cells. The rate of DNA synthesis begins to plateau after a 20- to 30-minute incubation. The rate cannot be increased by the addition of more dNTPs or ATP (Berger and Johnson, 1976). However, if fresh,

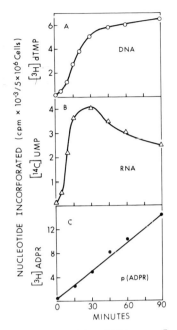

FIG. 2.   (A) Radioactivity incorporated into DNA from $[methyl\text{-}^3H]$dTTP (specific act 6 × 10⁴ cpm/nmole) in the presence of DNA synthesis mix and 5 × 10⁶ permeabilized cells. (B) Radioactivity incorporated into RNA from $[2\text{-}^{14}C]$UTP (specific activity 10 × 10³ cpm/nmole) in the presence of RNA synthesis mix and 5 × 10⁶ permeabilized cells. (C) Radioactivity incorporated into poly (ADPR) from $[adenine\text{-}2,8\text{-}^3H]$ NAD (specific activity 10 × 10³ cpm/nmole in the presence of poly (ADPR) synthesis mix and 5 × 10⁶ permeabilized cells (Berger and Johnson, 1976; Berger et al., 1978).

permeabilized cells are added to the reaction, the rate of DNA synthesis increases, indicating that the reaction components are not exhausted. RNA synthesis goes on for 15–20 minutes, after which the amount of labeled RNA begins to decrease, suggesting that RNA is being degraded or processed in the permeable cells. Poly (ADPR) is synthesized at a linear rate for 90 minutes. These studies demonstrate that the permeable cells are capable of continuing nucleic acid synthesis for at least 90 minutes. The cessation of DNA synthesis after a 20- to 30-minute incubation is probably due to the completion of active replicons and the absence of initiation of synthesis of new replicons in the permeable cells (Berger et al., 1977).

A double-label experiment using $[methyl\text{-}^3H]$dTTP and $[\alpha - {}^{32}P]$dTTP was performed to demonstrate that permeable cells incorporate deoxynucleotides directly into their DNA. If the nucleotides were hydrolyzed before they entered the cells and then rephosphorylated inside the cells, the ³H label on the pyrimidine base and the ³²P label would have been incorpo-

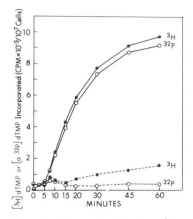

Fig. 3. Incorporation of radioactive nucleotide from [*methyl*-³H]dTTP (solid circles) and [α-³²P]dTTP (open circles) into DNA of permeabilized (solid line) and normal non-permeabilized cells. The complete DNA synthesis reaction mix contained [*methyl*-³H] dTTP and [α-³²P]dTTP, with the specific activity of each isotope equal to 23 × 10³ cpm/nmole (Berger and Johnson, 1976).

rated into DNA at different rates. If the nucleotides were incorporated directly into the DNA, the ³H and ³²P would have been incorporated at the same rate. Figure 3 demonstrates that intact cells incorporated the ³H label at a faster rate than the ³²P. The permeable cells incorporated both labels from dTTP at a much greater rate than the intact cells. More importantly, the permeabilized cells incorporated the ³H on the pyrimidine base and the ³²P at the same rate, confirming that the exogenously supplied nucleotides were incorporated directly into the DNA. Similar results were obtained with thymidine kinase–deficient cells in which it would have been impossible to rephosphorylate any hydrolyzed dTTP (Berger and Johnson, 1976).

B.   Requirements for DNA Synthesis

The amount of DNA synthesized by the permeable cells shows a linear dependence on the number of cells in the reaction between 1×10^6 and 1.5×10^7 cells (Berger and Johnson, 1976). Maximal DNA synthesis occurs in the presence of 70 mM NaCl or other monovalent cations such as K$^+$ or NH$_4$$^+$. Optimal DNA synthesis occurs at a MgCl$_2$ concentration of 9 mM and, as indicated in Table I, DNA synthesis is reduced by 77% in the absence of MgCl$_2$. In the absence of ATP, DNA synthesis is decreased by more than 90%. A similar requirement for ATP in replicative DNA synthesis has been noted in permeable prokaryotic and eukaryotic systems (Mordoh *et al.*, 1970; Moses and Richardson, 1970; Pisetsky *et al.*, 1972; Seki *et al.*,

TABLE I

REQUIREMENTS AND INHIBITORS OF DNA POLYMERASE AND POLY (ADPR) POLYMERASE
IN PERMEABLE CELLS[a]

Condition	Activity (% of control)	
	DNA	Poly (ADPR)
Complete	100	100
Minus mercaptoethanol plus N-ethymaleimide	4.6	6.3
Minus MgCl$_2$	23	20
Minus NaCl	57	—
Minus ATP	5.6	—
Minus dCTP	6.6	—
10 mM nicotinamide	96.4	4.4
10 mM 5-methyl nicotinamide	99.6	5.8
10 mM thymidine	81.2	4.1
0.1 mM arabinoside CTP	23.2	100.5
10 mM cytembena	12.0	104
10 mM phosphonoacetate	11.6	94.1
0.1 mM daunomycin	9.5	61

[a]The complete system for DNA synthesis measures incorporation from [^{3}H]dTTP into DNA in the presence of all the reaction components described for the DNA polymerase mix. The complete system for poly (ADPR) polymerase measures incorporation from [^{3}H]NAD into poly (ADPR) in the presence of all the reaction components described for the poly (ADPR) polymerase mix. All incubations were for 30 minutes, the DNA synthesis reaction was at 37°C, and the poly (ADPR) synthesis reaction took place at 30°C. All reactions are expressed as a percentage of the control complete reaction system (Berger and Johnson, 1976; Berger et al., 1978).

1975; Berger and Johnson, 1976) and in isolated nuclei (Friedman and Mueller, 1968). The ATP does not function by preventing hydrolysis of the other dNTPs but appears to be specifically required for DNA synthesis. 2-Mercaptoethanol is required for reproducible stability of DNA polymerase activity. In the absence of 2-mercaptoethanol, sulfhydryl binding reagents such as N-ethylmaleimide are effective inhibitors of DNA polymerase.

C.   Characteristics of DNA Synthesis

When permeable cells are allowed to synthesize DNA and then subjected to subcellular fractionation, 98% of the newly synthesized DNA is found in the nuclear fraction and 1% in the mitochondrial fraction (Berger and Johnson, 1976). To determine the nature of the newly synthesized DNA, L-cell DNA was prelabeled by growing cells in [^{14}C]dT. The cells were permeabilized and incubated with the complete DNA synthesis mix including [³H]dCTP and BrdUTP substituted for dTTP. The DNA was extracted and centrifuged to equilibrium in CsCl. As illustrated in Fig. 4, separation of the ¹⁴C-labeled normal-density DNA synthesized in the intact cells, and the dense ³H-labeled DNA synthesized by the permeable cells, demonstrates that DNA synthesis in the permeable cells is semiconservative (Berger and Johnson, 1976). In another experiment, the replication forks in intact cells were labeled with a brief exposure to [¹⁴C]BrdU. The cells were permeabilized and incubated in [³H]dTTP for 2 minutes and 10 minutes. DNA was extracted and sedimented to equilibrium in CsCl. The DNA synthesized in the first few minutes in the permeable cells sedimented along with the density-labeled material. DNA synthesized at subsequent time intervals was found at progressively lighter densities. Thus the initial DNA synthesized in the permeable cells occurred as extensions of replicons that were active *in vivo* (Berger *et al.*, 1977).

The size of the DNA synthesized in the permeable cells was analyzed on alkaline sucrose gradients. Cells were removed directly from the reaction and lysed on top of the gradients. Figure 5 demonstrates that the size of the bulk cell DNA was in the range of 150S. The DNA synthesized by the permeable cells had a heterogeneous distribution with peaks at 26 and 71S.

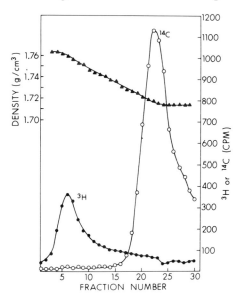

FIG. 4. Neutral CsCl equilibrium density gradient of newly synthesized, density-labeled DNA. Bulk cell DNA was prelabeled with [*methyl*-¹⁴C]dT. Following permeabilization, cells were allowed to synthesize DNA in complete reaction mix containing BrdUTP substituted for dTTP and [5-³H]dCTP (specific activity 6.5 × 10⁴ cpm/nmole). Incubation was for 30 minutes at 37 °C (Berger and Johnson, 1976).

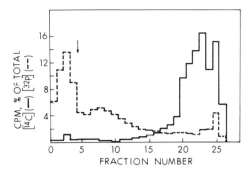

FIG 5. A 5–20% alkaline sucrose gradient pattern of DNA prelabeled with [¹⁴C]dT in intact cells (solid line), followed by labeling for 30 minutes at 37°C with [α − ³²P]dTTP after permeabilization (dashed line). The arrow indicates the position of a 40S lambda phage marker. The top of the gradient is at the left, and the direction of sedimentation is to the right.

The 71S material, corresponding to a molecular weight of 6.6×10^7, falls within the size range of eukaryotic replicons. In contrast to the size of the DNA synthesized in the permeable cells, isolated nuclei synthesized DNA which had a peak sedimentation coefficient of 26S. Very little of the newly synthesized DNA reached the size of eukaryotic replicons. In addition, as shown in Fig. 6, the rate of DNA synthesis in isolated nuclei is only 10% of that in the permeable cells. Isolated nuclei stop synthe-

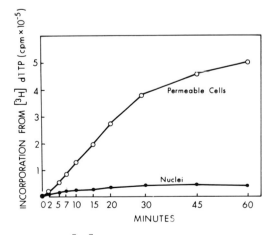

FIG. 6. Incorporation from [³H]dTTP (specific activity 43.7×10^6 cpm/nmole) into DNA by permeabilized L cells (open circles) or by isolated L-cell nuclei (solid circles). The complete reaction system contained high-specific-activity DNA synthesis mix and 1.25×10^6 permeabilized cells or 1.25×10^6 nuclei in a final volume of 75 μl.

sizing DNA after a 7- to 10-minute incubation at 37°C, whereas permeable cells continue DNA synthesis for at least 20–30 minutes. Thus, in comparison to isolated nuclei, DNA synthesis in permeable cells occurs at a faster rate, lasts for a longer period of time, and produces larger, repliconsized DNA intermediates. DNA synthesis in the permeable cells appears to be by elongation of chains that were initiated *in vivo* (Berger *et al.*, 1977). This system offers an advantage for studies of elongation in the absence of initiation. It may also be useful to reconstitute the initiation process.

D. Agents Affecting DNA Synthesis

The permeable cell technique can be used to evaluate drugs and other new agents that affect macromolecular synthesis to determine whether or not they have a direct effect on DNA polymerase acting on its native DNA template. Hydroxyurea, which inhibits DNA synthesis in intact cells by interfering with the production of deoxynucleotides (Reichard, 1972), has no effect on DNA synthesis in permeable cells where all dNTPs are supplied (Berger and Johnson, 1976; Berger and Weber, 1977). The nucleoside analog cytosine arabinoside has no effect on DNA synthesis in permeable cells, whereas the nucleotide, cytosine arabinoside triphosphate, has been demonstrated to cause competitive inhibition of DNA polymerase (Graham and Whitmore, 1970). Daunorubicin, which inhibits DNA synthesis by complexing the template (Kim *et al.*, 1968), causes prompt and total inhibition of DNA synthesis in permeable cells. Using this technique, we recently identified cytembena as a direct inhibitor of DNA synthesis (Berger and Weber, 1977). Phosphonoacetate has been found to inhibit

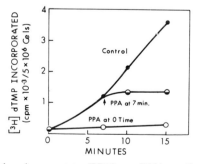

Fig. 7. Effect of phosphonoacetate (PPA) on DNA synthesis in permeabilized cells. The reaction system was composed of 0.1 ml of DNA synthesis reaction mix containing [³H]dTTP (specific activity 8 × 10⁴ cpm/nmole) and 0.2 ml containing 5 × 10⁶ permeabilized cells. The incorporation of radioactivity was measured at 37°C. ●, Control cells; ○, phosphonoacetate added just before reactions were started; ○, phosphonoacetate added after a 7-minute incubation at 37°C.

several types of eukaryotic polymerases (Leinbach *et al.*, 1976). Figure 7 demonstrates that the addition of phosphonoacetate to the permeable cells at the start of the reaction causes a marked inhibition of DNA synthesis. When phosphonoacetate is added after the reaction has been in progress for 7 minutes, there is a prompt reduction in the rate of DNA synthesis, demonstrating that this compound has rapid access to DNA polymerase functioning on its intrinsic template.

In the presence of 0.05% Triton X-100, DNA synthesis continues in a semiconservative fashion. The cells retain their cytoplasm, however, the cytoplasmic membrane is no longer intact and the permeability is increased so that exogeneously added macromolecules can gain rapid, direct access to the nucleus. This was demonstrated by first prelabeling cell DNA with [^{14}C]thymidine. The cells were then permeabilized, the reaction made 0.05% in Triton X-100, and DNase and venom phosphodiesterase added to the final incubation. The enzymes caused rapid degradation of the cell DNA, demonstrating that they reached the DNA within the first 2 minutes of the incubation period (Berger *et al.*, 1976). Added histones inhibited DNA synthesis in detergent-permeabilized cells, lysine-rich histones being the most effective and arginine-rich histones being the least effective (Berger *et al.*, 1976).

E. Characteristics of Poly (ADPR) Synthesis

Poly (ADPR) synthesis has been the subject of several reviews (Sugimura, 1973; Hayaishi and Ueda, 1977). When permeable cells are incubated with [^{3}H]NAD and Mg^{2+}, they synthesize the homopolymer poly (ADPR). As indicated in Table I, in the absence of Mg^{2+}, the reaction only achieves 20% of the maximal rate. The presence of NaCl, NaF, or ATP does not increase the reaction rate. When 2-mercaptoethanol is eliminated from the reaction, N-ethylmaleimide inhibits the poly (ADPR) polymerase activity. As demonstrated in Table I, the poly (ADPR) polymerase reaction can be clearly distinguished from the DNA polymerase reaction by the effect of nicotinamide, 5 methylnicotinamide, and thymidine, which totally inhibit poly (ADPR) polymerase activity while having a negligible effect on DNA polymerase activity. In contrast, cytosine arabinoside triphosphate, cytembena, and phosphonoacetate cause drastic inhibition of DNA polymerase but have a negligible effect on poly (ADPR) polymerase activity. The product of the poly (ADPR) polymerase reaction has been confirmed by examining its nuclease susceptibility. It is resistant to degradation by DNase and RNase but is totally degraded by venom phosphodiesterase (Berger *et al.*, 1978). Because of the problems in supplying precursors for poly (ADPR) polymerase in intact cells, it is difficult to define the physiological activity of this enzyme. The permeable cell technique should provide a useful tool for studies of the physiological activity and regulation of poly (ADPR) polymerase.

F. Physiological Changes in Nucleotide-Polymerizing Enzymes

As shown in Fig. 8, the permeable cell technique can be used to measure the changes that occur in the activity of nucleotide-polymerizing enzymes along with physiological changes in the cells. Plateau-phase cells at a density of $10 \times 10^5/\text{ml}$ were diluted to $1 \times 10^5/\text{ml}$ and allowed to reinitiate DNA synthesis and cell growth. Cells were then removed from culture at multiple time points during the course of the experiment: they were permeabilized and analyzed for DNA polymerase and poly (ADPR) polymerase activity. When the cells were at high density, there was a low level of DNA polymerase activity and a high level of poly (ADPR) polymerase activity. When the cells were shifted to fresh medium at a density of $1 \times 10^5/\text{ml}$, there were marked changes in the activities of both enzymes. The level of DNA polymerase reached four times its level in plateau-phase cells. The maximum level of activity occurred in cells in mid-log-phase growth and then gradually decreased as the cells left log-phase growth and entered the plateau phase. Chang *et al.* (1973) demonstrated that the L-cell DNA polymerase α fluctuated in a similar fashion with the replicative activity of these cells.

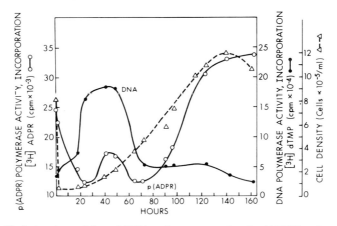

FIG. 8. Variation in activity of DNA polymerase and poly (ADPR) polymerase with cell growth. Plateau-phase cells were diluted to $10^5/\text{ml}$ in α modified Eagle's medium and 10% serum at time 0. Cell counts were performed at subsequent intervals (triangles). At each time point, cells were removed from culture, permeabilized, and assayed for activity of DNA polymerase (solid circles) and poly (ADPR) polymerase (open circles) (Berger *et al.*, 1978).

The activity of poly (ADPR) polymerase was highest in plateau-phase cells. When the cells were diluted to 1×10^5/ml, the activity fell to 20% of its plateau-phase level. The activity remained low during the period of active DNA synthesis and began to increase as the cells returned to the plateau phase. The permeable cell system has also proved useful in studying the changes in DNA polymerase and poly (ADPR) polymerase under other conditions which affect DNA synthesis, including a shift down to a glucose-deficient medium, infection with vaccinia virus, or following treatment with chemotherapeutic agents.

V. Summary

The technique described for rendering eukaryotic cells permeable to nucleotides has several advantages in the study of nucleic acid synthesis under near physiological conditions. The technique itself is simple; the cells remain in a monodisperse suspension and are easy to pipet quantitatively. The cells are freely permeable to phosphorylated compounds which gain rapid access to the nucleus. Nucleotides are incorporated into DNA without intermediate breakdown and rephosphorylation. DNA synthesis in permeable cells is semiconservative, the products are high-molecular-weight DNA intermediates, and DNA synthesis occurs as extensions of replication sites that were active *in vivo*. It is probable that all DNA synthesis in permeable cells is by elongation and that new replicon synthesis is not initiated. This offers the advantage of studying elongation in the absence of initiation. There is essentially no degradation of the cell DNA during the permeabilization process. The activity of the replicative DNA polymerase and poly (ADPR) polymerase have been found to undergo marked variations with the status of cell growth. This technique provides an assay system in which the polymerases function on their endogenous templates. It should be useful in determining whether agents that affect nucleic acid synthesis in intact cells exert their effects by direct action on the polymerase or on the template. The system should also be useful for studies of the intermediates in nucleic acid synthesis, and finally for studies of physiological changes in the activities of polymerases that occur with metabolic manipulation of the cells. Since the addition of Triton X-100 renders the cells permeable to exogenous proteins, this technique may also be useful in complementation studies of eukaryotic cells with genetic defects in DNA synthesis.

ACKNOWLEDGMENTS

I thank Elizabeth S. Johnson, Georgina Weber, Shirley J. Petzold, and Rebecca Minton for their contributions to the work, reported here. The research was supported by The Jewish Hospital of St. Louis.

REFERENCES

Berger, N. A., and Johnson, E. S. (1976). *Biochim. Biophys. Acta* **425**, 1–17.
Berger, N. A., and Weber, G. (1977). *J. Natl. Cancer Inst.* **58**, 1167–1169.
Berger, N. A., Erickson, W. P., and Weber, G. (1976). *Biochim. Biophys. Acta* **447**, 65–75.
Berger, N. A., Petzold, S. J., and Johnson, E. S. (1977). *Biochim. Biophys. Acta* **478**, 44–58.
Berger, N. A., Weber, G., and Kaichi, A. S. (1978). *Biochim Biophys Acta* (in press).
Boyse, E. A., Old, L. J., and Chouroulinkov, I. (1964). *Methods Med. Res.* **10**, 39–47.
Burgoyne, L. A. (1972). *Biochem. J.* **130**, 959.
Chang, L. M. S., Brown, M. K., and Bollum, F. J. (1973). *J. Mol. Biol.* **75**, 1–8.
Friedman, D. L., and Mueller, G. C. (1968). *Biochim. Biophys. Acta* **161**, 455–468.
Graham, F. L., and Whitmore, G. F. (1970). *Cancer Res.* **30**, 2636–2644.
Hayaishi, O., and Ueda, K. (1977). *Annu. Rev. Biochem.* **46**, 95–116.
Kim, J. H., Gelbard, A. S., Djerdjevic, B., Kim, S. H., and Perez, A. G. (1968). *Cancer Res.* **28**, 2437–2442.
Leinbach, S. S., Reno, J. M., Lee, L. F., Isbell, A. F., and Boezi, J. I., (1976). *Biochemistry* **15**, 426–430.
Mordoh, J., Hirota, Y., and Jacob, F. (1970). *Proc. Natl. Acad. Sci. U.S.A.* **67**, 773–778.
Moses, R. E., and Richardson, E. C. (1970). *Proc. Natl. Acad. Sci. U.S.A.* **67**, 674–681.
Pisetsky, D., Berkower, I., Wickner, R., and Hurwitz, J. (1972). *J. Mol. Biol.* **71**, 557–571.
Reichard, P. (1972). *Adv. Enzyme Regul.* **10**, 3–16.
Scholler, H., Otto, B., Nusslein, U., Huf, J., Hermann, R., and Bonhoeffer, F. (1972). *J. Mol. Biol.* **63**, 183–200.
Seki, S., LeMahieu, M., and Mueller, G. C. (1975). *Biochim. Biophys. Acta* **378**, 333–343.
Sugimura, T. (1973). *Prog. Nucleic Acid Res. Mol. Biol.* **13**, 127–151.
Tseng, B. Y., and Goulian, M. (1975). *J. Mol. Biol.* **99**, 317–337.

Chapter 2 2

Red Cell-Mediated Microinjection of Macromolecules into Mammalian Cells

ROBERT A. SCHLEGEL

*Department of Microbiology and Cell Biology, The Pennsylvania State University,
University Park, Pennsylvania*

AND

MARTIN C. RECHSTEINER

Department of Biology, University of Utah, Salt Lake City, Utah

I. Introduction

The direct microinjection of exogenous molecules into mammalian culture cells has proved to be a valuable technique (Graessmann *et al.*, 1974; Graessmann and Graessmann, 1976). However, the number of cells which can be injected is small because of the considerable manipulation required by this procedure. To overcome this limitation several laboratories have described a microinjection procedure based upon Sendai virus-mediated fusion of mammalian culture cells and red cells loaded with exogenous molecules (Furusawa *et al.*, 1974; Schlegel and Rechsteiner, 1975; Loyter *et al.*, 1975). By this procedure large numbers of culture cells can be rapidly microinjected with physiological quantities of macromolecules.

II.   General Methods

The following methods apply to the general microinjection procedure outlined by Schlegel and Rechsteiner (1975):

Red cells: Human blood (10 ml) is collected from the forearms of healthy volunteers and added to heparin (1000 units) to prevent coagulation. The serum and white cells are carefully removed by aspiration during four or five washes with saline.

Cell culture: Conventional methods of tissue culture are employed. Cells are grown in plastic tissue culture flasks (Falcon 3018) in F-12 medium (Ham, 1965) containing 5% fetal calf serum. Trypsin (0.2–1% in Ca^{2+}-, Mg^{2+}-free saline) is used to remove cells from the flasks, and cell numbers are determined by using a hemocytometer.

Sendai virus: Sendai virus is grown as described by Harris and Watkins (1965). The virus is titered by serial twofold dilutions in 0.5 ml of physiological saline. Two hours after adding 2 drops of a 4% suspension of human red cells, tubes are examined for agglutination. The reciprocal of the last dilution to show agglutination is taken to be the titer in hemagglutinating units (HAU) per milliliter. The virus is inactivated for 3 minutes by ultraviolet irradiation at a distance of 30 cm from a General Electric G8T5 bulb. With the use of this procedure identical levels of microinjection can be obtained with different preparations of Sendai virus using a single preparation of loaded red cells and culture cells.

Fusions: Typically, fusion mixtures consist of 10^7 culture cells, 10^9 loaded red cells, and 500 HAU of Sendai virus in 0.5 ml of Tris–saline (0.15 M NaCl, 20 mM Tris, pH 7.4) containing 0.2–2.0 mM $MnCl_2$. The components are combined in the order listed, mixed well, and kept on ice for 10 minutes prior to transfer to a reciprocating (120 strokes per minute) water bath at 37°C. After 30 minutes of incubation fusion is terminated by the addition of 5 ml of cold Tris–saline. At this point a sample of the fusion mixture is usually examined by phase microscopy. Large numbers of red cell–red cell fusion products and red cells that appear to be fusing with culture cells are taken as preliminary signs of successful microinjection (see Fig. 2 in Schlegel and Rechsteiner, 1975). The cells are then washed several times with culture media and plated for further study.

Separation of culture cells from red cells: If a microinjected molecule is to be examined by extraction from culture cells, residual loaded red cells must be removed from the microinjected culture cells. To accomplish this, fusion products are placed in plastic culture flasks or on glass microscope slides and incubated at 37°C long enough for the tissue culture cells to attach to the glass or plastic surface. Rinsing the attached cells several times frees them of red cells which do not attach to the surface.

Efficiency of microinjection: When culture cells are fused with ^{125}I-labeled bovine serum albumin ($[^{125}I]$BSA)-loaded red cells, the efficiency of microinjection is determined by autoradiography. Microscope slides to which the culture cells have attached are rinsed several times in saline to remove residual loaded red cells and then rinsed several times in methanol to fix the attached cells. The slides are coated with Kodak NTB-2 emulsion, exposed for several days, and developed as previously described (Schlegel and Rechsteiner, 1975). After staining with Giemsa, the fraction of labeled cells is determined by examining 500 cells in randomly chosen fields. A cell is considered microinjected if autoradiographic grains are found dispersed over the entire cytoplasm and the grain density is at least fivefold greater than the background density.

III. Red Cell Loading

A. Preswell Loading and Alternative Procedures

Hypotonic hemolysis occurs when red cells are suspended in sufficiently dilute salt solutions. Although the cells release soluble macromolecules, metabolites, and ions, they spontaneously regain impermeability to all these molecules. It has been demonstrated in several laboratories that macromolecules contained in the hypotonic lysing medium can enter red cells during hemolysis and remain trapped within them after the restoration of impermeability. Three general procedures have been described for trapping or loading macromolecules inside red cells. These procedures differ in the volume of hypotonic saline used to hemolyze the cells and the rate at which the cells swell before lysis. In descriptive terms these can be designated preswell loading, dialysis loading, and rapid-lysis loading.

In the preswell loading procedure (Rechsteiner, 1975) washed red cells prepared as described in Section II are centrifuged at setting 6 on an International clinical table-top centrifuge (IEC 428) for 15 minutes, which yields about 10^{10} packed red cells per milliliter. An equal volume of Hanks' solution is added to the packed red cells to make a 50% suspension. A 0.2-ml portion of this suspension is added to 11 ml of dilute Hanks' solution prepared by adding 5 ml of distilled water to 6 ml of Hanks' solution. The swollen red cell suspension is centrifuged in a 15-ml conical centrifuge tube at setting 6 for 10 minutes in the cold to produce a pellet of swollen red cells which is about 0.2 ml. The overlying dilute Hanks' solution is carefully removed with a pasteur pipet drawn to a fine tip. To the swollen red cell pellet is added with vortex mixing 0.1 ml of hypotonic buffer containing the macromolecules to be loaded. (Although 10 mM Tris, pH 7.4, is generally used, the pH of the buffer is not critical, since upon lysis the pH of the red cell contents dominates.) The lysing red cells are placed on ice for 2 minutes, and then 15 μl of 10-fold concentrated Hanks' solution is added during vortex mixing to shrink the red cells. Loading is then complete. However, the loaded red cells are incubated at 37°C in a water bath for 60 minutes before washing. This treatment is necessary to stabilize them against a loss of contents during washing and also during subsequent fusion.

Greater than 90% of input red cells are loaded when the preswell procedure just described is used and, as Fig. 1 illustrates, most red cells are loaded to the same extent. If more than the 10^9 red cells loaded by the above protocol are required, the loading procedure can be scaled up proportionately. All operations of the preswell procedure can easily be performed under sterile conditions.

The dialysis loading procedure is performed by dialyzing red cells and macromolecules in physiological saline against dilute saline (Seeman, 1967; Loyter et al., 1975). As in the preswell loading procedure, concentrated saline is added after loading is complete. Loyter and his colleagues include BSA or cytochrome during dialysis loading to produce loaded red cells which can be fused by Sendai virus.

Rapid-lysis loading is performed by diluting red cells into large volumes of hypotonic saline containing the molecules to be loaded (Hoffman, 1958; Seeman, 1967; Baker, 1967; Ihler et al., 1973). Hemolysis is evident within a few seconds, and the uptake of macromolecules is maximal by about 1 minute, at which time concentrated saline is generally added. Although, as in the other two procedures, macromolecules enter and are trapped in the red cells, the rapid-lysis loading procedure yields loaded cells which are relatively unstable and which fuse poorly (Peretz et al., 1974; M. Rechsteiner, unpublished results).

Both preswell loading and dialysis loading yield red cells which can be readily fused by Sendai virus. Both also allow the ratio of red cells to lysis buffer to be altered to conserve input macromolecules (Rechsteiner, 1975;

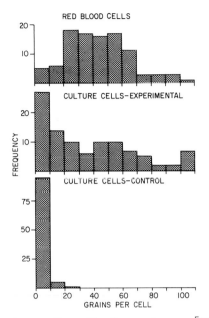

FIG. 1 Distribution of autoradiographic silver grains over [^{125}I] BSA-loaded red cells and recipient 3T3 cells following microinjection. The top panel shows the distribution of silver grains over red cells loaded with [^{125}I] BSA. The middle panel shows the distribution of silver grains over 3T3 cells after fusion to [^{125}I] BSA-loaded red cells. Cells in the bottom panel were fused to buffer-loaded red cells in the presence of [^{125}I] BSA free in the solution.

Wasserman *et al.*, 1976). Preswell loading, however, is perhaps more convenient, and in addition offers the advantage that culture cells need not be microinjected with cytochrome or BSA in red cells loaded by the dialysis procedure.

Both fresh blood used on the day of collection (Schlegel and Rechsteiner, 1975) and blood aged 3–6 weeks (Loyter *et al.*, 1975) have been used as a source of red cells. Although human red cells are generally used because of their high fusion capacity relative to other mammalian red cells (Furusawa *et al.*, 1974), rabbit red cells have also been successfully used (Kaltoft *et al.*, 1976).

B.   Properties of Loaded Red Cells

Loaded red cells after incubation at 37°C are washed several times in isotonic saline prior to fusion to remove untrapped macromolecules, released hemoglobin, and other cellular constituents. Examination of these cells by phase microscopy reveals dark-gray cells with high contrast compared to the highly refractile appearance of normal red cells and the virtual transparency of irreversibly lysed red cells. A variety of shapes including spheres, football shapes, cup shapes, and the biconcave disks typical of normal red cells are seen. It is not known whether or not the extent to which a cell is loaded is dependent upon its shape, but the spread in the distribution of autoradiographic grains over red cells loaded with [^{125}I]BSA (Fig. 1) is not inconsistent with this possibility.

Red cells loaded by the preswell procedure are relatively stable when stored in the cold in Tris–saline containing 2 mM Mn^{2+}. No decrease in efficiency of microinjection relative to freshly loaded red cells was observed using [^{125}I]BSA-loaded red cells which had been stored for 8 days at 4°C (R. Hanseen and M. Rechsteiner, unpublished results).

C. Mechanism of Loading

Careful electron microscope studies by Seeman (1967) have revealed 200- to 500-Å holes in the red cell membrane during hemolysis. These holes or pores were observed only from 15 to 25 seconds following the onset of hemolysis, in this case induced by the rapid-lysis procedure. Similarities in the partitioning of macromolecules are observed when either of the three loading procedures is used (see Section III,D). This suggests that the hemolytic events may be similar for all three procedures despite the differences in methodology. Thus a likely mechanism for the uptake of macromolecules by red cells is passive diffusion through transient pores in the red cell membrane.

If the concentration of macromolecules is able to equilibrate by passive diffusion during the brief period when the membrane pores are open, the percentage of input macromolecules trapped within red cells should equal the percentage of the total lysis volume the interior of the red cells occupies. With use of the preswell loading protocol described above, roughly one-half of the added macromolecules would be expected to be trapped. In actuality, approximately one-quarter of smaller macromolecules are trapped (Schlegel and Rechsteiner, 1975; Wasserman et al., 1976). Considering the volume between red cells, this value does not significantly deviate from the expected value. However, not all molecules load to this extent, as discussed in Section III,D.

Evidence that the loaded macromolecules have entered the red cell cytoplasm and have not merely become adsorbed to the membrane is considerable:

1. If the concentration of macromolecules in the lysis buffer remains constant but the ratio of lysis buffer volume to red cell volume is systematically varied, the amount of trapped macromolecule approximates that

expected for passive diffusion partitioning rather than adsorption (Hoffman, 1958; Rechsteiner, 1975; Wasserman *et al.*, 1976).

2. Relysis of loaded red cells by a variety of methods results in release of the loaded macromolecules (Rechsteiner, 1975).

3. When red cells loaded with fluorescent dye are examined in solution by a focused laser, the distribution of dye mirrors the volume of the loaded cell and not the distribution of the red cell membrane (J. Schlessinger, personal communication).

4. Electron microscopy of red cells loaded with ferritin reveals ferritin molecules throughout the interior of the red cell (Seeman, 1967; Loyter *et al.*, 1975).

D. Quantitative Aspects of Loading

A corollary to the mechanism of passive diffusion is that the percentage of total input macromolecules trapped within red cells remains constant even though the concentration of macromolecules is varied in a given volume of lysis buffer. This was found to be true over a concentration range of 28 μg/ml to 1.3 mg/ml for IgG (Rechsteiner, 1975) and 1–30 mg/ml for BSA (Wasserman *et al.*, 1976). This of course indicates that the amount of a macromolecule which can be trapped within red cells is proportional to its concentration in the lysis buffer. Red cells loaded in the presence of BSA at a concentration of 2–4 mg/ml contain on the order of 10^6 BSA molecules per cell (Schlegel and Rechsteiner, 1975; Wasserman *et al.*, 1976). If this number of molecules were transferred to individual culture cells during fusion, for most proteins physiologically significant levels would be attained within the microinjected cells.

The transient pores in the red cell membrane appear to produce a molecular sieving effect during loading. When Ihler *et al.* (1973) compared the size distribution of proteins from a crude *Escherichia coli* extract with that of proteins trapped in red cells during rapid-lysis loading, smaller proteins were found to be preferentially trapped, although β-galactosidase with a molecular weight of 540,000 could be loaded. Table I summarizes the loading of various macromolecules by the preswell procedure. The value of 1.0 corresponds to the partitioning described above in which approximately one-quarter of the added molecules are trapped. Although several smaller macromolecules partition similarly, ferritin and isolated chromatin nu bodies are partially excluded, while rRNA and bacteriophage T4 particles are entirely excluded. Cells loaded by the dialysis loading method have been reported to trap bacteriophage T4 particles and latex spheres with

TABLE I

Macromolecule	Relative entry
[^{125}I]BSA	1.0
[^{32}P]tRNA	1.0
[^{125}I]myoglobin	1.0
[^{125}I]IgG	1.0
[^{32}P]snRNA (4.5 and 5S)	1.0
[^{125}I]HMG 1 and [^{125}I]HMG 2	1.0
[^{3}H]histone I	1.0
[^{125}I]ferritin	0.3
[^{125}I]nu bodies (11S)	0.2
[^{32}p]rRNA (18S)	~0
[^{32}p]rRNA (28S)	~0
T$_4$ bacteriophage	~0

diameters up to 7000 Å (Loyter *et al.*, 1975). It may be that such particles can be loaded to the same extent by the preswell method but that the absolute number loaded in either case is small. The trapping of these particles by dialysis loading was detected by electron microscopy of sectioned loaded cells, and the number of particles trapped was not quantitated.

Besides the restrictions which may be imposed because of size, there are other factors which may affect the loading of particular macromolecules. Red cells possess a negative charge on their surface and thus tend to adsorb molecules which have a net positive charge in solution. When red cells are loaded with histone, as much as one-half of the histone associated with them after washing can be removed by trypsinization of the loaded cells. Present evidence indicates that the rest of the histone associated with the cells is internalized. There is, however, evidence suggesting that the presence of high concentrations of basic proteins during loading may somehow prevent normal resealing of the transient pores opened in the membrane during lysis (R. A. Schlegel and M. Rechsteiner, unpublished results). Other complications may arise. Proteins which are sensitive to oxidation may denature because of the high oxygen content within the red cells; other proteins may not be able to withstand the low ionic strength of the lysis buffer (although some protection may be afforded by not adding the macromolecule until immediately after lysis by the low-ionic-strength buffer); still other macromolecules may be adversely affected by the 37°C incubation period following loading. However, in general the efficient loading of neutral and acidic proteins with diameters less than 100 Å should be possible, and evidence is presented in Section IV,A that several types of small nucleic acids can also be effectively loaded.

IV.   Red Cell–Culture Cell Fusions

A.   Evidence for Microinjection

It is assumed that macromolecules gain entry into culture cells when the cytoplasm of loaded red cells and the cytoplasm of culture cells mix during fusion. However, some lysis of loaded red cells is always observed during fusion, which leaves open the possibility that transfer to culture cells may occur by the direct uptake of macromolecules from solution rather than via red cells. Control fusions are consequently performed routinely by some investigators, in which macromolecules are added free in solution during fusions of culture cells with red cells prepared by preswell lysis or dialysis lysis in the absence of exogenous macromolecules (Schlegel and Rechsteiner, 1975; Loyter et al., 1975). The results of such a control fusion are compared with an actual microinjection in Fig. 1. It is apparent that transfer was mediated by the fusion of loaded red cells and culture cells. T4 endonuclease V (Tanaka et al., 1975) and tRNA (Kaltoft et al., 1976) can, however, apparently enter culture cells directly from solution during Sendai virus fusion. The discrepancy between these results and the results of the control fusions outlined above may be due to the marked enhancement of fusion among culture cells when red cells are omitted.

When culture cells are microinjected with [^{125}I]BSA and examined autoradiographically, the number of grains over individual culture cells is very similar to the number of grains over individual loaded red cells (Fig. 1). This is consistent with the hypothesis that fusion generally takes place between one red cell and one culture cell. However, the presence of culture cells with a larger number of grains per cell indicates that fusions with more than one red cell do occur and that as many as four or more red cells may participate (Schlegel and Rechsteiner, 1975; Schlegel et al., 1976). The conclusion that the number of macromolecules transferred to a culture cell is roughly equivalent to the number of macromolecules loaded in each red cell was also reached by Wasserman et al. (1976) who compared the amount of radioactivity found in cells microinjected with [^{125}I]BSA with the amount of radioactivity in an equal number of loaded red cells.

Evidence that macromolecules are microinjected and not simply adsorbed to membranes is provided by the following observations:

1. Following red cell microinjection of [^{125}I]BSA, autoradiography reveals both labeled and unlabeled cells. If [^{125}I]BSA were adsorbed to membranes or entered the cells by direct uptake, all cells would be expected to be labeled.

2. Following red cell microinjection of fluoresceinated BSA, recipient culture cells exhibit uniform fluorescence and not the ring fluorescence

characteristic of plasma membrane-bound fluorescent molecules (Rechsteiner, unpublished results).

3. Autoradiography reveals that ^{125}I-labeled nonhistone chromosomal proteins (HMG 1 and HMG 2; Johns *et al.*, 1975) localize within HeLa cell nuclei following microinjection (M. Rechsteiner and L. Kuehl, unpublished results).

4. Electron microscopy reveals ferritin molecules dispersed throughout the cytoplasm of culture cells microinjected with ferritin (Loyter *et al.*, 1975).

Besides this cytological evidence there is considerable functional evidence for the cytoplasmic location of microinjected macromolecules:

1. Thymidine kinase–deficient cells incorporate significantly more [³H]-thymidine into nuclear DNA following red cell microinjection of thymidine kinase (Schlegel and Rechsteiner, 1975).

2. The half-life of [¹²⁵I]BSA microinjected into HeLa cells is 30 hours (Hanseen and Rechsteiner, 1977). This is considerably longer than the 1-hour half-life of [¹³¹I]BSA taken up by pinocytosis (Ryser, 1968).

3. Microinjected [³²P]tRNAs from both mammalian cells and *E. coli* exhibit half-lives identical to the endogenous mammalian tRNA prelabeled

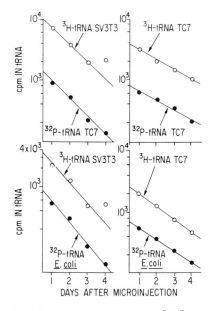

FIG. 2. Stability of microinjected tRNA molecules. [³²P]tRNAs isolated from *E. coli* or the African green monkey kidney line TC7 were microinjected into TC7- or SV40-transformed 3T3 cells whose tRNAs were prelabeled with [methyl-³H] methionine. Subsequent extraction and sucrose gradient analyses illustrate that microinjected tRNAs are as stable as internal tRNAs.

with [methyl-³H]methionine as demonstrated in Fig. 2 (Schlegel and Rechsteiner, 1977).

4. Microinjection of suppressor tRNAs has been used to identify nonsense mutations of the enzyme hypoxanthine-guanine phosphoribosyltransferase (HGPRT) in mouse L cells (Capecchi et al., 1977). HGPRT activity was restored in a CRM⁺ mutant upon red cell microinjection of ochre suppressor tRNA from yeast or E. coli. No restoration of activity was observed upon microinjection of amber or UGA suppressor tRNAs.

B. Optimal Conditions for Microinjection

As mentioned, Sendai virus causes lysis as well as fusion of red cells. Various cations have been employed to reduce this lysis, presumably by stabilizing the red cell membranes (Zakai et al., 1974; Peretz et al., 1974). In the initial report of Schlegel and Rechsteiner (1975) the divalent cation Mn^{2+} was added to fusion mixtures at 20 mM. Subsequently it was reported that high levels of Mn^{2+} during fusion markedly reduced protein synthesis in tissue culture cells (Wasserman et al., 1976; M. Capecchi, personal communication). As a result, lower concentrations of Mn^{2+} are now commonly employed, and recent results indicate that concentrations as low as 0.2 mM are effective in stabilizing red cells during fusion (Hanseen and Rechsteiner, 1977). Mn^{2+} was also initially used by Loyter et al. (1975), but subsequent studies by these workers indicated that La^{3+} is just as effective as Mn^{2+} in preventing lysis of loaded red cells (Wasserman et al., 1976). Furthermore, La^{3+} was found to be more effective than Mn^{2+} in promoting fusion and was found not to be detrimental to protein synthesis in microinjected culture cells.

In a series of experiments designed to determine optimal conditions for fusion of loaded red cells and HeLa cells, Hanseen and Rechsteiner (1977) made the following observations:

1. The efficiency of microinjection, defined as the fraction of culture cells which have been microinjected, is not affected by virus concentrations ranging from 500 to 2000 HAU per milliliter when a standard fusion is performed with 10^7 HeLa cells.

2. The efficiency of microinjection is not affected by reducing the fusion incubation period at 37°C from the usual 30 minutes to 15 minutes.

3. The efficiency of microinjection is dependent on the ratio of red cells to culture cells, but ratios as high as those used previously (2500:1 by Schlegel and Rechsteiner, 1975) are unnecessary. More than half of 10^7 HeLa cells are microinjected at a ratio of 50:1, and 90% are microinjected at 100:1.

4. The efficiency of microinjection at a fixed ratio of red cells to culture

cells is increased by reducing the volume in which the fusion is performed.

The last two observations indicate that the efficiency of microinjection is influenced by the number of collisions which occur between red cells and culture cells.

The microinjection process does not appear to discriminate among the phases of the cell cycle or among culture cell lines with different origins. Since 90% of a HeLa cell population in various phases of the cell cycle can be microinjected, there does not appear to be a marked correlation between the ability of a cell to be microinjected and the stage of the cell cycle. As further evidence, Colcemid-arrested mitotic cells can be readily micro-injected (R. Hanseen and M. Rechsteiner, unpublished results). A wide variety of tissue culture cell types has been microinjected. These include the mouse cell line 3T3-4E (Schlegel and Rechsteiner, 1975), rat hepatoma tissue culture cells (Loyter *et al.*, 1975), the presumed HeLa derivative D98/AH2 (Schlegel *et al.*, 1976), C$_3$H mouse cells (Kaltoft *et al.*, 1976), and the African green monkey kidney line TC7 and an SV40-transformed mouse line, SV3T3-B4 (Schlegel and Rechsteiner, 1977).

As indicated in Section II, culture cells can often be separated from red cells by virtue of the different adherence properties of the two cell types. Although this simple procedure has proven effective for D98/AH2 and 3T3-4E cells (Schlegel and Rechsteiner, 1975; Schlegel *et al.*, 1976), signi-ficant numbers of red cells are found attached but not fused to culture cells when TC7 cells are microinjected. This attachment is virus-mediated, since it does not occur when virus is omitted from the fusion mixture. The red cells can be removed, however, by treating the culture cells with a dilute trypsin solution once they have become firmly attached to a culture flask (Schlegel and Rechsteiner, 1977). Alternative approaches involve treating the attached cells with a dilute NH$_4$Cl solution to lyse the red cells (Kaltoft *et al.*, 1976) or to separate the red cells from the culture cells prior to plating by a fetal calf serum step gradient (Wasserman *et al.*, 1976).

C. Viability of Microinjected Cells

The applications of red cell microinjection would be very limited if the procedure proved lethal to the recipient cells. To assess the viability of microinjected cells Hanseen and Rechsteiner (1977) performed several experiments differing in their basic approach but all directed at this problem.

The cloning efficiency of culture cells following microinjection was examined in two experiments. In the first, red cells prepared by the preswell lysis procedure but not loaded with any exogenous macromolecules were fused with culture cells and plated into culture flasks containing growth medium. Other portions of the same culture cells were treated similarly, except that Sendai virus was omitted or red cells, virus, and Mn^{2+} were omitted. After 7 days the microinjected cells and control cells were fixed and stained with Giemsa, and the number of colonies containing more than eight cells was determined. No decrease in cloning efficiency was observed between the controls and the microinjected culture cells when mouse 3T3-4E cells were used. An observed reduction in the cloning efficiency of HeLa cells was attributed to exposure to Mn^{2+}, since a similar reduction occurred in the absence of Sendai virus.

The second cloning efficiency experiment was predicated on the finding that the efficiency of microinjection increases with higher ratios of red cells to culture cells. It was reasoned that, if microinjection were lethal, the cloning efficiency of culture cells would be progressively reduced as the ratio of red cells to culture cells was increased. In fact, cloning efficiency was found to be independent of the efficiency of microinjection over a 4–30% range of microinjection efficiency.

The question of culture cell viability after microinjection was also addressed by autoradiographic analysis of mitotic cells after the fusion of HeLa cells with $[^{125}I]$BSA-loaded red cells. If cells are unable to divide following microinjection, labeled mitotic cells should not be found. However, if microinjected cells enter mitosis at the same rate as nonmicroinjected cells, the fraction of labeled mitotic cells should equal the fraction of labeled interphase cells. An examination of mitotic spreads of culture cells 1 and 2 days after microinjection showed that at both times the ratio of labeled to unlabeled mitotic cells was equivalent to the ratio of labeled to unlabeled interphase cells.

In the final experiment designed to assess viability, culture cells microinjected with $[^{125}I]$BSA were plated onto microscope slides in growth medium, and the cells were fixed and examined autoradiographically after 12 hours and daily over the next 4 days. During this period only a slight decrease in the proportion of labeled cells was observed. Cell number, however, increased sixfold, indicating that microinjected cells must have divided.

V.   Concluding Remarks

Red cell microinjection permits the rapid transfer of macromolecules into large numbers of mammalian cells. Routinely, 10^7 culture cells can be microinjected in several hours. A wide range of macromolecules can be transferred subject only to the limitations of red cell loading discussed above. Individual culture cells have been microinjected with 10^6 BSA molecules. For comparison, many chromosomal proteins are represented at 10^3 to 10^5 copies per cell (Peterson and McConkey, 1976) and the cytoplasmic enzyme HGPRT is found at 2×10^5 copies per L cell (Hughes *et al.*, 1975). A similar comparison can be made for tRNA and other small nucleic acids. For example, it should be possible to microinject individual culture cells with 5×10^7 to 10^8 tRNA molecules. This level is similar to that found in various eukaryotic cell lines (Weinberg and Penman, 1968; Crook *et al.*, 1971). Thus it is clear that physiologically significant amounts of proteins and nucleic acids can be transferred. This ensures that the technique of red cell-mediated micro-injection can be used to elucidate the roles played by various macromole-cules in mammalian cells.

REFERENCES

Baker, R. F. (1967). *Nature (London)* **215**, 424
Capecchi, M., Vonder Haar, R., Capecchi, N., and Sveda, M. (1977). *Cell* **12**:371.
Crooke, S., Okada, S., and Busch, H. (1971). *Proc. Soc. Exp. Biol. Med.* **137**, 837.
Furusawa, M., Nishimura, T., Yamaizumi, M., and Okada, Y. (1974). *Nature (London)* **249**, 449.
Graessmann, A., Graessmann, M., Hoffman, H., and Niebel, J. (1974). *FEBS Lett.* **39**, 249.
Graessmann, M., and Graessmann, A. (1976). *Proc. Natl. Acad. Sci. U.S.A.* **73**, 366.
Ham, R. G. (1965). *Proc. Natl. Acad. Sci. U.S.A.* **53**, 290.
Hanseen, R., and Rechsteiner, M. (1977). Submitted for publication.
Harris, H., and Watkins, J. F. (1965). *Nature (London)* **205**, 640.
Hoffman, J. F. (1958). *J. Gen. Physiol.* **42**, 9.
Hughes, S., Wahl, G., and Capecchi, M. (1975). *J. Biol. Chem.* **250**, 120.
Ihler, G., Glew, R., and Schnure, F. (1973). *Proc. Natl. Acad. Sci. U.S.A.* **70**, 2663.
Johns, E. W., Goodwin, G., Walker, J., and Sanders, C. (1975). *Ciba Found. Symp., 1975* p. 95.
Kaltoft, K., Zeuthen, J., Engbäck, F., Piper, P., and Celis, J. (1976). *Proc. Natl. Acad. Sci. U.S.A.* **73**, 2793.
Loyter, A., Zakai, N., and Kulka, R. (1975). *J. Cell Biol.* **66**, 292.
Peretz, H., Toister, Z., Laster, Y., and Loyter, A. (1974). *J. Cell Biol.* **63**, 1.
Peterson, J., and McConkey, E. (1976). *J. Biol. Chem.* **251**, 548.
Rechsteiner, M. (1975). *Exp. Cell Res.* **93**, 487.
Ryser, H. (1968). *Science* **159**, 390.
Schlegel, R. A., and Rechsteiner, M. (1975). *Cell* **5**, 371.
Schlegel, R. A., and Rechsteiner, M. (1977). In preparation.
Schlegel, R. A., Darrah, L., and Rechsteiner, M. (1976). *In* "Molecular Mechanisms in the Control of Gene Expression" (D. P. Nierlich, W. J. Rutter, and C. F. Fox, eds.), pp. 465–470. Academic Press, New York.
Seeman, P. (1967). *J. Cell Biol.* **32**, 55.
Tanaka, K., Sekiguchi, M., and Okada, Y. (1975). *Proc. Natl. Acad. Sci. U.S.A.* **72**, 4071.
Wasserman, M., Zakai, N., Loyter, A., and Kulka, R. (1976). *Cell* **7**, 551.
Weinberg, R. A., and Penman, S. (1958). *J. Mol. Biol.* **38**, 289.
Zakai, N., Loyter, A., and Kulka, R. (1974). *J. Cell Biol.* **61**, 241.

Chapter 23

Selection of Somatic Cell Hybrids Between HGPRT⁻ and APRT⁻ Cells

Selection of Somatic Cell Hybrids Between
HGPRT⁻ and APRT⁻ Cells

R. MICHAEL LISKAY

Department of Molecular, Cellular and Developmental Biology,
University of Colorado, Boulder, Colorado

AND

DAVID PATTERSON

Eleanor Roosevelt Institute for Cancer Research, and Department of Biophysics and Genetics,
University of Colorado Medical Center, Denver, Colorado

I. Introduction

The analysis of mammalian somatic cell hybrids has provided new and fruitful approaches to the study of gene mapping, genetic complementation and dominance tests, differentiated cell functions, and malignancy (Ringertz and Savage, 1976). The induction of cell hybrids can be accomplished by several techniques (Ringertz and Savage, 1976; Davidson, 1976). The isolation of pure hybrid cell clones is more difficult and depends on the ability to select for growth of the hybrids while inhibiting growth of the parental cells. The well-known HAT system selects against both hypoxanthine-guanine phosphoribosyltransferase (HGPRT) mutants and thymidine kinase (TK) mutants while allowing growth of their hybrids (Littlefield, 1964). Mutants defective in adenine phosphoribosyltransferase (APRT) have become available in several cell lines (Astrin and Caskey, 1976; Chasin, 1974; Jones and Sargent, 1974). We describe a system utilizing a media called GAMA that can be used for the isolation of hybrids between HGPRT⁻ and APRT⁻ mutant cell lines.

II.   Techniques

A.   Cell Lines and General Procedures

The HGPRT⁻ line V79-HG⁻ is a derivative of Chinese hamster V79 fibroblasts (Robbins and Scharff, 1967) and was a gift from R. Klevecz. This line is resistant to the toxic guanine analog 6-thioguanine and lacks detectable HGPRT activity. The APRT-deficient line CAK-AP⁻ was derived from mouse CAK cells (Farber and Liskay, 1974) on the basis of its resistance to 2,6-diaminopurine, a toxic analog of adenine. Techniques for the isolation of HGPRT⁻ and APRT⁻ mutants have been described elsewhere (Astrin and Caskey, 1976; Chasin, 1974; Jones and Sargent, 1974; Thompson and Baker, 1973).

Cells were routinely cultured in Dulbecco's modified Eagle's medium plus 15% fetal calf serum (DMEM). Procedures for chromosome preparations (Farber and Liskay, 1974), Sendai-induced cell fusion (Ringertz and Savage, 1976), and HGPRT and APRT enzyme assays (Hughes *et al.*, 1975) have been described.

B.   Method for the Selection of Cell Hybrids

Following the fusion process (e.g., treatment with Sendai virus) cells are cultivated in DMEM for 2–3 days to allow nuclear fusion and the production of synkaryons. Cells are seeded into Falcon plastic flasks or dishes at a maximum cell density of 1.3×10^3 cells/cm² and then fed with the selective medium GAMA. This medium is comprised of DMEM supplemented with: 2×10^{-4} M guanine and 10^{-4} M adenine as purine sources; 10^{-5} M azaserine (Calbiochem), an inhibitor of endogenous purine biosynthesis (Patterson, 1975); and 6×10^{-6} M mycophenolic acid (Eli Lilly Laboratories), an inhibitor of the conversion of IMP to XMP (Franklin and Cook, 1969). Stock solutions of mycophenolic acid ($500\times$) are prepared in 100% *ethanol* as the solvent, as the drug is *not* readily soluble in aqueous solutions. Cells are subsequently refed with GAMA at 3-day intervals. Following a 10- to

20-day period in GAMA, surviving colonies are harvested individually and placed in GAMA for 4–7 days. This is to ensure that the clones are indeed hybrid in nature (discussed in Section III). Cells are then transferred to DMEM supplemented with adenine and guanine for 4–7 days (see Section III) and then can be cultured in DMEM and examined further.

III. Results, Discussion, and Some Precautions

A medium, referred to as GAMA, can be used to select somatic cell hybrids between HGPRT⁻ and APRT⁻ cell types. Basically, the rationale behind this selective system is to force cells to rely solely on added adenine and guanine for their purine nucleotides. This is accomplished by the addition of (1) azaserine to block endogenous purine synthesis, and (2) mycophenolic acid to block the AMP → GMP pathway. The ability to proliferate in GAMA should absolutely require that a cell have both HGPRT and APRT enzyme activity (see Fig. 1).

One hybridization experiment is presented to illustrate the use and effectiveness of GAMA selections. As controls, the parental lines, V79-HG⁻ and CAK-AP⁻, were fused to themselves, and a total of 10^6 cells was seeded into GAMA as described in Section II,B. In neither case did any colonies appear even after a 20-day period. However, when V79-HG⁻ was fused to CAK-AP⁻, colonies appeared after 12 days at a frequency of 1×10^{-4} to

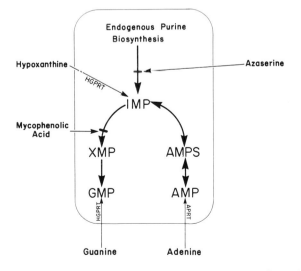

FIG. 1.   The steps of purine biosynthesis relevant to GAMA medium selections.

TABLE I[a]

KARYOTYPE ANALYSIS

Cell	Metacentrics and submetacentrics[b]	Acrocentrics[b]	Total[b]
Parents			
V79HG⁻	17 (16–18)	4 (3–5)	21 (19–22)
CAK-AP⁻	1	40 (38–41)	41 (39–42)
Hybrid clones			
HM-1	18 (16–19)	40 (34–44)	58 (52–61)
HM-2	18 (17–19)	41 (34–44)	59 (50–63)
HM-3	18 (17–19)	42 (38–44)	60 (56–63)
HM-4	19 (17–20)	41 (32–44)	60 (50–63)
HM-5	19 (18–19)	38 (35–45)	57 (53–63)

[a]The values shown represent the mean number of chromosomes as calculated from 12 to 15 metaphases scored for each cell line.

[b]Numbers in parentheses indicate the range of chromosome numbers observed in each case.

3×10^{-4}. Several of these colonies were isolated, grown in GAMA, then in DMEM plus adenine and guanine, and finally in DMEM. Karyotype analysis was performed on each parental line and their putative hybrid clones. These results as seen in Table I confirm the hybrid nature of all five clones examined. In general, chromosomal analysis is considered a good indication of whether or not the clones that survive GAMA selection are indeed true hybrids. The actual rationale behind GAMA was tested by

TABLE II

ACTIVITY OF HGPRT AND APRT IN PARENTAL
AND HYBRID CELLS

Cell	APRT	HGPRT
CAK-7	831	339
V79-8	811	307
CAK-AP⁻	0	338
V79-HG⁻	987	0
HM-1	932	196
HM-2	861	32
HM-3	977	188
HM-4	730	198
HM-5	993	221
-PRPP[b]	24	16

[a]Activities are given in counts per minute per 15 minutes per microgram of protein and represent averages of two separate experiments. CAK-7 is the wild-type parent of CAK-AP⁻. V79-8 is the wild-type parent of V79-TG⁻.

[b]Average of 34 determinations (controls).

assaying HGPRT and APRT enzyme activity in the parental and hybrid cells. These results are shown in Table II and support the notion that growth in GAMA requires the presence of both HGPRT and APRT enzyme activity. Therefore GAMA medium has proved successful in the isolation of Chinese hamster–mouse hybrids. Furthermore, hybrids of the following types have also been produced: Chinese hamster; Syrian hamster; Syrian hamster–mouse; and Chinese hamster–Chinese hamster (Liskay, 1978; Patterson, D., unpublished observations).

Several precautions concerning the use of GAMA should be noted. First, the amount of hypoxanthine present in the medium is important. The levels of hypoxanthine should be kept to a minimum because hypoxanthine can interfere with the effectiveness of the GAMA system by allowing the growth of APRT⁻ cells (see Fig. 1). Therefore media such as Dulbecco's that completely lack hypoxanthine should be used. It is true that Fetal calf serum (FCS) contains detectable levels of hypoxanthine. In fact, if APRT⁻ cells are placed in DMEM plus undialyzed FCS, azaserine and adenine cells survive because of their ability to convert hypoxanthine to AMP. Using dialyzed FCS or calf serum (which lack significant levels of hypoxanthine) can eliminate this problem. However, we have found that the high levels of guanine used in GAMA appear to eliminate the effect of hypoxanthine in undialyzed FCS. It is possible that this alternative (i.e., high levels of guanine) will not work in all circumstances. It is therefore advised that the investigator using GAMA test undialyzed FCS versus dialyzed FCS in the particular system.

Second, the cell density used in the initial plating of cells into GAMA is somewhat critical. Cell densities greater than $1.3 \times 10^3/cm^2$ can result in the undesirable appearance of colonies which are not purely hybrid in nature (i.e., they are contaminated with varying degrees of parental cells). This "protection" of the mutant cells is most likely due to metabolic cooperation (Subak-Sharpe et al., 1969). In view of this potential problem it is recommended that cell densities equal to or below $1 \times 10^3/cm^2$ be used. In addition, following clonal isolation, putative hybrids should be placed in GAMA for 4–7 hours to rid the cultures of any contaminating parental cells. We have found that the degree to which this problem occurs seems to depend on the particular mutant lines being used (Liskay, R. M. and Patterson, D., unpublished observations).

Third, cells should *not* be transferred directly from GAMA to DMEM but should go through a transitional cultivation in DMEM plus adenine and guanine. This serves as a precaution against possible purine starvation caused by residual azaserine and/or mycophenolic acid remaining in the cells.

Finally, the concentrations presented here for GAMA medium are intended to serve only as general guidelines. To ensure efficient selection of hybrids, variations in these conditions may be necessary for certain cell types. In fact, it appears that a particular clone of HGPRT⁻ CHO-K1 cells requires levels of mycophenolic acid 20–25% higher than that described in Section II,B (Patterson, D., unpublished observations).

In conclusion, the GAMA medium for the selection of hybrids between HGPRT⁻ and APRT⁻ cells should permit additional flexibility in the production of mammalian somatic cell hybrids.

ACKNOWLEDGMENTS

The authors thank Dr. R. Demars for his helpful suggestions concerning the work reported here. Dr. Liskay thanks Dr. David Prescott for the use of his laboratory for a portion of the research. This study was supported by NIH-NCI Fellowship Grant Award No. 6F32 CAO5203 to Dr. Liskay, by Grant No. VC-193 to Dr. Prescott from the American Cancer Society, and by grants from the National Foundation, March of Dimes (1–423), and the National Institute on Aging to Dr. Patterson. The authors thank Dr. Theodore T. Puck for a critical reading of the manuscript and Ms. Diane V. Carnright for excellent technical assistance.

This investigation is in part a contribution (no. 239) from the Eleanor Roosevelt Institute for Cancer Research and the Department of Biophysics and Genetics, University of Colorado Medical Center, Denver.

REFERENCES

Astrin, K. H., and Caskey, C. L. (1976). *Arch. Biochem. Biophys.* **176**, 397.
Chasin, L. A. (1974). *Cell* **2**, 37.
Davidson, R. L. (1976). *Somatic Cell Genet.* **2**, 271.
Farber, R. A., and Liskay, R. M. (1974). *Cytogenet. Cell Genet.* **13**, 384.
Franklin, L. J., and Cook, J. M. (1969). *Biochem. J.* **118**, 515.
Hughes, S. J., Wahl, G. M., and Capecchi, M. R. (1975). *J. Biol. Chem.* **250**, 120.
Jones, G. E., and Sargent, P. A. (1974). *Cell* **2**, 43.
Littlefield, J. W. (1964). *Science* **145**, 709.
Liskay, R. M. (1978). *Exptl. Cell Res.*, in press.
Patterson, D. (1975). *Somatic Cell Genet.* **1**, 91.
Ringertz, N. R., and Savage, R. E. (1976). "Cell Hybrids," Academic Press, New York.
Robbins, E., and Scharff, M. (1967). *J. Cell Biol.* **34**, 684.
Subak-Sharpe, J. W., Burke, R. R., and Pitts, J. D. (1969). *J. Cell Sci.* **4**, 353.
Thompson, L. H., and Baker, R. M. (1973). *Methods Cell Biol.* 210.

Chapter 24

Sequential Dissociation of the Exocrine Pancreas into Lobules, Acini, and Individual Cells

ABRAHAM AMSTERDAM

Department of Hormone Research,
The Weizmann Institute of Science, Rehovot, Israel

TRAVIS E. SOLOMON

Veterans Administration Center, Los Angeles, California

AND

JAMES D. JAMIESON

Section of Cell Biology
Yale University School of Medicine, New Haven, Connecticut

I.   Introduction

Intensive studies have been devoted to the understanding and characterization of the synthesis, intracellular transport, and discharge of the exportable proteins of the exocrine pancreas in tissue slices and in lobules from this gland (Jamieson, 1972; Scheele and Palade, 1975; Tartakoff *et al.*, 1975). Some aspects of regulation of the secretory process appear to depend upon the structural organization of the tissue and are best studied in preparations where the integrity of the tissue is preserved. Working with separated cells, however, offers certain theoretical and practical advantages in exploring the secretory process. In dispersed cells the entire cell surface is exposed to the external environment. Thus ready access of external medium to all the secretory cells is ensured. Moreover, the properties of the cell surface can be more easily studied in dispersed cells by the use of external probes (Maylié-Pfenninger *et al.*, 1975) visible by electron microscopy as a result of the elimination of diffusion barriers present in intact tissue.

There are basically three criteria for a satisfactory tissue dissociation procedure: (1) High yields of cells should be obtained; (2) the cells should remain structurally intact, and full function should be retained; (3) the technique should be simple and reproducible.

The dissociation of fully differentiated epithelial tissue, such as the exocrine pancreas, is particularly difficult, since the cells are firmly held together through their junctional elements (Farquhar and Palade, 1963), are encased basally by a basement membrane, and are enmeshed in a connective tissue (collagenous) stroma. In order to achieve complete separation of the exocrine pancreas into single cells, it is essential to treat the tissue with collagenase, to remove Ca^{2+} ions with chelating agents, and to apply mild shearing forces at the end of the enzymic treatment. Since each manipulation of the dissociation can damage the cells, conditions must be optimized at all stages in order to meet the criteria mentioned above. In our early studies (Amsterdam and Jamieson, 1972) crude collagenase was used, based on the procedure worked out for the liver (Berry and Friend, 1969), and led to good preparations of isolated cells, although the viability of the cells obtained fluctuated from batch to batch of commerical enzyme, possibly because of the presence of variable amounts of other proteases. In our later studies (Amsterdam and Jamieson, 1974a,b) we used chromatographically purified collagenase which contained only traces of other proteases, and added to the dissociation medium constant small amounts of chymotrypsin. The latter enzyme was chosen because it does not activate the zymogen forms of the endogenous proteolytic enzymes of the gland, especially trypsinogen. When this procedure was employed, exocrine pancreatic cells were obtained reproducibly and in high yield. Such cells retained the same structural polarity as in the intact tissue, as well as functional polarity with respect to the synthesis and transport of the exportable proteins. The responsiveness of these cells to secretagogues was, however, impaired, possibly because of partial damage to the hormone receptor sites exposed on the cell surface during the proteolytic stage of the dissociation process. Nevertheless, such preparations have been successfully employed in studying the effect of secretagogues on cyclic nucleotide levels and calcium transport in the exocrine cell (Gardner *et al.*, 1975, 1976; Robberecht *et al.*, 1976; Christophe *et al.*, 1976).

With the use of purified collagenase alone, without added proteases

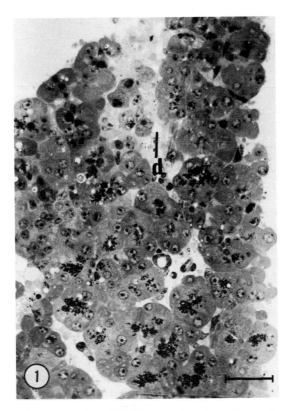

FIG. 1. Light micrograph of a thick Epon section through a portion of a minilobule prepared by the method described in the text. The preparation consists of large numbers of acini still attached to, and feeding into, an intralobular duct region (d). Marker = 20 μm. × 580.

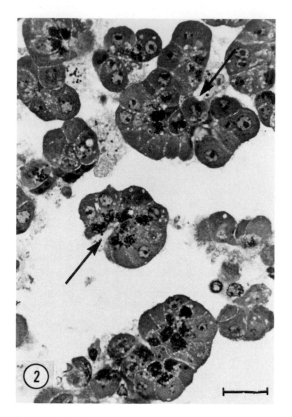

Fig. 2.   Light micrograph of a thick Epon section from an acinar preparation. The arrows indicate potential access to the lumenal front of the acinus by the incubation medium. Cell debris noted in several areas disappears from the preparation during subsequent washes. Marker = 20 μm. × 560.

and without Ca^{2+} ion chelation, the exocrine pancreas can be partially dissociated into small lobules or single acini (Gunther *et al.*, 1977). In such preparations the responsiveness to secretagogues is fully maintained, and therefore we also outline this procedure here, but will publish full details of the preparation separately.

II.   Materials and Methods

Dispersion of the pancreas into isolated single cells involves sequential enzymic digestion of stromal collagen and basement membrane, Ca^{2+} chelation to disrupt cell–cell junctions (i.e., desmosomes and tight junctions), and mechanical shearing to complete the separation of gap and tight junctions. Omission of the Ca^{2+} chelation step and variations in the other treatments result in less disassembly of the tissue, such that individual acini or groups of acini (minilobules) are produced (Figs. 1 and 2). Filtration and washing of the tissue digest yield a purified preparation free of cell debris, collagenase, and subcellular organelles liberated in the final steps of the procedure (Fig. 3). The major differences between the method presently used and that previously published (Amsterdam and Jamieson, 1972, 1974a,b) are the use of purified collagenase without hyaluronidase or exogenous proteolytic (chymotrypsin) activity and the newly developed technique of producing controlled tissue dissociation into groups of acini, single acini, or single cells.

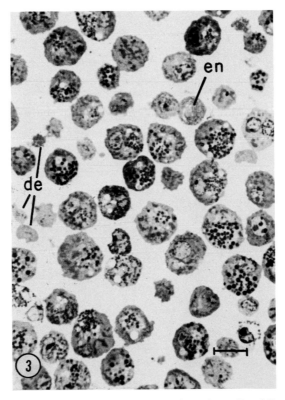

FIG. 3.   Light micrograph of a thick Epon section through a pellet of dispersed single cells obtained from the pancreas. The majority of the population consists of intact exocrine cells recognized by their population of zymogen granules. Endocrine cells (en) and residual cell debris (de) are indicated. Marker = 10 μm. × 840. From Amsterdam and Jamieson (1974a), courtesy of the *Journal of Cell Biology*.

Each step in the procedure is described in detail, along with the variations necessary to produce the above-mentioned three morphological units.

A. Animals

Guinea pigs, 250–350 gm in weight, are fasted overnight with free access to water. Younger animals are more satisfactory than older, heavier ones, because of their less dense collagen bundles and lack of stromal fat. The rat is another species which has been used with success. The animals are fasted to ensure the accumulation of zymogen granules for subsequent discharge assays and for the sake of experimental consistency. However, the effects of even brief fasting on pancreatic metabolism (Webster *et al.*, 1972) should be kept in mind when designing experiments concerned with protein secretion and metabolism.

B. Glassware

All vessels and pipets in contact with tissue are either plastic or heavily siliconized glass (Siliclad, Clay Adams, Parsipanny, N.J.).

C. Solutions

All solutions are based on Krebs–Ringer bicarbonate buffer equilibrated with 95% O_2 and 5% CO_2 to pH 7.4 (KRB). Additions present in all steps include 14 mM glucose, a complete L-amino acid supplement (Eagle, 1959) with 0.4 mM leucine, 0.2% bovine plasma albumin (BPA) (Fraction V, Armour Pharmaceutical Company, Phoenix, Ariz.), and 0.1 mg/ml chromatographically purified soybean trypsin inhibitor (STI) (Worthington Biochemical Corporation, Freehold, N.J.). Adjustments of Ca^{2+} and Mg^{2+} concentrations are described in Section F. The BPA is dissolved in and dialyzed against KRB (without amino acids, STI, Ca^{2+}, or Mg^{2+}) at a concentration of 35% (w/v). After dialysis against KRB at 4°C for 48 hours, the final concentration is calculated from the increase in volume of the solution. This stock is stored frozen. BPA at the concentration used is necessary to protect the cells agains blebbing and lysis during tissue dissociation; higher concentrations slow the dissociation process.

D. Enzyme

Chromatographically purified collagenase (*Clostridium histolyticum*), 450 U/mg, is obtained from Worthington Biochemical Corporation, Freehold, N.J. The importance of assaying both the collagenolytic and proteolytic activity of each batch of collagenase is discussed in Section H.

E. Analytic Procedures

DNA is determined by the method of Burton (1956). Cell counts are made using a hemocytometer.

F. Dissociation Procedure

Table I summarizes the solutions and procedures in each step.

1. *Removal of pancreas.* The animal is stunned and exsanguinated intrathoracically. The pancreas is removed and quickly trimmed free of connective tissue, fat, and lymph nodes as it floats in KRB at room temperature. It is weighed and trimmed to 0.8–0.9 gm wet weight.

TABLE I

SUMMARY OF CONDITIONS FOR DISSOCIATION

Step	Time	Solution
1. Gland collection	~ 5 minutes	10 ml KRB + 14 mM glucose, amino acids, 0.2% BPA, 0.1 mg/ml STI, 2.5 mM Ca^{2+}, 1.2 mM Mg^{2+}, room temperature
2. First digestion	Single cells, 20 minutes; minilobules, 30 minutes; single acini 45 minutes	5 ml KRB + 14 mM glucose, amino acids, 0.2% BPA, 0.1 mg/ml STI, collagenase 100 U/ml, 2.5 mM Ca^{2+}, 1.2 mM Mg^{2+}, 37°C
3. Ca^{2+} removal	5 minutes, twice (only for single cells); for minilobules and single acini, replace with fresh enzyme as in step 2 and proceed to step 5	2 × 5 ml KRB + 14 mM glucose, amino acids, 0.2% BPA, 0.1 mg/ml STI, no Ca^{2+} or Mg^{2+}, 2 mM EDTA, 37°C
4. Ca^{2+} replacement	3 minutes, twice (only for single cells)	As in step 1, 37°C
5. Second digestion	Single cells, ~ 35 minutes; minilobules, 30 minutes; single acini, 45 minutes	As in step 2; at the end of the second digestion, 1 mg/ml STI is added as a powder to the digestion mixture
6. Mechanical	Pipetting conditions to obtain minilobules, acini, and dispersed cells are given in the text	
7. Washing	~ 10 minutes	Single cells, minilobules, and acini as in step 1
8. Pulse	10 minutes	As in step 1 plus 20 μCi/ml [^{3}H] leucine, 37°C
9. Chase	—	As in step 1 plus 4.0 mM leucine, 37°C; volume discussed in the text

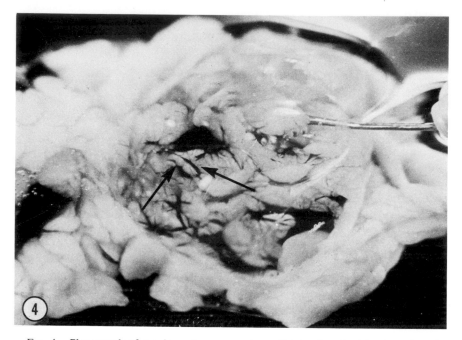

FIG. 4. Photograph of a guinea pig pancreas immediately after the injection of a collagenase-containing solution into the interstitial space. The arrows indicate groups of pancreatic lobules exposed by fluid dissecting through the interlobular connective tissue spaces. Groups of such lobules can be cut from the gland with fine scissors and have been used in previous studies on exocrine function (Scheele and Palade, 1975; Tartakoff *et al.*, 1975). × 3.

2. *Enzyme injection.* The pancreas is pinned to a wax plate and injected three to five times end to end with 5 ml of the enzyme solution using an angled 26-gauge needle on a 5-ml syringe. This distends the tissue and exposes it to the collagenase solution (Fig. 4).

3. *First digestion.* The distended pancreas and enzyme solution are transferred to a 50-ml erlenmeyer flask, gassed with 95% O_2 and 5% CO_2 and the flask tightly stoppered with a silicone stopper. As seen in Table I, the solution contains 100 U/ml collagenase (units determined by the method of Soberano and Schoellman, 1972), 2.5 mM Ca^{2+}, and 1.2 mM Mg^{2+}. The flask is incubated at 37°C at 120 oscillations per minute in a water bath. Every 10 minutes the flask is removed and swirled vigorously by hand to facilitate dissociation. The time of incubation depends on the desired final degree of dissociation. For singe cells 20 minutes is allowed; minilobules require 30 minutes, and acini 45 minutes.

4. *Ca^{2+} removal.* This step is necessary only for the production of single cells. The enzyme solution is removed by decanting and aspiration and

discarded. The tissue is exposed to two separate 5-ml portions of KRB (without Ca^{2+} or Mg^{2+}) containing 2 mM ethylenediamine tetraacetate (EDTA) for 5 minutes each. The flask is gassed, stoppered, and shaken as above. At this stage the tissue begins to loosen, and a sandy suspension of small tissue pieces develops. The supernatant is easily removed by aspiration after the suspension is allowed to settle.

5. *Ca^{2+} replacement.* For the single-cell procedure, two sequential 3-minute incubations in 5 ml of KRB containing 2.5 mM Ca^{2+} are carried out.

6. *Second digestion.* A fresh 5-ml portion of enzyme solution (as in step 3) is added, and incubation continued. Again, vigorous swirling by hand every 10 minutes accelerates disintegration of the tissue into a sandy suspension of small pieces. If the Ca^{2+} removal step has been used to produce single cells, the suspension becomes progressively cloudy as single cells are released. The duration of the second digestion is set at 30 minutes for minilobules and 45 minutes for single acini. For single cells the progress of dissociation is monitored by microscopy every 5 minutes after the first 20 minutes of the second digestion. A drop of the suspension is placed on a glass slide (after the slide and coverslip have been dipped in a 1% BPA solution and air-dried), and the degree of digestion determined by applying pressure to the coverslip with a pencil. The cells should be rounded up, and aggregates of cells should disperse with gentle pressure.

7. *STI quench.* To protect the tissue against the possibility of any active endogenous trypsin liberated in pipetting, filtering, and washing, 5 mg of STI is added at the end of the second digestion.

8. *Pipetting.* The final treatment in the dissociation procedure is dispersion of the tissue by mechanical shearing forces. For single cells, tight and gap junctions are ruptured by five passes through each of three siliconized glass pipets. This consists first of passage sequentially through a 10-ml TD Kimax pipet (No. 37033 with a tip diameter of 1.2 mm), second through a 10-ml-long-tip TD Corning pipet (No. 7064 with a tip diameter of 0.8 mm) and, finally, through a 9-inch disposable pasteur pipet with the tip flamed to 0.5 mm. Only the larger pipet is used for minilobule preparation, and five passes are sufficient. Acinar preparations are also obtained using all three pipets, but the extent of dispersion is limited by the presence of physiological levels of Ca^{2+} maintained during enzymic treatment.

9. *Filtering.* The minilobule preparation is not filtered. The acinar preparation is filtered through 200-μm Nitex gauze, while 26-μm gauze is used for the single-cell suspension (Nitex is available from Tetko, Inc., Elmsford, N.Y.).

10. *Washing.* After filtering the single-cell digest, the cells are separated from enzyme, subcellular particles, nuclei, and other debris by centrifuging for 2 minutes at 50 g in 12-ml conical glass tubes. The cell pellet is re-

suspended and centrifuged twice more using KRB containing 0.2% BPA. The pellet can be suspended by shaking or by forcefully blowing the KRB over its surface. For pulse-labeling the cells are resuspended in KRB containing the pulse label. After the pulse, the cells are centrifuged and washed once with chase medium. This procedure yields a final cell suspension almost totally free of debris, as judged by low levels of DNA, amylase, and ^{3}H-labeled protein in the medium (Fig. 3). Acini are washed after filtering, as for single cells. This procedure is repeated twice more, and the acini are resuspended in KRB with 0.2% BPA for pulse-labeling. Very few single cells are present in the final suspension (Fig. 2). Minilobules are washed three times in KRB containing 0.2% BPA with a short "up-and-down" procedure (accelerating to 50 g and switching the centrifuge off). The heavy minilobules are preferentially separated by this method (Fig. 1).

11. *Pulse-labeling and final suspension.* On centrifugation each of the above procedures yields about 0.6 ml of packed dispersed tissue or cells per gram wet weight of starting material. The DNA content varies from 1.5 to 2.0 mg, corresponding to a 40–60% yield. For single cells, about 10^8 cells are produced. Pulse-labeling of secretory proteins is accomplished by re-suspending the cells, acinar pellets, or lobules in 8–10 ml KRB containing 20 μCi/ml [^{3}H]leucine ([4,5-^{3}H]-L-leucine, 50–60 Ci/mmole) and incubating for 10 minutes at 37°C. The speed of oscillation is adjusted downward to a level which barely prevents sedimentation. Washing after pulse-labeling is done as described above, using chase medium. Final suspension is done in chase medium to a cell density of $\sim 10^6$ cell/ml or 50 ml total for acini and minilobules.

G.   Studies on Cell Function

The suspension is portioned into 1-ml aliquots in 15-ml, round-bottom, screw-capped tubes. After gassing, these tubes are incubated under the desired conditions. Following incubation, the tissue and medium are separated by centrifuging the incubation tubes at 1000 g for 5 minutes, and the supernatant is removed with a pasteur pipet and saved for further assays. Amylase and ^{3}H-labeled protein are analyzed as described previously (Amsterdam and Jamieson, 1974b).

H.   Assays for Enzyme Quality

As reported elsewhere (Gunther *et al.*, 1977), the quality of the collagenase used is essential for successful dissociation of the pancreas into structurally and functionally intact units. In this regard, it is essential that the investigator assay each batch of purified collagenase for specific collagenolytic activity and for potential detrimental contaminating enzymes which we have found to vary considerably from batch to batch of so-called purified collagenase and among suppliers. In our studies, the major offender has been clostripain activity, and in the article referred to above we describe simple specific assays for this activity and for collagenase, and define the upper limits of clostripain activity which can be tolerated by the tissue during dissociation. In addition, a simple economical method for preparing collagenase which exceeds commercial standards for purity is described.

III.   Results and Discussion

The protocol for tissue dissociation described here allows the investigator to control the extent of dissociation of pancreatic exocrine tissue into minilobules, single acini, and separated cells, and to analyze how the progression of dissociation affects cell structure and function. Treatment of the tissue with purified collagenase and with mild mechanical shearing forces was found to be sufficient to dissociate the tissue into minilobules (Fig. 1) and single acini (Fig. 2), whereas an additional step of Ca^{2+} chelation was found to be essential to break down the junctional elements between cells in order to obtain separated cells (Figs. 3 and 5).

A.   Enzymic Requirement for Tissue Dissociation

In our previous work, we showed (Amsterdam and Jamieson, 1974a,b) that the use of chromatographically purified collagenase with a low concentration of chymotrypsin yields a viable preparation of exocrine cells. Here we report that the chymotrypsin can be omitted, but proteases other than collagenase may be required for complete separation of cells, since the batches of highly purified collagenase used in our previous experiments still contain about 10% of the proteolytic activity of the crude enzyme. Studies under way, to be reported elsewhere (Gunther *et al.*, 1977), should clarify this and define tolerable limits of noncollagenolytic activity employed in tissue dissociation.

Carboxypeptidase B and leucyl amino peptidase have been tested by us and found to be ineffective in tissue dissociation (even at concentrations of 1 mg/ml), while the use of pronase yields cells with poor viability (A. Amsterdam and J. D. Jamieson, unpublished observations). Trypsin was

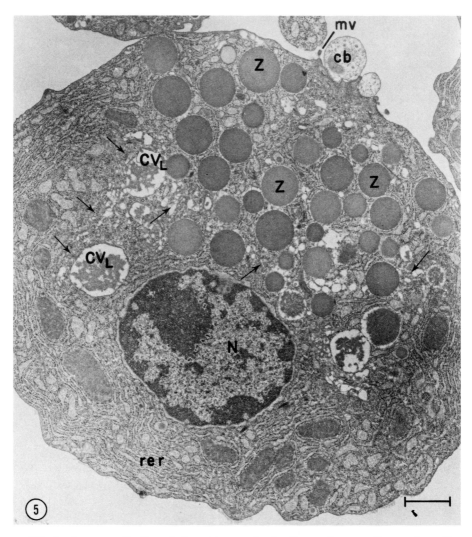

Fig. 5. Low-magnification electron micrograph of a dispersed pancreatic exocrine cell. Retention of the structural polarity of intracellular organelles is indicated by the location of fields of rough-surfaced endoplasmic reticulum (rer) and of the nucleus (N) at the former basal pole of the cell, the central distribution of elements of the Golgi complex (arrows) including its condensing vacuoles (CV_L), and the field of zymogen granules (Z) located at the former apical pole of the cell which is identified by the microvilli (mv) of the former lumenal plasmalemma. A cytoplasmic bleb (cb) torn from a neighboring cell remains attached to the apical pole of this cell, presumably by a tight junction. Marker = 1 μm. × 11,500. From Amsterdam and Jamieson (1974a), courtesy of the *Journal of Cell Biology*.

not used for dissociation of the pancreas, although it has been successfully employed, along with collagenase, to obtain acini from rat lacrimal glands, using a modification of our previous procedure (V. Herzog, personal communication).

B. Chelation of Calcium Ions

The removal of Ca^{2+} was found to be essential for complete separation of exocrine cells, since these ions are required to maintain the integrity of desmosomes and tight junctions between cells. Simply lowering the Ca^{2+} concentration from 2.5 mM to 0.1 mM is sufficient to split the desmosomes (Amsterdam and Jamieson, 1974a), whereas total removal of calcium ions by chelation is required for tight-junction breakdown (Gilula et al., 1977). EDTA was found to be the most suitable chelating agent for dissociation. Using up to 10 mM citrate results in a very poor yield of separated cells, while ethylene glycol bis (β-aminoethyl ether) N,N'-tetraacetic acid (EGTA) leads to complete cell lysis (A. Amsterdam and J. D. Jamieson, unpublished observations). The EDTA step was introduced midway during the dissociation process (see Table I), since this resulted in effective separation of well-preserved cells (Figs. 3 and 5). EDTA added after more prolonged enzymic treatment was equally effective in tissue dissociation, but cell viability was poor.

C. The Fate of Junctional Complexes during Tissue Dissociation

We showed earlier that, during tissue dissociation, desmosomes are disrupted, following which the hemidesmosomes are internalized (Amsterdam and Jamieson, 1974a). We also showed (Amsterdam and Jamieson, 1974a) that there is incomplete separation of opposing membranes in the region of gap and tight junctions and that remnants of the plasma membrane derived from adjacent cells remain attached to the separated cells (Fig. 5). However, it is clear from the freeze-fracture images of isolated cells that partial disruption of tight junctions with rearrangement and possible lateral movement of the disrupted junctions occurs (Fig. 6). In addition, it is clear from Fig. 7 that internalization of fragments of tight junctions also takes place after cell separation. Details of the fate of the junctional elements during tissue dissociation are given elsewhere (Gilula et al., 1977).

D. Yield of Cells

The yield of cells following tissue dissociation was found to be about 50% on a DNA weight basis. Comparable yields were observed following

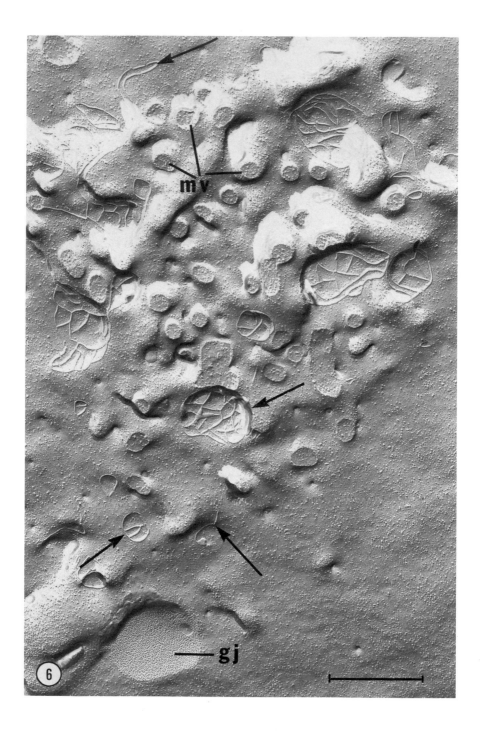

mv

mv

gj

6

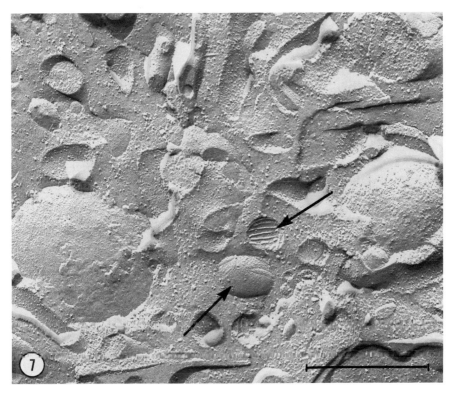

FIG. 7. Freeze-fracture replica of a dispersed isolated exocrine cell. Remnants of tight junctions are seen internalized in the walls of phagocytic vacuoles (arrows). The direction of shadowing is from the bottom of the micrograph. Marker = 0.5 μm. × 62,000.

dissociation into minilobules and acini. It is reasonable to believe that during the dissociation process cells to which the junctional elements remain attached should survive, whereas cells from which portions of the plasma membrane are sheared off at the junctional area should lyse. Since every acinar cell is attached to more than one neighboring cell which shares

FIG. 6. Freeze-fracture replica through the former apex of a dispersed single exocrine cell. The P fracture face shows isolated strands and circular islands of disrupted tight junctions, some of which are migrating laterally from the apical pole of the cell (arrows). This distribution of tight junctions is to be contrasted to the beltlike array of tight junctions surrounding the apical plasmalemma of intact tissue. Note also that the densities of intramembranous particles per unit surface area over the apical and lateral plasmalemma are approximately equal, in contrast to the usual lower density of particles in the apical plasmalemma of exocrine cells *in situ* (DeCamilli *et al.*, 1974). Cross-cleaved microvilli (mv) and a gap junction (gj) are noted. The direction of shadowing is from the bottom of the micrograph. Marker = 0.5 μm. × 48,000.

common junctional elements, a yield of 50% should be theoretically impossible for a single-cell preparation. It is also noteworthy that the temperature during mechanical dissociation (pipetting) is critical, since the yield of cells at 23 or 37°C was optimal, whereas at 4°C almost complete lysis of cells occurred. Taken together, the above considerations suggest that, following disruption of the plasma membrane in the junctional area during dissociation, rapid sealing of the membrane occurs, which is presumably a temperature-dependent process. After separation and healing are complete, cells can be maintained at 4°C.

E.   Preservation of Cell Polarity

The exocrine cell is a highly polarized cell which is involved in the vectorial transport and release of secreted material into the acinar lumina and subsequently into the duct system of the gland. The regional distribution of cell organelles (Figs. 1 and 2), as well as the disposition of microvilli only at the apical part of the plasmalemma, are well preserved in minilobules and single acinar preparations. Furthermore, even in fully separated cells which have rounded up, the regional distribution of the subcellular organelles (Amsterdam and Jamieson, 1974a, b and Fig. 5), as well as the local distribution of the microvilli at the former apical zone of the plasma membrane (Figs. 5 and 8), are preserved. Since this polarity is maintained despite disruption of the junctional elements, other cell constituents must be responsible for the maintenance of general regional topography of the plasma membrane. Nevertheless, it is clear that in the fully separated cells dramatic redistributions of individual membrane constituents occur, as indicated by fragmentation and basal migration of tight junctions and concomitant migration of intramembrane particles from the lateral plasmalemma to the apical plasmalemma, which in tissues with intact tight junctions has relatively few intramembranous particles (DeCamilli et al., 1974). As we describe elsewhere (Gunther et al., 1977), disruption of tight junctions and redistribution of intramembranous particles does not occur in our acinar preparation. Since full access of the luminal plasmalemma as well as the basal and lateral plasmalemma to tracers as large as ferritin is obtained in isolated acini, this preparation appears to be more suitable for studies on plasmalemmal structure and function compared to dispersed individual cells.

F.   Functional Studies

We previously showed that the overall rate of protein synthesis in single cells was not diminished compared to that in the intact tissue (Amsterdam and Jamieson, 1974b). Moreover, we demonstrated by autoradiography that

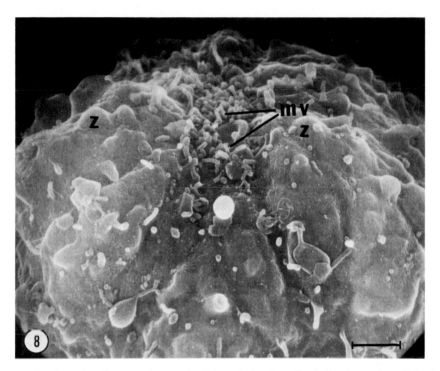

FIG. 8. Scanning electron micrograph of the apical region of an isolated exocrine cell fixed with glutaraldehyde, critical-point-dried, and gold-coated. Note microvilli at the former apical pole of the cell (mv) and humps under which presumably lie zymogen granules (z). Marker = 1 μm. $\times$ 12,000. Micrograph courtesy of M. F. Maylié-Pfenninger, P. Male, and J. D. Jamieson.

the intracellular transport of the exportable proteins from the endoplasmic reticulum to condensing vacuoles and subsequently to zymogen granules was not altered in separated cells (Amsterdam and Jamieson, 1974b). However, responsiveness to secretagogues (cholinergic agents and peptide hormones), as assayed by the release of amylase and of radiolabeled proteins to the medium was clearly impaired. In isolated cells, the minimally effective amount of secretagogue needed to elicit discharge was found to be one order of magnitude higher than in intact tissue, while the maximal rate of discharge was about 50% of that intact tissue (Amsterdam and Jamieson, 1974b; Fig. 9). In contrast, the preparations of single acini (Gunther *et al.*, 1977) and of minilobules showed that responsiveness to secretagogues was completely maintained. It is not clear whether the secretory defect in single cells result from the additional EDTA treatment step, which in combination with the enzymic treatment may alter receptor sites for secretagogues, or from a breakdown of the junctional elements, including gap junctions which may provide chemical or electrical coupling between the cells in the intact tissue. Such coupling may be essential for amplification of the stimulatory signal for discharge. The use of preparations of minilobules, single acini, and separated cells obtained from exocrine tissue may therefore provide a convenient model system for a direct approach to this problem.

ACKNOWLEDGMENTS

The studies reported above were supported in part by USPHS Grant AM-17389 and by grants from the National Cystic Fibrosis Research Foundation and The National Foundation March of Dimes awarded to J. D. Jamieson.

REFERENCES

Amsterdam, A., and Jamieson, J. D. (1972). *Proc. Natl. Acad. Sci. U.S.A.* **69**, 3028.

Amsterdam, A., and Jamieson, J. D. (1974a). *J. Cell Biol.* **63**, 1037.

Amsterdam, A., and Jamieson, J. D. (1974b). *J. Cell Biol.* **63**, 1057.

Berry, M. N., and Friend, D. S. (1969). *J. Cell Biol.* **43**, 506.

Burton, K. (1956). *Biochem. J.* **62**, 315.

Christophe, J. P., Frandsen, E. K., Conlon, T. P., Krishna, G., and Gardner, J. D. (1976). *J. Biol. Chem.* **251**, 4640

DeCamilli, P., Peluchetti, D., and Meldolesi, J. (1974). *Nature (London)* **248**, 245.

Eagle, H. (1959). *Science* **130**, 432.

Farquhar, M. G., and Palade, G. E. (1963). *J. Cell Biol.* **17**, 375.

Gardner, J. D., Conlon, T. P., Klaeveman, H. L., Adams, T. D., and Ondetti, M. A. (1975). *J. Clin. Invest.* **56**, 366.

Gardner, J. D., Conlon, T. P., and Adams, T. D. (1976). *Gastroenterology* **70**, 29.

Gilula, N. B., Amsterdam, A., and Jamieson, J. D. (1977). in preparation.

Gunther, G. R., Schultz, G. S., Hull, B. E., Alicea, H. A., and Jamieson, J. D. (1977). In preparation.

Jamieson, J. D. (1972). *Curr. Top. Membr. Transport* **3**, 273.

Maylié-Pfenninger, M. F., Palade, G. E., and Jamieson, J. D. (1975). *J. Cell Biol.* **67**, 333a.

Robberecht, P., Conlon, T. P., and Gardner, J. D. (1976). *J. Biol. Chem.* **251**, 4635.

Scheele, G. A., and Palade, G. E. (1975). *J. Biol. Chem.* **250**, 2660.

Soberano, M. E., and Schoellman, G. (1972). *Biochim. Biophys. Acta* **271**, 133.

Tartakoff, A. M., Jamieson, J. D., Scheele, G. A., and Palade, G. E. (1975). *J. Biol. Chem.* **250**, 2671.

Webster, P. D., Singh, M., Tucker, P. C., and Black, O. (1972). *Gastroenterology* **62**, 600.

Chapter 25

Bromodeoxyuridine Differential Chromatid Staining Technique: A New Approach to Examining Sister Chromatid Exchange and Cell Replication Kinetics

EDWARD L. SCHNEIDER,[1] RAYMOND R. TICE,[2] AND
DAVID KRAM[1]

[1] Laboratory of Cellular and Comparative Physiology, Gerontology Research Center, National Institute on Aging, National Institutes of Health, Public Health Service, Department of Health, Education and Welfare, Baltimore, Maryland.

[2] Medical Department, Brookhaven National Laboratories, Upton, New York. Supported by grant number 5-T-32CA09121 awarded by the National Cancer Institute, DHEW, and the U.S. Department of Energy.

I. Introduction

The 1970s have witnessed a technological explosion in the field of cyto-genetics, converting this discipline from a purely descriptive modality to a field capable of analyzing the mechanisms of DNA synthesis, damage, and repair. The initial part of this decade brought the advent of chromosomal banding techniques, based on differences in base composition along the vertical axis of the chromosome (Caspersson *et al.*, 1970; Arrighi and Hsu, 1971). These techniques coupled with somatic cell genetics have permitted detailed mapping of the human genome (McKusick, 1976).

The next major breakthrough was the differential staining of sister chro-matids by nonautoradiographic techniques. This was achieved by allowing cells to replicate in the presence of a nucleotide analog, bromodeoxyuridine (BrdU). Although differential labeling of sister chromatids had been achieved by Taylor (1958) by pulse labeling *Vicia faba* root tips with tritiated thymidine, followed by autoradiographic examination of metaphase cells, BrdU substitu-tion has the great advantage of simplicity. As discussed in more detail in Section V, there are several other advantages of BrdU over the combination of tritiated thymidine and autoradiography.

Zakharov and Egolina (1972) were the first to demonstrate that incorpo-rated BrdU alters chromatid structure. They demonstrated that incubation of cells with BrdU resulted in the production of chromosomes which had asymmetrical chromatids, one chromatid being more condensed than its sister chromatid. Latt (1973) examined the effect of several DNA dyes on BrdU-substituted chromosomes and found that the fluorescent dye Hoechst 33258 was particularly effective in producing differential staining of sister chromatids based on relative proportions of BrdU substitution (Fig. 1). Other fluorescent compounds such as acridine orange (AO) (Dutrillaux *et al.*, 1973; Franceshini, 1974; Kato, 1974b; Perry and Wolff, 1974) 4′,6-diamidino-2-phenylindole; (DAPI) (Lin and Alfi, 1976) were also employed to produce differential chromatid fluorescence in BrdU-substituted chromo-somes. As an alternative to fluorescent microscopy, Perry and Wolff (1974) found that Hoechst 33258, followed by Giemsa staining of BrdU-substituted chromosomes, produced "harlequin" staining which can be seen with normal light microscopy (Fig. 2). Wolff and Bodycote (1977) also obtained this differential Giemsa staining without pretreatment with Hoechst 33258

FIG. 1.   Differential chromatid fluorescence produced by Hoechst 33258 staining of a second replication cycle cell that has replicated twice in the presence of BrdU. Note that all chromosomes possess one bright and one dull staining sister chromatid. Points where the fluorescence switches from one sister chromatid to the other are SCEs (Tice *et al.*, 1975).

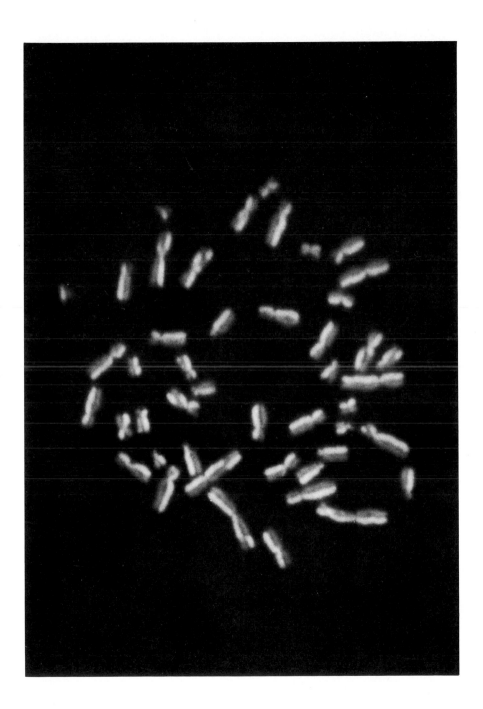

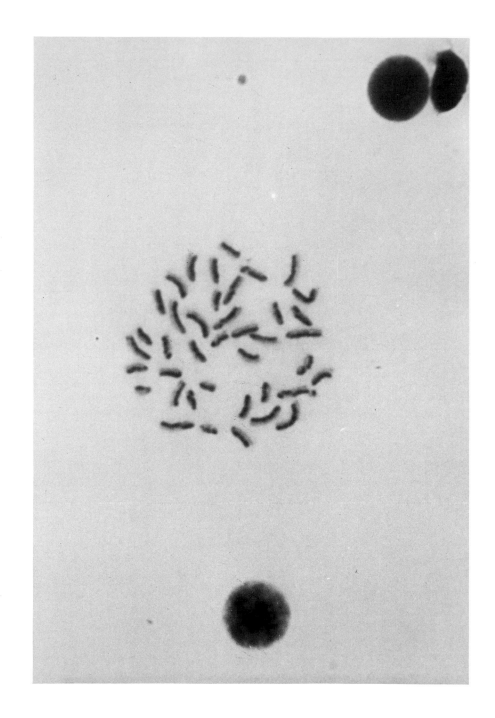

and conclude that differential staining is produced by alterations in chromosomal protein structure induced by incorporated BrdU rather than by reduced staining of the DNA–Giemsa complex.

In Fig. 3, a schema is presented to explain the mechanisms of differential staining of BrdU-substituted metaphase chromosomes (Tice *et al.*, 1975). It is based on three premises: that DNA replication is semiconservative in nature, that chromatid structure is moneneme (comprised of a single double helix), and that chromosomes segregate randomly at cell division. Although evidence to support these premises has come from a variety of investigations, studies of BrdU-substituted chromosomes further support them. After one replication cycle in the presence of BrdU, both chromatids contain one BrdU-substituted DNA strand (dashed lines) and one parental unsubstituted DNA strand (solid lines). These chromatids are biochemically identical and therefore have equal intensities when stained with either a fluorescent or a nonfluorescent dye. After two replication cycles in the presence of BrdU, one chromatid has its DNA unifilarly substituted with BrdU, while the sister chromatid is bifilarly substituted with BrdU. Chromatids containing bifilarly substituted DNA fluoresce less intensely than their unifilarly substituted sister chromatids when stained with a fluorescent dye such as Hoechst 33258, AO, or DAPI, and have diminished Giemsa staining with light microscopy. After three replication cycles in the presence of BrdU, half the chromosomal length has only bifilarly substituted DNA and fluoresces with equally diminished fluorescence or with equivalently decreased Giemsa staining. The remaining chromosomal length has chromatids which biochemically resemble second replication cycle chromosomes and therefore have similar differential staining properties. With further replication cycles, an increasing proportion of the chromosomal length will have bifilarly substituted DNA in both chromatids and will therefore stain with diminished intensity, fourth-division cells will have only one-quarter of their chromosomal length possessing differential staining properties, fifth-division cells will have one-eighth of their chromosomal length with these properties, etc. The problems of discriminating between third and subsequent replication cycle cells is discussed in Section V. However, this ability to identify unequivocally cells which have replicated one, two, three, or more times in the presence of BrdU has led to the use of this technique for cell cycle analysis (Tice *et al.*, 1976; Schneider *et al.*, 1977).

Of particular interest are second replication cycle cells, for it is in these cells that exchanges of staining intensity are most easily discerned. These

FIG. 2. Differential chromatid staining obtained in a second replication cycle cell with Giemsa staining of Hoechst-33258-pretreated slides. Again, SCEs can be easily distinguished as switches in light and dark staining along the chromosomal length.

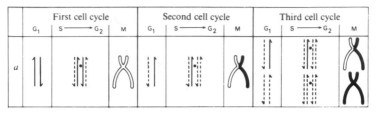

FIG. 3. Schema outlining the mechanism for the differential staining produced by Hoechst-33258 staining of BrdU-substituted metaphase chromosomes (based on a momeneme model of chromosome structure). The solid lines represent unsubstituted DNA strands, while the dashed lines indicate BrdU-substituted DNA. White metaphase chromatids represent bright fluorescence, while black chromatids represent relatively dull fluorescence.

points, called regions of sister chromatid exchange (SCE), were first described by Taylor *et al.* (1957) with tritiated thymidine. However, the ease of identifying these events with the BrdU differential staining technique has led to a proliferation of SCE studies. It has been shown by several investigators that SCEs are a sensitive indicator of DNA damage. The BrdU differential staining technology has also permitted examination of the nature of chromosomal structure (Holmquist and Comings, 1975; Tice *et al.*, 1975; Wolff and Perry, 1975). Last, the BrdU differential staining techniques can be utilized for the analysis of DNA replication patterns within specific chromosomes (Latt *et al.*, 1975b; Stubblefield, 1975). This chapter is equally divided between a detailed discussion of the methodology for BrdU differential staining, both *in vitro* as well as *in vivo*, and an examination of the many applications of this technique.

II.   Methodology

A.   *In Vivo* BrdU Labeling of Chromosomes

BrdU labeling of chromosomes has been achieved in a wide variety of species ranging from *V. faba* (Kilhman and Kronberg, 1975) to mud minnows (Kligerman and Bloom, 1976). In this section, we focus on the BrdU labeling of mammalian chromosomes *in vivo*. Several independent approaches to this area have been developed in parallel. They involve multiple subcutaneous injections (Pera and Mattias, 1976), multiple intraperitoneal injections (Allen and Latt, 1976; Vogel and Bauknecht, 1976), or continuous intravenous infusion (Schneider *et al.*, 1976) of BrdU. We prefer the last-mentioned approach, since it avoids multiple traumatization of the animal, permits reproducible serum concentrations of BrdU, and keeps fluctuations in BrdU

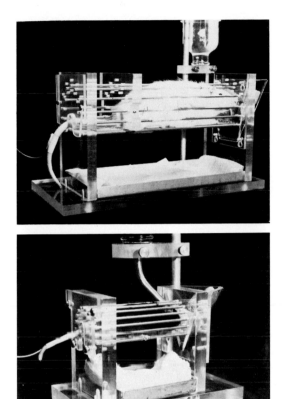

FIG. 4. Specially designed, modified Bollman restrainers for rats (above) and mice (below), which permit access to food and water while preventing the animals from interrupting intravenous infusions by tail vein (Schneider *et al.*, 1976).

levels to a minimum. Animals are kept in specially designed modifications of Bollman restrainers (seen in Fig. 4). These containment devices permit the animals to have access to food and water, while they prevent the interruption of continuous intravenous infusion by tail vein. The intravenous route also permits the addition of colchicine without further traumatization of the animal. Furthermore, if water-soluble mutagenic or carcinogenic compounds are to be tested, they can be administered as well by this infusion system.

Intravenous infusions are started while the rats are briefly anesthetized with ether, and the animals are then placed in the restrainers. Mice are placed in similarly designed but smaller restrainers, and the intravenous infusions are started without anesthesia. A 23-gauge needle attached to Clay Adams P.E. 50 tubing is used for cannulation of the rat tail vein, and a 27-

gauge needle attached to P.E. 80 tubing is utilized for mouse tail vein infusion. A Watson-Marlow 10-channel peristaltic pump with AutoAnalyzer tubing (Gamma Ent.) is used to deliver approximately 3 ml/24 hours to mice and 50 ml/24 hours to rats. Pump channels are calibrated prior to each experiment, so that precise amount of BrdU can be delivered to each animal. At the conclusion of the BrdU infusion, animals are sacrificed by air injection and bone marrow is removed from exposed femurs by washing with Dulbecco's phosphate-buffered saline, pH 7.4. The cells are then centrifuged at 200 g, and the resultant cell pellet is suspended in hypotonic KCl (0.06 M for rat bone marrow cells and 0.075 M for mouse bone marrow cells) for 10 minutes. The swollen cells are then centrifuged for 200 g and resuspended in methanol–acetic acid, 3:1. The last step is repeated twice, and the cells are then placed on slides.

In addition to bone marrow cells, thymocytes, splenic lymphocytes, spermatogonia, and macrophages are also amenable to *in vivo* BrdU labeling. Furthermore, since BrdU can pass the placental barrier, *in utero* labeling of fetal tissues is also possible with these techniques (Tice, 1978a). However, since BrdU can inhibit cell populations which are rapidly differentiating, not all cell populations *in vivo* may be amenable to this technique.

For *in vivo* labeling of animals without tails, or for animals not easily restrained, we recommend implantation of the delayed-release BrdU tablets recently developed by Latt (1977).

B. *In Vitro* BrdU Labeling of Chromosomes

Examination of the concentrations of BrdU used *in vitro* for differential chromatid staining reveals extreme variation, even for the same cellular system. BrdU concentrations reportedly used in phytohemagglutinin (PHA)-stimulated human lymphocyte cultures range from as high as 200 μg/ml (Dutrillaux *et al.*, 1974) to as low as 3.07 μg/ml (Latt *et al.*, 1975a). However, even when identical concentrations of BrdU are used, some investigators have also added additional compounds such as fluorodeoxyuridine (presumably to block endogenous thymidine synthesis), uridine (to counter the effect of fluorodeoxyuridine), and deoxycytidine (to counter the toxic effect of BrdU) (Latt, 1974, 1975).

Examination of cellular kinetics (see Section V) requires low concentrations of BrdU to ensure a lack of proliferative inhibition. Increasing BrdU concentrations are assessed for their ability to provide good differential chromatid staining without significant toxic effects. For PHA-stimulated human lymphocytes and human diploid fibroblasts [cultured in Eagle's minimal essential medium (MEM) with Earle's salts, nonessential amino acids, 1% L-glutamine, and 10% fetal calf serum], we have found that BrdU at a final concentration of 5–10 μg/ml produces satisfactory results. However, other cellular systems may yield their best results at other BrdU concentrations. We have avoided the addition of other compounds so as to simplify the nature of the dose–response curves.

An important consideration when dealing with *in vitro* studies is the presence of light. Ikushima and Wolff (1974) demonstrated that exposure to light significantly increased SCE frequencies in cells which had incorporated BrdU. This effect is dependent on the BrdU concentration and can occur with normal laboratory lighting. We therefore recommend minimizing the ambient light exposure of growing cultures either by using light-tight boxes or by wrapping flasks and dishes in aluminum foil.

C. Chromosome Preparations

Techniques which currently provide good morphology for chromosome banding techniques also yield good preparations for SCE examinations and/or cell cycle determinations. The reader is directed to a review on chromosome banding by Lubs *et al.* (1973) for a complete discussion on slide-making techniques.

D. Fluorescent Differential Staining Techniques

1. HOECHST 33258

The first technique for the direct demonstration of differential staining of sister chromatids involved use of the DNA dye Hoechst 33258 (Latt, 1973). We use a 100X stock of Hoechst 33258 solution in distilled water at 50 μg/ml, which appears to remain stable for several months. Prior to use, the solution is diluted 1:100 with distilled water, and slides are stained for 10–15 minutes followed by rinsing in running distilled water. A coverslip is then placed on the slides with citrate–phosphate buffer (0.05 *M* citric acid, 0.1 *M* dibasic sodium phosphate, pH 7.0), and the slides can then be examined by fluorescence microscopy. Alternatively, stained slides can be allowed to dry and subsequently reexamined within 2–4 days.

Hoechst 33258 fluorescence is pH-dependent (Latt, 1973), with a maxima at pH 7.0. The greatest problem with this fluorescent stain is its tendency to fade quickly on exposure to incident light. This fading quality necessitates the use of a fast film for photography and prohibits visual scoring of even moderate levels of SCE. Bloom and Hsu (1975) attempted to bypass this fading difficulty by using McIlvaine's pH-4.4 buffer as the mounting solution. This lowered pH results in a longer duration of differential fluorescence, but at considerably lower contrast.

While SCE analysis is difficult with Hoechst 33258, this dye permits extremely fast and accurate assessments of both the replicative history of metaphase cells and the percent of replicated interphase cells (i.e., the proportion of interphase cells which had undergone at least some proportion of DNA synthesis in the presence of BrdU). As shown by Latt (1974), interphase cells which have incorporated BrdU exhibit depressed fluorescence intensities proportional to the extent of BrdU incorporation. This depressed intensity can be used to distinguish cells which have replicated in the presence of BrdU from those which have not. Furthermore, cells in their first S phase exhibit a mottled appearance, while those that have undergone a complete period of DNA synthesis exhibit uniformly depressed fluorescence. This contrast in fluorescence intensity can be further enhanced by incubating stained slides at 5°C for ~10 minutes and/or using ice-cold mounting buffer.

2. ACRIDINE ORANGE

AO staining of BrdU-substituted chromosomes, independently introduced by Kato (1974a), Dutrillaux et al. (1973), and Perry and Wolff (1974), permits long-term examination of differential staining patterns. However, in contrast to Hoechst 33258, the differential staining patterns must "burn in" before good contrast occurs (Perry and Wolff, 1974). We have found AO staining to be the best differential staining technique for examining SCEs, because of its reliability and because of the contrast obtained between unifilarly and bifilarly substituted DNA. We use a modification of the technique of Kato (1974a). Slides are stained for 5–6 minutes in AO (0.125 mg/ml, pH 7.4, phosphate buffer), rinsed well with running distilled water, incubated in pH 7.4 phosphate buffer for 10 minutes, and mounted in the buffer. As with Hoechst 33258 staining, slides can be allowed to dry and subsequently used, or previously examined slides can be reexamined. Time in the buffer appears to be the only critical step. Too brief a time in the buffer, and good differential staining contrast requires more time to occur; too long a time, and the contrast is not as good. It is possible to examine the slides for quality under low magnification; slides not incubated long enough appear reddish, while slides left in the buffer too long appear green. The desired color contrast is greenish nuclei and reddish cytoplasm. Also, preexposure to enhance contrast can be accomplished by exposing the slides briefly to light from a mercury source, and fading can be decreased by inserting an exciter filter blocking light below 470 nm after contrast has been sufficiently enhanced.

3. 4',6-DIAMIDINO-2-PHENYLINDOLE

Recently, Lin and Alfi (1976) introduced a new stain for examining differential fluorescence. This stain, DAPI, appears to have the sensitivity of Hoechst 33258 for low levels of BrdU incorporation, while also resulting in greater photostability than either Hoechst 33258 or AO (Lin and Alfi, 1976). Slides are stained with DAPI (1 μg/ml in phosphate-buffered saline, pH 7.0) for 10 minutes, rinsed in distilled water, and mounted in 0.2 M sodium phosphate buffer, pH 11.0.

E. Giemsa Differential Staining Techniques

In contrast to fluorescent methods, Giemsa staining permits permanent chromosome preparations to be made. However, Giemsa staining is more dependent on incorporated BrdU levels, and the length of light treatment must be varied to produce optimal differential chromatid staining. For laboratories which do not alter BrdU concentrations, Giemsa staining can be invaluable. For those who frequently alter BrdU concentrations, these techniques can be tedious.

Perry and Wolff (1974) introduced a Giemsa staining technique based on prior staining with Hoechst 33258 (0.5 μg/ml in distilled water), followed by washing with distilled water. Mounted slides are then exposed to ambient light for 24 hours, treated with 2× standard saline citrate (SSC) (0.30 M NaCl, 0.30 M sodium citrate) at 62.5°C for 2 hours, and stained with 3% Giemsa. This technique (called the fluorescent plus Giemsa—FPG—technique) has subsequently undergone several modifications in regard to light exposure and SSC treatment (Wolff and Perry, 1974).

Sugiyama et al. (1976) and Goto et al. (1976) extensively examined the relationship between Hoechst 33258 prestaining, light exposure, and Giemsa staining. We have adapted their findings and developed a Giemsa technique which has provided extremely high-quality differential staining. Slides are stained with Hoechst 33258 (50 μg/ml) in distilled water for 10–15 minutes, rinsed, air-dried and, when convenient, mounted with Sorensen's citric acid–phosphate buffer, pH 7.0, exposed to light from two 15-W Westinghouse blacklight fluorescent tubes in a desk lamp from a distance of 2–3 cm for 15–20 minutes, and stained with Giemsa (Harleco, 2% in phosphate buffer, pH 6.8).

Pathak et al. (1976) demonstrated that techniques which gave G banding (trypsin, urea) could also give differential staining patterns. We have also found that other G-banding techniques can also result in differential staining patterns. Consequently, it can be concluded that almost any technique which results in G banding can, with appropriate modifications, produce differential staining of BrdU-substituted sister chromatids.

III. SCE Analysis for Examining Clastogenic Agents

SCEs were first observed by Taylor (1958) in *V. faba* root tips, using tritiated thymidine pulse-labeling and autoradiography. SCEs were seen as occasional symmetrical exchanges of autoradiographic grain distribution between sister chromatids.

The tritiated thymidine used for SCE detection was noted to increase SCE frequency in a dose-dependent manner (Gibson and Prescott, 1972). Similarly, the exposure of cultured cells to ultraviolet light immediately after labeling with tritiated thymidine also resulted in a severalfold increase in SCE (Kato, 1973). These early studies were difficult to conduct, and precise quantitation was not possible. Introduction of the BrdU differential staining technique has greatly facilitated studies on the nature of SCE in both *in vitro* and *in vivo* systems.

A. *In Vitro* Systems

An extensive study on the relationship of DNA damage and SCE was performed *in vitro* by Perry and Evans (1975). X-irradiation and 15 chemical agents were tested for their ability to induce SCE and chromosomal aberrations. Perry and Evans found a strong correlation between the ability of a chemical agent to produce chromosomal aberrations and its efficiency in inducing SCE. The most potent SCE inducers were alkylating agents, with bifunctional compounds, such as mitomycin C (MCC) and nitrogen mustard, several orders of magnitude more potent than monofunctional agents such as quinacrine mustard. These investigators concluded that SCE induction is a sensitive means of assessing DNA damage. Doses of mutagens which produced a twofold increase in SCE produced only minimal or even no increase in chromosomal aberrations.

The usefulness of this system for detecting mutagens and carcinogens has been confirmed by several studies. To date, a large variety of compounds has been tested for their ability to produce SCE (Vogel and Bauknecht, 1976; Solomon and Bobrow, 1975; Rudiger *et al.*, 1976; Stetka and Wolff, 1976a,b; Bayer and Bauknecht, 1977). In general, compounds found to be mutagenic, carcinogenic, or clastogenic in other assay systems also induce SCE.

A major problem of *in vitro* testing is that some known clastogens, such as cyclophosphamide and dimethylnitrosamine (Lilly *et al.*, 1975), do not induce SCE in tissue culture. These compounds require *in vivo* metabolism to attain their active form. To circumvent this difficulty, rat liver microsomes can be added to the cell culture test system. With this approach, cyclophosphamide treatment results in a dramatic increase in the frequency of

SCE (Stetka and Wolff, 1976a). The weakness of this system is that liver microsomes may not contain all the enzymes utilized *in vivo* for the conversion of drugs into their active form.

B. *In Vivo* Systems

The limitations of *in vitro* systems have led to the development of *in vivo* systems for examining DNA damage. Initial *in vivo* experiments were carried out in the root tips of the field bean *V. faba* (Kihlman, 1975). Although several compounds were found to induce SCE in this plant, extrapolation to animal cells is difficult. For example, the herbicide maleic hydrazide produces chromosomal aberrations in plants but has not been found to damage animal cells (Perry and Evans, 1975).

Several systems for *in vivo* analysis of SCE have been developed for use in mammals. Allen and. Latt (1976), employing multiple intraperitoneal BrdU injections, demonstrated that SCE could be induced by compounds requiring metabolic activation, such as cyclophosphamide.

An *in vivo* system has been developed by Stetka and Wolff (1976b) in which rabbits are treated with chemical agents and bled at timed intervals after exposure; SCEs are then measured in cultured lymphocytes. This system is sensitive to agents which require *in vivo* activation as well as those which do not.

In our laboratory, an *in vivo* system has been developed for assessing clastogens, in which rats and mice are continuously infused for 26 hours by tail vein with BrdU at 50 mg/kg wt per hour. Clastogens which are soluble in phosphate-buffered saline are injected intravenously following commencement of the BrdU infusion. Clastogens that are relatively insoluble are injected intraperitoneally.

The addition of MMC to animals 1 hour after the start of the infusion results in a linear dose–response curve for SCE induction (Fig. 5). At MMC concentrations of 6 mg/kg, over 100 SCEs per metaphase are observed in mice bone marrow cells. We have found that the induction of SCE in old animals is identical to that in young animals for MMC doses up to 2.5 mg/kg (Kram *et al.*, 1978). Above this concentration, the frequency of SCEs plateaus in old animals (both C57BL/6J mice and Wistar rats), while continuing to rise in a linear fashion in younger animals (Kram *et al.*, 1978). This diminished SCE formation at high MMC concentrations was accompanied by an increased frequency of chromosomal aberrations. We conclude that cellular aging appears to result in a diminished capacity to repair DNA damage by SCE formation.

We are presently conducting a survey of several strains of mice to determine the effect of genetic background on baseline and MMC-induced SCE

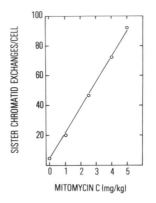

FIG. 5. Induction of SCEs by MMC *in vivo* in CBA mice. Animals are infused with a single MMC injection 1 hour after the commencement of BrdU infusion and 25 hours prior to sacrifice. Each point represents the mean of SCE measurements on 15 second replication cycle cells at each concentration of MMC. The line is derived by the method of least squares.

frequency. To date, we have found that the baseline frequency of SCEs is similar in all mouse strains tested. In the AKR strain, which features spontaneous development of lymphocytic leukemia, we found a decreased ability to form SCEs in response to high doses of MMC (Kram and Schneider, 1978a). It is hoped that these investigations will lead to the development of animal models for human genetic diseases, which display alterations in baseline and/or mutagen-induced levels of SCE.

C. Nature of SCE

In early experiments, Kato (1973) and Wolff, Bodycote, and Painter (1974) found that SCE could be induced by ultraviolet light in a dose-dependent manner. It was also observed that exposing cells in late S or G_2 did not increase SCE frequency, while exposure in early S resulted in a dramatic increase in SCE. This suggested that a round of DNA synthesis is required for SCE induction.

In another series of experiments, Kato (1973) found that ultraviolet damage could induce SCE several days after exposure. Kato concluded that all ultraviolet-induced damage was not eliminated in one cell cycle. Instead, damage seemed to persist and could produce SCE in cells six to eight cell cycles removed from the time of exposure.

These observations prompted Kato to study the effects of caffeine on SCE induction, since it has been shown that caffeine interferes with postreplicational repair (Trosko *et al.*, 1965). The exposure of ultraviolet-treated cells to caffeine resulted in a decreased induction of SCE and an increased frequency of chromosomal aberrations. These observations led to the hypothesis that SCE may represent a postreplication DNA repair process.

Later studies by Kato (1974a) showed that SCE, chromosomal aberrations, and DNA repair were not as simply related as had been thought. The drugs MMC and 4-nitroquinoline 1-oxide (4NQO) behaved similarly to ultraviolet light. SCEs were induced with an increasing concentration of the drug, and caffeine inhibited SCE formation and increased chromosomal aberrations. However, proflavine-induced SCEs and chromosomal aberrations were not affected by caffeine. Proflavine also differed in its ability to induce SCE at time points following exposure. Unlike ultraviolet light MMC, and 4NQO, proflavine induced no new SCEs 48 hours after acute exposure. Another drug, N-methyl-N-nitro-N-nitrosoguanidine (MNNG), induced SCEs very weakly but was efficient in causing chromosomal aberrations.

Our own work with MMC in $vivo$ suggests that DNA damage and/or repair may occur in a different fashion than in tissue culture (Kram and Schneider, 1977b). When MMC is given 26 hours prior to the initiation of BrdU infusion, only baseline levels of SCE are detected. Because the average cell cycle duration time under conditions of infusion is 13 hours, these data indicate that MMC-induced damage does not cause SCE after two cycles of replication.

Clearly, much more research will be necessary to determine the precise relationship between SCE, DNA damage, and DNA repair. The data presently available indicate that SCEs are induced by a variety of DNA-damaging agents having different mechanisms of action. Some agents which damage DNA are not efficient inducers of SCE. The data for caffeine sensitivity and chromosomal aberrations are inconclusive, since with some agents SCE behaves as a postreplicational repair mechanism, while with other agents it appears unrelated to DNA repair.

IV. Assessment of the Spontaneity of SCE

As detailed in the previous section, a wide variety of drugs can induce SCE. Among the agents known to induce SCE formation are the two chemicals that have been utilized for SCE detection, tritiated thymidine and BrdU. Since SCE cannot be presently detected without incorporation of one of these two agents, the only feasible approach to investigating this question is to examine the frequency of SCEs as a function of these agents. Brewen and

Peacock (1969) and Gibson and Prescott (1972) performed this experiment with tritiated thymidine and concluded that all SCEs were induced by the incorporated radioactivity and that these events do not occur spontaneously. However, evaluating SCEs at low levels of incorporated thymidine is extremely tedious, and the results must be interpreted with caution.

The advent of the Brdu differential staining technique permitted a more sensitive reevaluation of this question. Kato (1974b) first utilized this approach by measuring the frequency of SCEs in Chinese hamster cells at various BrdU concentrations. SCE frequency remained stable from the lowest BrdU concentrations permitting accurate scoring to BrdU concentrations 10 times higher. This plateau was followed by a linear increase in SCE frequency at higher BrdU concentrations. From these data, Kato concluded that some SCEs were spontaneous in nature.

Kato's conclusions were subsequently challenged by the results of a similar study of the frequency of SCEs as a function of BrdU concentration performed by Wolff and Perry (1975) on Chinese hamster cells. Instead of a plateau of SCE frequencies at low BrdU concentrations, these workers found a rapid increase in SCE frequency. Since they could not exclude the possibility that their dose–response curve could be extrapolated through zero, Wolff and Perry (1975) concluded that there was insufficient evidence to verify the spontaneous occurrence of SCEs.

One difficulty in these examinations of the spontaneity of SCEs is that they were performed under tissue culture conditions. Since it is extremely difficult to culture cells in the absence of light and since light has been demonstrated to incude SCE in BrdU-substituted chromosomes, it is difficult to exclude the possibility that the observed SCEs were induced by exposure to light. Adaptation of the BrdU differential staining techniques for the examination of cell populations *in vivo* permitted us to reexplore this question under conditions of no light exposure (Tice *et al.*, 1976). The results of these experiments, seen in Fig. 6, revealed no significant variation in SCE frequency from the minimum BrdU concentration which permitted unequivocal determination of differential fluorescence (1.9 mg/kg wt per hour) to an infusion level of 7.5 mg/kg wt per hour. Above this concentration of BrdU, SCE frequency increased significantly. We concluded from this study that, although BrdU can induce SCE formation, SCEs occur spontaneously at a rate of 1.5 SCEs per second replication cycle cell or 0.75 SCEs per cell per cell cycle. This observation leads to speculation on the mechanisms of spontaneous SCEs. Do they represent normal levels of DNA repair in response to normally occurring DNA damage? Are SCEs a normal consequence of DNA replication? Answers to these questions await elucidation of the mechanisms of SCE.

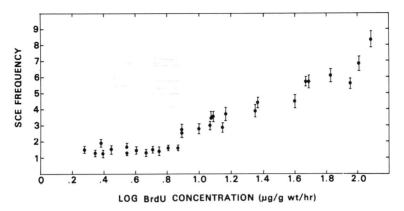

FIG. 6. Abscissa: BrdU concentration; ordinate: SCE frequency. The frequency of SCEs in rat bone marrow cells after two cell cycles in the presence of BrdU as a function of BrdU concentration. Each point represents the mean SCE frequency of 30 analyzed metaphase cells. Ranges indicate S.E.M. Note that, while BrdU induces SCEs at high concentrations, the frequency of SCEs plateaus below 10 mg/kg wt per hour, indicating the occurrence of spontaneous SCE.

V. Determination of Cellular Proliferation Kinetics

A. Introduction

Investigations into the proliferation kinetics of *in vitro* and *in vivo* cellular systems have chiefly utilized techniques based on the incorporation of radioactive precursors into DNA, followed by autoradiographic examination of labeling patterns as a function of time (Cleaver, 1967). The main disadvantages of these autoradiographic techniques include (1) the time and effort required, (2) perturbations in proliferation kinetics which may arise from the addition and/or removal of the radioactive precursor, (3) technical problems such as resolution, label dilution, or label reutilization, and (4) the insensitivity of the techniques in detecting cells with long intermitotic times (Cleaver, 1967; Steel, 1972; Shackney, 1975; Nicolini, 1975).

The ability to identify unequivocally the replicative history of metaphase cells based upon their BrdU differential staining patterns provides a feasible alternative to current autoradiographic techniques for assessing proliferation kinetics. Several investigators have independently used BrdU incorporation to examine various aspects of cellular proliferation (Kim, 1974; Craig-Holmes and Shaw, 1976; Dutrillaux and Fosse, 1976; Tice *et al.*, 1976). These investigations have been primarily restricted to examinations of the *in vitro* proliferation of PHA-stimulated human lymphocytes, and only recently have such studies been extended to *in vivo* cellular systems (Schneider *et al.*, 1977). We have designated the analysis of cellular kinetics based on the incorporation of BrdU into eukaryotic DNA and the subsequent utilization of specific differential staining patterns as BISACK (BrdU incorporation system for the analysis of cellular kinetics).

B.   Inhibition of Proliferation by BrdU

The validity of BISACK depends on demonstrating that examinations of cellular proliferation by this method are not conducted at concentrations of BrdU which inhibit cell replication. The concentrations of BrdU which permit the accurate assessment of proliferation kinetics depend on the cell origin as well as on the type of *in vitro* culturing system or the organ-specific BrdU levels *in vivo*. Consequently, for each cellular system under examination, a BrdU concentration–proliferation inhibition curve must be generated. Experimentally, this can be accomplished by comparing the relative proportions of different generation metaphase cells as a function of increasing BrdU concentration from the lowest concentration permitting accurate identification of the differential chromatid labeling patterns. This is best accomplished with a BrdU exposure time which results in cells in all three consecutive generations.

The BrdU concentration–proliferation inhibition curves we have gene-

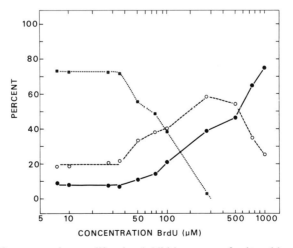

FIG. 7.   BrdU concentration–proliferation inhibition curve of cultured human peripheral lymphocytes. Distribution of metaphase cells among the first (solid circles), second (open circles), and third and subsequent (squares) generations as a function of BrdU concentration (Tice *et al.*, 1976).

rated experimentally exhibit two component parts (Fig. 7). At low BrdU concentrations, the relative proportions of cells in each generation remain relatively stable. This plateau is followed by the inhibition of cell replication by BrdU, as indicated by an exponential increase in the frequency of earlier generation metaphase cells. These observations strongly suggest that low BrdU concentrations, *in vivo* as well as *in vitro*, do not inhibit proliferation kinetics. In fact, intermitotic times of PHA-stimulated human lymphocytes *in vitro* (Tice, 1976) and of rat bone marrow stem cells *in vivo* (Schneider *et al.*, 1977) determined by BISACK are comparable to the most rapid times measured autoradiographically.

C. *In Vitro* Assessment of Cellular Proliferation

In vitro assessment of proliferation kinetics with BISACK involves the cultivation of cells at concentrations of BrdU demonstrated to not inhibit cell replication. At various time intervals after BrdU addition, Colcemid (or a comparable mitotic arrester) is added to accumulate metaphase cells. The cultures are then terminated as described in Section II, and the frequency of metaphase cells in each generation is determined. Both the length of Colcemid exposure and the interval between data points depend in a pragmatic manner on both the mitotic index and the expected rate of proliferation of the cellular system under examination. With high mitotic rates, only a brief Colcemid treatment is needed to ensure adequate sample size. With low mitotic indexes, longer Colcemid treatments are required. In a similar manner, the faster the proliferation rate, the shorter the time span must be between data points to provide accurate kinetic curves. With slower rates, longer time periods between data points should be employed.

In Fig. 8, a typical proliferation kinetic curve is generated from data obtained on PHA-stimulated human lymphocytes. This *in vitro* system was assayed at 4-hour intervals with a 1-hour exposure to Colcemid prior to each termination, beginning 40 hours after PHA administration and extending to 100 hours after PHA administration.

A more rapidly proliferating *in vitro* cellular system, such as the transformed Chinese hamster cell line V79, is probably best examined at 2-hour intervals with a 30-minute Colcemid block.

If only intermitotic times are being determined, the best results will be obtained by subculturing heavily confluent cell populations into BrdU-containing medium. This process effectively minimizes the number of cells actively undergoing DNA synthesis at the onset of BrdU addition. Metaphase chromosomes of cells undergoing DNA synthesis at the time BrdU is introduced will have discontinuous labeling patterns (Fig. 9).

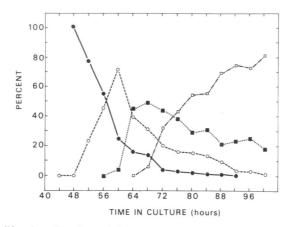

FIG. 8. Proliferation kinetics of PHA-stimulated human peripheral lymphocytes. Abscissa: time in culture; ordinate: percent first generation (solid circles), second generation (open circles), third generation (solid squares), and fourth and subsequent generation (open squares) metaphase cells as a function of time after PHA stimulation (Tice *et al.*, 1976).

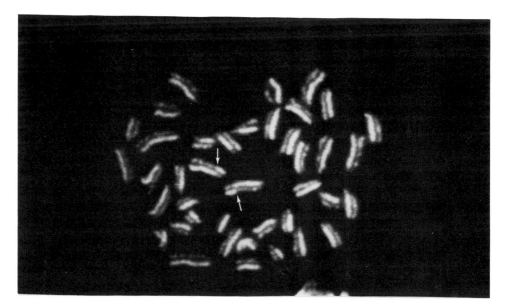

FIG. 9. Discontinuous labeling pattern (speckling) of a rat bone marrow metaphase cell that was in the S cell cycle phase at the onset of BrdU infusion (arrows indicate SCEs) (Schneider *et al.*, 1976).

In addition to determining intermitotic times, it is possible to assess the duration of G_1, S, and $G_2 + 1/2 M$ by appropriate modifying techniques. This can be accomplished by the addition and/or replacement of BrdU by thymidine at various intervals prior to cell harvesting, using a termination time which provides the maximum frequency of metaphase cells. Cells in G_2 at the time of thymidine addition exhibit no alteration in their labeling patterns. Cells undergoing active DNA synthesis show progressively decreasing proportions of bifilar BrdU substitution as they progress through the S phase. Cells in G_1 at the time of thymidine addition do not exhibit bifilar BrdU substitution or discontinuous labeling but show differential staining patterns.

Atlernatively, BrdU can be added to cultures in logarithmic growth, in which case discontinuous labeling patterns will occur for cells in the process of synthesizing DNA. The extent and timing of the appearance of these discontinuous labeling patterns can be used to determine the different phases of the cell cycle (Fig. 10).

Several investigators have grown cells in culture medium supplemented with various additional compounds such as fluorodeoxyuridine (to increase BrdU incorporation by blocking endogenous thymidine production) or deoxycytidine (to decrease the toxic effects of BrdU) (Madan et al., 1976).

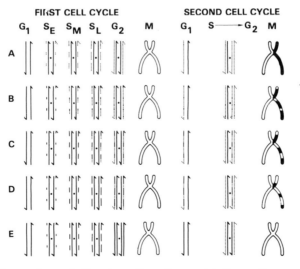

FIG. 10. Schema demonstrating how discontinuous labeling patterns can be utilized for the further examination of cell cycle kinetics. (A) BrdU labeling commencing in G_1; (B) early S phase; (C) middle S phase; (D) late S phase; and (E) G_2 phases of the first cell cycle. The solid lines represent unsubstituted DNA strands, while the dotted lines represent BrdU-substituted DNA. White metaphase chromatids indicate bright fluorescent regions, while black chromatid regions indicate diminished fluorescence.

However, Horn and Davidson (1976) showed that the effect of deoxycytidine is to decrease the levels of actual BrdU incorporation into DNA. Consequently, we feel that, at least for the examination of cellular kinetics, it is better to avoid the possible complicating effects of secondary nucleotide analogs or precursor compounds.

D.   *In Vivo* Assessment of Cellular Proliferation

In vivo methods of cellular proliferative kinetics closely parallel *in vitro* techniques. We feel that the infusion apparatus described earlier (Fig. 4) currently offers the most reliable means for assessing cellular proliferation kinetics *in vivo*. This system permits the constant infusion of known concentrations of BrdU, while avoiding both undue surgical stress or the technical difficulties inherent in multiple injections.

The proliferation kinetics of rat bone marrow stem cells determined by this technique are presented in Fig. 11. Because of the rapid proliferation of these cells, data points were obtained at 1.5-hour intervals. In this case, no effort was made to determine the duration of G_1, S, or G_2 + M phases of the cell cycle. Instead, metaphase cells were identified only on the basis of the number of S phases they fully or partially completed. Other *in vivo* cellular systems (e.g., thymus, spleen, testicular) appear to be equally amenable to investigation but not necessarily at the same BrdU levels.

In both *in vivo* and *in vitro* cellular systems, we believe that the examination and identification of 100 consecutive metaphase cells per culture or per animal tissue at each time point are sufficient for an accurate assessment of proliferation kinetics. However, we encourage the determination of at least triplicate values where possible. As long as metaphase cell identification is

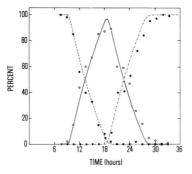

FIG. 11.   Proliferation kinetics of rat bone marrow cells *in vivo*. Distribution of first replication cycle (squares), second replication cycle (open circles), and third or subsequent replication cycle cells (solid circles) as a function of BrdU infusion time (Schneider *et al.*, 1977).

limited to first, second, and third or subsequent generation, good chromosome morphology is not critical. If an assessment of cell cycle phases is desired, excellent morphology is needed for the analysis of discontinuous labeling patterns. In the interest of expediency, it is important to maximize the number of metaphase cells in the microscope scanning area. We also recommend consecutive scoring to avoid possible bias; even highly contracted metaphase cells can be unequivocally identified and should be scored.

E.   Identification of Fourth and Subsequent Replication Cycle Metaphase Cells

With BISACK, unequivocal identification is limited to first and second replication cycle cells. Third and subsequent generation cycle cells are combined for this type of analysis. However, on a theoretical basis, it should also be possible to identify third, fourth, and perhaps subsequent generation metaphase cells. Their identification would be based on the random segregation of daughter chromatids and the resultant binomial distribution of differentially labeled chromosomes (i.e., containing DNA unifiliarly substituted with BrdU) among cells of each subsequent generation (Heddle, 1968). However, analyses are somewhat limited by the presence of SCEs which create technical difficulties by producing chromosomes with differentially labeled regions along only part of the chromosomal length. However, this technical difficulty can be overcome by limiting assessment to labeled centromeric regions (LCRs) (Heddle, 1968). Using LCR analysis, it appears that third generation metaphase cells can be identified and distinguished from fourth and subsequent generation cells, further increasing the scope of BISACK (Tice et al., 1976b). However, the ability to distinguish third, fourth, and fifth generation metaphase cells depends upon (1) the experimental generation of binomial distributions not significantly different from theoretically expected ones and (2) the degree of overlap between distributions for each generation. This overlap is actually a complex function related to the number and stability of the chromosomal complement and to the relative proportion of metaphase cells in particular generations. Consequently, in human cells, possible delineation is limited to third generation metaphase cells from subsequent generations and to estimations of relative proportions of metaphase cells based on simple 95% confidence limits, a rather simplified approach. Further research may result in the development of mathematical models which can analyze these complex interrelationships. These difficulties can be avoided, however, by limiting identification to metaphase cells of the first, second, and third and subsequent generations.

F.   Examination of Interphase Cells

With BISACK, the cumulative number of cells which have undergone DNA synthesis, as well as the replicative history of metaphase cells, can be examined. Latt (1974) demonstrated that interphase cells replicating in the presence of BrdU exhibit a greater than 40% loss in Hoechst 33258 fluorescence intensity by one-third completion of DNA synthesis. Further BrdU incorporation decreases nuclear fluorescence intensity in a linear fashion in proportion to the number of subsequent generations in the presence of the nucleotide analog (Fig. 12). The nuclei of cells exposed to BrdU for only part of one replication cycle appear speckled with regions of dull fluorescence. The contrast between "bright" and "dull" interphase nuclei can be enhanced by incubating Hoechst-stained slides at 5°C for 30 minutes.

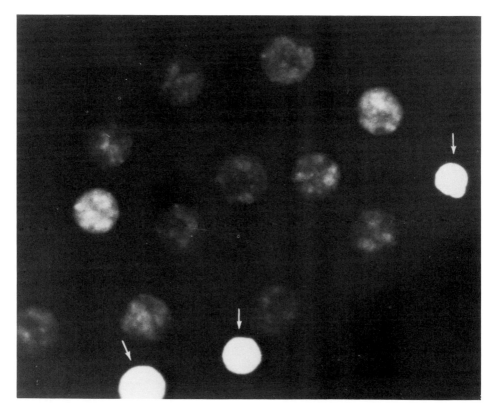

FIG. 12.   Peripheral human lymphocyte interphase nuclei 72 hours after PHA stimulation and BrdU addition. Bright nuclei (arrows) have not replicated in the presence of BrdU.

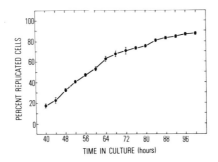

FIG. 13. Cumulative number of human peripheral lymphocytes which have replicated in the presence of BrdU as a function of time after PHA stimulation.

The increase in the cumulative proportion of PHA-stimulated human lymphocyte interphase cells exhibiting depressed fluorescence as a function of time is presented in Fig. 13. While this analysis is currently restricted to an all-or-none determination, it should be theoretically possible to use interphase cells for cell cycle determinations, either with microdensitometric measurements or with flow microfluorometric analysis. Both types of analyses require simultaneous measurements of cellular DNA content and BrdU-induced depressed fluorescence.

G.   Mathematical Analysis of BISACK-Derived Data

Considerable effort has been expended in the development of detailed mathematical techniques for the analysis of experimental results derived from autoradiographic measurements of proliferation kinetics (Barrett, 1966; Steel and Hanes, 1971; Brockwell et al., 1972; Gilbert, 1972). The importance of simulation methods for deriving accurate estimates of cell cycle times and the theoretical errors involved in the use of nonsimulation methods have been summarized by Mendelsohn and Takahashi (1971). Equally important is the development of simulation methods for BISACK-derived data.

The length of the cell cycle can be experimentally approximated from the interval between the appearance of second and third replication cycle cells. However, these values do not provide an estimate of cell cycle variance. We have developed mathematical simulation models for both G_1-arrested cell populations such as human peripheral lymphocytes, as well as for asynchronously cycling populations such as rat bone marrow cells. These models assume that cells, once they begin to cycle, continue cycling and that cell cycle durations follow a gamma distribution. The mathematics are presented in detail in another article (Thorne *et al.*, 1978). We feel that this is an important area in which research should be encouraged.

VI.   Examination of DNA Replication Patterns

Until recently, research directed toward the examination of DNA replication patterns within eukaryotic chromosomes had been largely restricted to autoradiographic methods. However, it was quickly observed that BrdU differential staining techniques offered direct light microscope visualization of DNA synthesis with greatly increased resolution (Latt, 1973).

Various exposure regimes and staining protocols were extensively used to examine different aspects of DNA replication. Late-replicating DNA regions were investigated in human (Grzeschik *et al.*, 1975; Epplen *et al.*, 1976), Chinese hamster (Stubblefield, 1975; Wang, 1976), and primary mouse (Madan *et al.*, 1976) cells. Several investigators have also examined replication patterns in human heterocyclic X and in X-autosome translocations (Latt, 1974, 1975; Willard and Latt, 1976). Early and middle S-phase DNA replication patterns have also been assessed with this technique (Stubblefield, 1975; Kim *et al.*, 1975).

For late-replicating regions, the method of choice has been to grow cells in the presence of BrdU throughout the first cell cylce and then to transfer them to thymidine-containing medium at various times prior to culture termination (Latt, 1974, 1975; Grzeschik *et al.*, 1975). This system, by demarcating late-replication regions with bright fluorescence or dark Giemsa staining, offers good visual contrast to the rest of the genome.

Alternatively, cells can be grown in thymidine-containing medium and then transferred to BrdU-containing medium at various times prior to culture termination (Latt, 1975; Kim *et al.*, 1975; Willard and Latt, 1976). This regime highlights the early-replicating DNA regions by depressing the fluorescence intensities of the regions still synthesizing DNA at the onset of BrdU incorporation and has the advantage of not requiring extensive synchronization to provide first generation metaphase cells.

Ideally, investigations into DNA replication patterns should involve thymidine pulsing of BrdU-substituted cells at various time frames throughout the S phase. This regime would result in intensely fluorescent chromosome regions at metaphase, pinpointing the regions involved in DNA synthesis during thymidine pulsing. If this approach were coupled with an examination of prophase-type chromosomes or G_2 prematurely condensed chromosomes, extremely fine band resolution should be possible.

VII. Conclusions

The applications of BrdU differential staining techniques have been discussed in the previous sections. Of these applications, the two most important areas will probably be the utilization of SCEs as an indication of DNA damage and repair and the employment of BISACK for examining cellular replication.

SCE appears to be an extremely sensitive measure of chemically induced chromosomal damage. The examination of SCEs is almost two orders of magnitude more sensitive than the previous approach to assessing clastogenic damage by the determination of chromosomal aberrations. Thus several laboratories utilize this technique for the screening of mutagenic and carcinogenic compounds. For compounds that do not require metabolic activation, a variety of cell cultures can be utilized. For substances that require metabolic activation, a microsomal fraction can be added to the cell culture or the examination can be performed *in vivo* in mice or other laboratory animals. Future adaptation of automated microdensitometry of BrdU differentially stained metaphases coupled with the appropriate computer software should provide the speed for a large-scale screening of compounds. Another potential approach involves flow microfluorometric analysis of BrdU differentially stained chromosomes isolated from treated cells.

Although bacterial screening systems may provide a simpler and more rapid approach to large-scale screenings of clastogenic compounds, SCE determination has the advantage that eukaryotic cells can be employed. Therefore SCE analysis could potentially detect an agent which may not be mutagenic for bacterial systems and yet be mutagenic, and perhaps even carcinogenic, for human cells.

SCE analyses have been employed to examine various human genetic disorders. The results of some of these studies are summarized in Table I. Alterations in baseline and induced SCEs have been found in several diseases characterized by either increased chromosomal aberrations or diminished ability to repair DNA damage. In Bloom's syndrome, an autosomal recessively inherited dermatological condition having an extremely high frequency of malignancy, a significant increase in baseline SCEs has been reported (Chaganti *et al.*, 1974). While baseline SCEs appear normal in Fanconi anemia, another autosomal recessive condition characterized by pancytopenia and an increased frequency of malignant tumors, the induced frequency of SCEs appears to be lower than that observed in control cells (Latt *et al.*, 1975a). Last, in xeroderma pigmentosum, a group of autosomal recessively inherited skin disorders, baseline frequencies appear normal while induced levels of SCEs are elevated (Wolff *et al.*, 1977). The last-

TABLE I

HUMAN GENETIC DISORDERS CHARACTERIZED BY ALTERATIONS IN SCE FREQUENCY

Human genetic disorder	Reference	Baseline SCE levels	Induced SCE levels
Bloom syndrome	Chaganti et al. (1974)	Markedly increased	—
Fanconi anemia	Latt et al. (1975)	Normal	Reduced
Ataxia telangectasia	Galloway and Evans (1975)	Normal	Normal
Xeroderma pigmentosum	Wolff et al (1975), (1977)	Normal	Increased
Down syndrome	Biederman and Bower (1976)	Normal	Increased
Cockayne syndrome	Fischman and Joy (1977)	Increased	—

mentioned disorders are of particular interest in that they comprise several genetically distinct conditions, each of which has an abnormality in different steps of excision repair and/or postreplication DNA repair.

The severity of these disorders makes prenatal diagnosis a reasonable choice for prospective parents who have previously had a child afflicted with one of these conditions. Baseline or induced levels of SCEs could be used for prenatal diagnosis and thus lead to the prevention of these disabling conditions. However, one must be cautious in this endeavor, since viral contamination (Brown and Crossen, 1976) of amniotic fluid cell cultures and potential mycoplasma contamination of these cells (Schneider et al., 1974) could lead to false positive results.

The system for analyzing cell replication kinetics (BISACK), described in Section V, should permit new insight into the nature of replicating cell populations both in vivo and in vitro. It has several advantages over previous autoradiographic techniques: (1) There is minimum perturbation of the cellular system; (2) difficulties with label dilution or reutilization are not important, since BrdU is present continuously; (3) the identification of first and second from third and subsequent replication cycle cells is unequivocal, and (4) analytic time is considerably shortened. We are currently employing BISACK in an examination of the decline in cell replication observed as a function of aging, and we anticipate that other investigators will utilize this technique for investigating diverse areas ranging from developmental biology to oncology.

We have intentionally saved the discussion of the nature of SCE for last, since the exact mechanism(s) of this event remain to be elucidated. SCEs are clearly a response to chromosomal damage since they can be induced by a wide variety of agents that damage DNA. Whether or not they represent a form of DNA repair has been the subject of considerable debate (Cleaver, 1977). It is clear from several studies that they do not appear to be involved in the classic pathways for repair of DNA damage such as excision repair. However, there is also considerable evidence suggesting that they represent a form of DNA repair. This includes the finding of chromosomal aberrations such as breaks at sites of incomplete exchange (Kato, 1977) and the observation of altered SCE response in disorders characterized by altered DNA repair (Chaganti *et al.*, 1974; Latt *et al.*, 1975; Wolff *et al.*, 1977). It has been recently suggested that SCE may represent more than one phenomenon, that one type of SCE may be intimately related to DNA replication while a second form may be a type of DNA repair (Tice, 1976; Kato, 1977). We find this explanation most appealing, since it would explain the increase in SCE frequency we have observed when inhibitors of DNA synthesis are added to cell cultures (Tice, 1978b), as well as the decreased frequency of SCEs and increased level of chromosomal aberrations we find as a function of cellular aging (Kram *et al.*, 1978; Tice *et al.*, 1978). SCE related to DNA replication would be responsible for the increase in the first case, while diminished SCE related to DNA repair would operate in the latter situation.

We anticipate that considerable attention will be focused in the future on the nature of SCEs. If they are confirmed to be involved in DNA repair, they may provide a new tool for investigating this important area. However, regardless of their nature, they will remain an extremely sensitive indicator of chromosomal damage.

REFERENCES

Allen, J. W., and Latt, S. A. (1976). *Nature (London)* **260**, 449–451.
Arrighi, F. E., and Hsu, T. C. (1971). *Cytogenetics* **10**, 81–86.
Barrett, J. C. (1966). *J. Natl. Cancer Inst.* **37**, 443–450.
Bayer, U., and Bauknecht, T. (1977). *Experientia* **33**, 25.
Biederman, B., and Bower, P. (1976). *Abstr., 15th Annu. Meet. Somatic Cell Genet.* p. 129.
Bloom, S. E., and Hsu, T. C. (1975). *Chromosoma* **51**, 261–267.
Brewen, J. G., and Peacock, W. J. (1969). *Mutat. Res.* **7**, 433–440.
Brockwell, P. J., Trucco, E., and Fry, R. J. M. (1972). *Bull. Math. Biophys.* **34**, 1–12.
Brown, R. L., and Crossen, P. E. (1976). *Exp. Cell Res.* **103**, 418–420.
Caspersson, T., Zech, L., Johansson, C., and Modest, E. J. (1970). *Chromosoma* **30**, 215–227.
Chaganti, R. S. K., Schönberg, S., and German, J. (1974). *Proc. Natl. Acad. Sci. U.S.A.* **71**, 4508–4512.
Cleaver, J. E. (1967). *In* "Thymidine Metabolism and Cell Kinetics." North-Holland Publ., Amsterdam.
Cleaver, J. E. (1977). *J. Supra. Mol. Biol.* **100**, 1978.
Craig-Holmes, A. P., and Shaw, M. W. (1976). *Exp. Cell Res.* **99**, 79–87.
Dutrillaux, B., and Fosse, A. M. (1976). *Ann. Genet.* **19**, 95–102.
Dutrillaux, B., Fosse, A. M., Prieur, M., and Lejeune, J. (1974). *Chromosoma* **48**, 317–340.

Dutrillaux, M. B., Laurent, C., Couturier, J., and Lejeune, J. (1973). *C.R. Hubd. Seances Acad. Sci.* **276**, 3179–3181.

Epplen, J. T., Bauknecht, T., and Vogel, W. (1976). *Hum. Genet.* **31**, 117–119.

Fischman, H. K., and Joy, C. (1977). *J. Supramol. Struct. Suppl.* **1**, 117.

Franceschini, P. (1974). *Exp. Cell Res.* **89**, 420–421.

Galloway, S. M., and Evans, H. J. (1975). *Cytogenet. Cell Genet.* **15**, 17–29.

Gibson, D. A., and Prescott, D. M. (1972). *Exp. Cell Res.* **74**, 397–402.

Gilbert, C. W. (1972). *Cell Tissue Kinet.* **5**, 53–63.

Goto, J., Akematsu, T., Shimazu, H., and Sugiyama, T. (1976). *Chromosoma* **53**, 223–230.

Grezeschik, K. H., Kim, M. A., and Johannsmann, R. (1975). *Humangenetik* **29**, 41–59.

Heddle, J. A. (1968). *Mutat. Res.* **6**, 57–65.

Holmquist, G. P., and Comings, D. E. (1975). *Chromosoma* **52**, 245–259.

Horn, D., and Davidson, R. L. (1976). *Somatic Cell Genet.* **2**, 469–481.

Ikushima, T., and Wolff, S. (1974). *Exp. Cell Res.* **87**, 15–19.

Kato, H. (1973). *Exp. Cell Res.* **82**, 383–390.

Kato, H. (1974a). *Exp. Cell Res.* **85**, 239–247.

Kato, H. (1974b). *Nature (London)* **251**, 70–72.

Kato, H. (1977). *Chromosoma* **59**, 179–191.

Kihlman, B. A. (1975). *Chromosoma* **51**, 11–18.

Kihlman, B. A., and Kronberg, D. (1975). *Chromosoma* **51**, 1–10.

Kim, M. A. (1974). *Humangenetik* **25**, 179–188.

Kim, M. A., Johannsmann, R., and Grzeschik, K. H. (1975). *Cytogenet. Cell Genet.* **15**, 363–371.

Kligerman, A. D., and Bloom, S. E. (1976). *Chromosoma* **56**, 101–109.

Kram, D., and Schneider, E. L. (1977a). *Hum. Genet.* (in press).

Kram, D., and Schneider, E. L. (1977b). In preparation.

Kram, D., Tice, R. R., and Schneider, E. L. (1978). *Exp. Cell Res.* (in press).

Latt, S. A. (1973). *Proc. Natl. Acad. Sci. U.S.A.* **70**, 3395–3399.

Latt, S. A. (1974). *J. Histochem. Cytochem.* **22**, 478–491.

Latt, S. A. (1975). *Somatic Cell Genet.* **1**, 293–321.

Latt, S. A. (1977). *J. Supramol. Biol.* (in press).

Latt, S. A., Stetten, G., Juergens, L. A., Buchanan, G. R., and Gerald, P. S. (1975a). *Proc. Natl. Acad. Sci. U.S.A.* **72**, 4066–4070.

Latt, S. A., Stetten, G., Juergens, L. A., Willard, H. F., and Scher, C. D. (1975b). *J. Histochem. Cytochem.* **23**, 493–505.

Lilly, L. J., Bahner, B., and Magee, P. N. (1975). *Nature (London)* **258**, 611–612.

Lin, M. S., and Alfi, O. S. (1976). *Chromosoma* **57**, 219–225.

Lin, M. S., Alfi, O. S., and Donnell, G. N. (1976). *Can. J. Genet. Cytol.* **18**, 545–547.

Lubs, H. A., McKenzie, W. H., Patil, S. R., and Merrick, S. (1973). *Methods Cell Biol.* **6**, 346–380.

McKusick, V. A. (1976). *Cytogenet. Cell Genet.* **16**, 6.

Madan, X., Allen, J. W., Gerald, P. S., and Latt, S. A. (1976). *Exp. Cell Res.* **99**, 438–444.

Mendelsohn, M. L., and Takahashi, M. (1971). *In* "The Cell Cycle and Cancer" (R. Baserga, ed.), pp. 6–26. Dekker, New York.

Nicolini, C. (1975). *J. Natl. Cancer Inst.* **55**, 821–826.

Pathak, S., Stock, A. D., and Lusby, A. (1976). *Experientia* **31**, 916–918.

Pera, F., and Mattias, P. (1976). *Chromosoma* **37**, 13–18.

Perry, P., and Evans, H. J. (1975). *Nature (London)* **258**, 121–125.

Perry, P., and Wolff, S. (1974). *Nature (London)* **251**, 156–158.

Rudiger, H. W., Kohl, F., Mangels, W., von Wichert, P., Bartram, C. R., Wohler, W., and Passarge, E. (1976). *Nature (London)* **262**, 290–292.

Schneider, E. L., Stanbridge, E. J., Epstein, C. J., Golbus, M., Abbo-Halbasch, G., and Rodgers, G. (1974). *Science* **184**, 477–479.

Schneider, E. L., Chaillet, J., and Tice, R. (1976). *Exp. Cell Res.* **100**, 396–399.

Schneider, E. L., Sternberg, H., and Tice, R. R. (1977). *Proc. Natl. Acad. Sci. U.S.A.*, **74**, 2041–2044.

Shackney, S. E. (1975). *J. Natl. Cancer Inst.* **55**, 827–829.

Solomon, E., and Bobrow, M. (1975). *Mutat. Res.* **30**, 273–278.

Steel, G. G. (1972). *Cell Tissue Kinet.* **5**, 87–100.

Steel, G. G., and Hanes, S. (1971). *Cell Tissue Kinet.* **4**, 93–105.

Stetka, D. G., and Wolff, S. (1976a). *Mutat. Res.* **41**, 333–342.

Stetka, D. G., and Wolff, S. (1976b). *Mutat. Res.* **41**, 343–350.

Stubblefield, E. (1975). *Chromosoma* **53**, 209–221.

Sugiyama, T., Goto, K., and Kano, T. (1976). *Nature (London)* **259**, 59–60.

Taylor, J. H. (1958). *Genetics* **43**, 515–529.

Taylor, J. H., Woods, P. S., and Hughes, W. L. (1957). *Proc. Natl. Acad. Sci. U.S.A.* **53**, 122–127.

Thorne, P., Schneider, E. L., and Tice, R. R. (1978). In preparation.

Tice, R. R. (1976). Ph.D. Thesis.

Tice, R. R. (1978a). In preparation.

Tice, R. R. (1978b). In preparation.

Tice, R. R., Chaillet, J., and Schneider, E. L. (1975). *Nature (London)* 251, 70–72.

Tice, R. R., Schneider, E. L., and Raiy, J. M. (1976). *Exp. Cell Res.* **102**, 232–236.

Tice, R. R., Kram, D., and Schneider, E. L. (1978). In preparation.

Trosko, J. E., Chu, E. H. Y., and Carrier, W. L. (1965). *Radiat. Res.* **24**, 667–671.

Vogel, W., and Bauknecht, T. (1976). *Nature (London)* **260**, 448–449.

Wang, H. C. (1976). *Chromosoma* **58**, 225–261.

Willard, H. F., and Latt, S. A. (1976). *Am. J. Hum. Genet.* **28**, 213–227.

Wolff, S. (1977). *J. Supramol. Struct., Suppl.* **1**, 100.

Wolff, S., and Perry, P. (1974). *Chromosoma* **48**, 341–353.

Wolff, S., and Perry, P. (1975). *Exp. Cell Res.* **93**, 23–30.

Wolff, S., Bodycote, J., and Painter, R. B. (1974). *Mutat. Res.* **25**, 73–81.

Wolff, S., Bodycote, J., Thomas, G. H., and Cleaver, J. E. (1975). *Genetics* **81**, 349–355.

Wolff, S., Rodin, B., and Cleaver, J. E. (1977). *Nature (London)* **265**, 347–349.

Zakharov, A. F., and Egolina, N. A. (1972). *Chromosoma* **38**, 344–365.

Chapter 26

Preparation and Characterization of Mitochondria and Submitochondrial Particles of Rat Liver and Liver-Derived Tissues

PETER L. PEDERSEN, JOHN W. GREENAWALT[1]
BALTAZAR REYNAFARJE, JOANNE HULLIHEN,
GLENN L. DECKER, JOHN W. SOPER, AND
ERNESTO BUSTAMENTE[2]

Laboratory for Molecular and Cellular Bioenergetics,
Department of Physiological Chemistry,
Johns Hopkins University School of Medicine,
Baltimore, Maryland

[1] Deceased.

[2] Present address: Department of Physiological Sciences, Universidad Peruana Cayetano Heredia, Lima, Peru.

I. Purpose

This chapter deals specifically with techniques for the preparation and characterization of mitochondrial and submitochondrial fractions from liver and liver-derived tissues. It was written initially so that all incoming students and postdoctoral fellows at The Johns Hopkins School of Medicine interested in bioenergetics would have immediate access to the exact procedure for mitochondrial preparation and characterization developed for routine day-to-day use in our laboratories. After writing this chapter, however, we felt that the procedures included herein might be useful to other investigators in the field as well. To the best of our knowledge this article represents the most detailed account of preparative procedures available for mitochondria and submitochondrial particles of rat liver.

We believe this chapter will be useful to a broad spectrum of investigators. First, it should be useful not only to new students of bioenergetics but to senior investigators as well. It is hoped that such investigators will realize that mitochondrial and submitochondrial preparations from liver are frequently more intact, more well defined, and more directly applicable to certain biochemical studies than mitochondrial preparations from other tissues. Second, it should be useful to laboratory technicians. Methods are deliberately described in a straightforward, stepwise manner which we believe leaves very little to the imagination. Moreover, flow diagrams are included with many of the methods, so that one can obtain an overall view of a given procedure and what it entails prior to carrying out the individual steps.

II. Advantages of Using Rat Liver Mitochondria for Studying Mitochondrial Biochemistry and Bioenergetics

Except in cases in which it may be desirable to purify large quantities of a mitochondrial component, rat liver mitochondrial systems offer numerous advantages over other types of mitochondria for studying mitochondrial bioenergetics. Some of the more important advantages are:

1. Availability. Rats are readily available from numerous animal farms throughout the world. They are easy to maintain and can be kept near the laboratory so that fresh liver mitochondria can be prepared at any time.

2. Rapidity of preparation. Mitochondria can be prepared from rat liver in high yield within 1–2 hours. The use of meat grinders, proteases, and other potentially mitochondrially damaging techniques is not required to break down the plasma membrane of liver, as is the case for many other tissues.

3. A fairly homogeneous population of liver mitochondria can be obtained by standard techniques. It is not necessary to separate a large light layer from a heavy layer, as is the case for heart mitochondria.

4. Intactness. Mitochondria prepared from rat liver are morphologically and biochemically intact. The outer and inner membranes suffer no obvious damage during preparation, and such mitochondria exhibit high acceptor control ratios (ACRs).

5. Ease of subfractionation. The four components of mitochondria (outer membrane, inner membrane, intracristal space, and matrix) are prepared readily.

6. Availability of reliable assays. Details of many enzymic and energy-linked assays have been well worked out for rat liver mitochondria.

7. Applicability as controls for neoplastic mitochondria. In studies of neoplastic tissue, liver mitochondria serve as ideal controls for hepatoma mitochondria.

8. Applicability as controls for other rapidly growing tissues. The rat can be subjected to a partial hepatectomy which results in rapid liver regeneration. Therefore mitochondria prepared from regenerating rat liver can be used as controls for mitochondrial studies involving rapidly proliferating tissue. As a case in point, alterations found in hepatoma mitochondria may be either a consequence of the normal-to-neoplastic cell transformation process or of rapid cell proliferation. Studies of mitochondria isolated from regenerating liver cells aid the investigator in distinguishing between these two possibilities.

9. Applicability to studies involving alterations in metabolic state. In studies in which alterations in metabolic state are induced the rat and the liver therefrom frequently provide the ideal animal and tissue of choice, respectively. For example, the rat can be readily utilized to study the effects on animals of starvation, diabetes, hormones, drugs, alcohol, carcinogens, radiation, and various toxic substances. Moreover, since the liver is a generalized organ providing the principal site of metabolic "crossroads," the possibility that these conditions or substances may alter the biochemistry of liver subcellular organelles such as mitochondria warrants serious consideration.

10. Applicability to tissue culture studies. When it is experimentally desirable, liver cells and hepatoma cells can be readily grown in tissue culture. Therefore direct effects of substances on the cells and on their mitochondrial properties can be observed without having to deal with the whole organism.

The rat liver system has been avoided in the past for many molecular studies involving the mechanisms of ATP synthesis and utilization and the regulation of these processes, partly because the mitochondrial ATPase complex was not available in purified form from this tissue. This shortcoming has now been overcome, and in recent years reliable procedures have been worked out for preparing from rat liver mitochondria F_1 ATPase in purified form (Catterall and Pedersen, 1971; Lambeth and Lardy, 1971), ATPase inhibitor peptide(s) in purified form (Chan and Barbour, 1976; Cintron and Pedersen, 1978), and a partially purified preparation of oligomycin-sensitive ATPase or OS-F_1 (Soper and Pedersen, 1976).

III.   Introductory Comments regarding Methods for Preparing Rat Liver Mitochondria

Today most investigators studying properties of rat liver mitochondria use one of two types of mitochondrial preparations: (1) mitochondria prepared by differential centrifugation, (2) mitochondria prepared by differential centrifugation and further purified either by gradient centrifugation or by mild detergent treatment.

Investigators carrying out studies on the energy-linked functions of mito-chondria usually work with mitochondria of type 1. Such mitochondria can be prepared in a short period of time by simple differential centrifuga-tion techniques. Under the electron microscope positively stained pre-parations of these mitochondria appear intact. The inner and outer membranes are continuous, and the matrix appears dense. However, mitochondria prepared in this way are by no means homogenous. They are often contaminated with lysosomes, endoplasmic reticulum, plasma mem-brane, and perhaps with small amounts of other cellular components as well. They should be used with caution in metabolic and quantitative studies involving lipids, nucleic acids, and carbohydrates.

Investigators interested in biochemical aspects of mitochondria not directly involving energy-linked functions frequently prefer to work with type-2 mitochondria. Type-2 mitochondria are especially recommended for studies in which there may be some question about whether a given process, enzyme, or component is localized in mitochondria; for studies involving the quantification of such components; and for studies of metabolic pro-cesses that do not take place exclusively in the mitochondria, e.g., protein and nucleic acid synthesis.

The first part of this chapter summarizes several methods for isolating rat liver mitochondria. Some of these methods yield type-1 mitochondria (partially purified but intact), and some yield type-2 mitochondria (highly purified).

IV.   Mitochondrial Isolation Media

To date, investigators working in the field of mitochondrial bioenergics and biochemistry have not agreed upon an isolation medium which is superior to all others. Depending upon the source from which the mito-chondria are to be prepared the choice of an isolation medium may vary. This section deals only with isolation media which have been used to prepare rat liver mitochondria. As noted in Table I, some isolation media are simple and contain only sucrose or buffered sucrose. Other media are more complex. The only common property of the various media is the tonicity. With the exception of the 0.88 M sucrose medium, they are all isotonic or nearly isotonic with the mitochondrial matrix which has been estimated to be 230–300 mOsM (Goyer and Krull, 1969).

For much of our mitochondrial work we have abandoned the traditionally favored 0.25 M sucrose solution introduced by Schneider (1948) and

TABLE I

Preparation Media for Rat Liver Mitochondria

Medium	pH[a]	Comments	Reference
0.88 M Sucrose	Unbuffered	The sucrose concentration is hypertonic relative to the tonicity of the mitochondrial matrix. Some damage to enzyme complexes involved in fatty acid oxidation results. Higher centrifugal forces are required to sediment mitochondria than when isotonic sucrose is used.	Schneider and Hogeboom, 1950
0.25 M Sucrose	Unbuffered	The sucrose concentration is isotonic (or nearly isotonic) relative to the tonicity of the mitochondrial matrix. Less damage results to large enzyme complexes, and lower centrifugal forces can be used to sediment mitochondria than when 0.88 M sucrose is used. An apparent problem with the preparation method using this media is that a large fluffy layer sometimes results.	Schneider, 1948; Schneider and Hogeboom, 1950
0.25 M Sucrose, 3.4 mM Tris–Cl, 1.0 mM EGTA	7.4 at 20°C	EGTA chelates Ca^{2+}, an uncoupler of oxidative phosphorylation. EGTA is advised over EDTA because EDTA extracts Mg^{2+} from the mitochondrial membrane, rendering it leaky to K^+ ions.	Chappell and Hansford, 1972
0.25 M Sucrose, 2.0 mM HEPES, 1.0 mM EGTA	7.4 at 0°C	Basically the same as the medium above except HEPES is preferred as a buffer. Mitochondria prepared in media buffered with Tris have been shown to exhibit lower ACRs than mitochondria prepared in media buffered with HEPES. This may be because Tris, unlike negatively charged buffers, is taken up by mitochondria in response to the membrane potential.	Good et al., 1966, Spencer and Bygrave, 1972

416

	pH		
Solution A (0.33 M sucrose, 1.0 mM EDTA, 0.02 M Tris–Cl) Solution B (0.175 M KCl, 1.0 mM EDTA, 0.02 M Tris–Cl buffer)	7.4 7.4	The first low-speed centrifugation step and the first mitochondrial centrifugation step are carried out in solution A. All subsequent mitochondrial washes are carried out in solution B.	Skidmore and Catravas, 1970
0.2 M D-Mannitol	CO_2-free solution	None	Johnson and Lardy, 1974 Schnaitman and Pedersen, 1968; Schnaitman and Greenawalt, 1968; Pedersen and Schnaitman, 1969; Chan et al., 1970; Pedersen et al., 1970; Pedersen and Morris, 1974
220 mM D-Mannitol, 70 mM sucrose, 2.0 mM HEPES, 0.5 gm/liter BSA (H-medium)	7.4 with KOH at 0°C	This medium is ideal for preparing intact mitochondria from liver, regenerating liver, and hepatoma tissues. Also, a more intact mitoplast (inner membrane–matrix) fraction is obtained when H-Medium is used than when 0.25 M sucrose media are used.	

[a]It should be noted that in the range of pH from 7 to 8 the pH of Tris buffers changes much more with temperature than the pH of HEPES buffers. For example, a 50 mM solution of Tris–Cl of pH 7.4 at 25°C has a pH of about 7.8 at 0°C. HEPES buffer stays essentially constant over this temperature range.

Schneider and Hogeboom (1950) and use instead what we refer to as H-medium (Schnaitman and Pedersen, 1968; Schnaitman and Greenwalt, 1968), a HEPES-buffered medium containing sucrose, mannitol, and defatted albumin (Table I). Our main reason for favoring this medium is its wide applicability to liver and liver-derived systems. Thus we have found that liver (Schnaitman and Pedersen, 1968; Schnaitman and Greenawalt, 1968), regenerating liver (Pedersen and Morris, 1974), and many hepatoma mitochondria (Pedersen and Morris, 1974; Pedersen *et al.*, 1970) prepared in H-medium appear ultrastructurally and morphologically intact and exhibit high ACRs (See Section XVI,B,1). In addition, more biochemically and ultrastructurally intact mitoplast fractions can be prepared in our hands when H-medium is the isolation medium than when 0.25 M sucrose is used (Schnaitman and Pedersen, 1968; Schnaitman and Greenawalt, 1968; Pedersen and Schnaitman, 1969; Chan *et al.*, 1970; Decker and Greenawalt, 1977).

No claim is made here that H-medium yields more physiologically intact mitochondria than other media, or that H-medium is the best medium to use for all studies involving mitochondria. In fact, we recommend that each investigator decide by direct comparison which preparation medium yields mitochondria which optimally suit his or her specific studies. If this recommendation were followed we might ultimately obtain some indication of the preparation medium which yields what might more appropriately be regarded as physiologically intact mitochondria. In addition, we might obtain clearer insight into the rationale underlying preferences for a certain medium in a specific study.

V.   Preparation of Rat Liver Mitochondria

A.   In 0.25 M Sucrose

Comment: This procedure is patterned after the original procedure described by Schneider and Hogeboom (1950). All operations are carried out at 0–4°C. The yield of mitochondria prepared by this method is 200–250 mg/10 gm wet weight liver. Such mitochondria have an ACR in the presence of succinate of about 5 when a respiration medium containing KCl, KP_i, and HEPES is used (see Table III). See Fig. 1 for a flow diagram.

Equipment: Potter–Elvehjem glass homogenizer and Teflon pestle with a diameter 2.5 to 3 mm smaller than the internal diameter of the tube; Sorvall RC 2-B centrifuge with a SS-34 rotor (10.8-cm radius), or equivalent centrifuge.

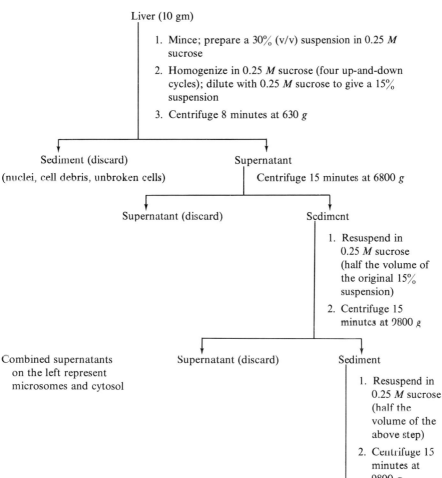

FIG. 1.   Flow diagram for the preparation of rat liver mitochondria by differential centrifugation in 0.25 M sucrose.

Solution: Freshly prepared ice-cold sucrose.

Method:

1. Sacrifice a rat (150–200 gm) by cervical decapitation and allow it to bleed.

2. Remove the liver (10–15 gm) and place it in a beaker containing ice-cold sucrose (about 30 ml). Rinse once or twice with small volumes of sucrose and leave the beaker on ice. Estimate the weight of the liver during this step.

3. Mince the liver with scissors as finely as possible and make a 30% suspension with the ice-cold 0.25 *M* sucrose. Transfer the suspension to the homogenizer.

4. Attach the pestle to a motor and homogenize by applying four complete cycles, where one cycle equals the movement of the pestle completely down and up through the suspension. (The pestle rotates at about 400 rpm.)

5. Dilute the homogenate to twice the initial volume (∼ 15% v/v) and centrifuge at 2250 rpm (630 *g*) for 8 minutes.

6. Carefully decant the supernatants so as to avoid including particulate material. (It is advisable to use a pasteur pipet.).

7. Centrifuge the supernatant at 7500 rpm (6800 *g*) for 15 minutes. Decant and discard the supernatant and most of the fluffy (very loosely packed) material.

8. Add 0.5 ml of sucrose to the sediment and gently homogenize with a "cold finger" (a test tube filled with ice). Resuspend the sediment in half the volume used in step 5 (the same volume as that used in step 3).

9. Centrifuge at 9000 rpm (9800 *g*) for 15 minutes. Discard the supernatant and resuspend the sediment in half the volume of sucrose used in step 3.

10. Centrifuge at 9000 rpm (9800 *g*) for 15 minutes. Discard the supernatant and resuspend the sediment to give a stock suspension containing 50 mg protein per milliliter in 0.25 *M* sucrose.

B. In H-Medium

1. Fast, Low-Yield Procedure

Comment: The following method is essentially identical to the procedure described by Schnaitman and Greenawalt (1968). The yield of mitochondria prepared by this procedure is 150–200 mg/10 gm wet weight liver. Such mitochondria have ACRs in the presence of succinate of 4 to 8 when the respiration medium described by Schnaitman and Greenawalt (1968) is employed (see Table III). For many studies it seems important to discard preparations with an ACR of < 4. Under the electron microscope mitochondria prepared in this fashion appear morphologically and ultra-structurally intact (Fig. 16A). All operations are carried out at 0–4°C. See Fig. 2 for a flow diagram.

Equipment: Same as for the 0.25 *M* sucrose procedure.

Solution: H-Medium, ice-cold. (Commercial sources of ingredients are those currently used in our laboratories. Components from other sources may work equally well.) Adjust pH to 7.4 at 0°C with KOH.

Component	Grams/liter	Concentration
Sucrose (Baker)	23.9	0.7 *M*
D-Mannitol (Schwarz/Mann)	38.3	0.21 *M*
HEPES (Sigma)	0.5	0.002 *M*
BSA (ICN, Fraction V, defatted)	0.5	0.05% (w/v)

Method:

1. Sacrifice a rat by cervical decapitation. After the cessation of bleeding, remove the liver (10–15 gm), dissect off the fat, and immediately put it into a small beaker containing ice-cold H-medium (2 ml/gm liver). Decant most of the medium and leave the beaker on ice.

2. Mince the liver into small pieces with scissors and rinse two times with H-medium as above.

3. Make a 30% (v/v) suspension (1 volume liver mince and 2 volumes H-medium).

4. Homogenize in a glass homogenizer (55-ml capacity), applying four up-and-down cycles with a Teflon pestle having multiple radial serrations. The pestle is attached to a motor and operates at ∼1500 rpm. (This step is extremely critical. Usually, when mitochondria are prepared with low acceptor control, it is because the homogenization process has been "overdone".)

5. Dilute the homogenate to 10% v/v (1 volume liver mince and 9 volumes H-medium) and immediately centrifuge 15 minutes at 660 *g* in the Sorvall SS-34 rotor (RC 2-B centrifuge) or in a comparable centrifuge.

6. Decant the supernatant very carefully into ice-cold Sorvall tubes so that no loosely packed material is decanted. (Removing the supernatant with a pasteur pipet is slower but leads to less nuclear contamination in the final mitochondrial sediment.)

7. Centrifuge for 15 minutes at 6800 *g* in the Sorvall SS-34 head. Decant the supernatant.

8. Gently homogenize the sediments into a paste with a "cold finger" (a test tube filled with ice) and a small amount of H-medium. Resuspend in

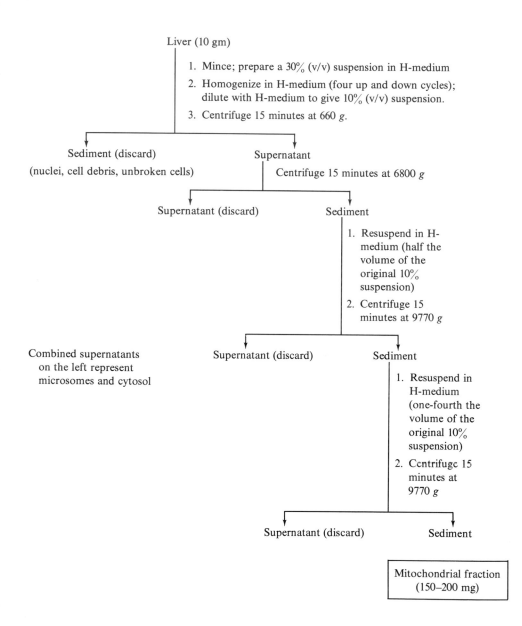

Liver (10 gm)

1. Mince; prepare a 30% (v/v) suspension in H-medium
2. Homogenize in H-medium (four up and down cycles); dilute with H-medium to give 10% (v/v) suspension.
3. Centrifuge 15 minutes at 660 *g*.

Sediment (discard)
(nuclei, cell debris, unbroken cells)

Supernatant
Centrifuge 15 minutes at 6800 *g*

Supernatant (discard)

Sediment

1. Resuspend in H-medium (half the volume of the original 10% suspension)
2. Centrifuge 15 minutes at 9770 *g*

Combined supernatants on the left represent microsomes and cytosol

Supernatant (discard)

Sediment

1. Resuspend in H-medium (one-fourth the volume of the original 10% suspension)
2. Centrifuge 15 minutes at 9770 *g*

Supernatant (discard)

Sediment

Mitochondrial fraction
(150–200 mg)

FIG. 2. Flow diagram for the preparation of rat liver mitochondria by differential centrifugation in H-medium (fast, low-yield procedure).

half the original volume used in step 5 and centrifuge in the Sorvall SS-34 head for 15 minutes at 9770 g.

9. Decant the supernatant and carefully rinse off the fluffy layer. Resuspend in one-fourth the original volume used in step 5. Again centrifuge in the Sorvall SS-34 rotor for 15 minutes at 9770 g. Decant the supernatant.

10. Suspend the mitochondria in H-medium using a loose-fitting homogenizer. Dilute if necessary to give exactly 100 mg/ml biuret protein.

2. Fast Procedure Involving Nagarse

Comment: This procedure is essentially identical to the procedure described by Kaschnitz *et al.* (1976). It avoids the use of a homogenizer and relies upon the protease Nagarse (Nagase and Company, Ltd., Osaka, Japan) to break down the cell membranes. The yield of mitochondria with this method is ~10 mg protein/gm liver. Such mitochondria have ACRs considerably greater than 10 when succinate or β-hydroxybutyrate is the substrate, and measurements are made in the medium described by Schnaitman and Greenawalt (1968). All operations are carried out at 0–4°C. See Fig. 3 for a flow diagram.

Solutions: H-medium (220 mM mannitol, 70 mM sucrose, 2 mM HEPES, pH 7.4, at 25°C with KOH); H-medium with bovine serum albumin (BSA) (1 and 2% w/v) (BSA is Cohn Fraction V for Sigma). [Other sources of BSA may be suitable. This source noted is the one preferred by Kaschnitz *et al.* (1976a).]

Method:

1. Anesthetize two to four rats with ether.

2. Remove the livers 10–15 gm each) and wash several times with H-medium.

3. Mince the tissue thoroughly with scissors, suspend at 2 ml/gm tissue in H-medium, and pulp with three strokes in a loosely fitting homogenizer.

4. Add 1 mg Nagarse per gram of tissue. Stir vigorously for 20 minutes in the cold using a magnetic stirrer. (*Note*: Prolonged treatment of liver tissue in the presence of Nagarse may result in mitochondrial damage. Therefore it is important to adhere to a 20-minute (or less) time period.)

5. Add 2 ml/gm tissue H-medium containing 2% w/v BSA.

6. Centrifuge the homogenate at 3200 g for 1 minute (average) and discard the sediment.

7. Centrifuge the supernatant at 17,000 g for 2 minutes.

8. Rinse the sediment thoroughly with H-medium containing 1% BSA to remove any fluffy layer and resuspend *by hand* in a glass homogenizer in 2 ml of the same solution per gram of sediment.

9. Centrifuge the suspension and repeat the rinsing procedure.

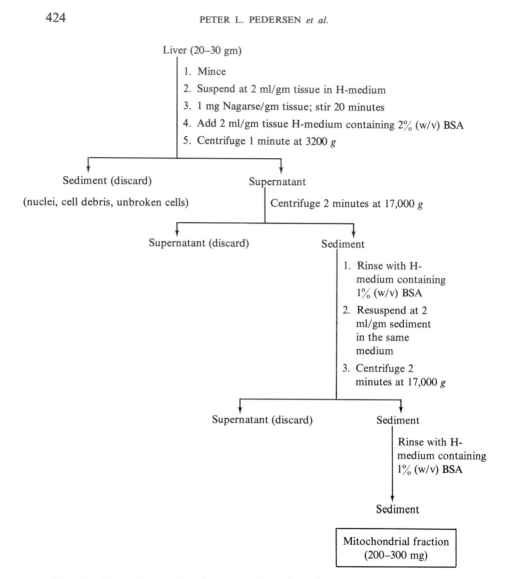

FIG. 3. Flow diagram for the preparation of rat liver mitochondria by differential centrifugation in H-medium [fast procedure of Kaschnitz *et al.* (1976) involving Nagarse].

10. Suspend the mitochondria at 20–50 mg/ml in H-medium containing 1% BSA.

3.   HIGH-YIELD PROCEDURE

Comment: This method can be used when it is desirable to obtain maximal yields of mitochondria from a given amount of starting liver. The procedure is designed for preparing large amounts of mitochondria for subsequent enzyme purification work and is essentially identical to the procedure described by Bustamante *et al.* (1977). The yield of mitochondria obtained is about 360–400 mg/10 gm wet weight liver. This method results in a higher yield of mitochondria than the procedures described in Section V,A and B,1 and 2 because the nuclear fraction is resuspended and washed several times in analogy to the method reported earlier by Johnson and Lardy (1967). Under the electron microscope these mitochondria appear untrastructurally intact and seem to be free of obvious contamination by extramitochondrial fractions. Similar to the low-yield preparation described in Section V,B,1, they have ACRs with succinate of 4 to 8 when measurements are made in the medium described by Schnaitman and Greenawalt (1968) (see Table III). All operations are carried out at 0–4°C. See Fig. 4 for a flow diagram.

Equipment and solution: Same as for the procedure in Section V,B,1.

Method:

1. Sacrifice a retired breeder rat ($\sim$700 gm per rat) by cervical decapitation. (Retired breeder rats provide a source of large livers.) After bleeding, remove the liver (15–20 gm) and place it in ice-cold H-medium.

2. Remove fat and connective tissue, blot dry, and weigh in a tared glass beaker.

3. Immediately cover the liver with H-medium, mince, and homogenize in a total volume of about three times the weight of the liver.

4. Centrifuge the homogenate in 50-ml tubes for 3 minutes at 3000 rpm (1100 g) in the Sorvall SS-34 rotor (RC 2-B centrifuge).

5. Save the supernatant, resuspend the sediment to the original volume, and centrifuge as in step 4.

6. Repeat step 5 two additional times.

7. After the fourth centrifugation step, pool all the supernatants and discard the sediment.

8. Centrifuge the supernatant fractions at 7500 rpm (6780 g) for 15 minutes in 50-ml tubes.

9. Resuspend the sediment in half the volume recorded in step 7. Centrifuge for 10 minutes at 13,000 rpm (20,200 g) in 50-ml tubes.

10. Resuspend the sediment from step 9 in one-fourth the volume recorded in step 7. Centrifuge at 13,000 rpm (20,200 g) for 15 minutes in 50-ml tubes.

11. Resuspend the sediment from step 10 in one-eighth the volume recorded in step 7. Centrifuge at 5000 rpm (3000 g) for 3 minutes in 50-ml tubes. Resuspend the sediment in one-sixteenth the volume recorded in step 7. Repeat the process. Discard the final low-speed sediment.

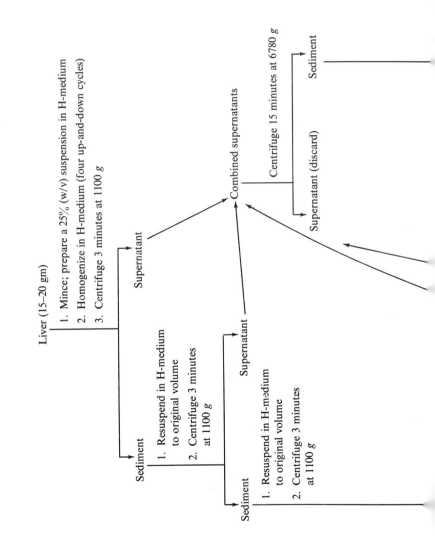

Liver (15–20 gm)

1. Mince; prepare a 25% (w/v) suspension in H-medium
2. Homogenize in H-medium (four up-and-down cycles)
3. Centrifuge 3 minutes at 1100 g

Supernatant

Sediment

1. Resuspend in H-medium to original volume
2. Centrifuge 3 minutes at 1100 g

Supernatant

Sediment

1. Resuspend in H-medium to original volume
2. Centrifuge 3 minutes at 1100 g

Combined supernatants

Centrifuge 15 minutes at 6780 g

Supernatant (discard)

Sediment

426

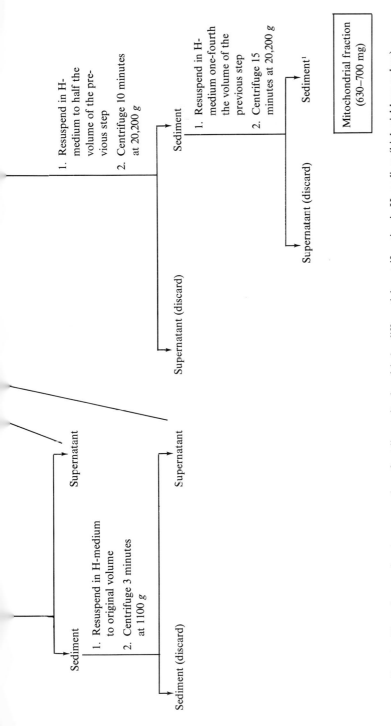

Sediment

1. Resuspend in H-medium to original volume
2. Centrifuge 3 minutes at 1100 g

Sediment (discard)

Supernatant

1. Resuspend in H-medium to half the volume of the previous step
2. Centrifuge 10 minutes at 20,200 g

Supernatant (discard)

Sediment

1. Resuspend in H-medium one-fourth the volume of the previous step
2. Centrifuge 15 minutes at 20,200 g

Supernatant (discard)

Sediment[1]

Mitochondrial fraction (630–700 mg)

FIG. 4. Flow diagram for the preparation of rat liver mitochondria by differential centrifugation in H-medium (high-yield procedure).

[1]If it is desirable to remove the fluffy layer, steps 11–14 in the text should be carried out.

427

12. Centrifuge the supernatants from step 11 for 20 minutes at 13,000 rpm. Gently rinse off the fluffy layer with H-medium. Resuspend in one-eighth the volume recorded in step 7.

13. Centrifuge the suspension from step 12 at 13,000 rpm (20,200 g) for 20 minutes. Gently rinse off the fluffy layer with H-medium and discard.

14. Suspend the mitochondria at 50 mg protein per milliliter in H-medium.

VI.   Additional Purification of Mitochondria Prepared by Differential Centrifugation

A.   Treatment with low Concentration of Digitonin

Comment: This procedure is essentially that of Lowenstein *et al.* (1970). It takes advantage of the fact that very low levels of digitonin preferentially lyse lysosomes. Subsequent to mild digitonin treatment the mitochondria can be sedimented, presumably leaving the lysosomal membranes (and lysosomal matrix material) in the supernatant. Greater than 90% of the acid phosphatase activity is removed from the mitochondrial fraction by this procedure. The ACR of digitonin-treated mitochondria is unaffected compared to that of untreated mitochondria. All operations are carried out at 0–4°C.

Equipment: Any refrigerated centrifuge which will sediment mitochondria at 20,000 g.

Solution: Ice-cold 0.25 M sucrose, 1 mM Tris–Cl, pH 7.5.

Method:

1. Prepare mitochondria by any of the procedures previously described in this chapter. (Record the weight of the starting liver.)

2. Suspend the mitochondria at 20 mg/ml in 0.25 M sucrose, 1 mM Tris–Cl, pH 7.5.

3. Add 1 ml sucrose–Tris (the same concentration as in Step 2) containing 1.75 mg digitonin per gram wet weight of original liver.

4. Shake for 2 minutes in an ice bath.

5. Add 3 volumes sucrose–tris solution.

6. Centrifuge the mitochondria at 19,000 g for 2 minutes and wash twice with the same medium.

7. Resuspend at 50 mg/ml in the sucrose–Tris medium.

B. Isopycnic Sedimentation in a Continuous Sucrose Gradient

Comment: This method (E. Bustamante and P. L. Pedersen, unpublished results) can be used for further purification of certain types of hepatoma (e.g., Morris 5123 and 7800) mitochondria prepared by differential centrifugation. It results in the removal of about 10–20% of nonmitochondrial protein [as assessed by succinic dehydrogenase assays (Schnaitman and Greenawalt, 1968)] and yields mitochondria which retain acceptor control of respiration [ACR = 3–5 when succinate is the substrate and the respiration medium described by Schnaitman and Greenawalt (1968) is employed]. This procedure, however, does not completely remove lysosomes. The technique can also be used to purify liver mitochondria, however, under these conditions liver mitochondria, unlike certain hepatoma mitochondria, do not retain acceptor control of respiration. The greater stability of certain hepatoma mitochondria during sedimentation has been discussed by Wattiaux (Section VI,F).

Equipment: Device for preparing gradients; Büchler peristaltic pump; Beckman preparative ultracentrifuge with SW-50.1 rotor; cellulose nitrate tubes (5 ml); apparatus for making density gradients; refractometer; 18-gauge needle.

Solutions: 20 and 55% (w/w) sucrose; mineral oil.

Method:

1. Using an appropriate setup for mixing and layering continuous gradients, carefully make a 20% (top) to 55% (bottom) sucrose gradient in 5-ml centrifuge tubes. [This can be done by starting with a mixing chamber containing 2.3 ml 55% (w/w) sucrose in one compartment and 2.3 ml 20% (w/w) sucrose in another compartment. The two compartments are initially separated from one another by a stopcock. The compartment containing the 55% sucrose contains a small stirring bar for rapid mixing and is connected with Tygon tubing to a Buchler peristaltic pump which pumps the sucrose into 5-ml Beckman cellulose nitrate tubes. With the stirring apparatus and the pump on, the stopcock is opened to allow the 20% sucrose to mix with the 55% sucrose, the resultant mix of which flows into the 5-ml cellulose nitrate tubes.]

2. Layer on top of the gradient 200 µl (~ 20 mg/ml) of mitochondrial suspension.

3. Layer mineral oil on top of the mitochondrial layer to the top of the tube to balance the tubes and prevent evaporation.

4. Place the cap on, and centrifuge at 45,000 rpm (230,000 g) in the Spinco Model L SW-50.1 rotor for 3 hours at 4°C.

5. Puncture the bottom of the tube with an 18-gauge needle and collect

about 10 equal fractions of 0.5 ml. [A very useful readily made device for puncturing tubes has been described by Englund (1971)].

6. Determine the density of sucrose by refractometry in each fraction.

7. Dilute each fraction very slowly to 10 ml with H-medium and sediment the mitochondria at 10,000 *g* in the Sorvall centrifuge. Resuspend the mito-chondrial sediment in H-medium to the protein concentration of choice. The purpose of this step is to remove the high-density sucrose and to equalize the tonicity of all fractions prior to assay. (Rapid dilution may cause osmotic shock to the mitochondria, resulting in membrane rupture and loss of matrix proteins. This can be avoided by slow dilution, e.g., by the addition of small aliquots of H-medium with mixing.)

8. Assay for succinic dehydrogenase (Schnaitman and Greenawalt, 1968) or cytochrome oxidase (Schnaitman *et al.*, 1967) activity to localize the fraction with the highest purity of mitochondria.

C. Sedimentation in a Discontinuous Sucrose Gradient

Comment: Glew *et al.* (1973) have used this procedure to further purify mitochondria prepared by differential centrifugation. The resultant mito-chondria have been used to study the binding of concanavalin A to rat liver mitocondria.

Equipment: Beckman preparative ultracentrifuge; Beckman cellulose nitrate tubes (1 × 3 inches); 18-gauge needle.

Solutions: Sucrose (54% w/w, 45% w/w, 39% w/w, and 20% w/w).

Method: The washed mitochondrial sediment obtained by differential centrifugation is resuspended in homogenizing medium (approximately 30 mg of protein per milliliter), and 1 ml is layered onto a discontinuous sucrose density gradient formed in the Beckman cellulose nitrate tubes (1 × 3 inches) by layering successively the following sucrose solutions: 3 ml of 54% w/w (density 1.25 gm/ml), 8.3 ml of 45% w/w (1.21 gm/ml), 8.3 ml of 39% w/w (1.17 gm/ml), and 8.3 ml of 20% w/w (1.08 gm/ml) and centrifuged at 63,500 *g* for 2 hours. This procedure results in the resolution of three cytochrome oxidase-containing fractions. Routinely, 70% of the cyto-chrome oxidase activity applied to the gradient is recovered from the 1.21 gm/ml shelf. Discontinuous sucrose gradients are collected by puncturing the bottom of the tube with an 18-gauge needle [or with the device described by Englund (1971)] and collecting 20 equal fractions.

D. Sedimentation in a Continuous Ficoll Gradient

Comment: This method has been used by Brown (1968) to separate mito-chondria, lysosomes, and peroxisomes. Similar to the digitonin method, it is a direct method which avoids the preliminary injection of Triton WR-1339 into rats in order to alter the banding density of the lysosomes (Wattiaux et al., 1963).

Equipment: B-XV, B-XXIII, or B-XXIX zonal rotors and appropriate centrifuge.

Method: A 95-ml sample of a 5% rat liver homogenate is applied to a 1-liter, 0–15% (w/v) linear Ficoll gradient containing 0.25 M sucrose throughout, with a 250-ml cushion of 45% (w/v) sucrose. [Ficoll contributes less than half of 1% of the total osmotic pressure. It is a high-molecular-weight polymer synthesized by the cross-linking reaction of epichlorohydrin and sucrose and can be obtained from Pharmacia (Piscataway, N.J.).] Centrifugation is carried out for 15 minutes at 10,000 rpm in either the B-XV, B-XXIII, or B-XXIX zonal rotor.

E.   Sedimentation in a Continuous Metrizamide Gradient

Comment: This method has been used by Aas (1973) to separate mitochondria and lysosomes. Metrizamide [2-(3 acetamido-5-N-methylacetamido-2,4,6-triodobenzamido)-2-deoxy-D-glucose) is very soluble in water or dilute buffer solutions and forms dense solutions of relatively low viscosity. (Metrizamide can be obtained from Nyegaard and Company, A/S, Oslo, Norway.) It is chemically inert and nonionic and is therefore potentially useful for isopycnic banding of unfixed biological macromolecules.

Method: Complete details of the method are not given in the preliminary reference (Aas, 1973). Centrifugation is carried out in a Spinco SW-27 rotor at 25,000 rpm. Mitochondria band at an isopycnic density between 1.20 and 1.25 gm/ml and require several hours to reach equilibrium. Lysosomes, however, reach equilibrium after about 1 hour and band at an isopycnic density between 1.15 and 1.20 gm/ml.

F.   Precautionary Measures to Prevent Mitochondrial Damage during Isopycnic Centrifugation

Wattiaux and his colleagues (1971; Wattiaux, 1974) have reported recently that, when mitochondria are subjected to isopycnic centrifugation, they may undergo disruption at high speeds (~39,000 rpm in the studies of Wattiaux et al. 1971; Wattiaux, 1974). Apparently, the hydrostatic pressure becomes too high for mitochondrial structure to remain intact. The centrifugation speed at which disruption occurs varies for different mitochondrial types. Some hepatoma (e.g., Morris hepatoma 16) mitochondria appear to be less susceptible to disruption by high centrifugal forces than rat liver mitochondria. Although the speed of centrifugation seems to be the primary factor to consider when carrying out gradient centrifugation of mitochondria, the reader should take note of the following statement of Wattiaux: "Our results illustrate that several (additional) factors are to be taken into account in isopycnic centrifugation experiments: The kind of rotor, the limits of the gradient, the temperature of centrifugation, and the tissue from which the organelles originate" (Wattiaux, 1974).

VII.   Preparation of Rat Hepatoma Mitochondria

A.   Procedures for Solid Tumors

Comment: The procedures described here yield intact mitochondria from a variety of Morris hepatomas (Pedersen and Morris, 1974; Pedersen *et al.*, 1970, 1971; Kaschnitz *et al.*, 1976, 1978), and from Novikoff hepatomas (Greenawalt, unpublished observations). Such mitochondria usually have ACRs of > 4 when succinate is the substrate and the assay medium described by Schnaitman and Greenawalt (1968) is used (see Table III). Although there has been some question as to whether or not these two procedures yield functionally identical mitochondria (Kaschnitz *et al.*, 1976), it now appears that mitochondria prepared by both techniques are quite similar (Kaschnitz *et al.*, 1978; P. L. Pedersen, unpublished results). They both exhibit acceptor control of respiration and high uncoupler-stimulated ATPase activity. Interestingly, hepatoma mitochondria frequently require the addition of ATP before the uncoupler to catalyze maximal ATPase activity (Kaschnitz *et al.*, 1976, 1978). This is not the case with normal or host liver mitochondria, or with mitochondria from regenerating liver (Pedersen and Morris, 1974).

Method:

a. Isolation of hepatomas.   Morris hepatomas are allowed to grow in the hind legs of Buffalo strain rats (Simson Laboratories, Gilroy, Calif.) for the following time periods after inoculation: hepatoma 9618A (6–9 months), hepatoma 16 (6 months), hepatoma 7794A (1.5 months), hepatoma 7800 (1 month), hepatoma 7777 (2 weeks), and hepatoma 3924A (1–2 weeks). Solid Novikoff hapatomas are grown intraperitoneally for 5–6 days in 150 gm Holtzman rats (Holtzman Company, Madison, Wis.). Immediately after the decapitation of two to four hepatoma-bearing rats, livers and hepatomas are removed. Hepatomas are cleaned of fat, connective tissue, and necrotic areas. Livers and hepatomas are placed on ice until all operations are complete and then transferred to breakers and rinsed twice with

100 ml of a medium containing 220 mM D-mannitol, 70 mM sucrose, 0.50 mg/ml defatted BSA, and 2 mM HEPES, pH 7.4.

b. Preparation of mitochondria.   Mitochondria can be prepared from Morris hepatomas by following one of the two methods previously discussed in this chapter: (1) Section V,B,1, the Schnaitman and Greenawalt procedure (1968); (2) Section V,B,2, the Kaschnitz *et al.* procedure (1976, 1978).

c. Additional notes.   For other techniques and general precautions involved in the preparation of mitochondria from certain solid hepatomas and other tumors, the reader should also consult the articles of Devlin (1967) and Sordall and Schwartz (1971). The latter investigators employ a modified Schneider–Hogeboom method (1950) in which the isolation medium consists of 0.25 M sucrose, 1 mM Tris–Cl, 1 mM EDTA, and 1% BSA, pH 7.2–7.4.

B.   Procedures for Ascites Tumors

1.   MITOCHONDRIA FROM ASCITES HEPATOMAS (As-30D AND NOVIKOFF)

Comment: This procedure was developed for the preparation of mitochondria from AS-30D ascites hepatoma cells but can also be applied for the preparation of mitochondria from Novikoff ascites cells. Although this method employs Nagarse to break down the cell wall, a Dounce homogenizer works equally well. Mitochondria prepared from AS-30D cells have ACRs of 3 to 5 when assayed in the medium of Schnaitman and Greenawalt (1968) (see Table III). See Fig. 5 for a flow diagram.

Animals: The AS-30D tumor cell line, an azo dye-induced line, was originally induced and carried intraperitoneally in Sprague-Dawley male albino rats (Smith *et al.*, 1970). In our hands this line is being carried in the Sprague-Dawley version of the Charles River Company. Animals of this strain are referred to as Charles River®rats. Two-month-old animals are preferred for easy handling.

Equipment: Swinging-bucket, low-speed centrifuge (e.g., International Model PR-1 centrifuge); centrifuge for sedimentation at forces as high as 25,000 g.

Solutions: NKT buffer (0.15 M NaCl, 5 mM KCl, 10 mM Tris–Cl, pH 7.4); STE buffer (0.25 M sucrose, 5 mM Tris–Cl, 1 mM EDTA, pH 7.4); H-medium (see Table I).

Method:

a. Isolation and transplantation of cells

1. Fluid collection.   Sacrifice by decapitation an animal that has been carrying the tumor for 7 days. With the aid of a 5-ml syringe and a 16-gauge

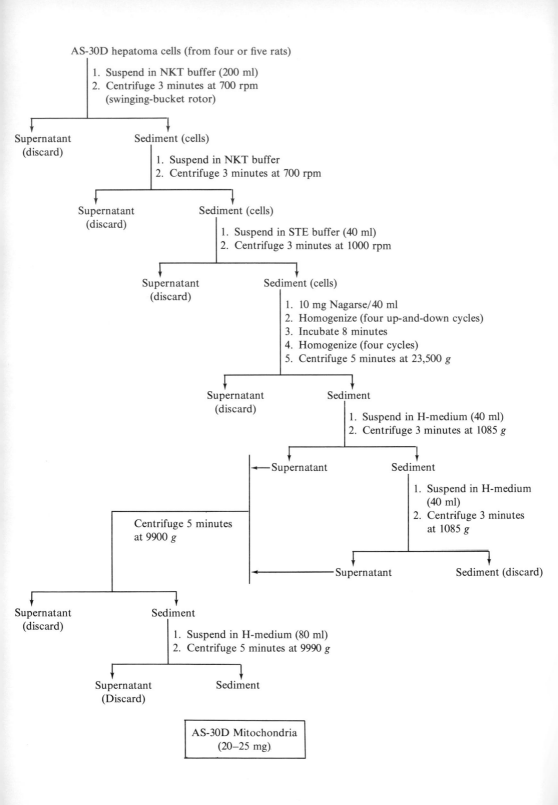

needle, extract about 10 ml of ascites fluid from the peritoneal cavity. Collect the fluid in a small container (heparin is not necessary).

2. *Fluid injection.* The injection is made with a 1-ml syringe and a 22-gauge needle. Inject 0.3 ml ascites fluid into the peritoneal cavity of each rat. It is recommended that five animals be injected each week, since the mortality rate of rats bearing this tumor is quite high after 7 days.

3. *Important notes.* It is necessary to check as frequently as possible that the animal's water supply is sufficient. This is essential because water is necessary for "normal" growth of this tumor. The rate of tumor growth is retarded in the absence of an adequate water supply.

b. *Preparation of mitochondria*

1. Collect the ascites fluid from four or five animals as indicated in step 1 of the isolation procedure.

2. Place in a flask containing 3 mg heparin.

3. Add ice-cold NKT buffer to 200 ml. (All additional operations are carried out at 0–4°C.)

4. Centrifuge for 3 minutes at 700 rpm in six 50-ml Sorvall tubes in a swinging-bucket rotor.

5. Remove the supernatant. The cells remain in the sediment.

6. Resuspend the cells in 200 ml NKT buffer.

7. Centrifuge as in step 4 and save the sediment (cells).

8. Combine the contents of the six tubes into two tubes, using STE buffer (~ 40 ml).

9. Centrifuge for 3 minutes at 1000 rpm in a swinging-bucket rotor.

10. Remove the supernatant and save the cells.

11. Add 10 mg Nagarse per 40 ml total volume of STE buffer.

12. Homogenize with a Potter–Elvehjem homogenizer with a loose-fitting pestle (four up-and-down cycles).

13. Allow to stand for 8 minutes.

14. Repeat the homogenization.

15. Centrifuge for 5 minutes at 14,000 rpm (23,500 g) in the SS-34 Sorvall rotor.

16. Remove the supernatant. The sediment is nuclei plus mitochondria.

17. Resuspend the sediment in 40 ml H-medium.

18. Centrifuge for 3 minutes at 3000 rpm (1085 g) in the SS-34 Sorvall rotor. Collect and save the supernatant.

19. Suspend the sediment in 40 ml H-medium.

20. Centrifuge at 3000 rpm (1085 g) as in step 18. Collect and save the supernatant.

FIG. 5. Flow diagram for the preparation of mitochondria from AS-30D hepatoma cells. The procedure can also be applied to Novikoff ascites hepatoma cells.

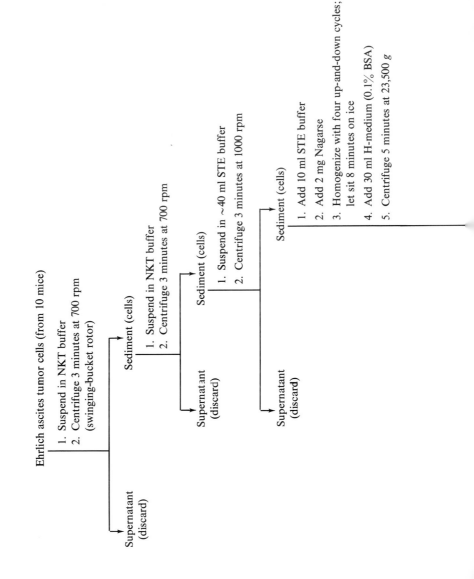

Ehrlich ascites tumor cells (from 10 mice)

1. Suspend in NKT buffer
2. Centrifuge 3 minutes at 700 rpm (swinging-bucket rotor)

Supernatant (discard)

Sediment (cells)

1. Suspend in NKT buffer
2. Centrifuge 3 minutes at 700 rpm

Supernatant (discard)

Sediment (cells)

1. Suspend in ~40 ml STE buffer
2. Centrifuge 3 minutes at 1000 rpm

Supernatant (discard)

Sediment (cells)

1. Add 10 ml STE buffer
2. Add 2 mg Nagarse
3. Homogenize with four up-and-down cycles; let sit 8 minutes on ice
4. Add 30 ml H-medium (0.1% BSA)
5. Centrifuge 5 minutes at 23,500 g

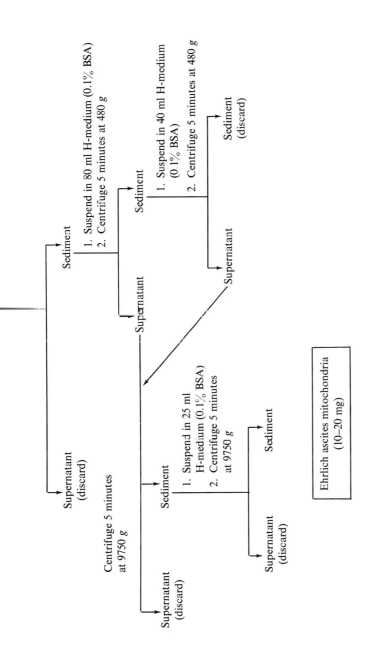

FIG. 6. Flow diagram for the preparation of mitochondria from Ehrlich ascites tumor cells.

21. Centrifuge the collected supernatants for 5 minutes at 9000 rpm (9990 g) in the SM-24 Sorvall rotor (outer row).

22. Resuspend the sediment in the same volume of H-medium.

23. Centrifuge as in step 21. Suspend the sediment in H-medium at 20 mg protein per milliliter or greater.

2. MITOCHONDRIA FROM EHRLICH ASCITES TUMOR CELLS

Comment: Although Ehrlich ascites cells are not derived from liver hepatocytes, the properties of mitochondria prepared from such cells are frequently compared to the properties of liver mitochondria. For this reason a procedure is included in this article for preparing mitochondria from Ehrlich ascites cells. The method described is essentially identical to the procedure described by Pedersen and Morris (1974) and should yield mitochondria with ACRs of 3.5 to 6 when succinate is the substrate and the respiration medium of Schnaitman and Greenawalt (1968) is used (see Table III). The yield of mitochondria from cells isolated from 10 mice is 10–20 mg. See Fig. 6 for a flow diagram.

Animals: Mice, CDI strain, males (20–25 gm) obtained from Charles River Breeding Laboratories, Wilmington, Mass.

Equipment: Swinging-bucket centrifuge for low-speed centrifugation (International Model PR-1 centrifuge); centrifuge which will sediment material at 25,000 g or greater (e.g., Sorvall RC-2B centrifuge).

Solutions: NKT buffer (150 mM NaCl, 5 mM KCl, 10 mM Tris–Cl, pH 7.3); STE buffer (0.25 M sucrose, 5.0 mM Tris–Cl, 1.0 mM EDTA, pH 7); H-medium (regular, see Table I); H-medium containing 0.1% defatted BSA, pH 7.4.

Method:

1. Remove the cells from ~10 mice, 7 days after inoculation, by cutting a small hole in the peritoneal cavity. Hold the mouse up with a pair of forceps and remove the cells with a pasteur pipet. Rinse the cavity with NKT buffer.

2. Place the cells from 10 mice in a 250-ml flask containing 3 mg heparin. Add NKT buffer to about 200 ml. Stir frequently to prevent the blood from clotting. (To transplant, inject 0.1 ml cells into each of 15 mice and allow to grow for 7 days.)

3. Place the cell suspension in 50-ml tubes and centrifuge in an International centrifuge (swinging-bucket rotor) at 700 rpm for 3 minutes.

4. Remove the supernatant via suction, leaving the cells at the bottom undisturbed.

5. Resuspend the cells in NKT buffer and centrifuge as in step 3 (700 rpm for 3 minutes).

6. Remove the supernatant as in step 4.

7. Combine six tubes into two tubes with STE buffer (~40 ml) and sediment at 1000 rpm for 3 minutes in the centrifuge used in step 3.

8. Remove the supernatant via suction, leaving the cells at the bottom undisturbed.

9. Combine the cells from 10 mice in ~10 ml STE buffer plus 2 mg Nagarse. Homogenize with a Potter–Elvehjem homogenizer (four up-and-down cycles) to mix them; allow the cells to sit 8 minutes on ice, add 30 ml H-medium containing 0.1% BSA, and homogenize (five up-and-down cycles).

10. Centrifuge at 14,000 rpm (23,500 g) for 5 minutes in the Sorvall SS-34 rotor.

11. Suspend the sediment in H-medium containing 0.1% BSA (80 ml).

12. Save the supernatant.

13. Suspend the sediment in 40 ml H-medium containing 0.1% BSA and centrifuge at 2000 rpm (480 g) for 5 minutes in the Sorvall SS-34 rotor. Save the supernatant.

14. Combine the supernatants from steps 12 and 13 and spin at 9000 rpm (9750 g) for 5 minutes in the Sorvall SS-34 rotor. Discard the supernatant and resuspend the sediment in approximately 25 ml H-medium containing 0.1% BSA.

15. Centrifuge again at 9000 rpm (9750 g).

16. Suspend the final sediment in 0.5 ml regular H-medium.

3. MITOCHONDRIA FROM L1210 ASCITES TUMOR CELLS

Comment: Although L1210 ascites cells are derived from white blood cells, the properties of their mitochondria are frequently compared to the properties of liver mitochondria. For this reason a procedure is included in this chapter for making mitochondria from L1210 cells. The method outlined is essentially identical to the method described by Pedersen and Morris (1974) and should yield mitochondria with ACRs of 3.5 to 6 when succinate is the substrate and the respiration medium of Schnaitman and Greenawalt (1968) is used (see Table III). See Fig. 7 for a flow diagram.

Animals: Mice, BDFI strain, females (20–25 gm), 6–8 weeks old, obtained from Buckberg, N.Y. or Cumberland View Farms, Clinton, Tenn.

Equipment: Swinging-bucket centrifuge for low-speed centrifugation (International PR-1 centrifuge); centrifuge which will sediment material at 25,000 g or greater (e.g., Sorvall RC-2B centrifuge).

Solutions: Saline solution (0.15 M NaCl); STE buffer (0.25 M sucrose, 5 mM Tris–Cl, 1 mM EGTA, pH 7.4); H-medium (see Table I);

Method:

1. Cells are harvested after 4 days of growth in the peritoneal cavity. To remove the cells from the peritoneal cavity of a mouse, decapitate the

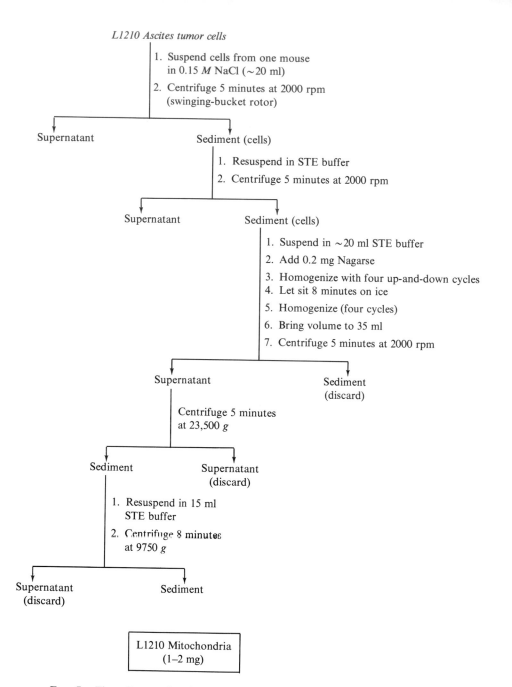

FIG. 7. Flow diagram for the preparation of mitochondria from L1210 ascites tumor cells. *Note*: This procedure is based on cells from one mouse. To obtain higher yields of mitochondria, it is necessary to commence the procedure with cells from more animals. As noted in the text, cells from 20 mice provide 15–20 mg mitochondria.

mouse with a pair of scissors and inject 2 ml of saline solution; shake the mouse; insert a 19-gauge needle into the cavity and allow the cells to drip out into a small test tube. Do this for 20 mice.

2. Dilute the cells 10-fold with saline solution. (At this time additional mice should be injected with 0.1 ml cells with a 26-to 28-gauge needle.)

3. Centrifuge the cells for 5 minutes at 2000 rpm in a swinging-bucket rotor (International centrifuge).

4. Resuspend in the same volume of STE buffer.

5. Repeat the centrifugation step in 3.

6. Suspend the cells in about 20 ml STE buffer and add 4 mg Nagarse. Homogenize lightly (four up-and-down cycles) with a Potter–Elvehjem homogenizer. Allow to sit for 8 minutes. Homogenize (four cycles) again and bring the volume to approximately 35 ml with STE buffer.

7. Centrifuge 5 minutes at 2000 rpm (480 g) in the Sorvall SS-34 rotor. Discard the sediment.

8. Centrifuge the supernatant 5 minutes at 14,000 rpm (23,500 g) in the same rotor.

9. Resuspend the sediment in about 15 ml STE buffer and centrifuge in the same rotor at 9000 rpm (9750 g) for 8 minutes.

10. Resuspend the sediment in 0.5 ml H-medium.

VIII. Preparation of Mitochondria from Regenerating Liver

Comment: To obtain regenerating liver the partial hepatectomy procedure described by Higgins and Anderson (1931) is followed. This entails first tying three lobes of the liver, excising these three lobes (two lobes should remain intact), suturing the animal, and allowing the liver to regenerate. Normally the half-life of a liver hepatocyte is about 1 year, but after three lobes have been removed the cells in the other two lobes start to divide rapidly. In about 28 hours many of the old cells have been replaced by new cells. The system is back to near normal in 3–6 days.

Animals: Sprague-Dawley rats.

Equipment: Sorvall RC-2B or equivalent centrifuge.

Materials and solutions: Cotton; forceps; scissors; thread, nos. 30 and 70; needles, no. 18, triangle "1"; nose bags (gauze, shaped to cover a rat's nose); anesthetic ether; 10% glucose; H-medium (see Table I).

Method:

a. *Operation for the induction of liver regeneration (partial hepatectomy)*

1. Place a rat (~250 gm, preferably male) in a container (e.g., dessicator)

in which the air is saturated with ether. (Saturate a paper towel or cotton with ether and place it in the container to allow the air to become saturated with ether.) *Note*: *No smoking.*

2. When the animal's head starts to droop, remove it from the container and place a nose bag (saturated with ether) close to or over its nose and check respiration. Do not allow the animal to go too far under the ether when in the container, or it may die. It must be taken out when its head starts to droop.

3. Locate the xyphoid cartilage near the diaphragm (by touch). Shave the rats abdomen, about 3 cm below the xyphoid cartilage, with a small pair of scissors.

4. Make about a 3-cm incision in the skin just below the xyphoid cartilage and retract the skin.

5. Make an identical incision in the muscle tissue below. Do not cut the diaphragm, but only up to and a little above the xyphoid cartilage. The liver should be obvious and exposed.

6. Excise the xyphoid cartilage at the base.

7. By gently squeezing protrude the three visible lobes of the liver (Fig. 8).

8. Tie these three lobes (ligature) with no. 30 thread (reef knot: left over right, then right over left).

9. Excise these three lobes with a pair of scissors. If part of the three lobes remain after this, remove it with a pair of forceps. Use cotton to absorb any blood.

10. Suture the muscle (no. 70 thread and triangular needle) starting at the xyphoid cartilage. Make a knot first, then suture, and then make another knot. After suturing, use cotton to absorb any blood on the surface of the muscle.

11. Suture the skin in same way.

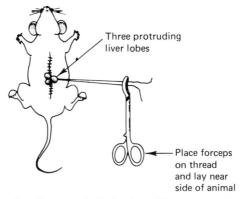

FIG. 8.   Operation for induction of liver regeneration.

12. For a control, repeat the operation without removing the liver (sham operation).

13. Place the animals in a cage and supply them with food and water. The first day after the operation the water should contain 10% glucose (w/v). The temperature in the room should be a little higher than normal.

b. *Preparation of mitochondria.* Mitochondria with high ACRs can be prepared by the procedure of Schnaitman and Greenawalt (1968) exactly as described in Section V,B,1.

IX. Introductory Comments regarding Submitochondrial Particle Preparations of Rat Liver

Submitochondrial particle preparations of rat liver can be divided into four main categories (see Fig. 9).

1. *Those in which all or a large part of the matrix protein remains.* These particles are usually prepared by a procedure (such as water lysis or limited digitonin treatment) which results in fairly selective removal of the outer mitochondrial membrane. The types of submitochondrial particles in this class are *ghosts* and *mitoplasts* (inner membrane plus matrix).

2. *Those in which the bulk of the matrix protein has been removed.* These particles are usually prepared by subjecting intact mitochondria or mitoplasts to sonic oscillation (with or without prior swelling and contraction) or to treatment with a nonionic detergent such as Lubrol WX, followed by a high-speed centrifugation step. The types of submitochondrial particles in this class are *sonic particles*, *Lubrol particles*, *sonic inner membranes*, and *Lubrol inner membranes*.

3. *Those in which the bulk of the outer membrane, matrix, and ATPase knobs have been removed.* These particles are prepared by allowing inner membrane particles to stand at 0–4°C for 2–3 days. They are referred to as *ATPase-deficient particles*.

4. *Those in which the inner membrane particles are enriched in ATPase.* These particles are prepared by subjecting inner membrane vesicles prepared by the digitonin–Lubrol method to extensive washing with a high-ionic-strength P_i buffer. They are referred to as *ATPase-enriched particles*. They have been referred to in our laboratory as *Pi-washed membranes* or *light membranes* (Soper and Pedersen, 1976).

Particles of class 1 are probably for the most part "right side in"; that is, the inner mitochondrial membrane has the same orientation ("sidedness") as in intact mitochondria. Certainly, this is the case for mitoplasts prepared

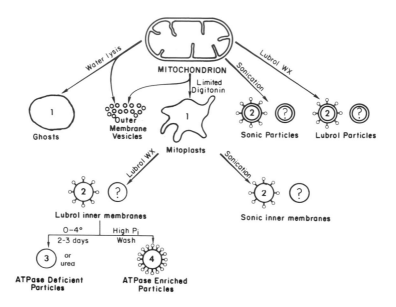

Fig. 9. Submitochondrial preparations of rat liver. Class-1 particles are those in which the outer compartment has been removed with retention of most of the inner membrane and matrix space. Class-2 particles are those in which the bulk of the matrix and intermembrane space have been removed with retention of the inner membrane and in some cases the outer membrane. Class-3 particles are those in which only the inner mitochondrial membrane remains but which is devoid of 90-Å ATPase molecules. Class-4 particles consist of the inner mitochondrial membrane enriched in 90-Å ATPase molecules.

Sonication and Lubrol treatment are depicated as producing two classes of particles, some inside out and some of unknown orientation (see text). The bulk of the particles are probably inside out, but a small but significant fraction of right-side-out particles cannot be excluded, nor can particles which have been stripped of the ATPase molecules by the negative stain.

by the digitonin procedure of Schnaitman and Greenawalt (1968). Less information is available on the sidedness of ghost preparations (Caplan and Greenawalt, 1968); however, the electron microscope image, the retention of phosphorylation capacity, and the latency of matrix enzymes, as well as ion transport, contractibility, and osmotic properties (Caplan and Greenawalt, 1968; Caplan, 1966), strongly suggest they are "right side out."

Particles in the remaining three classes are probably for the most part "inside out"; that is, the inner membrane orientation is inverted relative to intact mitochondria. Investigators are admonished, however, not to implicitly assume that a given submitochondrial preparation is 100% inside out. In fact even today, there remains controversy concerning the sidedness of submitochondrial preparations (Wainio, 1970; Malviya *et al.*,

1968). Some investigators believe that sonication of mitochondria results in the "pinching off" of cristae, giving inside-out particles with the F_1 ATPase on the surface (Lee and Ernster, 1968). Indeed, the bulk of the membrane vesicles in many submitochondrial preparations appear to have the F_1-ATPase molecules lining the surface when such preparations are negatively stained with phosphotungstic acid (PTA) and examined under the electron microscope. However, when such preparations are examined closely, membrane vesicles devoid of F_1-ATPase molecules may be seen as well (Chan et al., 1970). Whether such vesicles are right side out and not penetrated by the negative stain, whether they represent "noncristate" inner membrane vesicles (inner boundary membrane) devoid of ATPase molecules, or whether they represent a small fraction of the total population which has been damaged by the negative stain remains to be established.

There is probably no single test which can unequivocally reveal whether a given mitochondrial preparation consists of right-side-in vesicles or inside-out vesicles, or a mixture of these two populations. However, there are several tests (Chan et al., 1970; Fernández-Morán, 1962; Lee, 1963; Hackenbrock and Hammon, 1975; Fessenden and Racker, 1966; Racker et al., 1970) which when taken together can give a fairly reliable indication of the sidedness of submitochondrial preparations. These involve negative staining (Fernández-Morán, 1962), antibody studies (Fessenden and Racker, 1966; Racker et al., 1970), studies to assess the effect of Mg^{2+} on ATPase activity (Chan et al., 1970), and studies to assess the effect of cytochrome-c addition on respiration (Lee, 1963; Hackenbrock and Hammon, 1975). For details the reader should consult Table II.

The second part of this chapter is concerned specifically with methods for preparation of the four classes of submitochondrial particles described above and illustrated in Fig. 8. This chapter does not include the earlier procedure for preparing submitochondrial particles by using high concentrations of digitonin [referred to in the literature as digitonin particles (Devlin and Lehninger, 1958)], nor does it deal with procedures for the preparation of mitoplasts by the swelling-sonication technique of Sottocasa et al. (1967). Methods for preparing these submitochondrial particles are summarized in detail elsewhere (Devlin and Lehninger, 1958; Sottocasa et al., 1967). Today very few laboratories use classically prepared digitonin particles (Devlin and Lehninger, 1958), and it seems that many investigators prefer the Schnaitman and Greenawalt procedure (1968) [over the Sottocasa et al. method (1967)] for the preparation of mitoplasts. The Schnaitman and Greenawalt method (1968), in contrast to the Sottocasa method (1967), yields mitoplasts which contain ~90% of the matrix enzymes and exhibit acceptor control of respiration. Moreover, Decker and Greenawalt (1977) recently devised a procedure whereby "intact" mitoplasts with

TABLE II

Experimental Tests That Can Be Used to Determine Whether a Given Submitochondrial Preparation is Right Side Out or Inside Out

Test for sidedness[a]	Anticipated result if particles are completely inside out	Comment	Reference
Stimulation of ATPase activity by Mg^{2+}	The ATPase activity of submitochondrial particles should be activated maximally by $MgCl_2$. Addition of Lubrol or other detergents should not activate Mg^{2+}-ATPase activity beyond the level observed with Mg^{2+} alone.	This test is probably most applicable to rat liver mitochondria, since as isolated they exhibit a low Mg^{2+}-ATPase activity and do not take up Mg^{2+} unless special conditions are employed.	Chan et al., 1970
Negative staining with PTA	Under the electron microscope all submitochondrial particles should have 90-Å knobs on their periphery to be considered completely inside out.	There is controversy over whether or not the negative stain interacts with the ATPase particles and "pulls" them out of the membrane. However, even if it were shown that PTA popped the 90-Å particles out of the membrane, it is unlikely that this would negate the PTA staining test for sidedness. This is because in inside-out particles the stain would still be predicted to pull the 90-Å knobs to the outer surface, whereas in right-side-out particles they should be pulled in the other direction,	Fernández-Morán, 1962; Chan et al., 1970

Test	Basis	Comment	Reference
Failure to observe a stimulation of respiration by cytochrome c	Since cytochrome c is normally localized on the exterior surface of the mitochondrial inner membrane, it is localized on the interior surface in inside-out particles. The respiration of such particles should therefore not be stimulated by the addition of cytochrome c.	It is implicit in this test that, as a control, the investigator shows in some right-side-out cytochrome-c-depleted system that addition of cytochrome c stimulates repiration.	Lee, 1963; Hackenbrock and Hammon, 1975
Effectiveness of an F_1-ATPase antibody to inhibit Mg^{2+}-ATPase activity, and the ineffectiveness of a cytochrome-c antibody to inhibit respiration.	The F_1-ATPase antibody should inhibit (completely or nearly completely) Mg^{2+}-ATPase activity, and the cytochrome-c antibody should have no effect on repiration supported by succinate or NAD^+	This test assumes that interaction of the antibody with the antigenic sites of the proteins involved also inhibits their activity. Results obtained by Racker and his colleagues tend to suggest that this assumption is valid.	Fessenden and Racker, 1966; Racker et al., 1970

[a]There are many tests that can be used to give an indication of whether the majority of particles in a given submitochondrial particle preparation are inside out. However, these tests only give an indication of a majority, which could be as low as 60% of the population. Among these tests are oxidation by NAD^+, net alkalization of the external medium when respiration is initiated, net uptake of lipophilic anions such as tetraphenylboron, and nonspecificity of Mg^{2+}-ATPase activity for nucleoside triphosphates.

acceptor control of respiration can be prepared utilizing a French pressure cell, which circumvents the use of digitonin.

X. Preparation of Sonic Particles

Comment: This method yields submitochondrial vesicles which are maximally or near maximally activated with respect to Mg^{2+}-ATPase activity (Pedersen and Morris, 1974). Moreover, such particles retain their respiration and oxidative phosphorylation capacities. Although the majority of particles may be assumed to be inside out, this has not been rigorously proven. All operations are carried out at 0–4°C. See Fig. 10 for a flow diagram.

Equipment: Sorvall RC 2-B or equivalent centrifuge; Beckman Spinco Model L or equivalent centrifuge; Bronwill Biosonik sonicator equipped with a small probe.

Solutions: H-medium (see Table I).

Method:

1. Suspend freshly prepared mitochondria at 20 mg/ml in H-medium.

2. Place 1 ml of this mitochondrial suspension in a small Pyrex test tube. After clamping, place the tube in an ice bath; put a small magnetic stirrer in the tube and place the ice bath containing the tube on a magnetic stirring base.

3. Place the small probe of the Bronwill Biosonik sonicator just beneath the surface of the mitochondrial suspension.

4. Sonicate for 2 minutes at 15-second intervals; i.e., sonicate for 15 seconds, turn off for 15 seconds, sonicate for 15 seconds, etc. Sonication should be done at 20% of maximal intensity when using the small probe of the Bronwill sonicator.

5. Sediment the resultant suspension at 105,000 g for 1 hour at 0–4°C in the Spinco Model L centrifuge. (If it is desirable to rule out all possibilities of mitochondrial contamination, the sonicate in step 4 should be centrifuged for 10 minutes at 7500 g prior to the 105,000 g spin.] The supernatant from this step is then centrifuged at 105,000 g. This modification will result in a decrease in yield, because the larger submitochondrial vesicles will sediment at 7500 g.

6. Discard the supernatant. Rinse the sediment with H-medium and discard the rinse. Resuspend the sediment in H-medium at a protein concentration of 20 mg/ml.

Note: When larger preparations of sonic particles are desired, the pro-

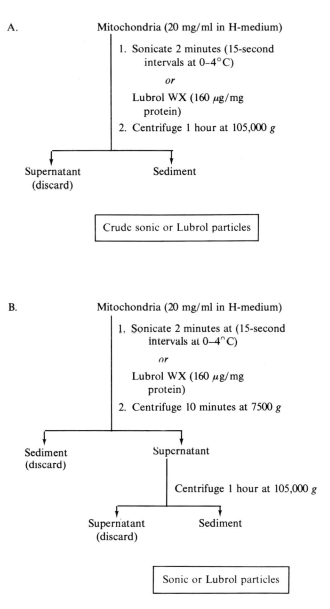

FIG. 10. Flow diagram for the preparation of sonic or Lubrol particles of rat liver mitochondria. The procedure summarized in (A) provides a crude sonic or Lubrol particle preparation which may be contaminated by mitochondria or partially ruptured mitochondria. The procedure in (B) provides a mitochondrial-free sonic or Lubrol particle preparation.

cedure can be modified. In such cases mitochondrial suspensions are placed in stainless-steel containers of about 50–to 100-ml capacity, and the large probe of the Bronwill Biosonik sonicator is used.

XI.   Preparation of Lubrol Particles

Comment: Lubrolization as described here appears to result in essentially complete disruption (membrane lysis) of rat liver mitochondria and to stimulate Mg^{2+}-ATPase activity maximally (Chan *et al.*, 1970; Pedersen *et al.*, 1971). The resultant particles respire well with succinate and ascorbate–N,N,N',N'-tetramethyl-*p*-phenylenediamine (ascorbate-TMPD). They do not respire with externally added β-hydroxybutyrate. Oxidative phosphorylation is observed at site I I (but not at sites I and III) provided assays are conducted at low ionic strength, low protein concentration (≤ 0.5 mg/ml), and in the presence of defatted BSA. About 60–70% of the total mitochondrial protein (mostly matrix and intracristal space protein) is solubilized by Lubrol WX. All operations are carried out at 0–4°C. See Fig. 10 for a flow diagram.

Equipment: Beckman Spinco Model L or equivalent centrifuge.

Solution: H-medium (see Table I) Lubrol WX (19mg/ml).

Method:

1. Treat freshly prepared rat liver mitochondria in H-medium (25–50 mg/ml) with a volume of Lubrol WX (19 mg/ml) such that the final suspension contains 0.16 mg Lubrol WX per milligram of mitochondrial protein. Gently mix.

2. Allow the lubrolized suspension to stay on ice for 15 minutes.

3. Dilute the suspension with isolation medium so that the final volume is twice the original volume of the mitochondrial fraction.

4. Centrifuge at 105,000 g for 1 hour in the Spinco Model L centrifuge.

5. Carefully remove the supernatant with a pasteur pipet.

6. Suspend the sediment to the original mitochondrial volume in H-medium and centrifuge again as in Step 4.

7. Discard the supernatant and resuspend the sediment in H-medium.

Note: Another procedure has been described by Winkler and Lehninger (1968) for preparing Lubrol particles in which the lubrolization process and resultant resuspensions are carried out in buffered 0.25 M sucrose. Particles prepared in buffered 0.25 M sucrose (in contrast to particles prepared in H-medium) have been reported by Winkler and Lehninger (1968) to have no sucrose-impermeable space and therefore to be completely permeable to small molecules.

XII. Preparation of Mitochondrial Ghosts

Comment: The procedure described here is essentially that described by Caplan and Greenawalt (1966). It involves subjecting mitochondria to water lysis under well-defined conditions. The particles (ghosts) obtained are thought to be devoid of much of the outer membrane and about 50–60% of the matrix space. The carry out respiration and oxidative phosphorylation but have no acceptor control of respiration. One gram of rat liver normally yields 10–15 mg of ghost protein. See Fig. 11 for a flow diagram.

Equipment: Beckman Spinco Model L or equivalent centrifuge.

Solution: 0.25 M sucrose.

Method:

1. Prepare rat liver mitochondria in 0.25 M sucrose (see Section V,A).

2. Dislodge the mitochondrial sediment obtained with 10 gm of rat liver with the bottom of a cold test tube and dilute to a total volume of 10 ml with ice-cold distilled water. (The mitochondria are drawn into and expelled from a 10-ml pipet several times until a uniform suspension is obtained.)

3. Centrifuge the suspension in a Spinco No. 40 rotor at 40,000 rpm (105,000 g) for 30 minutes.

4. Decant the supernatant fraction from the first centrifugation and disperse the sediment with a cold stirring rod.

5. Dilute the sediment to 10 ml with cold water, resuspend with a pipet, and centrifuge as in step 3.

6. Repeat steps 4 and 5 twice more.

7. Decant the final supernatant.

8. Suspend the final sediment in 10 ml of water. [Final suspensions are based on the gram (wet weight) of starting liver; a 1:1 ratio is always maintained.]

XIII. Subfractionation of Mitochondria

A. Mitoplasts (Inner Membrane plus Matrix)

1. DIGITONIN PROCEDURE

Comment: The mitoplast fraction prepared in the manner described here is essentially identical to the method described by Schaitman and Greenawalt (1968). Only a few operational modifications have been made. The mitoplast fraction prepared in this manner should:

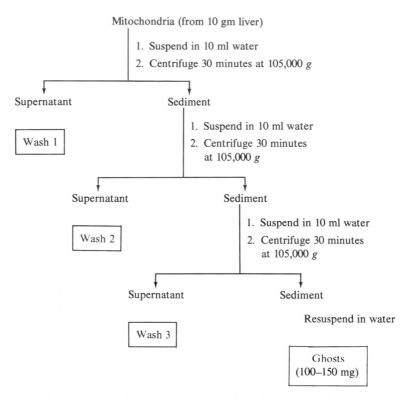

FIG. 11.   Flow diagram for the preparation of mitochondrial ghosts.

1. Exhibit ACRs in the absence of added Mg^{2+} of 1.7 to 2.5 when succinate is the substrate and the measurements are made in the medium described by Schnaitman and Greenawalt (1968). [Table III]

2. Have less than 5% of the total monoamine oxidase and adenylate kinase activity of intact mitochondria (Pedersen and Schnaitman, 1969; Chan *et al.*, 1970).

3. Exhibit oxidative phosphorylation, ADP–ATP exchange, P_i–ATP exchange, and Ca^{2+} uptake with H^+ ejection when measurements are made in the absence of added Mg^{2+} (Schnaitman and Greenawalt, 1968).

4. Contain citric acid cycle enzymes and the electron transport chain.

5. Exhibit pseudopodia-like projections when observed under the electron microscope (in the absence of Mg^{2+}) (Schnaitman and Greenawalt, 1968; Chan *et al.*, 1970) (see Fig. 17B).

6. Undergo orthodox to condensed conformational changes in the absence of added Mg^{2+} (Greenawalt, 1969).

All operations are carried out at 0–4°C. See Fig. 12 for a flow diagram.
Equipment: Sorvall RC 2-B or equivalent centrifuge.

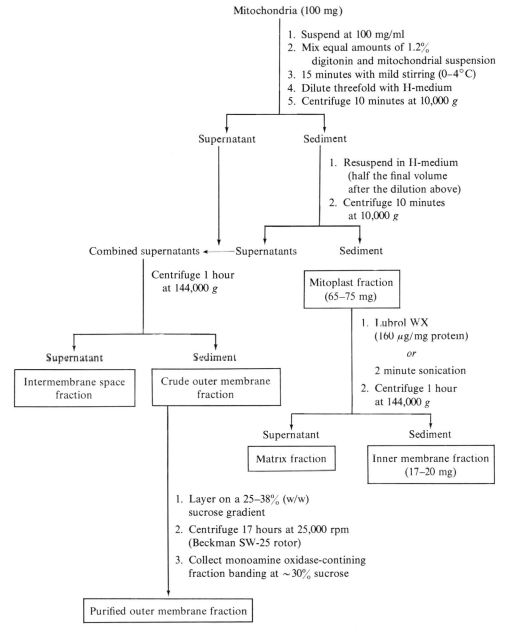

FIG. 12. Flow diagram for the subfractionation of mitochondria into their four components: outer membrane, intermembrane space, inner membrane, and matrix. Steps 2 and 3 in the initial subfractionation involving digitonin treatment can be replaced by using a French pressure cell (see Section XIII,A,2).

Solutions: H-medium (see Table I).

Method:

1. Use rat liver mitochondria freshly prepared by the procedure summarized in Section V, B, 3 or the methods outlined in Figs. 2 and 4.

2. Make a 1.2% digitonin solution (digitonin, A grade, can be obtained from Calbiochem or Gibco). Weigh out 60 mg digitonin and place in a small erlenmeyer flask. Add 5.0 ml H-medium (minus BSA), which has been heated almost to boiling. Swirl on a hot plate for about 1 minute to dissolve the digitonin. Allow to cool and then add 0.05 ml BSA (50 mg/ml).

3. Surround a small cylindrical vial ($\sim$20 ml in capacity) with ice by placing the vial in a beaker. Add to the vial a small magnetic stirring flea and place the beaker containing the vial on a magnetic stirring base. (This operation should be done while the digitonin is being dissolved in step 1.)

4. Add 4 ml of the mitochondrial suspension (100 mg/ml) to the vial and start to stir slowly. Now add an equal volume (4 ml) of the 1.2% digitonin solution.

5. Incubate with gentle stirring for 15 minutes.

6. Add 24 ml H-medium and immediately centrifuge at 10,000 g in the Sorvall SS-34 head for 10 minutes. Carefully pipet off the supernatant and save it. Leave the fluffy layer with the sediment. (In good preparations there is little fluffy layer.)

7. Resuspend the sediment with a "cold finger" (a test tube to which ice has been added) in one-half the original volume of H-medium (the volume after dilution in step 6). Centrifuge at 10,000 g for 10 minutes in the Sorvall SS-34 head. Remove the supernatant with a pasteur pipet and combine with the supernatant in step 6. The combined supernatants are used in the preparation of outer membrane and intermembrane space (see Section XIII, B).

8. Suspend the sediment (mitoplasts) in 8 ml H-medium.

Comment: If it is desirable to avoid the use of digitonin in preparing mitoplasts of rat liver mitochondria, the procedure described here, which employs a French pressure cell, can be used. This technique, developed recently by Decker and Greenawalt (1977), yields mitoplasts similar in biochemical and morphological properties to mitoplasts prepared by the digitonin procedure (see Fig. 17). All operations are carried out at 0–4°C.

Equipment: French pressure cell (American Instruments Company, Model 4-3398); Sorvall RC 2-B or equivalent centrifuge.

Solutions: H-medium (see Table I) 2 $\times$ medium (twice the weights of H-medium components in the same volume, pH 7.4).

Method:

1. Resuspend mitochondria from three rat livers in 2–3 ml of 2× medium and dilute the suspension to 25–50 mg protein per milliliter (8–10 ml if the yield is normal). Save a small aliquot for protein and enzyme assays. Cover and keep on ice.

2. Load the diluted mitochondrial suspension into a prechilled French pressure cell and apply 1,500 psi. Upon reaching constant pressure in the hydraulic press, open the needle valve slowly to obtain continuous flow from the pressure cell into a cold collecting tube (usually ~1.5 ml is lost in the press).

3. Pipet a known volume (6–8 ml) of French-pressed mitochondria into a cold 40-ml Sorvall centrifuge tube containing equal volumes of 2X medium. Stir gently by swirling and then centrifuge in the RC 2-B Sorvall SS-34 rotor at 10,000 rpm (12,100 g) for 10 minutes.

4. *Carefully* remove the supernatant with a pasteur pipet and *save*. (The loose fluffy layer should be removed with the supernatant).

5. Muddle the sediment (mitoplasts) into a smooth paste with a "cold finger" (a test tube filled with ice). Continue to muddle adding 2X medium dropwise until a fluid suspension is obtained. Finally, resuspend in 2X medium to the original volume of pressed mitochondria selected for washing. Repeat the centrifugation in step 3. The resulting sediment is the mitoplast fraction. Resuspend in cold H-medium to about 50 mg/ml and store on ice. Combine the supernatant with the supernatant in step 4 and save for preparation of the outer membrane and intermembrane space fractions.

B. Intermembrane Space and Outer Membrane

1. INTERMEMBRANE SPACE

The combined supernatants from steps 6 and 7 in the digitonin procedure or steps 4 and 5 in the French pressure cell procedure are centrifuged at 144,000 g for 1 hour in the Spinco No. 40 rotor. The supernatant is removed with a pasteur pipet. This fraction is called the *intermembrane space* or *intracristal space* fraction. It contains essentially all the original adenylate kinase activity of the mitochondria.

2. OUTER MEMBRANE

The sediment from step 1 contains the outer mitochondrial membrane. However, since the outer membrane constitutes only about 4–5% of the rat liver mitochondria, small amounts of contamination by the inner membrane significantly contribute to the total amount of this fraction. In fact, there is more inner membrane protein in this fraction than there is outer membrane fraction. To obtain a relatively pure outer membrane fraction carry out the following density gradient centrifugation step described by Schnaitman (1969).

Layer the preparation on a 25–38% (w/w) sucrose gradient in 10 mM HEPES buffer, pH 7.4, and centrifuge in a Beckmen SW-25 rotor for 17 hours at 25,000 rpm. Monoamine oxidase is a marker enzyme for the outer membrane. The highest activity of this enzyme is observed in material banding at 30% sucrose (w/w).

C. Inner Membrane and Matrix

Comment: The inner membrane fraction prepared in the manner described here is referred to as the *Lubrol inner membrane fraction*. This fraction should:

1. Consist of vesicles ($\leq 0.4 \mu$m diameter) which when negatively stained with PTA exhibit 90-Å particles on their periphery (Chan *et al.*, 1970).

2. Exhibit no acceptor control (Chan *et al.*, 1970).

3. Respire in the presence of either succinate or TMPD plus ascorbate but not in the presence of either β-hydroxybutyrate or NADH (Chan *et al.*, 1970).

4. Phosphorylate in the presence of succinate as substrate at a low protein concentration (Chan *et al.*, 1970).

5. Be essentially free of most matrix, intermembrane space, and outer membrane activity (Chan *et al.*, 1970).

6. Exhibit about 21 to 23 stainable bands after SDS gel electrophoresis in 10% polyacrylamide (Pedersen *et al.*, 1971). All operations are carried out at 0–4°C. See Fig. 12 for a flow diagram.

Equipment: Beckman Spinco Model L or equivalent centrifuge.

Solutions: H-medium (see Table I); Lubrol WX (19 mg/ml).

Method:

1. Place 4 ml of a freshly prepared mitoplast suspension (30–35 mg/ml) in a calibrated centrifuge tube.

2. Add a volume of Lubrol WX (19 mg/ml) such that the final suspension contains 0.16 mg Lubrol WX per milligram of mitoplast protein. (Lubrol WX can be obtained from Grand Island Biological Company, Grand Island, N.Y.)

3. Let stand at 0°C for 15 minutes.

4. Dilute to 8 ml with H-medium.

5. Centrifuge for 1 hour at 144,000 *g* in the Spinco Model L No. 40 rotor.

6. Remove the supernatant (the matrix) with a pasteur pipet.

7. Rinse the sediment (the Lubrol inner membrane) twice with a small amount of H-medium and suspend the sediment in H-medium at a concentration of $\sim$10 mg/ml.

Note: To prepare what we refer to as *sonic inner membranes* (see Figs. 9 and 12) suspend the mitoplast fraction at 20 mg/ml in H-medium. Carry out sonication, centrifugation, and resuspension exactly as described in Section X. The supernatant fraction is the matrix, and the sediment fraction represents the sonic inner membranes. Sonic inner membranes prepared by this procedure should have properties similar to those of sonic particle preparations of mitochondria (see Section X).

This technique can be refined as described by Hackenbrock and Hammon (1975) to yield submitochondrial vesicles which appear to be essentially completely inside out (i.e., respiration is not affected by either cytochrome-c addition or by the addition of an antibody to cytochrome c). To prepare these inside-out vesicles 50 mg of mitoplasts is suspended in 25 ml distilled water after which the inner membrane is centrifuged at 10,000 g for 15 minutes. The membranes are then resuspended in 1 ml distilled water and sonicated for 2 minutes (at 15-second intervals) with the microtip of a Branson Model W185 sonifer. The sonicated preparation is then centrifuged at 10,000 g for 10 minutes to remove any large univerted vesicles. The small inverted vesicles ($\sim$300 nm) are sedimented by centrifugation at 100,000 g for 1 hour. See Fig. 13 for a flow diagram.

XIV. Preparation of an Inner Membrane Fraction Deficient in ATPase Activity

A. Urea Procedure

Removal of F_1 ATPase from Inner Membrane Vesicles Followed by Rebinding of the Purified Enzyme

This procedure can be used to remove F_1-ATPase particles from inner membrane vesicles prepared by the method of Hackenbrock and Hammon (1975; see Fig. 13). One milliliter of 6 M urea is added to 60 mg inner membrane vesicles in 1 ml H-medium (final urea concentration 3M). The mixture, after sitting on ice for 5 minutes, is diluted to 10 ml with H-medium and centrifuged at 48,000 rpm for 30 minutes in the Spinco No. 60 rotor. The sediment is suspended in 8.0 ml isolation medium and centrifuged again at 48,000 rpm. The final sediment is then suspended in 1 ml H-medium. Significantly, these F_1-depleted urea particles can be used in reconstitution

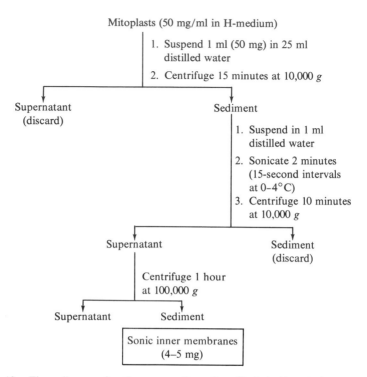

Fig. 13. Flow diagram for the preparation of purified, inside-out, inner membrane vesicles by the procedures of Hackenbrock and Hammon (1975). The procedure has been modified slightly in our laboratories by suspending the initial sediment in H-medium rather than in distilled water (J. P. Wehrle and P. L. Pedersen, unpublished).

studies in which it is of interest to rebind the homogeneous F_1 ATPase to the membrane and reconstitute either oligomycin-sensitive ATPase activity, oxidative phosphorylation, or both. To rebind homogeneous liver F_1 ATPase to urea particles the following reconstitution protocol is used:

The following components are incubated in a final volume of 0.1 ml for 2 hours at 25°C: 0.5 mg urea particles, 150 μg purified F_1 ATPase (Catterall and Pedersen, 1971), 1 mg defatted BSA, 88 mM D-mannitol, 28 mM sucrose, 0.8 mM HEPES, 100 mM KP$_i$, 2 mM EDTA, pH 7.5. After centrifuging the mixture in the Spinco No. 60 rotor (0–4°C) (at 48,000 rpm for 30 minutes the sediment (reconstituted membranes) is resuspended in 0.1 ml H-medium (0–4°C). Under these conditions 85–95% of the original ATPase activity of the membrane is restored, and between 20–30% of the oxidative phosphorylation capacity. To obtain maximal reconstitution of oxidative phosphorylation ($\geq$75% of the original activity) it is essential to include either 2 mM ATP or 2 mM AMP–PNP in the reconstitution assay.

B. Cold-treatment Procedure

Comment: This procedure is essentially that described by Chan *et al.* (1970). The inner membrane fraction prepared by the simple procedure described here has lost most of its original Mg^{2+}-stimulated ATPase and ATP–P_i exchange activities. Most of the ATPase, knoblike, 90-Å particles are also missing from the periphery of the membrane.

Equipment: Beckman Spinco Model L or equivalent centrifuge.

Solution: H-medium (see Table I).

Method:

1. Suspend the inner membrane fractions prepared by the digitonin–Lubrol procedure at 20 mg/ml in H-medium.

2. Allow the preparation to sit on ice for 2–3 days. (The preparation should be monitored from day to day for ATPase activity. Step 3 is carried out only when 80% or more of the ATPase activity is lost.)

3. Centrifuge for 1 hour at 104,000 *g*.

4. Discard the supernatant and suspend the sediment in the original volume.

XV. Preparation of an Inner Membrane Fraction Enriched in ATPase Activity ("Light" Membranes)

Comment: The procedure described here represents a slight modification of the procedure described by Soper and Pedersen (1976). It yields an inner membrane fraction enriched in ATPase activity (specific activity in Tris–bicarbonate buffer of greater than 30 μmoles ATP hydrolyzed per minute per milligram). The high specific ATPase activity of this fraction can be compared with that of normal Lubrol or sonic inner membrane preparations which have a specific activity of 2–5 μmoles ATP hydrolyzed per minute per milligram in Tris–bicarbonate buffer [2 mM ATP, spectrophotometric assay (Soper and Pedersen, 1976)]. The procedure described here also yields a "heavy" inner membrane fraction with an ATPase specific activity in Tris–bicarbonate of 8–10 μmoles ATP hydrolyzed per minute per milligram. See Fig. 14 for a flow diagram.

Equipment: Sorvall RC 2-B or equivalent centrifuge; Beckman Spinco Model L or equivalent centrifuge.

Solutions: H-medium (see Table I); H-medium lacking BSA; Lubrol WX (10% w/v); PE buffer (300 mM KP$_i$, 50 mM EDTA, pH 7.9). PA buffer (300 mM KP$_i$, 2 mM ATP, 10% ethylene glycol, 5 mM EDTA, and 0.5 mM dithiothreitol, pH 7.9); PE buffer containing 10% ethylene glycol; PA buffer containing 50% glycerol (w/w).

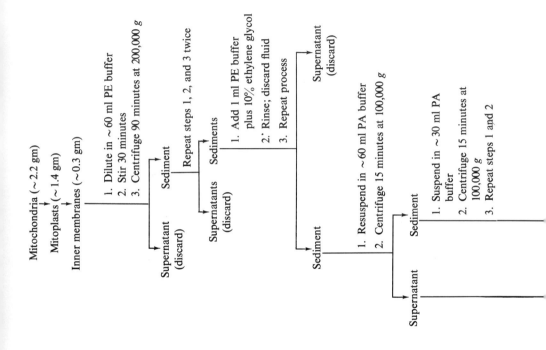

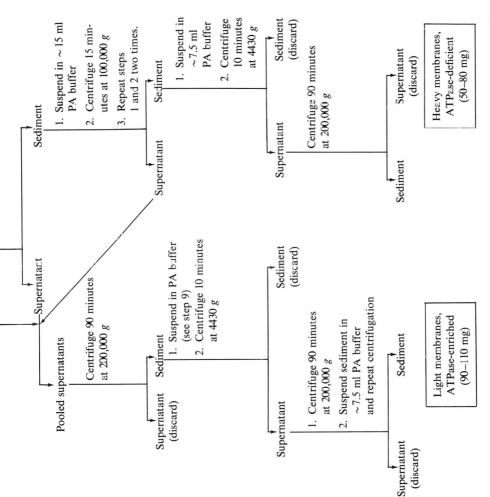

Supernatant

Sediment

1. Suspend in ~ 15 ml PA buffer
2. Centrifuge 15 minutes at 100,000 g
3. Repeat steps 1 and 2 two times.

Sediment

1. Suspend in ~7.5 ml PA buffer
2. Centrifuge 10 minutes at 4430 g

Sediment (discard)

Supernatant

Supernatant

Centrifuge 90 minutes at 200,000 g

Supernatant (discard)

Sediment

Heavy membranes, ATPase-deficient (50–80 mg)

Pooled supernatants

Centrifuge 90 minutes at 200,000 g

Supernatant (discard)

Sediment

1. Suspend in PA buffer (see step 9)
2. Centrifuge 10 minutes at 4430 g

Sediment (discard)

Supernatant

1. Centrifuge 90 minutes at 200,000 g
2. Suspend sediment in ~7.5 ml PA buffer and repeat centrifugation

Supernatant (discard)

Sediment

Light membranes, ATPase-enriched (90–110 mg)

FIG. 14. Flow diagram for the subfractionation of the mitochondrial inner membrane into an ATPase-enriched and an ATPase-deficient fraction.

Method:

Preparation of Mitoplasts

1. Commence this preparative procedure with 20–30 ml of a mitochondrial suspension of between 70 and 100 mg/ml in H-medium. Weigh an amount of digitonin equal to 0.12 × the total milligrams of mitochondrial protein. Dissolve the digitonin in a volume of H-medium (lacking BSA) equal to half the volume of the mitochondrial suspension. Add 0.5 mg defatted BSA per milliliter.

2. Quickly pour the digitonin solution into the mitochondrial suspension contained in an ice-cold erlenmeyer flask equipped with a magnetic stirrer. Mix and begin timing. Add sufficient additional H-medium to adjust the protein concentration to 50 mg/ml. (Add this first to the cylinder containing digitonin and then to the flask.)

3. After 20 minutes immediately dilute five-to eightfold with H-medium. Centrifuge for 20 minutes at 14,000 rpm (23,500 g) in the Sorvall SS-34 rotor.

4. Resuspend the sediments from step 3 in half the original volume and centrifuge for 40 minutes at 14,000 rpm (23,500 g) in the Sorvall SS-34 rotor.

5. Resuspend the sediment from step 4 to about one-half the volume of the original mitochondrial digitonin suspension. Determine protein by the biuret procedure. Record the suspension volume.

Preparation of Inner Membrane

1. Add to a graduated cylinder sufficient 10% Lubrol WX to give 0.16 mg Lubrol per milligram of mitoplasts. Add mitoplast suspension (40–50 mg/ml). Adjust the protein concentration to 35 mg/ml with H-medium. Mix immediately, place on ice, and incubate for 20 minutes.

2. Dilute twofold with ice-cold PE buffer. Centrifuge in the Spinco No. 65 rotor at 48,000 rpm (~20,000 g).

3. Resuspend the sediments from *step* 2 in PE buffer to a volume equal to the total volume of eight Spinco tubes (~60 ml). Wash by stirring in a capped, 50-ml, ice-cold erlenmeyer flask for 30 minutes. Use a stirring speed just below that needed to cause foaming. Centrifuge the washed membrane suspension for 90 minutes at 48,000 rpm (~ 200,000 g) in the Spinco No. 65 rotor (wash 1).

4. Resuspend the sediments from step 3 in PE buffer to a volume equal to the total volume of four Spinco tubes (~ 30 ml). Centrifuge the suspension for 10 minutes at 6000 rpm (4430 g) in the outer rim of the Sorvall SM-24 rotor. Resuspend the sediment in the same volume and centrifuge again at 6000 rpm. Discard the sediment. Wash the pooled supernatants and sediment them as in step 3 (wash 2).

5. Resuspend the sediments from step 4 in PA buffer containing 50% (w/w) glycerol. Carry out resuspension with a minimal amount of this buffer. Freeze and store in liquid nitrogen.

Washing of Lubrol Inner Membranes to Give the ATPase-Enriched or Light Membrane Fraction

1. Dilute the membranes from step 5 to a volume equal to that of eight spinco tubes (~60 ml) with PE buffer. Wash exactly as in step 3. Centrifuge 90 minutes at 48,000 rpm (~200,000 g) in the Spinco 65 rotor (wash 3).

2. Resuspend the sediments to the original volume with PE buffer, wash, and centrifuge exactly as in step 3 (wash 4).

3. Repeat step 2 (wash 5). Carefully pour off the supernatant.

4. Gently add 1 ml of PE buffer containing 10% ethylene glycol to each tube. Swirl the tubes and tap lightly to rinse off the small amount of white material on the tops of the sediments. Discard the fluid. Repeat this process.

5. Resuspend the sediment from step 4 in PA buffer to a volume equal to that of eight Spinco tubes (~60 ml). Centrifuge in the Spinco No. 65 rotor for 15 minutes at 33,000 rpm (100,000 g).

6. Save the supernatant (light membranes); resuspend the sediments in PA buffer to a volume equal to that of four Spinco tubes (~30 ml). Centrifuge as in step 5 and repeat the process (spins 2 and 3).

7. Resuspend the sediments in PA buffer to a volume equal to that of two Spinco tubes (~15 ml). Centrifuge as in step 5 and repeat the process two more times (spins 4, 5 and 6). Resuspend the sediment (heavy membranes) in PA buffer to a volume equal to that of one Spinco tube (~7.5 ml). Hold for step 9.

8. Pool 16 of the 22 light membrane supernatants and sediment the membranes by centrifuging at 48,000 rpm (~200,000 g) for 90 minutes in the Spinco No. 65 rotor.

9. Resuspend the light membrane sediments in the remaining light membrane supernatant (6 of the 22 tubes remaining). Centrifuge the light and heavy membranes separately at 6000 rpm (4430 g) for 10 minutes in 12-ml Sorvall tubes (~8 ml per tube) using the SM 24 rotor. Discard the sediments and centrifuge the heavy and light supernatants at 48,000 rpm (~200,000 g) for 90 minutes.

10. Resuspend the light membranes in PA buffer to a volume equal to that of one Spinco tube (~7.5 ml). Centrifuge as in step 9.

11. Resuspend the heavy membranes in the minimal volume of PA buffer containing 50% glycerol and freeze in liquid nitrogen.

12. Resuspend the light membranes in the minimal amount of PA buffer containing 50% glycerol. Determine the biuret protein and adjust the protein concentration to 25–30 mg/ml.

13. Determine the weight of 1 ml of the light membrane suspension and calculate its density (~1.1 gm/ml). (One cannot readily pipet the viscous suspension. Therefore, in order to transfer accurately an aliquot of known protein concentration to a freezer vial, it is necessary to determine the

density of the suspension.) Now calculate the weight of suspension that contains ~ 15 mg of membrane protein (0.5–0.7 gm of suspension).

14. Put into freezer vials and place in liquid nitrogen. Store until used.

Note: Light and heavy membranes may represent two different types of inner membranes. The fact that they can be isolated and shown to have different ATPase activities raises the question whether the inner mitochondrial membrane is homogeneous with respect to the ATP-synthesizing system.

XVI.   Introductory Comments Regarding the Characterization of Mitochondria and Submitochondrial Particles for Intactness

Prior to carrying out biochemical studies on mitochondria and submitochondrial particles it is important to assess the quality of the preparations involved. The two most common tests used to determine the quality of mitochondrial and submitochondrial preparations involve an examination of their respiratory characteristics in a small chamber equipped with an oxygen electrode, and an examination of their structural and morphological features under the electron microscope.

Respiration measurements are carried out to determine whether or not the preparation in question exhibits the biochemical characteristics referred to as *respiratory control*. There are two types of respiratory control, one of which involves control of respiration by ADP and is referred to as *acceptor control* (Chance and Williams, 1955), and one which involves control of respiration induced by oligomycin, an inhibitor of ATP synthesis. The latter type of respiratory control is referred to as *oligomycin-induced respiratory control* (Lee and Ernster, 1965).

Intact mitochondria and mitoplasts should exhibit acceptor control of respiration. That is, they should be able to turn on their respiration when ADP is present, i.e., when there is a demand for ATP synthesis, but to turn off their respiration when ADP is absent. Normally, in intact mitochondria respiration proceeds very slowly when an oxidizable substrate and phosphate are present (state IV respiration), and very rapidly when an oxidizable substrate, phosphate, and ADP are present (state III respiration) (Fig. 15A). If limiting amounts of ADP are employed, ADP addition to intact mitochondria respiring in state IV should result in a burst of respiration (state III) followed by a return to the state IV rate (Fig. 15A). During this period of time (state IV → state III → state IV) coupled ATP synthesis takes place. The mitochondria are said to have acceptor control of respiration, and an ACR is defined as state III rate/state IV rate.

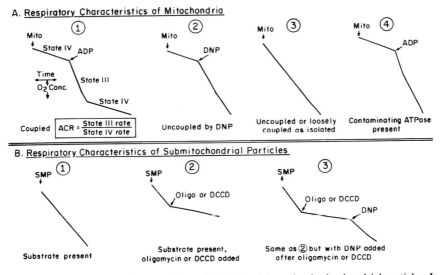

FIG. 15.   Respiratory characteristics of mitochondria and submitochondrial particles. In all cases it is assumed that substrate and P_i are present. Mitochondria in (A1) exhibit what is called acceptor control of respiration (see text) and are said to be coupled. For rat liver mitochondria the ACR usually ranges from 4 to 10 depending on the substrate and the intactness of the mitochondria. In (A2) the effect of the uncoupler DNP is seen on the respiratory rate. (A3) is an example of an isolated mitochondrial preparation that may be uncoupled because of structural damage (irreversible uncoupling) or uncoupled by an endogenous uncoupler like a fatty acid. (Mitochondria contain a Ca^{2+}-activated phospholipase in their outer compartment. Lysosomes, which frequently contaminate mitochondria, also contain phospholipases.) The addition of serum albumin binds many endogenous uncouplers, especially fatty acids, and may reverse this type of uncoupling effect. (A4) is an example of an isolated mitochondrial preparation that may be contaminated by a nonmitochondrial membrane ATPase. The ATPase would recycle any ATP formed to ADP and P_i and thus cause respiration to continue at an uncoupled rate. Submitochondrial particles in (B) are said to exhibit oligomycin-induced respiratory control.

The exact details as to why intact mitochondria exhibit acceptor control of respiration is beyond the scope of this chapter. Suffice it to say here that respiratory control is thought to be related to an electrochemical gradient across the mitochondrial inner membrane, which involves the movement of protons and perhaps other ions as well (Mitchell, 1966). The mitochondrial electron transport chain drives the formation of the electrochemical gradient. Normally, this gradient is discharged at the level of mitochondrial ATPase when ADP and P_i are both present. The energy for the discharging process is utilized to drive the synthesis of ATP. Thus respiration proceeds slowly when only substrate and P_i are present, presumably because the electrochemical gradient is in equilibrium with electron transport (state IV respiration). It is important to note that in the absence of both ADP and P_i the membrane is highly impermeable to protons. However, when ADP is added to the "charged" system, the gradient is dissipated to convert the ADP and P_i to ATP (at the level of the mitochondrial ATPase), and a sudden burst of respiration ensues (state III respiration), presumably to reestablish the preexisting electrochemical gradient. The gradient may also be dissipated by "uncoupling" agents such as 2,4–dinitrophenol (DNP) and fatty acids which make the membrane permeable to protons. Uncoupling agents, in analogy with ADP, stimulate respiration but, unlike ADP, they are never "used up." Thus respiration proceeds until essentially all the oxygen in the system has been consumed (Fig. 15A).

Because of the nature of the electrochemical gradient and its relationship to ATP synthesis it should be immediately clear, functionally speaking, that mitochondria are very fragile. If they are handled very "roughly" during isolation, they may become leaky to protons and other ions (even though their membranes may appear intact under the electron microscope). The net result will be a loss of acceptor control of respiration. Examples of respiratory traces of mitochondria which are either *uncoupled* or *loosely coupled* are shown in Fig. 15A.

Most submitochondrial preparations do not exhibit acceptor control of respiration (mitoplasts are an exception). As isolated, they are usually leaky to protons. Despite their uncoupled nature, however, if such particles are vesicular, they can frequently be "recoupled" in the sense that the proton leak can be blocked by agents such as oligomycin and dicyclohexylcarbo-diimide (DCCD). Thus, when such particles are supplied with a respiratory substrate, they respire at a high or uncoupled rate comparable to the state III rate observed with intact mitochondria. However, when oligomycin or DCCD is added, this rate is reduced to a level comparable to the state IV rate of intact mitochondria. The addition of DNP, which "reopens" the membrane, rendering it leaky again to protons, results in the stimulation of respiration back to the state IV rate (Fig. 15B).

From the above discussion it can be inferred that an assessment of the respiratory characteristics of mitochondria and submitochondrial particles provides a more sensitive test for intactness than *electron microscopy*. This is not always true. For example, the outer membrane of mitochondria can suffer damage or even be removed without the loss of acceptor control of respiration (Schnaitman and Greenawalt, 1968). Certainly, the electron microscope can reveal this damage. Moreover, submitochondrial particles can lose part of their ATP-synthesizing system (F_1 ATPase) without a pronounced effect on oligomycin-induced respiratory control (Ernster *et al.*, 1974). Again, the electron microscope can reveal this damage, since the F_1-ATPase molecules appear as 90-Å particles when submitochondrial preparations are stained with PTA. Thus, for rigorous quality control, it seems essential to employ not only respiration measurements as an index of the intactness of mitochondria and submitochondrial particles, but electron microscopy as well. Moreover, when purity is of concern, an electron microscope analysis also provides a rough index of the degree of contamination of a given mitochondrial or submitochondrial preparation.

A. Respiratory Control as an Index of Intactness

1. SUMMARY OF ACCEPTOR CONTROL MEDIA USED FOR LIVER MITOCHONDRIA

Table III summarizes various media which have proven suitable for carrying out acceptor control measurements. Similar to isolation media for mitochondria (see Table I), acceptor control media are designed to be isotonic or nearly isotonic with the mitochondrial matrix. Aside from their tonicity, however, acceptor control media may vary markedly in composition depending upon the type of mitochondria being assayed (liver or hepatoma) and depending upon what additional properties the investigator may wish to measure under coupled conditions.

For investigators whose primary concern is to obtain maximal ACRs in liver mitochondria or in mitochondria from liver-related systems the following suggestions may prove valuable.

1. Use mannitol or mannitol plus sucrose-containing media. There is no obvious rationale for this other than the preference of several investigators (Chance and Hagihara, 1960; Johnson and Lardy, 1967; Schnaitman and Greenawalt, 1968) for mannitol-containing media.

2. Use HEPES (or similar anionic buffers) rather than Tris buffers. Tris is a cation which may be taken up by mitochondria and thus lower the membrane potential thought to be essential for maximal coupling.

3. Use $MgCl_2$ in the medium. For reasons that are unclear $MgCl_2$ makes the "turn-off" rate (state III → state IV transition) sharper, resulting in a lower state IV rate and a higher ACR.

4. Use defatted BSA and/or EGTA. The net result of both agents is to lower free fatty acid levels which can uncouple mitochondria. BSA binds free fatty acids, whereas EGTA binds Ca^{2+}, an activator of certain phospholipases.

2. MEASUREMENT OF THE ACR OF MITOCHONDRIA

Comment: The procedure described here is one of the methods we routinely use in our laboratory to assess the acceptor control properties

TABLE III

ACCEPTOR CONTROL MEDIA THAT CAN BE USED FOR MITOCHONDRIA FROM RAT LIVER AND LIVER-RELATED TISSUES

Medium	Substrate(s)	pH	Comment	Reference
130 mM KCl, 2 mM KP$_i$, 3 mM HEPES	5.0 mM NAD-linked plus 1.0 mM malate	7.2	Mitochondria prepared from both heart and liver exhibit acceptor control of respiration in this medium. Addition of MgCl$_2$ results in higher ACRs for liver mitochondria and lower values for heart mitochondria. The medium is a "compromise" medium which allows certain properties of heart and liver mitochondria to be compared, especially Ca^{2+} uptake and H$^+$ ejection which would be difficult or impossible to assay in more complex media.	Brand et al., 1976
	0.4 mM pyruvate alone or 5.0 mM pyruvate plus 1.0 mM malate	7.2		
	5.0 mM succinate	7.2		
0.25 M Sucrose	5.0 mM of one of the following: glutamate, β-hydroxybutyrate, succinate, α-ketoglutarate	7.4	Mitochondria from certain mouse tumors have been reported to exhibit acceptor control in this simple medium. The system might provide more nearly optimal conditions if Tris were replaced with HEPES. As discussed in Table I, Tris$^+$ is permeable and may lower the membrane potential thought to be necessary for acceptor control of respiration.	Sordahl and Schwartz, 1971
100 mM Sucrose, 50 mM KCl, 10 mM KP$_i$, 2 mM MgCl$_2$·6 H$_2$0, 1 mM EGTA, 15 mM HEPES	10 mM succinate plus 0.5 µM rotenone	7.4	This is basically the same medium as that used by Thorne and Bygrave (1973) for measuring acceptor control of mitochondria from Ehrlich ascites tumor cells. The medium is also applicable to liver and hepatoma mitochondria.	Thorne and Bygrave, 1973; Reed, 1972
75 mM KCl, 50 mM Tris–Cl, 12.5 mM KP$_i$, 1 mM EDTA, 5 mM MgCl$_2$	5.0 mM of one of the following: glutamate, β-hydroxybutyrate, succinate, α-ketoglutarate	7.4	Mitochondria isolated from liver and from certain hepatomas have been shown to exhibit acceptor control in this medium. More optimal values might be obtained if Tris were replaced with HEPES for reasons discussed above.	Sordahl et al., 1969

Medium	Substrate	pH	Comments	References
70 mM Sucrose, 220 mM mannitol, 2 mM HEPES, 2.5 mM KP$_1$, 0.5 mM EDTA, 2.5 mM MgCl$_2$, 1 mg/ml BSA	10 mM succinate or 10 mM β-hydroxybutyrate	7.4	Mitochondria from liver, regenerating liver, hepatomas, and ascites tumor cells exhibit optimal or near optimal ACRs in this medium (or in minor variations of it). Mitoplasts also exhibit acceptor control in this medium when MgCl$_2$ is omitted and the EDTA concentration is increased to 1.0 mM.	Schnaitman and Greenawalt, 1968; Pedersen and Morris, 1974; Kaschnitz et al., 1976, 1978.
70 mM Sucrose, 220 mM mannitol, 2 mM HEPES, 2.5 mM KP$_1$, 0.5 mM EDTA, 10 mM MgCl$_2$, 1 mM malate, 1 mM ATP, 1 mM carnitine, 0.05 mM CoA, 0.7 mg/ml BSA (defatted)	8 µM sodium palmitate dissolved in absolute ethanol	7.4	The medium used for measuring oxidation of fatty acids in such a way that acceptor control can be obtained is complex. Several ingredients are required to "activate" the fatty acid. Malate is necessary to "charge" the TCA cycle. BSA, which binds fatty acids, is probably necessary to maintain a low concentration of the fatty acid substrate (an uncoupler at high concentrations) and thus prevent uncoupling.	P. L. Pedersen, unpublished observation

of rat liver and hepatoma mitochondria. The medium designed by Schnaitman and Greenawalt (1968); see Table III used in this procedure is one we have found to almost always provide optimal ACRs. ACRs obtained with rat liver mitochondria prepared in H-medium are usually greater than 4 when succinate is the substrate and greater than 6 when β-hydroxybutyrate is the substrate.

Equipment:

Three-milliliter chamber: This chamber must accommodate an oxygen electrode. Such a chamber is easy for any instrument shop to construct. It can be made of glass or plastic. The only requirements are that it have a flat base so that stirring is possible, and that it have two openings, one in which the oxygen electrode fits tightly and one for adding solutions. The latter should have a cap, so that during the oxygen uptake process the system is closed and back-diffusion of oxygen does not occur. Oxygen electrode chambers can be obtained from Oxygraph, Gilson Medical Electronics, Middleton, Wis.

Oxygen electrode: This can be obtained from Yellow Springs Instrument Company, Yellow Springs, Ohio.

Voltage box: The oxygen electrode operates at a voltage of 0.8 V. The voltage box contains a flashlight battery and a rheostat to adjust the voltage.

Sargent or equivalent recorder.

Overall setup: The oxygen electrode is contained within the 3-ml chamber and is connected to the voltage box. The voltage box in turn is connected to the recorder.

Solutions: 2X medium, containing, in 100 ml, 8.01 gm mannitol (440 mM), 4.7 gm sucrose (140 mM), 68 mg KH_2PO_4 (5.0 mM), 95.3 mg HEPES (4 mM), and 37.2 mg EDTA (1.0 mM), pH 7.4 (with KOH); succinate (1.0 M, pH 7.4) or β-hydroxybutyrate (1.0 M, pH 7.4); $MgCl_2$ (100 mM); ADP (10 mM, pH 7.4).

Method:

1. Place 1.4 ml 2X medium, 1.3 ml water, 15 μl succinate or 30 μl β-hydroxybutyrate, 50 μl $MgCl_2$, and 40 μl defatted BSA (50 mg/ml) into the 3-ml chamber equipped with a small stirring bar.

2. Adjust the voltage to ~0.8 V on the voltage box.

3. Calibrate the recorder so that full scale represents 100% oxygen in the system. This is done by shorting out the oxygen electrode set up at the level of the recorder and first adjusting the zero adjust so that the pen is at 0% full scale. The circuit is then reopened, and the pen is set at 100% full scale by adjusting the voltage on the voltage box.

4. Add 50 μl mitochondria (50 mg/ml) and allow the recorder to record the oxygen uptake process for about 1 minute (state IV respiration) and then return to the state IV rate.

5. Add 45 μl ADP. Respiration should increase rapidly (state III respiration) and then return to the state IV rate.

6. The ACR (see Fig. 14) is defined as

$$ACR = \frac{\text{state III rate of respiration}}{\text{state IV rate of respiration}}$$

By convention the state IV rate taken for the calculation is the rate occurring after ADP addition, i.e., the turn-off rate.

Note: From this type of experiment the P/O ratio of mitochondria can be calculated also. If it is assumed that all added ADP is converted to ATP, then the P/O ratio is calculated as

$$P/O = \frac{\text{nmoles ADP added}}{\text{natoms of oxygen consumed}}$$

The natoms of oxygen consumed is equal to the amount of oxygen consumed between the time ADP is added (initiation of state III respiration) and the time ADP is used up (the cessation or turn-off of state III respiration). By knowing that 1 ml of solution at room temperature contains approximately 480 natoms of oxygen, and that full scale on the recorder is equal to 100% oxygen in the system (480 $\times$ number of milliliters), the amount of oxygen consumed in any given time period can be calculated. For β-hydroxybutyrate the theoretical P/O ratio is 3.0; for succinate it is 2.0.

3. MEASUREMENT OF OLIGOMYCIN-INDUCED RESPIRATORY CONTROL IN SUBMITOCHONDRIAL PARTICLES

Comment: The procedure described here has been used in our laboratories for demonstrating respiratory control in sonic inner membrane particles (see Section XIII, C) of rat liver mitochondria. Respiratory control can be induced in these particles not only with oligomycin, but with DCCD and venturicidin as well. See Fig. 15B for an example of oligomycin-induced respiratory control.

Equipment: Oxygen electrode setup described in Section XVI, A, 2.

Solutions: Same solutions as for acceptor control measurements (see Section XVI, A, 2). Also 10 mM DNP and 100 μg/ml oligomycin.

Method:

1. To a 3-ml oxygen electrode chamber add 1.4 ml 2X medium, 1.3 ml H_2O, 15 μl succinate, and 50 μl $MgCl_2$.

2. Calibrate the recorder so that full scale is equivalent to 100% oxygen in the system.

3. Add 0.5 mg sonic inner membranes and allow respiration to proceed for 1–2 minutes.

4. Add 1.25 nmole oligomycin and allow respiration to proceed for another 1–2 minutes. Respiration should slow down dramatically if the

particles are vesicular and leaky to protons. (Oligomycin blocks the proton leak.)

5. Add 100 nmole DNP. Respiration should increase to at least the original rate and perhaps somewhat higher. (See Fig. 15B.)

B.   Electron Microscopy of Mitochondria and Submitochondrial Particles

1.   PREPARATION OF RAT LIVER MITOCHONDRIA AND SUBMITOCHONDRIAL
       PARTICLES FOR ELECTRON MICROSCOPY—CHEMICAL FIXATION,
       DEHYDRATION, THIN–SECTIONING, POSITIVE AND NEGATIVE STAINING

Comment: As indicated earlier, the major reason for viewing mitochondria in the electron microscope is to determine if they are ultrastructurally intact, i.e., whether or not the inner and outer membranes are unbroken and whether or not the matrix space stains densely. Also, the degree of gross contamination by other organelles can be estimated. In addition, submitochondrial preparations are often viewed to establish optimal conditions for disruption and thus to determine the extent of membrane vesiculation. If the membrane fragments are vesicular, what is the size range of vesicles and which side of the membrane is exposed to the external environment?

The techniques used most commonly to determine the above properties of mitochondrial and submitochondrial preparations are thin-sectioning and/or negative staining. Each of these procedures is carried out for a specific purpose. The object of the former technique is to preserve ultrastructural aspects of the specimen in the lifelike state. Preservation is primarily accomplished by the action of glutaraldehyde and/or OsO_4 on the biological specimen; these are widely used fixatives in electron microscopy (Glauert, 1974; Sabatini *et al.*, 1963).

The procedures leading to thin-sectioning as a method of ultrastructural analysis include chemical fixation, dehydration, infiltration, embedding, polymerization, and positive poststaining.

Negative staining (See Fig. 16), which can be used with or without fixation, is a relatively fast and simple technique which entails embedding the specimen in heavy-metal stains which dry as a glass to form a background of greater electron density than the unstained specimen and thus create a negative image of the specimen much like that of a bacterial suspension in india ink viewed in the light microscope (Brenner and Horne, 1959; Valentine and Horne, 1962). Positive staining (see Fig. 16), used in conjunction with a heavy-metal fixative (OsO_4) and thin-sectioning, yields the opposite image; i.e., the biological specimen acquires greater electron density than the embedding medium. Electron micrographs of mitochondrial

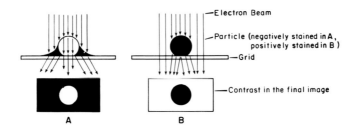

FIG. 16. The two general ways in which mitochondria or submitochondrial particles can be stained for viewing under the electron microscope. (A) Negative staining. Note that the stain surrounds or embeds the particles within an electron-dense material. (B) Positive staining. Note that the particles combine with the fixative and/or stain, usually containing a heavy metal of high atomic number In (A) the electron beam penetrates the particle itself but not the surrounding stain, whereas in (B) the electron beam penetrates the surroundings but not the particle. The contrast in the final image is completely reversed in the two cases. [After R. W. Horne (1967)]

and/or submitochondrial preparations subjected to positive or negative staining techniques described here are summarized in Fig. 17.

The methodology involved in preparing mitochondria for evaluation in the electron microscope is summarized here, together with a few statements about what each step involves and why it is carried out; the exact, detailed method we use then follows. Many variations of these basic approaches for thin-sectioning or negative staining can be obtained from the literature (Glauert, 1974; Kay, 1967; Hayat, 1970). These procedures have proved successful over many years' trial in our hands.

Chemical fixation: This step is carried out to stabilize the three-dimensional structure of the mitochondria. Double-fixation with glutaraldehyde followed with OsO_4 (postfixation) is most commonly used today. Glutaraldehyde forms inter- and intramolecular bonds between proteins Hayart, 1970) and probably between N-containing bases of phospholipids. It is in general, however, a poor lipid fixative (Glauert, 1974). OsO_4, however, strongly interacts with lipids (Hayat, 1970). In general, osmium fixation largely retards the extraction of lipids during subsequent dehydration, thereby increasing the contrast of the specimen against the background. Ca^{2+} (1–3 mM) can be added to a glutaraldehyde and/or osmium fixative to minimize lipid extraction (Hayat, 1970).

Dehydration: This step removes water from the specimen and is generally necessary because the specimen is later infiltrated with an epoxy resin which is relatively water-insoluble. The water is largely removed in a stepwise manner by exposing the fixed mitochondrial sample to a graded series (50–100%) of alcohol or acetone.

Infiltration, embedding, thin-sectioning:

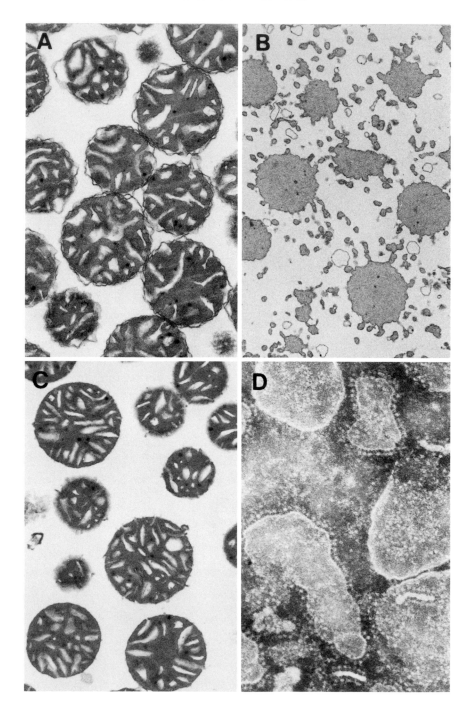

This step is the means of getting the epoxide into and around the mito-chondria. A block of uniform composition and hardness is obtained after embedding and polymerization, so that thin sections (400–900 nm) can be cut for examination under the electron microscope.

Poststaining (positive staining): A variety of stains and staining procedures is available. We generally use saturated *uranyl acetate* in 50% ethanol followed by *lead citrate* (Reynolds, 1963) which enhances the density of cellular structures, membranes, ribosomes, etc. (Hayat, 1970). Stains can be used separately or in combination. Lead citrate particularly gives a high-contrast image to membranes, while uranyl salts exhibit specificity toward nucleic acids. Within certain limits the enhancement of contrast increases resolution.

Equipment: Screw-capped vials or shell vials (45 × 15 mm) with no. 4 corks; Beckman Model 152 microfuge; Beckman 314326 micro test tubes; Beem capsules and holders, and pencil-written labels; Weck razor blades; pasteur pipets (long); graduated cylinders; beakers; flasks; brown, heavy-walled, ground-glass-stoppered bottles for fixatives; Parafilm; stirring apparatus; dessicator; silica gel; vial racks; Ladd Sonicor or equivalent; oven (with a controlled temperature of 60°C).

Solutions:

Stocks: Buffer A (0.2 *M* NaH$_2$PO$_4$); buffer B (0.2 *M* Na$_2$HPO$_4$); 23 ml of A plus 77 ml of B = pH 7.2 7.3. *Note*: Use at final concentration of 0.1 *M*.

Fixatives (1) Glutaraldehyde (3.0% in 0.1 M NaPO$_4$ buffer): To 58.3 ml of 0.2 *M* NaPO$_4$ buffer add distilled water to 100 ml then transfer quanti-tatively 5.0 ml of 70% Ladd glutaraldehyde (bring to final volume of 116.6 ml with distilled water) pH should remain 7.2–7.3. Store in the cold. (2) OsO$_4$ (2% in 0.1 *M* NaPO$_4$ buffer): Carefully break a 1/2 gm vial in a heavy-walled, ground-glass-stoppered reagent bottle *in a hood.* (*OsO$_4$ can cause permanent damage to the eyes and mucous membranes.*) Add 25 ml of 0.1 *M* NaPO$_4$ buffer. Seal tightly with Parafilm and store in the cold.

Fig. 17.   (A) Mitochondria isolated in H-medium and then fixed with glutaraldehyde followed by OsO$_4$. Poststaining was accomplished with uranyl acetate and lead citrate. (See text.) Note the dense, *positively stained* matrix and membranes. ×25,000. (B) Mitoplasts derived from digitonin treatment of mitochondria. Fixation and poststaining as in (A). Loss of typical cristal membrane profiles is apparent. ×25,000. (C) Mitoplasts prepared mechanic-ally by the passage of osmotically contracted mitochondria through a French pressure cell. Fixation and poststaining as in (A). The configuration and matrix density of the inner com-partment are very similar to those of intact mitochondria. × 25,000. (D) Submitochondrial vesicles *negatively stained* with potassium phosphotungstate at pH 6.8. Note the direction of the 90-Å particles (F$_1$-ATPase molecules) on the inner surface on the cristae. × 120,000.

Solubilization takes overnight on standing or 5 minutes with prolonged sonication in a Ladd Sonicor or equivalent.

Wash and storage medium (0.2 M sucrose in 0.1 M NaPO$_4$ buffer): Use between glutaraldehyde and OsO$_4$ fixations to remove unbound glutaraldehyde. Store in the cold.

Complete resin mixture (for embedding): Hardness and therefore cutting properties vary with composition. The mixture is altered accordingly for cutting sections with glass versus diamond knives as indicated in the accompanying table.

Chemicals for complete resin mixture	For glass knife	For diamond knife (Luft's 1:1)
Epoxy resin (Epon 812) (Luft, 1961)	50 ml	45 ml
Dodecenyl succinic anhydride (DDSA)	30 ml	30 ml
Nadic methyl anhydride (NMA)	20 ml	25 ml
Benzyldimethylamine (BDMA)	1 ml	—

Measure in the order DDSA, Epon, and NMA into a graduated cylinder. Mix immediately and thoroughly with a stirring bar. Add BDMA dropwise (slowly) while stirring. Avoid bubbles. If the solution is mixed 3 minutes with adequate stirring no "swirl" lines will be visible. (*Avoid contact with the skin*. If contact occurs wash the affected area at once with soap and water.) It is advisable to use a fume hood during this procedure. Both the complete resin mixture and the solvent–resin mixture (see below) can be stored frozen for long periods without an accelerator; with an accelerator present they can generally be stored for not more than 1 month. Let these mixtures warm to room temperatures before opening to avoid condensation of water vapor. Return them to the freezer as soon as possible.

Solvent-resin mixture: 1:1 mixture of complete resin and 100% acetone.

Aqueous acetone (for dehydration): 50%, 70%, 95%, 100%. Have one stock of 100% acetone stored *dry* or open a fresh container. Store cold, with an additional 100% stock bottle kept at room temperature.

Method:

a. Fixation

1. To obtain a microfuge pellet about 1/2 mm thick, place 10 μl of 10 mg protein per milliliter (i.e., a 0.1-mg sample of mitochondrial protein) or a corresponding amount of mitochondrial suspension in a precooled microfuge tube containing 300 μl of cold glutaraldehyde fixative. (Do not significantly dilute the fixative with the sample volume.) If preferred, micropellets can be made before the addition of fixative by spinning for 30 seconds to 1 minute. In either case fix at 0–4°C for 2 hours.

Note: Submitochondrial membrane preparations may require the following special treatment. Sediment in microfuge tubes placed in capped Spinco tubes (12 ml) containing medium of same density as that in the microfuge tubes (isolation medium can generally be used). Add medium to the Spinco tube until the microfuge tube just floats. Balance the tubes carefully. Centrifuge at 40,000 rpm in the Spinco No. 40 rotor. Remove, the supernatant fluid and add fixative to the pellets.

2. Discard the glutaraldehyde by shaking (leaving a small amount in the tube to keep the specimen wet) and replace it with two changes of 0.2 M sucrose in 0.1 M NaPO$_4$ (wash and storage medium) and store overnight in the last change at 0–4°C. If the pellet loosens, centrifuge for an additional 15 seconds in the microfuge.

3. Remove the wash and storage medium and replace it with OsO$_4$ fixative; fix for 2 hours at 0–4°C. *Beware of fumes and contact with the skin—work in a hood.* Wash the skin thoroughly with water upon contact with OsO$_4$. *Note*: Always discard OsO$_4$ in tightly capped waste containers for dangerous materials.

b. Dehydration

1. Remove the OsO$_4$ with a pasteur pipet and dispose of it as noted above. Rinse the tube and the surface of the pellet with 50% cold acetone and remove it at once with a pasteur pipet to avoid a residue of osmium.

2. Add cold 70% acetone two times for 10 minutes each time.

3. Discard the acetone by shaking and replace it with cold 95% acetone twice (10 minutes each time). Discard by shaking. *Be careful not to shake out the specimens at this point.* (At this time remove solvent resin mix from the freezer and bring it to room temperature.) Also, prepare glass vials with labels, and label the covers and cork stoppers. Fill the vials with 100% acetone (dry and at room temperature) and stopper.

4. Add cold 100% acetone and remove the specimen from ice; allow it to come to room temperature (15 minutes).

5. To remove micropellets from tubes, cut the solid bottom end from the microfuge tube with a razor blade, leaving a thin wall of plastic at the bottom between the tube and the specimen pellet. While inverting the tube, use a blunt object to press on the residual, thin, plastic wall and dislodge the pellet so that it moves about a centimeter from the bottom. Now cut off the bottom of the tube far enough from the bottom so that an opening is formed through which the pellet can pass. Return the tube to an upright position. Allow the pellet to settle to the opening; squeeze it out in the first drop into the vial filled with dry, 100% acetone and let stand for 15 minutes.

c. Infiltration. Pour off the 100% acetone and replace it with the solvent–resin mixture. Place the specimen to be infiltrated in a dessicator overnight with dessicant but without vacuum. Mark the level of the liquid in the vials and remove the corks.

When half-volume in the vials has been reached, most of the solvent is gone. Replace it with fresh, complete resin mixture prepared the previous day. (Bring the resin to room temperature before opening.) Allow the mitochondrial specimen to stand in the resin mixture for approximately 1 hour.

d. Embedding. Prepare the Beem capsules with labels and resin. Transfer the specimen to a capsule with large-bore pasteur pipet in the minimum amount of resin mixture coming from the vial. (A pasteur pipet can be used most effectively if first scored with a diamond marking pencil and then the tip removed up to a sufficiently large diameter to permit easy pipetting of the resin.) Place the specimen in a 60°C oven overnight (~22 hours) to polymerize the resin. *Note*: If the Beem capsule holder is tilted approximately 45°, many of the micropellets will come to rest in such a manner as to permit the entire pellet to be sectioned approximately parallel to the axis of sedimentation.

2. PREPARATION OF SUBMITOCHONDRIAL PARTICLES FOR VIEWING 90-Å ATPase PARTICLES—NEGATIVE STAINING WITH POTASSIUM PHOSPHOTUNGSTATE WITHOUT FIXATION

Comment: Submitochondrial particles are frequently viewed in the electron microscope after negative staining with potassium phosphotungstate. Under such conditions the investigator can obtain some indication as to whether the membrane vesicles have the 90-Å particles on the outer or inner surface, i.e., whether the vesicles are right side out or inside out. Since the negative stain surrounds or embeds the vesicles in an electron-dense material, the contrast of the final image is reversed (Fig. 17) when compared with that of positively stained preparations (Fig. 17A–C). The extent to which negative stains interact with or change mitochondrial membranes is not clear; there is some evidence that potassium phosphotungstate may actually react with the ATPase particles and tend to "pull" them out of the membranes (Catterall and Pedersen, 1970, G. L. Decker, J. W. Greenawalt, and P. L. Pedersen, unpublished observations). Nevertheless, the technique is still valuable as a method to determine in which direction the ATPase particles are oriented. In addition, several heavy-metal negative stains have been used effectively in the study of viral (Valentine and Horne, 1962; Horne and Wildy, 1963) and bacterial (Glauert *et al.*, 1963; Weigand *et al.*, 1973) substructure and of purified proteins, e.g., enzymes (Horne and Greville, 1963).

Method:

1. Dilute a mitochondrial stock suspension (50 mg protein/ml) in water to an optical density of 0.6–1.0 at 520 nm and then mix 1 : 1 with 2% potassium phosphotungstate. The pH of the phosphotungstate should be adjusted to 6.8 with KOH prior to dilution.

2. Add a drop of the diluted specimen to a Formvar- or carbon-coated grid. (We have used both and find no great advantage of carbon over Formvar for *most* routine purposes.)

3. Allow a droplet of the suspension to remain on the grid for about 2 minutes in order for the particles to settle and to adhere to the surface of the grid. (*This is a critical step*; if too short a time is used, uniform spreading of the sample on the grid does not occur.)

4. Touch the edge of the grid with a piece of Whatman No. 1 filter paper, drawing off most of the droplet but leaving a uniform thin film on the grid.

5. Let the grid air-dry at room temperature for a few minutes. The preparation is ready for viewing in the electron microscope.

6. View in the electron microscope at a magnification of about ×40,000 to observe projected ATPase particles. (A particle 100 Å in diameter is ∼0.4 mm on the final viewing screen at this magnification.)

ACKNOWLEDGMENTS

This article was written while the principal author (P.L.P.) was supported by NIH Grant CA-10951 from the National Cancer Institute and NSF Grant PCM-7611024. Other grant support, without which this article would not have been possible, includes NIH Grant NS-12016 to J. W. Greenawalt, Contract NO1-45610 from the National Cancer Institute to Albert L. Lehninger, and NIH Grant CA 10729 to Harold P. Morris. Ernesto Bustamante is a fellow on leave from the Universidad Peruana Cayetano Heredia (Lima, Peru), who was supported by fellowships from the Lilly Research Laboratories, the DuPont Company, and the Ford Foundation. John W. Soper was supported by NIH Fellowship CA 05055.

The authors also acknowledge Dr. Martin Brand for his help in supplying us with details of the digitonin procedure used for removing lysosomal contamination from mitochondria, Dr. Terence Spencer for information about mitochondrial isolation and acceptor control media, Dr. Janna Wehrle for information about the properties of submitochondrial particles, and Dr. Youssef Hatefi for information pertaining to the isolation of rat liver mitochondria by the Nagarse procedure.

REFERENCES

Aas, M. (1973). *Proc. Int. Congr. Biochem.*, *9th*, *1973* Abstract, p. 31.

Brand, M. D., Reynafarje, B., and Lehningar, A. L. (1976). *Proc. Natl. Acad. Sci. U.S.A.* **73**, 437.

Brenner, S., and Horne, R. W. (1959). *Biochim. Biophys. Acta* **34**, 103.

Brown, D. H. (1968). *Biochim. Biophys. Acta* **16**, 152.

Bustamante, E., Soper, J. W., and Pedersen, P. L. (1977). *Anal. Biochem.* **80**, 401–408.

Caplan, A. I. (1966). Ph.D. Thesis, Johns Hopkins School of Medicine, Baltimore, Maryland.

Caplan, A. I., and Greenawalt, J. W. (1966). *J. Cell Biol.* **31**, 455.

Caplan, A. I., and Greenawalt J. W. (1968). *J. Cell Biol.* **36**, 15.

Catterall, W. A., and Pedersen, P. L. (1970). *Biochem. Biophys. Res. Commun.* **38**, 400.

Catterall, W. A., and Pedersen, P. L. (1971). *J. Biol. Chem.* **246**, 4987.

Chan, S. H. P., and Barbour, R. L. (1976). *Biochim. Biophys. Acta* **430**, 426.

Chan, T. L., Greenawalt, J. W., and Pedersen, P. L. (1970). *J. Cell Biol.* **45**, 291.

Chance, B., and Williams, G. R. (1955). *J. Biol. Chem.* **217**, 383.

Chance, B., and Hagihara, B. (1960). *Biochem. Biophys. Res. Commun.* **3**, 6.

Chappell, J. B., and Hansford, R. G. (1972). *In* "Subcellular Components Preparation and Fractionation" (G. D. Birnie, ed.), 2nd ed., p. 77. Cambridge Univ. Press, London and New York.

Cintrón, N. and Pedersen, P. L. (1978). In "Methods in Enzymology" (in press).

Decker, G. L., and Greenawalt, J. W. (1977). *Ultrastruct. Res.* **59**, 44–56.

Devlin, T. M. (1967). *In* "Methods in Enzymology" (R. W. Estabrook and M. E. Pullman, eds.), Vol. 10, p. 110. Academic Press, New York.

Devlin, T. M., and Lehninger, A. L. (1958). *J. Biol. Chem.* **233**, 1586.

Englund, P. T. (1971). *Anal. Biochem.* **40**, 490.

Ernster, L., Nordenbrand, K., Chude, O., and Juntti, K. (1974). *In* "Membrane Proteins in Transport and Phosphorylation" (G. F. Azzone *et al.*, eds.), p. 29. North-Holland Publ., Amesterdam.

Fernández-Morán, H. (1962). *Circulation* **26**, 1039.

Fessenden, J. M., and Racker, E. (1966). *J. Biol. Chem.* **241**, 2483.

Glauert, A. M., ed. (1974) "Practical Methods in Electron Microscopy," Vol. 3. Am. Elsevier, New York.

Glauert, A. M., Kerridge, D., and Horne, R. W. (1963). *J. Cell Biol.* **18**, 327.

Glew, R. H., Kayman, S. C., and Kuhlenschmidt, M. S. (1973). *J. Biol. Chem.* **248**, 3137.

Good, N. E., Winget, G. D., Winter, W., Connolloy, T. N., Izawa, S., and Singh, R. M. M. (1966). *Biochemistry* **5**, 467.

Goyer, R. A., and Krall, M. (1969). *J. Cell Biol.* **41**, 393.

Greenawalt, J. W. (1969). *Fed. Proc., Fed. Am. Soc. Exp. Biol.* **28**, 663.

Hackenbrock, C. R., and Hammon, K. M. (1975). *J. Biol. Chem.* **250**, 9185.

Hayat, M. A., ed. (1970). "Principles and Techniques of Electron Microscopy; Biological Application." Vol. 1. Van Nostrand-Reinhold, Princeton, New Jersey.

Higgins, G. M., and Anderson, R. M. (1931). *Arch. Pathol.* **12**, 186.

Horne, R. W. (1967). *In* "Techniques for Electron Microscopy" (D. H. Kay, ed.), p. 328. Davis, Philadelphia, Pennsylvania.

Horne, R. W., and Greville, G. D. (1963). *J. Mol. Biol.* **6**, 506.

Horne, R. W., and Wildy, P. (1963). *Adv. Virus Res.* **10**, 101.

Johnson, D., and Lardy, H. A. (1967). *In* "Methods in Enzymology" (R. W. Estabrook and M. E. Pullman, eds.), Vol. 10, p. 94.

Kaschnitz, R. M., Morris, H. P., and Hatefi, Y. (1976) *Arch. Biochem. Biophys.* **449**, 224.

Kaschnitz, R. M., Hatefi, Y., Pedersen, P. L., and Morris, H. P. (1978). ·*In* "Methods in Enzymology" (in press).

Kay, D., ed. (1967). "Techniques for Electron Microscopy." Blackwell, Oxford.

Lambeth, D. O., and Lardy, H. A. (1971). *Eur. J. Biochem.* **22**, 355.

Lee, C. P. (1963). *Fed. Proc., Fed. Am. Soc. Exp. Biol.* **22**, 527.

Lee, C. P., and Ernster, L. (1965). *Biochem. Biophys. Res. Commun.* **18**, 523.

Lee, C. P., and Ernster, L. (1968). *In* "Regulation of Metabolic Processes in Mitochondria" (J. M. Tager *et al.*, eds.), p. 218. Adriatic Editrice, Bari.

Loewenstein, J., Scholte, H. R., and Wit-Peeters, E. M. (1970). *Biochim. Biophys. Acta* **223**, 432.

Luft, J. H. (1961). *J. Biophys. Biochem. Cytol.* **9**, 409.

Malviya, A. N., Parsa, B., Yokaiden, R. E., and Elliot, W. B. (1968). *Biochim. Biophys. Acta* **162**, 195.

Mitchell, P. (1966). "Chemiosmotic Coupling in Oxidative and Photosynthetic Phosphorylation." Glynn Res., Bodmin, Cornwall, England.

Pedersen, P. L., and Morris, H. P. (1974). *J. Biol. Chem.* **249**, 3327.

Pedersen, P. L., and Schnaitman, C. A. (1969). *J. Biol. Chem.* **244**, 5065.

Pedersen, P. L., Greenawalt, J. W., Chan, T. L., and Morris, H. P. (1970). *Cancer Res.* **30**, 2620.

Pedersen, P. L., Eska, T., Morris, H. P., and Catterall, W. A. (1971). *Proc. Natl. Acad. Sci. U.S.A.* **68**, 1079.

Racker, E., Burnstein, C., Loyter, A., and Christiansen, R. O. (1970). *In* "Electron Transport and Energy Conservation" (J. M. Tager *et al.*, eds.), p. 235. Adriatica Editrice, Bari.

Reed, K. C. (1972). *Anal. Biochem.* **50**, 206.

Reynolds, E. S. (1963). *J. Cell Biol.* **17**, 208.

Sabatini, D. D., Bensch, K., and Barrnett, R. J. (1963). *J. Cell Biol.* **17**, 19.

Schnaitman, C. A. (1969). *Proc. Natl. Acad. Sci. U.S.A.* **63**, 412.

Schnaitman, C. A., and Greenawalt, J. W. (1968). *J. Cell Biol.* **38**, 158.

Schnaitman, C. A., and Pedersen, P. L. (1968). *Biochem. Biophys. Res. Commun.* **30**, 428.

Schnaitman, C. A., Erwin, V. G., and Greenawalt, J. W. (1967). *J. Cell Biol.* **32**, 719.

Schneider, W. C. (1948). *J. Biol. Chem.* **176**, 259.

Schneider, W. C., and Hogeboom, G. H. (1950). *J. Biol. Chem.* **183**, 123.

Skidmore, W. D., and Catravas, G. N. (1970). *Armed Forces Radiobiol. Res. Inst. Rep.* **AFRRI SR-70-12**, 3.

Smith, D., Walborg, E., and Chang, J. (1970). *Cancer Res.* **30**, 2306.

Soper, J. W., and Pedersen, P. L. (1976). *Biochemistry* **15**, 2682.

Sordahl, L. A., and Schwartz, A. (1971). *Methods Cancer Res.* **6**, 159.

Sordahl, L. A., Blailock, Z. R., Liebelt, A. G., Kraft, G. H., and Schwartz, A. (1969). *Cancer Res.* **29**, 2002.

Sottocasa, G. L., Kuylenstierna, B., Ernster, L., and Bergstrand, A. (1967). *In* "Methods in Enzymology" (R. W. Estabrook and M. E. Pullman, eds.), Vol. 10, p. 448. Academic Press, New York.

Spencer, T., and Bygrave, F. L. (1972). *Biochem. J.* **129**, 355.

Thorne, R. F. W., and Bygrave, F. L. (1973). *Cancer Res.* **33**, 2562.

Valentine, R. C., and Horne, R. W. (1962). *Symp. Int. Soc. Cell Biol.* **1**, 263.

Waino, W. W. (1970). "The Mammalian Mitochondrial Respiratory Chain," p. 15. Academic Press, New York.

Wattiaux, R. (1974). *Mol. Cell. Biochem.* **4**, 21.

Wattiaux, R., Wibo, M., and Baudhuin, P. (1963). *Lysosomes, Ciba Found. 1963* p. 176.

Wattiaux, R., Wattiaux-De Conick, S., and Ronveaux-Dupal, M.-F. (1971). *Eur. J. Biochem.* **22**, 31.

Weigand, R. A., Holt, S. C., Shively, J. M., Decker, G. L., and Greenawalt, J. W. (1973). *J. Bacteriol.* **113**, 433.

Winkler, H. H., and Lehninger, A. L., (1968). *J. Biol. Chem.* **243**, 3000.

Chapter 27

The Use of Complement-Mediated Cytolysis to Analyze Specific Cell Populations of Eukaryote Cells in Cultures

S. M. CLARKE, P. F. KOHLER, AND L. M. FINK

Departments of Pathology and Medicine, University of Colorado Medical Center, Denver, Colorado

I.   Introduction

A.   Background

The use of antibodies against specific cell surface antigens for cell identification has been practiced for many years. The technique was first used by
immunologists to type red blood cells. Subsequently, it was found that antibodies could be tagged with fluorescent molecules, enzymes, and ferritin.
These generated colored or electron-dense moieties so that cells with unique
antigens could be identified by fluorescence, light, or electron microscopy.
After the introduction of techniques for culturing cells *in vitro*, it became
important to be able to isolate and propagate types of cells from mixtures
of cells. The complement-mediated antibody cytotoxicity assay has been
used in studies on the changes in cell surface antigens during neoplastic
transformation, the differentiation of mammalian cells, the chromosomal
localization of specific genes for cell surface antigens, the removal of
specific types of cells from mixtures so that biochemical analysis can be
performed, and the detection and isolation of specific populations of
lymphocytes. In essence, the development of antibody cytolysis techniques
allows any investigator, who wishes to prepare antibodies against cell
surface antigens, the capability to sort cells from populations which are
quantitatively or qualitatively heterogeneous for the antigen under study.
This can be done without extremely complex and expensive equipment.
The ensuing discussion briefly reviews the mechanism by which activation
of complement is thought to lyse cells, the techniques for the use of
complement-mediated antibody cytotoxicity to analyze populations of
neoplastically transformed and untransformed cells, and how other
investigators have used these techniques.

B.   The Mechanism of Antibody–Complement Cytolysis

Complement was first recognized in 1898 by Bordet who found that a
heat-stable antibody against *Cholera vibrio* could not lyse this organism
without the participation of a heat-labile serum cofactor. This heat-
sensitive serum cofactor was called complement, a designation currently
recognized as a collective term for a complex system of sequentially inter-
acting serum proteins and inhibitor proteins. Eleven proteins participate
in the classic reaction sequence leading to cytolysis. The nomenclature
and properties of the classic human complement proteins are shown in
Table I.

The entire mechanism of antibody–complement cytolysis is not completely understood (for a review, see Kohler, 1977). The overall reaction

TABLE I

PROPERTIES OF HUMAN COMPLEMENT SYSTEM PROTEINS

Classic component	Molecular weight	Mobility	Number of chains	Serum concentration ($\mu g/ml$)
C1q	400,000	γ	18	200
C1r	190,000	β	2	100
C1s	86,000	α	1	110
C2	117,000	β	1	30
C3	180,000	β^2	2	1500
C4	204,000	β^2	3	525
C5	180,000	β^1	2	75
C6	95,000	$\beta^?$	1	75
C7	110,000	β^2	1	55
C8	163,000	γ^1	1	80
C9	79,000	α^1	1	230

can be very simply described as a specific cell surface antigen reacting with the antibody made against it to form a surface antigen–antibody complex. Complement recognizes this complex, and through a series of reactions the cell is lysed (Fig. 1). The antigen can be any surface molecule capable of eliciting an antibody response. The classic pathway of immune cytolysis is triggered when a single IgM or two adjacent IgG antibody molecules interact with surface antigen.

The action of the 9 components of complement (11 proteins) can be best understood by grouping them into three functional units. These units are reversible macromolecular complexes in their native form but become transformed into functional units on cell membranes upon activation. The first unit is composed of C1 which is three proteins designated C1q, C1r, and C1s. This unit recognizes the immune complexes and forms the activated initiating enzyme C1. The second unit consists of C4, C2, and C3. These serum proteins are responsible for activation and amplification by assembly of the C3 and C5 converting enzymes. The last functional unit

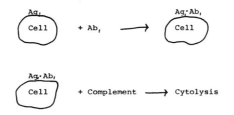

FIG. 1.   Overall reaction of complement-mediated cytolysis.

(C5b, C6, C7, C8, and C9) is the terminal membrane-attacking unit and is formed by noncovalent and nonenzymic assembly.

The interaction of the complement components with the cell antigen–antibody complex is illustrated in Fig. 2. The interaction of one IgM molecule or two adjacent IgG molecules with a membrane antigen establishes the recognition site for C1q binding. C1q becomes the molecular bridge which links C1r and C1s subunits to the IgM or IgG molecules. On binding it is thought that C1q undergoes a conformational change which converts C1r into an active internal enzyme. C1r then cleaves C1s to generate a serine protease (C̄1s). C4 and C2 are natural substrates for C̄1s. C4 is cleaved into C4a and C4b fragments. C4b has a nascent binding region for membrane sites that is separate from the antibody-combining site and attaches in a random manner on the cell membrane. C2 is also cleaved into C2a and C2b fragments. C2a attaches to C4b, establishing the C3 splitting enzyme, C3 convertase, which cleaves the C3 molecule into C3a and C3b. When C3b binds in close proximity to the C4b2a membrane-bound complex, the final enzyme, C5 convertase, is formed. C5 is cleaved by C5 convertase and attaches with C6 and C7 to a third topographically distinct membrane site. This complex forms a receptor for C8 and C9. The hydrophobic polypeptide chains from C5b and C7 subunits become inserted into the phospholipid bilayer of the cell membrane. Subsequent reactions with C8 and as many as six molecules of C9 are believed to open a channel across the membrane with the passage of water into the cell, resulting in osmotic lysis. In general, the cytolytic activity is close to the antigen–antibody site because the activated complexes of complement are extremely unstable and/or susceptible to inhibitors. However, reactive lysis can occur if innocent bystanders (cells not containing Ag–Ab complexes) are extremely close to the specific target cells.

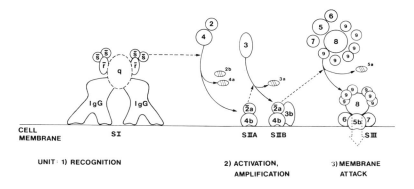

FIG. 2.   A model for complement activation by Ag–Ab surface complexes.

II.   Preparation of Reagents for Complement-Mediated Cytolysis

A.   Antibody Preparation

In the preparation of an antibody, it is ideal to use a pure antigen as the immunogen. However, cell surface antigens can be very difficult to obtain in pure form. This is especially true if one is interested in membrane proteins. Cell surface antigens have been obtained by extraction of cells with certain solvents or treatment with enzymes. For example, 0.2–1 M urea (Yamada and Weston, 1974; Yamada et al., 1975) and immobilized papain (Ruoslahti et al., 1973) have been used to obtain antigens from fibroblasts in culture. These techniques frequently result in the removal of more than one antigen from the cell surface. Either further purification prior to immunization or selective adsorption of the immune serum is often necessary for meaningful studies.

A useful laboratory technique in studying membrane components has been sodium dodecyl sulfate polyacrylamide gel electrophoresis (SDS-PAGE). This procedure keeps the proteins solubilized and separates them according to molecular weight (with the exception of those that contain large amounts of carbohydrate or are extremely hydrophobic). Since protein separation is possible, we suggest this as an alternative method for obtaining antigens when conventional chemical techniques are not successful. The primary objection to this method of obtaining antigens for injection is the possibility that more than one protein will migrate at the same rate. Nonetheless, we have found this to be a valuable method of obtaining antigens for the production of antibodies against very small amounts of specific cell surface components.

A typical procedure is to run a preparative SDS–PAG, loading as much of a cell lysate or membrane lysate as possible and still obtaining good protein separation. The gel is stained with 1% coomassie blue in 50% trichloroacetic acid for 1 hour at 37°C, followed by destaining in 5% methyl alcohol and 10% glacial acetic acid. The specific protein band, representing the antigen of interest, is cut out of the gel, emulsified in phosphate-buffered saline (PBS), and diluted 1:1 with Freund's complete adjuvant. After the desired band is cut out of the first electrophoresis gel, further purification can be obtained, if necessary, by rinsing it in electrophoresis buffer and placing the gel band on top of another gel. Warm 1% agarose is used to seal the connection, and the gel is rerun. The protein band is restained, isolated, and prepared for injection.

The adjuvant and sample protein are mixed until a very stiff sample is obtained. Small aliquots (approximately 0.1 ml) are injected subcutaneously into a rabbit along both sides of the back. Fourteen days later a second preparative gel is run and stained and the same protein band removed. This is emulsified in PBS, diluted 1:1 with Freund's incomplete adjuvant, and injected in a similar manner. On days 21, 22, and 23, after the initial injection of pure protein, increasingly larger amounts of crude protein (e.g., whole-cell extracts) are injected intravenously into the ear vein. This allows the animal to respond anamnestically to the first antigen. The rabbit is exsanguinated on day 30 before it can respond to the crude proteins injected. Intravenous saline profusion and canulization of the aorta give the best yield. Before any injections are made, it is important to bleed the animal for control serum. Table II gives the injection schedule.

TABLE II

Day	Injection material
0	Band from SDS-PAG with Freund's complete adjuvant injected subcutaneously at multiple sites along the back.
14	Band from SDS-PAG with Freund's incomplete adjuvant injected subcutaneously at multiple sites along the back.
21	Crude protein containing approximately 100 μg of the protein of interest injected intravenously
22	Crude protein containing approximately 1 mg of the protein of interest injected intravenously
23	Crude protein containing approximately 2 mg of the protein of interest injected intravenously
30	Exsanguination of the rabbit

B.   Antibody Testing

Testing the antiserum for specificity can also be a problem when insoluble proteins that are difficult to purify have been used as the antigen. Immunoprecipitation techniques, as well as most other standard immunological procedures for testing specificity, usually cannot be used because of nonspecific precipitation caused by solubility problems. When reagents such as SDS or urea are used to keep the proteins in solution, the antigen–antibody reaction may be inhibited.

The Ouchterlony immunodiffusion technique utilizing 1% agarose with

0.05% SDS is sometimes useful (Ouchterlony and Nilsson, 1973). The antigen usually is solubilized in no greater than a 1 M urea solution or 0.05% SDS. However, some investigators have reported using 8 M urea with retention of antigenicity (Ruoslanti and Vaheri, 1974). Nonionic detergents in concentrations of 1–2% have also been used to solubilize some surface proteins. The sensitivity of this reaction can be increased by utilizing radioactive labeled antigens (surface and/or metabolic labels) and autoradiographing the dried slide. The same applies for crossed immunoelectrophoresis.

Indirect immunofluorescent staining of polyacrylamide gels, as described by Stumph *et al.* (1974), can be useful in looking at the antiserum specificity of insoluble antigens. The antigen is electrophoresed in PAG, and rabbit antiserum is placed over the gel. The gel is then rinsed well, and a fluorescent goat anti–rabbit IgG is put on the gel. The fluorescent conjugate combines with any antigen–antibody complexes and thus localizes them by fluorescence. The disadvantages are the large amounts of antiserum necessary to cover the gel adequately and the problem of antigens diffusing out of the gel before antigen—antibody precipitation occurs.

Affinity chromatography utilizing cyanogen bromide-activated Sepharose columns may be useful in picking out the antigen(s) that react with the antibodies bound to the column. However, when the specimen to be tested contains thousands of proteins (e.g., whole-cell extracts), some of which may be large and/or sticky, the column may become overloaded and lose its specificity. This problem can be partially alleviated by making membrane preparations. A second problem associated with this method of testing is elution of the bound material. We have found that 3 M sodium thiocyanate, 10 mM NaH$_2$PO$_4$, pH 6, is superior to the glycine—HCl buffer, pH 2.5. However, it may still be difficult to release all bound material.

If the antigen is known to be present on one cell type and not on another, adsorption coupled with indirect fluorescent staining of the cells can be done. Cells having the surface antigen should fluoresce when treated with the antiserum followed by fluorescein-conjugated goat anti–rabbit IgG (Pollack and Rifkin, 1975). Cells without the antigen and cells treated with normal rabbit serum should show no fluorescence. When the antiserum is incubated 1 hour at 37°C with the cells containing antigen (0.5 ml antiserum per 25 × 10⁶ cells), its staining ability, as well as its cytotoxicity, should be decreased or lost completely (Oda and Puck, 1961).

In summary, testing for specificity of antiserum made against cell surface antigens is difficult. The best approach is to use several methods and compile all the data obtained.

C. Complement Preparation

Guinea pig, rabbit, and chicken complement have been used in cytotoxicity assays. In our testing with rabbit antibodies we find that normal rabbit serum, though occasionally toxic, is a superior source of complement because of its higher titer.

Normal rabbits are bled from the marginal ear vein or intracardiac puncture, and the blood is permitted to clot at 22°C. The serum is collected, filtered for sterility, divided into small aliquots, and frozen at below − 20°C. This type of handling decreases the loss of activity which occurs with thawing and refreezing. Frozen complement can be stored for a maximum of 6 weeks.

After filtration, each lot of rabbit serum is tested for its nonspecific cytotoxicity by plotting the percent of complement used versus the number of colonies obtained (Table III). It may frequently take several animals to find a satisfactory source of complement. Lots that show high toxicity at low concentrations are discarded. If very high concentrations of complement are required for cytotoxicity, small amounts of naturally occurring cytotoxic antibodies can be removed by adsorbing the serum to be used as a complement source with the test cells at 0–4°C in the absence of calcium (Boyse *et al.*, 1970). The cytotoxicity assay is standardized by using different concentrations of complement with various dilutions of antiserum (Table IV). This allows the investigator to know which conditions allow for optimum killing.

TABLE III

COMPLEMENT CYTOTOXICITY[a]

Complement (%)	Number of colonies
0.1	61
0.5	69
1.0	42
2.0	12
3.0	2

[a]Two thousand Syrian hamster embryo fibroblasts were plated in 1 ml MEM plus 5% fetal calf serum in 35-mm petri dishes and incubated 24 hours at 37°C before the addition of complement. Colonies were stained and counted after 7 days of incubation.

D. Test Cells Preparation

The cells to be tested can be maintained in suspension or monolayer using the medium best suited for their growth. At the time of testing they

TABLE IV

STANDARDIZATION OF CYTOTOXICITY ASSAY[a]

Complement concentration (%)	Final antiserum dilutions					
	1:1600	1:1200	1:800	1:400	1:200	0
0	25	20	21	24	20	0
0.1	21	25	24	19	5	0
0.5	17	9	4	2	1	6
1.0	2	5	3	1	0	2
2	1	0	0	0	6	0
3	0	0	0	0	0	0

[a]Colony counts were obtained with various concentrations of antiserum and complement incorporated with the medium in plates seeded with 2000 Syrian hamster embryo fibroblasts in 35-mm petri dishes.

are counted and placed in petri dishes at the density desired. If the surface antigen is sensitive to trypsinization, ethylene glycol–tetracetic acid (EGTA, 0.25 mM) or ethylenediamine tetracetate acid disodium salt (EDTA, 0.1 mM) can be used for cell dispersion from monolayer growth. If renewal of the cell surface antigen occurs after trypsinization, the assay can be carried out at times after protease treatment when the antigen is restored to the cell surface.

III. Cytotoxicity Assay

A. Sample Protocol Used for Assaying for the Presence of a 220,000-Molecular-Weight Cell Surface Protein

The test cells are trypsinized, counted, and plated in 35-mm petri dishes. The initial density depends on the plating efficiency of the cells used. For Syrian hamster embryo fibroblasts with a plating efficiency of 1–2%, 2×10^3 cells are plated in 1 ml of Eagle's minimal essential medium (MEM) supplemented with twice the concentration of amino acids and vitamins plus 5% fetal calf serum. The cells are incubated at 37°C for 24 hours to allow for attachment and regeneration of surface proteins lost by trypsinization. The dishes are divided into five sets and treated by adding 10 μl complement to the first set, 50 μ heat-inactivated antiserum diluted 1:40 in PBS to the second, 50 μl heat-inactivated normal rabbit serum (1:40 dilution) plus 10 μl complement to the third, and 50 μl antiserum

(1:40 dilution) plus 10 μl complement to the fourth. Nothing is added to the fifth set. The plates are mixed well and incubated at 37°C, 5% CO_2 for 5–7 days. The colonies are fixed with methanol for 5 minutes, stained with 10% Giemsa for 5 minutes, rinsed, and counted macroscopically.

B.   Other Methods for Quantitating Cytotoxic Effects

Another method used in determining cell survival is ^{51}Cr release. In this procedure, semiconfluent cells are washed well in fresh medium without serum and then exposed to approximately 10 μCi ^{51}Cr for 90 minutes prior to exposure to the antiserum. The cytotoxic effect of the antiserum is measured by the release of ^{51}Cr. Aliquots are taken at various time intervals and cooled; Bray's fluid is added, and the sample counted on a liquid scintillation spectrometer with adjustments similar to those used for tritium measurements (Rieber and Romano, 1976).

Since antibody cytotoxicity of cells in the complement-mediated reaction occurs within several hours, the uptake of vital dyes, such as trypan blue, can also be used to monitor the killing effect. It is possible to use a light-activated cell counter to count total cells and dye-containing cells (nonviable), and a cell sorter to separate cells containing dye from those not containing dye (Horan and Kappler, 1976). These methods can give rapid analysis of the antigenic character of the surface of cells in mixed populations.

C.   Standardization Curves

In order to obtain the replicate plating efficiency (RPE), cells are plated with complement and different concentrations of antiserum. The amount of complement to be used can be determined from the complement cytotoxicity (Table III), but is usually about 1% of the final concentration. Control plates with and without complement are set up. The plate containing complement only is designated as having 100% efficiency. From the replicate plating efficiency data, survival curves can be made by plotting the natural log of the RPE versus the antiserum concentration.

D.   Cytotoxic Effects of a Cell Surface Antibody on   Untransformed and Transformed Cells

An antiserum was made against an untransformed Syrian hamster cell surface protein that migrates in the 220,000-molecular-weight range on 7.5% SDS-PAG, using the procedure previously described. This protein

TABLE V

EFFECT OF ANTI–220,000-MOLECULAR-WEIGHT ON TERTIARY SYRIAN
HAMSTER CELLS[a]

Treatment of cells	Number of surviving colonies[b]
Media alone	19[c]
Rabbit complement	25[c]
Antiserum	23[c]
Complement and normal rabbit serum	23[c]
Complement and antiserum	1[d]

[a]The protocol described in Section III, A was used for these experiments.

[b]Colonies macroscopically scored after a 7-day incubation. Initial plating was 2000 cells per 35-mm plate. The average of 12 separate experiments performed in duplicate is given.

[c]Not significantly different ($P < .05$).

[d]Significantly different ($P < .05$).

was absent or decreased in amount on the surface of several chemically transformed Syrian hamster cells, as shown by [^{125}I]lactoperoxidase labeling of cell surface proteins. By use of the cytotoxicity assay it was found that this antiserum, undiluted and in titers up to 1:30 dilution (the final dilution of antiserum in the assay was 1:600), killed 100% of the untransformed hamster embryo fibroblasts. Complement alone, antiserum alone, and normal rabbit plus complement showed negligible effects on the cells

TABLE VI

COMPARISON OF THE CYTOTOXICITY OF ANTI–
220,000-MOLECULAR WEIGHT CELL PROTEIN ON
UNTRANSFORMED AND TRANSFORMED SYRIAN
HAMSTER CELLS[a]

Cells treated	Killing (%)[b]
Untransformed	96
AT$_1$(Cl)$_3$	25
P-MCA-75	17
A-6	59

[a]The protocol described in Section III, A was used for these experiments.

[b]Percent of killing = 1 − (Number of surviving colonies in the presence of antiserum and complement)/(number of surviving colonies in the presence of normal rabbit serum and complement) × 100.

(Table V). Transformed cells showed varying amounts of cytotoxic response to the antiserum and complement. When the antiserum was used at a 1:40 dilution (the final dilution of antiserum in the assay was 1:800), the difference in cytotoxic effect between transformed and untransformed cells was optimized. Using this dilution of antiserum, the cytotoxicity effect on three different chemically transformed Syrian hamster cell lines was compared to the effect on untransformed Syrian hamster cells. Table VI shows the results. These differences in sensitivity to the antiserum indicate a difference in availability and/or synthesis of this protein. Another possible explanation for the cytotoxicity data is that certain transformed cells lack receptors for the C5b-9 complexes.

IV.   Applications of Antibody–Complement Cytolysis

A.   The Detection of Membrane Changes in Neoplastic Cells

Complement-mediated antibody cytotoxicity has been used to study the surface antigens of neoplastically transformed cells in culture. The test has been used to measure tumor-specific cell surface antigens (TSSAs) which are believed to be an *in vitro* correlate of tumor-specific transplantation antigens (TSTAs). Antibody cytotoxicity has been used mostly in studies of cells transformed by tumor viruses (for reviews, see Terethia and Rapp, 1977; Bauer, 1974).

In studies on avian RNA tumor viruses it has been found that mouse immune serum, prepared from mice bearing Rous sarcoma virus-induced tumors, contains antibodies which are cytotoxic to chicken myeloblasts transformed with avian myeloblastosis virus. Thus there is a common antigen on the surface of cells derived from different species, and this antigen is similar in cells infected with either a leukosis or sarcoma virus. Cytolysis experiments have been performed with antiserum directed against specific polypeptides from purified virus in order to determine whether or not the virion proteins are responsible for TSSA changes in tumor virus-infected cells.

Studies on mammalian RNA tumor viruses have shown that cytotoxic antibodies directed against Gross lymphoma cells can be produced by immunization with Gross murine leukemia virus-infected lymphoma cells. In other studies, antibodies against the structural virion proteins p31 and p15 have been shown to be cytotoxic to cells infected by murine tumor viruses. Certain cytotoxic antibodies, which kill virally transformed leukemia cells, do not react with virion proteins. Studies are in progress to determine which TSSAs are virally coded, which are coded for by the host cell, and the relationship between the other phenotypic characteristics of the transformed cells and the expression of TSSAs.

DNA tumor virus-infected and transformed cells have also been studied for alternations in cell surface antigens, using cytotoxic antibodies. SV40-transformed cells have specific antigens on their cell surface, which can be recognized by humoral antibodies. There appear to be both SV40-specific changes and other host cell-directed changes such as the expression of carcinoembryogenic antigens on SV40-transformed cells. At present there is no evidence that any of the SV40-induced cell surface antigenic changes are caused by the insertion of virion proteins. Cytotoxic antibodies are found in animals with Shope virus-induced fibromas, and the regression of these tumors is correlated with the titer of this antibody.

B.   The Detection of Antigenic Changes in Cells during Differentiation

Preparations of antiserum from animals immunized with primitive mouse teratocarcinoma cells have been used in complement-mediated cytotoxicity assays on differentiated cells derived from the teratocarcinoma cells and on fertilized eggs and cleavage-stage embryos. The antibodies generated reacted with several different lines of primitive teratocarcinoma but not with the differentiated cells generated from these primitive lines. This antiserum reacts with testis cells and with mouse morula cells at specific stages during development (Artzt et al., 1973). These types of studies are being used to map the appearance and disappearance of cell surface antigens during development, and immunological techniques similar to this are being used to study development in mutants where there is an arrest at specific times in tissue genesis.

Another example of the use of antibody complement cytolysis has been in monitoring expression of the H-Y male antigen. In this study, female mice immunized with male cells produced antiserum capable of killing sperm in the presence of complement. After adsorption with spleen cells, the antiserum was tested for its effect on mouse embryos of varying gestational age. These studies showed that the H-Y antigen is present prior to implantation of the embryo, that the Y chromosome is necessary for expression of this antigen, and that expression of the H-Y antigen occurs before extensive male differentiation and testosterone production (Krco and Goldberg, 1976).

C.   The Chromosomal Mapping of Cell Surface Antigens

Antibody cytotoxicity assays have been used to map certain cell surface antigens to specific human chromosomes. One of the procedures is to make somatic cell hybrids between human cells and Chinese hamster ovary (CHO) cells with subsequent isolation of hybrids containing only one human chromosome. These clones are used to immunize rabbits, and the immune antiserum is then adsorbed with intact CHO cells to remove antibodies against the CHO cell surface antigens. The adsorbed serum is tested for its cytotoxic effect on the human–CHO hybrid clones and on human cells. The clones which are sensitive to killing are designated as having the human chromosome containing the information for the human antigens to which the rabbit has made antibodies. Karyotypic analysis is performed to designate which chromosome codes for the antigens being studied (Moore, 1976). In further studies adsorption of this antisera with specific human tissues is being used to determine in which tissues these cell surface antigens are being expressed.

D.   Immunosurgery

Cytotoxic antibodies can be used to remove specific populations of cells from a mixed population. This technique of immunosurgery has been used on mouse blastocysts which apparently are impermeable to antibodies (Solter and Knowles, 1975). Mouse blastocysts were exposed to rabbit antimouse antibodies, washed, and then exposed to complement. This treatment kills the trophoblastic cells and leaves the inner cell mass. Biological and biochemical studies can then be performed on the inner cell mass as it differentiates in culture.

E.   The Detection and Isolation of Specific Populations of Lymphocytes

Antibody–complement-mediated cytolysis has probably been most extensively used by immunologists studying the maturation and classification of lymphocytes. Lymphocytes are subjected to killing by different antibody preparations made against specific cell surface antigens (the antibodies are frequently made in congenic animals and cells from donor animals are injected into animals which are genetically different in only one gene locus). The surviving populations have been studied for their immunological and physiological properties. Examples of subclasses of lymphocytes which have been identified using these techniques include $\theta+$ and $\theta-$, Ly+ and Ly−, and Ia+ and Ia− lymphocytes (Niederhuber et al., 1976; Cantor and Boyse, 1975; Chan et al., 1970).

REFERENCES

Artzt, K., Dubois, P., Bennett, D., Condamine, H., Babinet, C., and Jacob, F. (1973). *Proc. Natl. Acad. Sci. U.S.A.* **70**, 2988.

Bauer, H. (1974). *Adv. Cancer Res.* **20**, 275.

Boyse, E. A., Hubbard, L., Stockert, E., and Lamm, M. E. (1970). *Transplantation* **10**, 446.

Cantor, H., and Boyse, E. A. (1975). *J. Exp. Med.* **141**, 1376.

Chan, E. L., Mishell, R. I., and Mitchell, G. F. (1970). *Science* **170**, 1215.

Horan, P. K., and Kappler, J. W. (1976). *J. Cell Biol.* **70**, 148a (abstr.).

Kohler, P. F., (1977). *In* "Immunologic Diseases" (M. Samter, ed.), 3rd ed. (in press).

Krco, C. J., and Goldberg, E. H. (1976). *Science* **193**, 1134.

Moore, E. E. (1976). Ph.D. Thesis, University of Colorado Medical Center, Boulder.

Niederhuber, J. E., Frelinger, J. A., Dine, M. S., Shaffner, P., Dugan, E., and Shreffler, D. C. (1976). *J. Exp. Med.* **143**, 372.

Oda, M., and Puck, T. T. (1961). *J. Exp. Med.* **113**, 599.

Ouchterlony, O., and Nilsson, L. A. (1973). *Hanb. Exp. Immunol., 2nd Ed.* (D. M. Weir, ed.) Vol. 1, p. 19–17.

Pollack, R., and Rifkin, D. (1975). *Cell* **6**, 495.

Rieber, M., and Romano, E. (1976). *Cancer Res.* **36**, 3568.

Ruoslahti, E., and Vaheri, A. (1974). *Nature (London)* **248**, 789.

Ruoslahti, E., Vaheri, A., Kuusela, P., and Linder, E. (1973). *Biochim. Biophys. Acta* **322**, 352.

Solter, D., and Knowles, B. (1975). *Proc. Natl. Acad. Sci. U.S.A.* **72**, 5099.

Stumph, W. E., Elgin, S. C. R., and Hood, L. (1974). *J. Immunol.* **113**, 1752.

Tevethia, S. S., and Rapp, F. (1977). *Contemp. Top. Immunobiol.* **6** (in press).

Yamada, K. M., and Weston, J. A. (1974). *Proc. Natl. Acad. Sci. U.S.A.* **71**, 3492.

Yamada, K. M., Yamada, S. S., and Pastan, I. (1975). *Proc. Natl. Acad. Sci. U.S.A.* **72**, 3158.

In "Affinity Chromatography Principles and Methods," 1976, p. 6–15. Pharmacia Fine Chemicals, Uppsala, Sweden.

Chapter 28

Microelectrophoresis of Enzymes in Animal Cell Colonies

WILLIAM C. WRIGHT

Human Tumor Cell Laboratory, Walker Laboratory,
Sloan-Kettering Institute, Rye and New York, New York

I. Introduction

Methods of electrophoresis of enzymes and other proteins have been developed extensively over the past 2 decades, and many interesting problems have been studied in depth. The amount of time and resources required to produce ample quantities of tissues or cultured animal cells have not been serious limiting factors in many studies. However, the analysis of cultured cells can be impeded by such limitations if: (1) the cells grow slowly, as is the case for some human tumor cells *in vitro*, or (2) the time required to obtain a result is limited to days or only a few weeks, as in cases of prenatal diagnosis using cells obtained by amniocentesis, or (3) it is not possible to grow large clonal populations from single primary cells because division potential is exhausted and senescence occurs, or (4) there is heterogeneity among the cells in a culture. For these reasons it is desirable to be able to perform microelectrophoresis on colonies of a few thousand cells down to the level of a single cell.

Observations and methods that suggest the feasibility of microelectrophoresis of enzymes in cell colonies came from histochemistry (Pearse, 1968, 1972). Enzyme activity was demonstrated in small groups of cells and in individual cells in histological sections for alkaline and acid phosphatases, esterases, dehydrogenases, peptidases, and other enzymes. Further encouragement came with the adaptation from the macrolevel to the microlevel of quantitative assays for a wide variety of enzymes (Lowry, 1957). An interesting and practical example of this was determination of the levels of activity of hypoxanthine-guanine phosphoribosyltransferase (HGPRT) and adenine phosphoribosyltransferase (APRT) in the cells of single hair follicles plucked from the human scalp (Gartler et al., 1971). By assaying separately several follicles per person it was possible to distinguish individuals that were normal, heterozygous or hemizygous at the HGPRT or Lesch-Nyhan HGPRT deficiency disease locus which is X-linked. HGPRT activity was also assayed in cell colonies (Richardson and Cox, 1973) and in samples of several and single hybrid cells (Hösli et al., 1974). The microassay of β-galactosidase and glucose-6-phosphate dehydrogenase (G6PD) in colonies or single cultured human fibroblasts was described (Galjaard et al., 1974a,b). Cells were grown and freeze-dried on thin plastic foil. Under the microscope a given area of foil with the required number of cells was cut away and placed in a microcuvet where the kinetics of an enzyme reaction were determined by microphotometry or microfluorometry.

Success in microelectrophoresis is more likely with microsamples having high concentrations of enzyme activity which can be achieved by harvesting cells in as small a volume as possible, storing samples at -20 to $-80°C$ until used, reducing the time required for electrophoresis by increasing voltages while maintaining adequate cooling, and increasing the sensitivity of the specific enzyme stains. For many enzymes microelectrophoretic methods have already been devised by adapting macromethods using starch, cellulose acetate, and polyacrylamide gels (PAGs). To present the range of technical possibilities a microsample is considered here to consist of a few thousand cells down to a single cell obtained from cultured cells, small tissue or organ fragments, or whole organisms of small species.

II.   Starch Gel Microelectrophoresis

Potato starch gels have been used extensively and successfully as an anti-convection and molecular sieving matrix for the electrophoresis of many enzymes (Smithies, 1955, 1959; Smith, 1968; Harris and Hopkinson, 1976). Starch gel microelectrophoresis of enzymes in animal cell colonies as small as 1000 mouse A9 cells grown *in vitro* was previously described (Wright, 1973). The enzymes microelectrophoresed were: Phosphohexose isomerase (PHI, EC 5.3.1.9), malate dehydrogenase (MDH, EC 1.1.1.37), lactate dehydrogenase (LDH, EC 1.1.1.27), adenylate kinase (AK, EC 2.7.4.3), and nucleoside phosphorylase (NP, EC 2.4.2.1). The micromethods described here for G6PD (EC 1.1.1.49), malic enzyme (ME, EC 1.1.1.40), and esterase D (ESD, 3.1.1.1) include only a few minor modifications or variations.

The demonstration of microelectrophoresis described here was carried out as part of a human tumor cell line characterization project in the laboratory of Jørgen Fogh, Sloan-Kettering Institute, New York, N.Y. Cell colonies were derived from a human pigmented malignant melanoma, Ti 6833, passed twice through nude mice, and then cultured *in vitro* in minimal essential medium with 26% fetal bovine serum, penicillin, and streptomycin (Fogh *et al.*, 1976). Portions of tumor taken from the nude mice were set aside for electrophoresis of enzymes for which humans and mice are known to have different mobilities. One of the objectives of the *in vitro* cultivation of human tumors passed through nude mice was to establish human tumor cell lines free of mouse cells and possible hybrid cells. Since cell colonies in this case grew slowly over a period of 6 months, it was desirable to be able to determine the species of the cell colonies in culture without sacrificing too many cells. Of the two types of colonies observed, the predominant, compact colonies had a dense melanotic center with indistinct cell morphology and an outer rim of spindle to polygonal cells relatively clear of pigment. These colonies grew slowly for 6 months, producing pigmented monolayer cultures. A few diffuse colonies, observed in the first few months, had cells that varied in size and shape from an epithelial to a fibroblast-like morphology. These cells had heavy pigmentation of the cytoplasm and one or two nuclei. Of three colonies of each type subcultured into Leighton tubes, the compact colonies grew, but the diffuse colonies failed to metabolize and grow.

In some studies cultures were terminated after harvesting microsamples, so that strict sterile technique was not required, only clean technique. In other experiments it was desirable that a portion of each harvested colony and other colonies not harvested be able to continue to grow for later study. In this case sterile technique was maintained.

Two types of glass instruments with microtips were made for harvesting and handling colony microsamples. One type, used for clean harvesting, was made from capillary tubes, and the other type, used for sterile harvests, was made from pasteur pipets and autoclaved before use. The middle of a capillary tube or the narrow part of a pasteur pipet, about 4 cm from the tip, was heated in a small bunsen flame until soft and then removed and immediately drawn into a fine capillary tube with a 30° angle to the main shaft of the capillary tube or pipet. The fine capillary was cut with a diamond pencil, leaving a 5- to 15-mm-long microtip.

In order to control carefully the inflow and explusion of fluid and cells in the microtip instruments when harvesting and later when loading into starch gels, microtip instruments were fitted with a plastic or rubber tube long enough to reach the operator's mouth when the glass instrument was held by hand. A hard plastic mouthpiece was fitted into the oral end of the tubing.

Cells can be grown in and harvested from a variety of glass or plastic flasks or petri dishes, depending on the requirements of a particular experiment and a laboratory's preferences and standards. When glass flasks, or petri dishes and plastic flasks with colonies to be saved for further culturing, were harvested, sterile microtip pasteur pipets were used to pick off the colonies individually. When a culture in a plastic flask was being terminated after harvest, the top of the flask was cut away with a flame-hot scalpel blade. This permitted the same easy access to colonies as for petri dish cultures that were terminally harvested with clean microtip capillary tubes.

The harvest operation was begun by noting colony morphology, size, and location in the culture vessels. The medium was drained off, and the colonies washed three times with saline A or Hanks' solution. A thin film of washing medium was left covering the colonies to prevent dessication. About 15 μl of sterile distilled water was taken into the microtip instrument by capillary action and controlled suction. The microtip was then guided by eye over the colony to be harvested and then, with microscope monitoring, the water was carefully expelled. The cells were scratched off the surface of the culture vessel into the water and drawn into the microtip by capillary action. To obtain high concentrations of enzyme the cells were collected in the smallest practical volume of water (5–10 μl). Until used for microelectrophoresis, the harvested colonies in labeled microtip instruments were stored at -20 or $-80°$C. The colonies remaining for continued cultivation were then washed three times with medium before feeding to remove loosened cells, so as to reduce the chances of contamination by floating cells. An alternative way of preventing such contamination was to subculture individual colonies into small Leighton tubes using the microtip harvesting method described.

Gels of 12% starch and 6 mm thickness (those of 3 mm thickness were also used) were loaded with control macro- and microsamples and experimental microsamples. A strip of clear cellulose (3 × 20 cm) was used as a temporary label across the width of the gel to indicate, with dots, dashes, and names, the location of each sample. This label remained in position while gels were loaded, examined, and photographed. Control macrosamples were extracts, diluted one-half to one-quarter with water, of tissues or cells with known isozyme patterns. A macrosample of 15 μl was usually loaded onto a 8 $\times$ 4 mm piece of Whatman No. 3 chromatography paper that was inserted with the aid of a razor blade into the gel. Control samples were also loaded into microsample holes in the gel.

Microsample holes in starch gels were made with pasteur pipets attached to tubing through which suction was applied orally. The tip of the pipet, lubricated with distilled water, was pushed vertically into the gel at designated positions. The pipet was slowly withdrawn while oral suction was applied to remove the solid cylinder of starch gel in the pipet. Any water of starch remaining in the hole was removed by suction. Care must be taken not to suck too hard, so as not to make a large hole in the gel. Pasteur pipets have a bore diameter of 1 mm at the end and, since gels are 6 mm thick, holes with a volume of about 5 μl were obtained. When larger holes were required, the pipet was cut with a diamond pencil proximally where the bore was wider.

Cell colonies harvested and stored in microtip pipets or capillaries were freeze-thawed three times to lyse the cells and then loaded directly from the microtip into holes in the starch gel. Sample fluid was carefully blown out of the microtip so that the hole was filled from the bottom up, thereby reducing the chance of obtaining air bubbles in the hole. Care was also taken to prevent the microsample from overflowing onto the gel surface. This was avoided by having a sufficiently wide microtip to allow easily controlled outflow of sample fluid and could be corrected by suking up the excess. After electrophoresis for 10 minutes to allow enzymes to enter the matrix, the gel was carefully covered with film (Saran wrap, Melinex, or Parafilm) to prevent desiccation and for electrical insulation; Melinex plastic was generously donated by Imperial Chemical Industries United States, Incorporated, Stratford, Connecticut.

A.   G6PD Microelectrophoresis

G6PD is an X-linked enzyme that is polymorphic in Negro populations in which equally fast electrophoretic type-A and type-A $^-$ isozymes and a slow type-B isozyme were observed (Kirkman, 1971). Only type-B was found in Caucasians. This ethnic difference has been the basis of the exoneration of more than 169 human tumor cell lines from the suspicion of contamination by the G6PD type-A HeLa cell line derived from a Negro cervical tumor (Fogh et al., 1977). Clonal cell colonies in vitro and tumors from type-AB Negro women have been useful in demonstrating lyonization of the X chromosome (Davidson et al., 1963) and the clonal or nonclonal origin of various types of cancer (McCurdy, 1967; Beutler et al., 1967). Mouse G6PD migrates faster toward the anode than human type-A and

-B isozymes, permitting identification of the species of different cell colonies in human-mouse mixed or hybrid cell cultures. The following demonstration of microelectrophoresis of G6PD in colonies has extended the usefulness of this important analytical genetic maker.

Starch gels at 12% were made and run using the Tris–EDTA–boric acid (TEB) buffer system at pH 8.6 (Ruddle and Nichols, 1971). After electrophoresis at 5 V/cm for 22 hours between cooling plates the gels were sliced horizontally and stained for G6PD (Harris and Hopkinson, 1976). The pattern (Fig. 1) in a control of mouse brain extract, macrosample (channel 1, C-1) and microsample (C-9), consisted of three isozymes with activity decreasing cathodally with only a trace of the most cathodal isozyme. Human G6PD type-B controls, macrosample (C-3) and microsample (C-5 and -10), have three isozymes with the middle isozyme predominant. Human type-A controls, macrosample (C-4) and microsample (C-11), have a similar but faster pattern. An extract of the pulp and capsule of Ti 6833 melanoma, taken directly from a nude mouse host was run as a macrosample (C-2). Both human type-B and mouse isozymes were present. Hybrid isozymes with intermediate mobility were absent, suggesting that human–mouse hybrid cells were not present or not frequent enough to be detected. Portions of colony microsamples loaded into C-6, -7, and -8 had cell lysates equivalent to about 2.2–3.5 mm^2 of confluent cells showed only human G6PD type-B isozyme was present in these colonies.

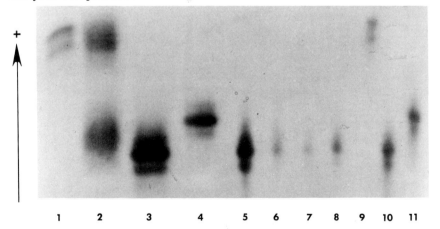

+

1   2   3   4   5   6   7   8   9   10   11

FIG. 1. G6PD microelectrophoresis. Mouse brain concentrated extract, 15-μl, macrosample, loaded on a Whatman No. 3 paper insert (C-1), and 5-μl microsample (C-9). Human HeLa, type-A control macrosample (C-4) and microsample (C-11). Human Detroit 562, type-B control, macrosample (C-3) and microsample (C-5 and -10). Concentrated extract of pulp and capsule of Ti 6833 human melanoma, as harvested from a nude mouse host, macrosample (C-2). Microsample of Ti 6833 melanoma compact colonies *in vitro* 6 months (C-6, -7, and -8). Origin not shown.

B. ME Microelectrophoresis

ME was electrophoresed under the buffer conditions described by Harris and Hopkinson (1976), except that 5 mg of NADP in 1 ml of water was added to 250 ml of 12% starch solution before pouring into molds (Cohen and Omenn, 1972). Gels were run at 7.5 V/cm for 17.5 hours and then stained (Harris and Hopkinson, 1976). Control extracts (Fig. 2) of mouse brain, macrosample (C-1) and microsample (C-8), had the anodally faster mitochondrial ME isozyme and the much slower soluble or cytoplasmic ME isozyme (Povey *et al.*, 1975). Concentrated human brain extract controls, macrosample (C-3 and -10) and microsample (C-4 and -9), showed the fast soluble ME and the slower mitochondrial ME isozymes, and the latter in

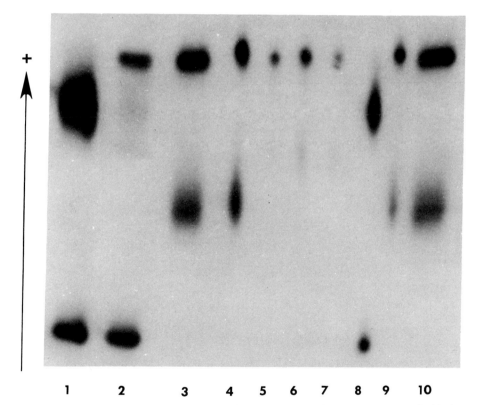

FIG. 2. ME microelectrophoresis. Mouse brain concentrated extract, macrosample (C-1) and microsample (C-8). Human brain control, concentrated extract, macrosample (C-3 and -10) and microsample (C-4 and C-9). Concentrated extract of pulp and capsule of Ti 6833 human melanoma, taken directly from a nude mouse host, macrosample (C-2). Microsamples of Ti 6833 compact melanoma colonies *in vitro* 6 months (C-5, 6, and -7). Origin not shown.

this case was known to be phenotype 1. An extract of the pulp and capsule of Ti 6833 melanoma, macrosample (C-2), had the fast human and slow mouse soluble ME isozymes of each species but little evidence of the mitochondrial ME isozymes of either species. Portions of microsamples of colonies loaded into C-5, -6, and -7 had strong activity for human soluble ME (and possibly a trace of human mitochondrial ME, phenotype 2). No mouse ME isozymes were observed in these colonies.

C. ESD Microelectrophoresis

Electrophoresis for ESD was carried out with the bridge and gel buffers used for AK with runs at 9 V/cm for 4 hours (Fildes and Harris, 1966; Harris and Hopkinson, 1976). The stain, applied as a paper overlay, consisted of 0.01% 4-methylumbelliferyl acetate in 0.5 M acetate buffer, pH 5.2

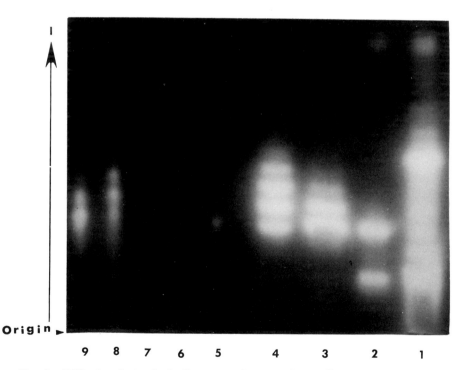

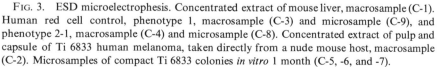

Fig. 3. ESD microelectrophesis. Concentrated extract of mouse liver, macrosample (C-1). Human red cell control, phenotype 1, macrosample (C-3) and microsample (C-9), and phenotype 2-1, macrosample (C-4) and microsample (C-8). Concentrated extract of pulp and capsule of Ti 6833 human melanoma, taken directly from a nude mouse host, macrosample (C-2). Microsamples of compact Ti 6833 colonies *in vitro* 1 month (C-5, -6, and -7).

(Hopkinson *et al.*, 1973). A concentrated human red cell lysate control (Fig. 3) for phenotype 1, macrosample (C-3) and microsample (C-9), showed the typical slow primary isozyme and two faster secondary isozymes usually seen in red cells but not in cultured cells (Hopkinson *et al.*, 1973). Red cell controls for phenotype 2-1, macrosample (C-4) and microsample (C-8), had the three primary isozymes and 2 anodally faster secondary isozymes usually seen in red cells but not in cultured cells. A control extract of a Swiss mouse liver, macrosample (C-1), had a complicated isozyme pattern. However, the slowest isozyme was observed in cultured mouse fibroblasts (Hopkinson *et al.*, 1973). This slowest of the mouse isozymes and the single primary band of human ESD, type 1, were observed in an extract from the pulp and capsule of Ti 6833 human melanoma, harvested directly from a nude mouse host (C-2). Microsamples (C-5, -6, and -7) of compact Ti 6833 melanoma cell colonies harvested after 6 weeks *in vitro* were found to have the human type-1 ESD isozyme and no murine isozymes.

Microelectrophoretic analysis of other colonies of Ti 6833 melanoma gave the following results. At 3 weeks one diffuse colony was found to have mouse G6PD and one compact colony to have human type-B G6PD. At about 6 months in culture, four compact colonies microelectrophoresed for soluble ME and five colonies for G6PD were found to have human but no mouse isozymes. The overall result of this brief analysis of isozymes of cultured cell colonies of Ti 6833 human melanoma, was that a few diffuse, nondividing mouse colonies containing melanin were initially observed but disappeared in 1–2 months. The predominant colony type throughout the culture period were compact human melanoma colonies that grew rather slowly. After 10 months a mass culture large enough for conventional electrophoresis was obtained. The results confirmed that murine isozymes were absent and that the human isozymes, including the mitochondrial ME, phenotype 2, had been characterized correctly in earlier microsamples. After 10 months a mass culture large enough for conventional electrophoresis was obtained. The results confirmed that murine isozymes were absent and that the human isozymes, including the mitochondrial ME_m2 phenotype, had been characterized correctly in earlier microsamples.

Another method of starch gel microelectrophoresis was used to analyze serum proteins (Daams, 1963). A starch solution at 1.25 times the concentration used in macroelectrophoresis was poured onto coverslips to a final thickness of 1 mm. Small slits in the gel surface were made with a razor blade, and 0.2-μl samples of serum were loaded into the slits. Electrophoresis at 85 V/cm for 15 minutes was performed in the type of apparatus used for agar gel electrophoresis. Gels stained for protein gave results comparable to those for macroelectrophoresis gels. A similar method using microscope slides on which 1.5-mm-thick starch gels were poured was used for the electrophoresis of esterases of calf duodenum (Ramsey, 1963).

A special apparatus and technique were designed for microelectro-phoresis of enzymes in 0.2- to 1.0-ml-thick starch gels or PAGs (Sviridov *et al.*, 1971). Single *Drosophila* larvae and microsamples of mouse material were tested to demonstrate isozymes of esterases, LDH, and alcohol dehydrogenase using starch microgels.

III. Cellulose Acetate Microelectrophoresis

Thin strips of cellulose acetate in the nongel form and in the technically preferred gel form (Cellogel) have been used for the electrophoresis of enzymes in macrosamples (Khan, 1971; Van Someren *et al.*, 1974). There were several apparent advantages to using cellulose acetate strips. Cell lysate samples with volumes of a few microliters or less could be loaded with skill onto the surface of the strip. Overheating during electrophoresis was not a problem even at room temperature, which permitted the use of higher voltages for shorter periods of time, thereby minimizing diffusion and heat-inactivation of the enzymes. When a specific enzyme stain was added to the gel after electrophoresis, it quickly penetrated the thin gel so that all the enzyme immediately began to react with the substrate. Thus rapid development of the isozyme pattern consisting of narrow bands was achieved.

Cellogel microelectrophoresis was demonstrated for hexosaminidase (HEX) A and B using colonies of amniotic fluid cells and skin fibroblasts cultured on microtest plates (Richardson and Cox, 1973). The samples of 0.5 μl, loaded onto the gel with a drafting pen, contained a lysate equivalent to 400 cells. After electrophoresis at 30 V/cm for 1 hour the gels were stained using a sensitive fluorescent substrate. The results showed that it was usually possible to distinguish colonies of cells from patients with Tay-Sachs disease lacking HEX A from colonies of cells from normal persons having this isozyme. HEX B was evident in microsamples from both affected and normal individuals.

IV. Polyacrylamide Gel Microelectrophoresis

The theory and practice of polyacrylamide gel electrophoresis (PAGE) were developed from 1959 to 1964 (Ornstein, 1964; Davis, 1964; Raymond, 1964). In the method of disc electrophoresis of Ornstein and Davis, polyacrylamide was poured into tubes about 5 mm in diameter to gel. A lower gel of 5% or more polyacrylamide had pores the size of protein molecules, so that by the process of molecular sieving during electrophoresis the proteins were separated according to molecular weight as well as charge. Above the molecular sieving gel were stacking and sample-containing gels of 3% polyacrylamide. Thus a *dis*continuous concentration of PAG was formed in the tube. The 3% gel had a large pore size with little molecular sieving potential. At the beginning of electrophoresis using a *dis*continuous buffer system, proteins in the sample gel rapidly migrated and stacked as thin concentrated *discs* at the interface between the stacking and sieving gels. The mobility of proteins entering the molecular sieve gel decreased considerably, but then varied significantly, resulting in separation. After electrophoresis the gels were extruded from the tubes and stained for proteins observed as colored *discs* that could be quantified with a densitometer. Another PAGE method was devised using vertical slabs of PAG in which several samples could be run in parallel, permitting easy comparison of enzyme or protein staining patterns between samples (Raymond, 1964). Sample slots were made in the upper end of the gel. Glucose was added to the samples before loading, making them denser than the buffer in the upper electrode tank, so that when samples were loaded they stayed at the bottom of the slots and upward convection into the buffer was prevented. The use of 3% sampling and stacking gels was thereby avoided.

Micro-PAGE adaptations were seen made for the analysis of proteins at the nanogram level (Grossbach, 1965). Capillary tubes with 0.2- to 0.45-mm diameters were used instead of the 5-mm tubes used in macro-PAGE, reducing the cross-sectional area by 1/123 to 1/625 times. This effectively kept the microsamples concentrated. Serum albumin was electrophoresed within 5 minutes at 40 V/cm. The capillary tubes were frozen, and the gels pushed out, fixed, and stained for protein that was quantified with a microdensitometer.

Neuhoff and colleagues (1970; Neuhoff 1972, 1973) applied and developed micromethods, including micro-PAGE, for the manipulation and chemical analysis of microsamples in the fields of neurophysisology and biochemistry. For micro-PAGE, capillary tubes were filled with PAG at concentrations usually greater than those used in macro-PAGE. Electrophoresis was carried out in minutes to about an hour. Proteins or enzymes were then stained and quantified by scanning with a microdensitomer. By repeating scans once a minute, kinetic data for individual isozymes of G6PD were obtained in addition to electrophoretic mobility data (Cremer *et al.*, 1972). A modification of these methods is microisoelectrofocusing using ampholine, pH 3–10, in the PAG to separate proteins (Quentin and Neuhoff, 1972). By this method the variation in the proportions of the five LDH isozymes was determined in different regions of the rabbit brain, using a few micrograms of tissue per sample.

Similar micro-PAGE methods have been used in entymological embryology. Esterase isozymes of single ganglia of housefly larvae were microelectrophoresed (Odintsov and Petrenko, 1970). Developmental changes in isozyme patterns in organs of *Drosophila* were analyzed quantitatively for acid and alkaline phosphatases and esterases; malic, glucose-6-phosphate, lactic, alcohol, and xanthine dehydrogenases; and peroxidase, cytochrome oxidase, and α-hydroxy acid oxidase (Pasteur and Kastritsis, 1971, 1973, 1974).

In the rat, small groups of cells along differentiating crypts and villi of the small intestine were removed by microdissection, electrophoresed through PAG in 5-mm capillaries, and stained for nonspecific esterases. Increased esterase activity during differentiation in the lower villus was shown to be caused by a uniform increase in the activity of all esterase isozymes and not specific isozymes (DeBoth et al., 1974).

V. Microelectrophoresis of Single Cells

Molluscan neurophysiologists have used micro-PAGE to analyze proteins in single large neurons that can be routinely identified and characterized by electrical activity in ganglia of *Aplysia californica*, a sea hare or slug, and *Otala lactea*, a land snail (Wilson, 1971; Wilson and Berry, 1972; Gainer, 1971, 1972a,b). Ganglia were dissected from the animals and cultured and labeled *in vitro* with radioactive leucine. Single neurons were isolated and prepared with sodium dodecyl sulfate for electrophoresis in capillary tubes loaded with PAG. A unique protein of 5000 molecular weight was found in endogenous bursting pacemaker neurons in *Aplysia*, neuron R15, and in *Otala*, neuron 11, but not in nonpacemaker neurons, R2 and 12, respectively (Gainer, 1971, 1972a). During the winter hibernation of *Otala* both the pacemaker activity and the 5000-molecular-weight protein were absent from neuron 11 (Gainer, 1972b). A portion of the total extract of a single mouse ovum was employed to separate G6PD isozymes, using micro-PAGE methods similar to those described above (Cremer et al., 1972).

A novel electrophoresis method was devised to observe different hemoglobin variants in single cells during the metamorphosis of *Rana catesbeiana*, a bullfrog (Dan, 1970). Erythrocytes were added to polyacrylamide that was then allowed to gel on microscope slides. After microelectrophoresis at high voltages (1800 V) between the electrodes for 40–70 seconds, gels were fixed and stained for protein. Erythrocytes examined microscopically were found to have one or two adjacent bands, suggesting that single erythrocytes may contain one or two types of hemoglobin.

Single erythrocytes of *R. catesbeiana* and human umbilical cord blood were analyzed by micro-PAGE in capillary tubes for hemoglobin, LDH, and G6PD (Rosenberg, 1970). Single cells were manipulated with glass micropipets fitted to a micromanipulator, with monitoring through a microscope. In single cells from metamorphic tadpoles only a single band of either tadpole or adult hemoglobin was observed. In human cord blood, red cells had fetal, adult, or both hemoglobin bands. Isozymes of LDH and G6PD were observed using stains 20 times the standard concentrations. Gels were photographed and scanned with a microdensitometer. Proportions of LDH isozymes varied significantly from cell to cell in *R. catesbeiana* and in human cord blood. Three G6PD isozymes were found in *R. catesbeiana* cells, but some cells had only one of the three isozymes while the remaining cells had the other two isozymes. When the hemoglobin and enzymes were detected with specific antisera, similar banding patterns were obtained.

Microelectrophoresis of carbonic anhydrase isozymes, particularly CA II, of the pigtailed macaque was demonstrated with techniques that should encourage development of single cell microelectrophoresis of other enzymes (DeSimone *et al.*, 1971). Macaque red cells and rabbit antihuman carbonic anhydrase II serum were suspended in 1% agar, poured onto microscope slides, and overlaid with a coverslip. The slides were electrophoresed for 1 minute at 70 V/cm during which the cells lysed and enzyme moved out of the cell and was immunoprecipitated. The slides were stained so that immunoprecipitation granules could be observed and counted microscopically. Other enzymes could likely be examined for electrophoretic and immunological variants with these technical approaches, using single cells and possibly other kinds of microsamples.

VI.   Concluding Remarks

The variety of microelectrophoretic methods available requires that the experimenter select a suitable approach for his or her particular problem. Some of the factors involved in this choice are likely to be the size of the microsample, the concentration of enzyme, the sensitivity of the particular enzyme stain, the satisfactory resolution of isozymes under selected gel and buffer conditions, whether or not the isozymes are to be quantified by microdensitometer, the cost of materials, and the methodological preferences of the worker. With technical perseverance many fields of research could be furthered by the microelectrophoresis of enzymes in animal cell colonies.

ACKNOWLEDGMENTS

I thank Dr. Jørgen Fogh and other colleagues at the Human Tumor Cell Laboratory, Sloan-Kettering Institute, and my wife, Dr. Morya Smith, Division of Medical Genetics, Department of Pediatrics, Mount Sinai Hospital, New York, N.Y., for their encouragement and assistance.

REFERENCES

Beutler, E., Collins, Z., and Irwin, L. E. (1967). *Clin. Res.* **15**, 110.
Cohen, P. T. W., and Omenn, G. S. (1972). *Biochem. Genet.* **7**, 289.
Cremer, T., Dames, W., and Neuhoff, V. (1972). *Hoppe-Seyler's Z. Physiol. Chem.* **353**, 1317.
Daams, J. H. (1963). *J. Chromatog.*, **10**, 450.
Dan, M. (1970). *Exp. Cell Res.* **63**, 436.
Davidson, R. G., Nitowsky, H., and Childs, B. (1963). *Proc. Natl. Acad. Sci. U.S.A.*, **50**, 481.
Davis, B. J. (1964). *Ann. N.Y. Acad. Sci.* **121**, 404.
DeBoth, N. J., Van Dongen, J. M., Van Hofwegen, B., Keulemans, J., Visser, W. J., and Galjaard, H. (1974). *Dev. Biol.* **38**, 119.
DeSimone, J., Daufi, L. M., and Tashian, R. E. (1971). *Exp. Cell Res.* **67**, 338.
Fildes, R. A., and Harris, H. (1966). *Nature (London)* **209**, 261.
Fogh, J., Goodenow, M., Loveless, J., and Fogh, H. (1976). *In* "In Vitro Methods in Cell Mediated and Tumor Imunity" (B. R. Bloom and J. R. David, eds.), p. 677. Academic Press, New York.
Fogh, J., Wright, W. C., and Loveless, J. D. (1977). *J. Natl. Cancer Inst.* **58**, 209.
Gainer, H. (1971). Anal. Biochem. **44**, 589.
Galner, H. (1972a). *Brain Res.* **39**, 369.
Gainer, H. (1972b). *Brain Res.*, **39**, 387.
Galjaard, H., Van Hoogstraten, J. J., De Josselin De Jong, J. E., and Mulder, M. P. (1974a). *Histochem. J.* **6**, 409.
Galjaard, H., Van Hoogstraten, J. J., De Josselin De Jong, J. E., and Mulder, M. P. (1974a). *Histochem. J.* **6**, 491.
Gartler, S. M., Scott, R. C., Goldstein, J. L., and Campbell, B. (1971). *Science* **172**, 572.
Grossbach, U. (1965). *Biochim. Biophys. Acta* **107**, 180.
Harris, H., and Hopkinson, D. A. (1976). "Handbook of Enzyme Electrophoresis in Human Genetics." North-Holland Publ., Amsterdam.
Hopkinson, D. A., Mestriner, M. A., Cortner, J., and Harris, H. (1973). *Ann. Hum. Genet.* **37**, 119.
Hösli, P. DeBruyn, C. H. M. M., and Oei, T. L. (1974). *Adv. Exp. Med Biol.* **41B**, 881.
Khan, P. M. (1971). *Arch. Biochem. Biophys.* **145**, 470.
Kirkman, H. N. (1971). *Adv. Hum. Genet* **2**, 1.
Lowry, O. H. (1957). *In* "Methods in Enzymology" (S. P. Colowick and N. O. Kaplan, eds.), Vol. 4, p. 366. Academic Press, New York.
McCurdy, P. R. (1967). *Clin. Res.*, **15**, 65.
Neuhoff, V. (1972). *Int. J. Neurosci.* **4**, 93.
Neuhoff, V., ed. (1973). "Micromethods in Molecular Biology." Springer-Verlag, Berlin and New York.
Neuhoff, V., Schill, W.-B., and Sternbach, H. (1970). *Biochem. J.* **117**, 623.
Odintsov, V. S., and Petrenko, V. S. (1970). *Dokl. Biochem. (Engl. Transl.)* **193**, 242.

Ornstein, L. (1964). *Ann. N.Y. Acad. Sci.* **121**, 321.
Pasteur, N., and Kastritsis, C. D. (1971). *Dev. Biol.* **26**, 525.
Pasteur, N., and Kastritsis, C. D. (1973). *Wilhelm Roux' Arch. Entwicklungsmech. Org.* **173**, 346.
Pasteur, N., and Kastritsis, C. D. (1974). *Arch. Zool. Exp. Gen.* **115**, 185.
Pearse, A. G. E. (1968). "Histochemistry: Theoretical and Applied," 3rd ed., Vol. 1. Little, Brown, Boston, Massachusetts.
Pearse, A. G. E. (1972). Histochemistry: Theoretical and Applied, 3rd ed., Vol. 2. Williams and Wilkins, Baltimore, Maryland.
Povey, S., Wilson, D. E., Jr., Harris, H., Gormley, I. P., Perry, P., and Buckton, K. E. (1975). *Ann. Hum. Genet.* **39**, 203.
Quentin, C.-D., and Neuhoff, V. (1972). *Int. J. Neurosci.* **4**, 17.
Ramsey, H. A. (1963). *Anal. Biochem.* **5**, 83.
Raymond, S. (1964). *Ann. N.Y. Acad. Sci.* **121**, 350.
Richardson, B. J., and Cox, D. M. (1973). *Clin. Genet.* **4**, 376.
Rosenberg, M. (1970). *Proc. Natl. Acad. Sci. U.S.A.* **67**, 32.
Ruddle, F. H., and Nichols, E. A. (1971). *In Vitro* **7**, 120.
Smith, I., ed. (1968). "Chromatographic and Electrophoretic Techniques," 2nd ed. Vol. II. Wiley (Interscience), New York.
Smithies, O. (1955). *Biochem. J.* **61**, 629.
Smithies, O. (1959). *Biochem. J.* **71**, 585.
Sviridov, S. M., Korochkin, L. I., and Matveeva, N. M. (1971). *Sov. J. Dev. Biol. (Engl. Transl.)* **2**, 352.
Van Someren, H., Van Henegouwen, H. B., Los. W., Wurzer-Figurelli, E., Doppert, B., Vervloet, M., and Khan, P. M. (1974). *Humangenetik* **25**, 189.
Wilson, D. L. (1971). *J. Gen. Physiol.* **57**, 26.
Wilson, D. L., and Berry, R. W. (1972). *J. Neurobiol.* **3**, 369.
Wright, W. C. (1973). *Exp. Cell Res.* **82**, 303.

SUBJECT INDEX

CONTENTS OF PREVIOUS VOLUMES

Volume I

Volume VII

Volume X

Volume XI

Volume XII

Volume XIII

Volume XV

Volume XVII

Volume XVIII

Volume XIX

A B C D E F G H I J
8 9 0 1 2 3 4 5 6